SPECIATION
AND THE
RECOGNITION
CONCEPT

SPECIATION AND THE RECOGNITION CONCEPT

THEORY AND APPLICATION

Edited by

David M. Lambert
Evolutionary Genetics Laboratory
School of Biological Sciences
University of Auckland

Hamish G. Spencer
Department of Zoology
University of Otago

SETON HALL UNIVERSITY
WALSH LIBRARY
SO. ORANGE, N.J.

The Johns Hopkins University Press

Baltimore and London

© 1995 The Johns Hopkins University Press
All rights reserved. Published 1995
Printed in the United States of America
on acid-free paper
04 03 02 01 00 99 98 97 96 95   5 4 3 2 1

The Johns Hopkins University Press
2715 North Charles Street
Baltimore, Maryland 21218-4319
The Johns Hopkins Press Ltd., London

A catalog record for this book is
available from the British Library.

Library of Congress Cataloging-in-Publication Data

Speciation and the recognition concept : theory and
application / edited by David M. Lambert, Hamish G.
Spencer.
    p.   cm.
Includes index.
ISBN 0-8018-4740-0 (hc : alk. paper). —
ISBN 0-8018-4741-9 (pbk. : alk. paper).
1. Species. 2. Sex recognition (Zoology). I. Lambert,
David M. II. Spencer, Hamish G.
QH380.S64   1994
575.1'3—dc20   94-27980
                                            CIP

Contents

Foreword, by P. H. Greenwood, F.R.S. *ix*

Preface *xi*

Introduction *xiii*

Ideas about the Nature of Species

ELISABETH S. VRBA
1. Species as Habitat-Specific, Complex Systems   3

BERNARD MICHAUX
2. Species Concepts and the Interpretation of Fossil Data   45

ALAN TURNER
3. The Species in Paleontology   57

KIYOHIKO IKEDA AND ATUHIRO SIBATANI
4. Kinji Imanishi's Biological Thought   71

RICHARD H. HUNT AND MAUREEN COETZEE
5. Mosquito Species Concepts: Their Effect on the Understanding of Malaria Transmission in Africa   90

JACK P. HAILMAN
6. Toward Operationality of a Species Concept   103

HAMPTON L. CARSON
7. Fitness and the Sexual Environment   123

Consequences of the Recognition Concept for Speciation, Ecology, and Evolution

HAMISH G. SPENCER

8. Models of Speciation by Founder Effect: A Review *141*

MICHAEL G. RITCHIE AND GODFREY M. HEWITT

9. Outcomes of Negative Heterosis *157*

MAUREEN COETZEE, RICHARD H. HUNT, AND DEBRA E. WALPOLE

10. Interpretation of Mating between Two Bedbug Taxa in a Zone of Sympatry in KwaZulu, South Africa *175*

GIMME H. WALTER

11. Species Concepts and the Nature of Ecological Generalizations about Diversity *191*

CHRISTOPHER A. GREEN

12. Heterogeneity in Distribution of Fixed Chromosomal Inversions in Ancestor and Descendant Species in Two Groups of Mosquitoes: Series Myzomyia and Neocellia of *Anopheles* (*Cellia*) *225*

DAVID M. LAMBERT

13. Biological Function: Two Forms of Explanation *238*

Properties of Specific-Mate Recognition Systems

CHARLES E. LINN, JR., AND WENDELL L. ROELOFS

14. Pheromone Communication in Moths and Its Role in the Speciation Process *263*

CHRIS S. WHITE, DAVID M. LAMBERT, AND STEVEN P. FOSTER

15. Chemical Signals and the Recognition Concept *301*

ROGER BUTLIN
16. Genetic Variation in Mating Signals and Responses   *327*

DAVID HENRY LANE
17. The Recognition Concept of Species Applied in an Analysis of Putative Hybridization in New Zealand Cicadas of the Genus *Kikihia* (Insecta: Hemiptera: Tibicinidae)   *367*

MARTIN VILLET
18. Intraspecific Variability in SMRS Signals: Some Causes and Implications in Acoustic Signaling Systems   *422*

JERONE DEN HOLLANDER
19. Acoustic Signals as Specific-Mate Recognition Signals in Leafhoppers (Cicadellidae) and Planthoppers (Delphacidae) (Homoptera: Auchenorrhyncha)   *440*

NILES ELDREDGE
20. Species, Selection, and Paterson's Concept of the Specific-Mate Recognition System   *464*

Author Index   *479*

Subject Index   *493*

Foreword

IT IS A GENUINE PLEASURE to introduce this collection of essays brought together as a pragmatic and analytical tribute to the ideas of Hugh Paterson.

Hugh and I have known one another since the late 1940s, when we were students at the University of the Witwatersrand. In those early days of our careers I am sure that, like most undergraduates of today, we "knew" exactly what a species was and gave little but passing thought to the processes involved in the origin of species. Later, and I suspect deeply influenced by Dobzansky's *Genetics and the Origin of Species* and Mayr's *Systematics and the Origin of Species*, we became increasingly intrigued by and involved in the well-named "species problem," Hugh as an entomologist and myself as an ichthyologist fascinated by the evolution and speciation of cichlid fishes in Africa.

I have to admit that when Hugh first outlined what was to become his new concept of a species, I was somewhat skeptical. But that skepticism passed as I gave the idea more thought, and particularly as I applied it to the species problems encountered in my own research.

Seemingly, such a change of mind was not experienced by Coyne, Orr, and Futuyma who (*Systematic Zoology 37:* 1988) posed the question, aimed at the Paterson school, "Do we need a new species concept?" In the course of their extended and negative reply to that question, Coyne et al. demonstrated, as far as I was concerned, a very real need to revive and rethink, as Paterson had done, the whole question of what a species is and how it originates. Indeed, in my view, Coyne et al. actually answered their query with a definite *Yes,* despite their obvious intentions of not so doing.

Apart from the intrinsic value of Paterson's Recognition Concept, its criticism of the Isolation Concept has played an important role in reawakening interest in a fundamental but rather dormant aspect of biology. Before the advent of Paterson's ideas the Isolation Concept seemed to have attained the status of received wisdom and thus a frightening degree of immunity from critical thought. Possibly my experience has been atypical, but I have a distinct impression that among many practicing biologists at that time, even among some taxonomists and systematists, little thought was given either to the need for a deeper understanding of what a species is or to the possibility that there could be both biological and logical flaws in the then generally accepted Biological Species Concept. Indeed, in some minds it would appear that the species category and the species taxon were confused and even conflated, while for many students the "species problem" existed only as the battles they had fought with seemingly intractable identification keys.

In part that situation can be blamed on the rather widespread and still current disappearance from many university curricula of courses in the more intellectual, and exciting, aspects of taxonomy and systematics. Its improvement, on the other hand, certainly lay in the minds of people like Hugh Paterson, not only for the novelty and incisiveness of his thoughts, but for the controversy those thoughts engendered. Even in the lofty climate of academia there is nothing like a scrap to excite and promote action, be it in thought or in the field and laboratory. For those reasons alone, and there are many others, it is a delight to welcome the appearance of this volume.

In a somewhat narrower perspective, I would welcome these collected essays as the first comprehensive and multifaceted review of Paterson's Recognition Species Concept, its consequences, and its ramifications as seen through the critical eyes of a variety of workers drawn from different backgrounds, both philosophical and biological, as well as neontological and paleontological. Apart from broadening the context in which the Recognition Concept can and should be viewed, their contributions also put paid to the not uncommon opinion that the Recognition and Isolation concepts are merely different sides of the same coin. They are indeed fundamentally different concepts of the products of speciation, as I have learned from applying both concepts to the situation presented by the complex species-flocks of African cichlid fishes. There, it is clear that the premises underlying the Recognition Concept make it, and not the Isolation Concept, the truly Biological Species Concept.

Recently, Hugh Paterson expressed an opinion (one by no means his alone) that it is unproductive to defend the status quo of the evolutionary synthesis by denying that objections to it are, in fact, really objections. "It is surely time," he wrote, "to cease shoring up the currently inadequate synthesis and begin its reconstruction." With the appearance of this volume I believe that the reconstruction has begun, a belief founded on the book's wide and varied coverage and its dissection and reexamination of the several issues arising from Paterson's formulation of the Recognition Concept.

P. H. Greenwood, F.R.S. ☐

Preface

WHEN WE STARTED OUT to produce this book, we had little idea of the amount of work it would entail or of the diverse array of people who would assist us in achieving our goal. First among these people must be Judith Masters, who was with us at the beginning and who acted as an editor for the first three years. This book would have been much the poorer were it not for her valuable input; indeed the book may never have existed without it. We thank her sincerely for her hard work.

We also owe a great debt to our contributors. They have been invariably patient, and they have promptly, for the most part, revised their essays in light of reviewers' comments and suggestions. Above all, they have produced what we believe is a stimulating and diverse set of papers, which explore the implications of the Recognition Concept in many areas of evolutionary biology. We hope they are pleased with the final result.

We are especially grateful to the numerous people who critically reviewed each one of the essays for us. Their efforts have resulted in numerous improvements both major and minor. These people include Prof. T. C. Baker, University of California at Riverside; Dr. N. H. Barton, University College, London; Prof. M. F. Claridge, University of Wales, Cardiff; Dr. R. Cooper, D.S.I.R., Lower Hutt, New Zealand; Prof. J. Cracraft, University of Illinois at Chicago; Prof. G. Davidson, London School of Hygiene and Tropical Medicine; Dr. R. L. Dorit, Harvard University; Dr. J. Dugdale, D.S.I.R., Auckland, New Zealand; Prof. L. Ehrman, State University of New York at Purchase; Dr. M. T. Ghiselin, California Academy of Sciences, San Francisco; Dr. M. T. Gillies, Whifield, Sussex; Prof. B. C. Goodwin, The Open University, Milton Keynes, U.K.; Assoc. Prof. J. A. Grant-Mackie, University of Auckland, New Zealand; Prof. B. Halstead, Imperial College, London; Dr. K.G.A. Hamilton, Agriculture Canada, Ottawa; Dr. M. O. Harris, Massey University, Palmerston North, New Zealand; A. J. Hughes, New Zealand AIDS Foundation, Auckland; Assoc. Prof. P. F. Jenkins, University of Auckland, New Zealand; Prof. K. Kaneshiro, University of Hawaii at Manoa; Dr. J. Lawton, University of York, U.K.; Prof. M. Littlejohn, University of Melbourne, Australia; Prof. S. Loevtrup, Muséum National d'Histoire Naturelle, Paris; Dr. C. Lofstedt, University of Lund, Sweden; Dr. R. J. Mahon, C.S.I.R.O., Black Mountain, Australia; Dr. B. H. McArdle, University of Auckland, New Zealand; Prof. B. Mishler, Duke University; Dr. G. Nelson, American Museum of Natural History, New York; Dr. K. Newberry, Veterinary Research

Institute, Onderstepoort, South Africa; Dr. J. G. Ollason, University of Aberdeen; Dr. C. Patterson, British Museum (Natural History), London; Dr. E. D. Penny, Massey University, Palmerston North, New Zealand; Dr. M. Potter, Massey University, Palmerston North, New Zealand; Dr. D. M. Rand, Harvard University; Dr. A. G. Rodrigo, University of Auckland, New Zealand; Dr. D. D. Shaw, Australian National University, Canberra; Prof. D. Simberloff, Florida State University, Tallahassee; Dr. E. Slooten, University of Otago, Dunedin, New Zealand; Prof. L. Stebbins, University of California at Davis; Dr. I. Tattersall, American Museum of Natural History, New York; Prof. P. J. Taylor, Cornell University; Dr. S. Telford, University of Cape Town, South Africa; Prof. A. Templeton, Washington University, St. Louis, Mo.; Dr. G. P. Wallis, University of Otago, Dunedin, New Zealand; Dr. L. Werdelin, Swedish Museum of Natural History, Stockholm; Prof. G. C. Williams, State University of New York at Stony Brook; and Prof. P. G. Williamson, Harvard University.

We are indebted to Peter Ritchie and Judith Robins for their assistance with editing. We would also like to thank the staff of Johns Hopkins University Press for their help in publishing this book. In particular, Richard O'Grady, their science editor, whose unstinting enthusiasm for the project ensured its completion, has our sincere gratitude. It has been a rare pleasure working with him. We also appreciate Sherry Hawthorne's copy editing skills.

David Lambert thanks Craig Millar, Judith Robins, Laurel Walker, and Garth Cooper for considerable assistance with the construction of the index.

Last (and clearly not the least), we would like to thank Hugh Paterson for being the inspiration behind the Recognition Concept and thus indirectly behind this collection. It is our firm belief that the Recognition Concept is a major advance in evolutionary thinking, and we hope that this book in some way reflects this conviction.

Introduction

BIOLOGISTS of whatever ilk recognize and attempt, in their individual ways, to explain the ordered diversity and variation of life on earth. Diversity is not seamless, but discontinuous, and this discontinuity occurs at many levels. Individual organisms are different from each other, all populations differ, and each species is distinct from every other species. But although there are differences at all levels, species have traditionally occupied a unique place in studies of diversity, even before Darwin's *Origin of Species*. To most biologists, species represent an important aspect of diversity at the populational level of biological organization. Indeed, species are populational units, groups of organisms that share something. The long-standing question is, simply, what do they share?

If species represent important units of biological diversity, then an obvious question is, What is the relationship between evolutionary processes and those units? Biologists from the latter half of the twentieth century have answered the question by asserting that species represent the direct outcome of adaptive evolution. They have been created, and are maintained, by forces that are actively responsible for them. The implicit view, then, is that species are a proper outcome of such adaptive processes. Hence their integrity will be protected by mechanisms which are responsible not only for their creation but for their maintenance also. Such isolating mechanisms became part of the established language and thought processes of evolutionary biology from the 1940s.

This view was challenged in a series of papers published from the mid-1970s in which Hugh Paterson rejected not simply the language but the underlying framework of the Isolation Concept of species. The language was inadequate, he argued, because species' integrities were not protected by "mechanisms" in the sense of Williams (1966). That is, they were not the direct products of selection for that "purpose" or "function." The only scenario under which "isolating mechanisms" were truly mechanisms was that of speciation by reinforcement. Paterson, however, pointed out a number of problems with this model. For example, the population genetics of heterozygote disadvantage suggests that a more likely result of such a situation is the elimination of the genetic cause of the disadvantage, rather than reinforcement. Paterson asserted that species need to be understood in terms of the underlying biological processes whereby conspecific males and females recognize each other. "Isolating mechanisms" simply represent an effect (*sensu* Williams 1966) or consequence of the existence of distinct mate recognition systems.

The Recognition Concept is a concept of species and is important in contemporary biology because it represents a distinct approach to the resolution of the problem of biological diversity. Paterson's concept is based upon a quite different analysis of the logic and language of species (see, e.g., MacNamara and Paterson 1984). Consequently, and not surprisingly, it offers significant implications for ideas about the origin of species.

The Recognition Concept has acted as a fillip in a range of diverse studies. The first part of this volume concerns the way Paterson's arguments have influenced people's ideas about just what species are. In the first chapter, Vrba argues for the reform of the language we use when dealing with biological diversity. She believes that because species represent complex systems, utilizing a unique combination of resources, and because speciation is the irreversible branching of lineages, the term *species* should be reserved for *recognition species*. This is the only concept to which the above criteria properly apply.

The chapters by Michaux and Turner both explore the implications of the Recognition Concept for species studies within the fossil record. There are a number of parallels between these studies. Michaux uses the Recognition Concept to investigate the relationship between extant and extinct species, using the molluscan genus *Amalda* as an example. He argues that species' delimitations of fossil material need to be based on data that are independent of the record itself. They need, he claims, to be based on biologically relevant information about extant species.

Like that of Michaux, Turner's essay deals with the relationship between the Recognition Concept and ideas about punctuated equilibrium. Turner argues for a decoupling of the concepts of within- and between-species change and consequently argues that the rate of change within a species actually offers no answer to questions about the speed at which it became that species or the processes that gave rise to any other. The error that leads to a conflation of the two sorts of change stems from the inadequate concept of species that underpinned the theory of speciation inspiring punctuated equilibrium.

An issue of interest to those concerned with the Recognition Concept has been the relationship between Paterson's ideas and those of the antiselectionist Japanese biologist Kinji Imanishi. The contribution by Ikeda and Sibatani explicates the central tenets of Imanishi's views on species and evolution and simultaneously provides an opportunity for western biologists to obtain insights into some contemporary ideas in Japanese science. The authors suggest that Imanishi believed that "each species is governed by its own laws which bind its individuals," a view with apparent parallels to the Recognition Concept. Nevertheless, they conclude that despite the superficial similarities between Paterson's and Imanishi's ideas they are conceptually distinct.

It is important to realize that many authors who use the Recognition Concept work in applied areas of biology. The chapter by Hunt and Coetzee illustrates this by discussion of the effects of Paterson's views on the understanding of malaria transmission in Africa. These authors review a range of genetic studies that have revealed the presence of cryptic species in *Anopheles* mosquitoes.

Jack Hailman uses his essay to further develop his long-held view that biology needs to be more operational. He outlines an operational view of species by rejecting the "nonoperational culprits" within the Biological Species Concept (i.e., the Isolation Concept), namely, "interbreeding" and "potentially." He explores a pedigree approach to the operational definition of interbreeding by defining organisms as interbreeding if they can be connected by parent-offspring links within a time since the last speciation event.

Although believing that the orthodox Isolation Concept is inadequate in the study of speciation, Hampton Carson suggests that individual recognition encompasses more than the evaluation of a potential mate's specific status. He argues that a mate recognition system also enables an assessment of the individual's fitness. Consequently he asserts that sexual selection must play a dominant role in the evolution of Specific Mate Recognition Systems (SMRSs).

Part Two of this volume concerns itself with the consequences of the Recognition Concept for speciation, ecology, and evolution itself. In the first chapter, Spencer reviews a range of models of speciation concerned with the founder effect. Paterson argued that these models are the only ones for which compelling evidence for their actual occurrence exists. Spencer concludes that there are in fact only two conceptually distinct models in this group: Mayr's genetic revolution (with its population genetic theory rectified) and Carson's founder-flush (which includes flush-crash).

Ritchie and Hewitt's chapter is prompted by Paterson's criticisms of simplistic assertions that cases of reduced hybrid fitness are likely, or certain, to result in the direct selection of isolating mechanisms. Reinforcement is only one of the four possible outcomes, they contend, together with maintenance of the hybrid zone, extinction of one taxon, and what they refer to as "amelioration." This last suggestion is similar to that of Spencer et al. (1986) that there could be selection of modifier genes which would diminish the strength of any deleterious heterozygous effects and which would be directly selected for.

Again from an applied perspective, Coetzee, Hunt, and Walpole, discuss their findings on two taxa of bedbugs that bite humans. Crosses between the taxa are usually unproductive, although cross-mating in at least one direction is known to occur frequently. The authors consider conflicting consequences of the Isolation Concept and the Recognition Concept in assessing the evolutionary status of these animals.

The chapter by Walter is a further development of his idea that the Recognition Concept implies a need in contemporary biology for an alternative theory of ecology. Such a theory would be centrally dependent upon the likely properties of species that have arisen via the divergence of their respective SMRSs in small populations.

Green's cladistic analysis of chromosomal inversions in *Anopheles* mosquitoes is another instance of applied research that has been influenced by ideas about the Recognition Concept. He examines two groups of *Anopheles* mosquitoes from the Afrotropical and Oriental regions that differ dramatically in the relative proportions of inversions unique to single species and those shared by two or more species. He discusses this pattern and the likely properties of the mutational events that produce inversions and the population-genetic events that fix inversion alternatives.

Lambert investigates a range of implications of changes in the meaning of *function* as it is used in evolutionary biology. He extends Williams's suggestion that adaptation is an onerous concept and argues that while the *function/effect* distinction is important, it is virtually impossible to precisely locate the function of any character, because the functions are simply lost in history. Indeed, from his perspective the historical approach itself is problematical, and he argues that these views represent a logical extension of Paterson's original distinction between function and effect in relation to ideas about species.

Part Three of this volume is concerned with the nature of mate recognition systems themselves. Linn and Roelofs begin with an extensive consideration of the literature on pheromone communication. They consider the factors in a noisy environment which, they suggest, might be involved in the formation of new SMRSs. The nature of such changes and the possible role of intraspecific competition are discussed, and they conclude that moth pheromone systems provide excellent opportunities to investigate some of these issues.

Similarly, White, Lambert, and Foster also discuss the SMRSs that employ pheromones and use the opportunity to analyze the properties of systems themselves and SMRSs in particular. They consider the vexed question of the significance of variability in components of the SMRS and assert that the concept of variation is applicable only to the components, and not to the system itself. Thus, although the SMRS will not be the same in all individuals, it is the components of the system that exhibit variation (along with almost all biological characters). The authors also contend that the nature of systems means that systems can change only in a discontinuous fashion, and they review information that indicates that changes in SMRSs occur in this way.

There are a number of authors who have quite fundamentally different ideas about the issues of stability and variation. Butlin, for example, reviews the evidence for variation in mating signals and responses. He concludes that there is widespread variation in the former and argues that this finding is more in keeping with what he calls the "competition" view (emphasizing lability), as opposed to the "recognition" view (emphasizing stability).

Lane's chapter exhibits the heuristic value of the Recognition Concept in his study of a possible subalpine hybrid zone between two taxa of New Zealand cicadas belonging to the genus *Kikihia*. In contrast to previous studies that have used the Isolation Concept as their framework, Lane shows that acoustic and morphological data are consistent with the existence of two distinct species. Significantly, the patterns of acoustic signals important in the SMRS are "remarkably stable" throughout the geographical ranges of the respective species.

The issue of variability is important to Villet, who deals with similar material and remarks that the degree of variation that would be expected in SMRS acoustic signals has not been explicitly detailed by proponents of the Recognition Concept. He asserts, however, that empirical studies of such acoustic components of the SMRS show more variation than he would have expected. He concludes by modifying Paterson's model of how SMRS signals change.

The analysis by den Hollander of the acoustic signals in leafhoppers also uses the Recognition Concept. He argues that the SMRS can be used to resolve the specific status of populations where other taxonomic methods cannot or in cases where there are conflicting results. Den Hollander illustrates a number of cases in which substrate-transmitted acoustic signals have been used to resolve species problems.

Finally, Eldredge discusses the contribution of Paterson's SMRS to our understanding of species and of speciation. He contrasts this with "Mayr's basic model of speciation" in which any genetically based change in allopatry may lead to speciation. In contrast, he contends, Paterson's view is that change in reproductive adaptations of a species is both necessary and sufficient for speciation to occur. Consequently, he suggests that "Paterson's Recognition Concept of species opens the door to a more general consideration of the relation between reproductive and economic stasis and change in phylogenetic history."

It should be clear from the above outlines that this book does not set out to present a comprehensive or definitive statement of the Recognition Concept; Paterson's own papers are more than adequate for that job (Paterson 1993). Nor is this book an attempt at a manifesto of the Recognition Concept. In fact, the book reviews some clear disagreements among contributors, notably with respect to sexual selection and issues of variability and stability. Anyone looking to this book for a unanimous statement of the principles and implications of the Recognition Concept, therefore, will be disappointed. Instead we have assembled a collection of papers by scientists who have been strongly influenced by the Recognition Concept, and these authors have written about the consequences of this idea for their own studies. We think it is hardly surprising that ideas of great consequence strike people differently. Indeed, we believe that the diversity of opinion expressed in this book is yet another measure of the timeliness and importance of Paterson's contribution to evolutionary biology.

References

Paterson, H.E.H. 1993. Evolution and the Recognition Concept of species: Collected writings. S.F. McEvey, ed. Johns Hopkins University Press, Baltimore.

Macnamara, M., and H.E.H. Paterson. 1984. The Recognition Concept of species. S. Afr. J. Sci. 80:312–319

Spencer, H.G., B.H. McArdle, and D.M. Lambert. 1986. A theoretical investigation of speciation by reinforcement. Am. Nat. 128:241–262.

Williams, G.C. 1966. Adaptation and natural selection. Princeton University Press, Princeton, N.J.

*Ideas about the
Nature of Species*

1

Species as Habitat-Specific, Complex Systems

ELISABETH S. VRBA

Department of Geology and Geophysics
Yale University

THE SPECIES CONCEPT has often been construed as an epistemological "chameleon" able to take on whatever meaning happens to be convenient in a given subfield. This tradition is reflected by the many species definitions bearing descriptive prefixes, like **chronospecies, morphospecies,** or **taxospecies** (terms defined in the glossary at the end of this chapter). The contrary view, which I share, is that all species should be the same kind of ontological entity and that all speciation should denote one particular kind of process. I will compare the major ontological concepts that have been debated recently (e.g. Otte and Endler 1989; Kimbel and Martin 1992): the **Isolation** (= "Biological," Mayr 1963), **Recognition** (Paterson 1978, 1985), **Phylogenetic** (Cracraft 1983), and **Cohesion** (Templeton 1989) species concepts. Table 1.1 compares the concepts in terms of the character divergence regarded as necessary for speciation.

It is largely a semantic matter which kind of entity should bear the venerable name *species* and what the names for other kinds of organismal groups should be. But the conceptual distinctions should be made and honored by different names of some kind. I will explore four questions that I regard as fundamental to the natural organization above the level of organisms:

1. Does a given concept require a species to have the *status of a complex system* (see Simon's [1962] definition in the next paragraph), by virtue of consistent, heritable among-organismal interactions? Or does it approach species and species' characters, respectively, as no more than aggregates of organisms and of organismal characters related by descent?

2. Is each species required to have a unique **habitat specificity**, that is, a unique combination of **resources** that it can use?

Table 1.1. Character divergence required for speciation under the Recognition (Paterson 1985), Isolation (= "Biological" [Mayr, 1963]), Cohesion (Templeton 1989), and Phylogenetic (Cracraft 1983) species concepts

Species concept	Required character divergence							
	Prefertilization system		Postfertilization system		Habitat adaptation		Any other	
Recognition	X							
Isolation	X	or	X					
Cohesion	X	or	X	or	X			
Phylogenetic	X	or	X	or	X	or	X	

3. Does a concept require speciation to result in *irreversible lineage branching*? Or does it allow species to be ephemeral with species-specific characters that either disappear or merge with those of the sister-species when the two come into secondary sympatry?

4. Is each species required to be a *monophyletic group* (*sensu stricto*, Nelson 1971) of organisms?

The concepts are characterized according to these four criteria in Table 1.2. I also discuss the operational problems of using each of the concepts to diagnose species.

Should Species Be Construed as Complex Systems?

One definition of a dynamic system is that of Caswell et al. (1971:4): "a collection of objects, each behaving in such a way as to maintain behavioral consistency with its environment which may include other objects in the system." This includes such a broad range of phenomena that "species" under

Table 1.2. Comparison of the implications of the Phylogenetic, Cohesion, Isolation, and Recognition species concepts

Species concept	Do species require:			
	Complex system status?	Unique habitat specificity?	Irreversible branching?	Monophyly?
Phylogenetic	No	No	No	Yes
Cohesion	No	No	No	No
Isolation	Yes	No	No	No
Recognition	Yes	Yes	Yes	No

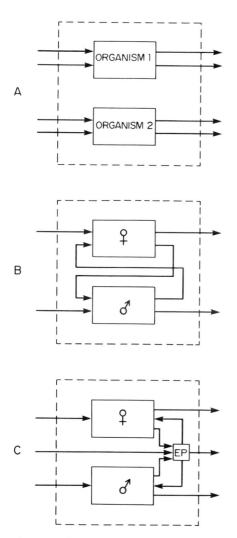

Fig. 1.1. Block diagrams representing organisms as subsystems within two kinds of organismal groups, or systems. a, The simple asexual case: organisms do not interact with each other, only with the environment. b, The complex biparental sexual system: males and females not only each have inputs from and outputs to the environment, but also interact with each other. c, Male-female interactions result in an emergent property, the Specific-Mate Recognition System (Fig. 1.2), that interacts with the environment.

all concepts qualify as systems. I will focus more narrowly on the complex system discussed by Simon (1962:86) as "a system made up of a large number of parts that interact in a nonsimple way. In such systems, the whole is more than the sum of the parts, not in an ultimate metaphysical sense, but in the important pragmatic sense that, given the properties of the parts and the laws of their interaction, it is not a trivial matter to infer the properties of the whole. In the face of complexity, an in-principle reductionist may be at the same time a pragmatic holist." Figure 1.1a represents a system, the subsystems (here organisms) of which interact only with the environment and not with each other. I am concerned here with complex systems (Fig. 1.1b,c) in which the causal influences acting between environmental inputs and outputs involve not only the subsystem states, but also emergent properties of the system resulting from particular interactions between subsystems.

Living entities clearly include complex, hierarchically structured systems. In fact, one of the most striking and quintessential attributes of life is its **hierarchy** of subsystems (or stable subassemblies, *sensu* Simon 1962) nested and interacting within larger systems, and so on, at levels of increasing complexity. If there is any biotic phenomenon, besides reproduction and resulting genealogy, that definitely needs representation in biological formulations and concepts, it is this one. Of course, the complex-system hierarchy *is* prominently acknowledged by the most basic concepts we have. A "gene" made up of certain base pairs is different from a collection of those same base pairs, because the gene consists of those base pairs *and* a particular arrangement and interaction of those base pairs relative to one another. "Cell" and "metazoan organism," referring to two kinds of complex systems each of interacting subsystems, are also keystone concepts, which are real to all biologists and for the most part unambiguously and uniformly defined.

Groups of Organisms as Systems

When it comes to entities arising from organization *among* organisms the meanings of the few terms we have are hotly disputed. This failure with respect to systems *above* the organism level is unprecedented for biology. I suggest that the overwhelming reality and importance of hierarchy in biology is one good argument to reserve the familiar term *species* for a kind of complex system existing above the organism level.

Many of the actions organisms do by which they stay alive are purely or largely direct transactions with the physical environment. For instance, when oryx antelopes of the species *Oryx gazella* thermoregulate during the heat of a day in the African deserts, by panting that sets up countercurrent cooling of arterial blood on its way to the brain (Taylor 1969), in one sense it is true that the *species* is thermoregulating. But it is true only in the sense of a simple

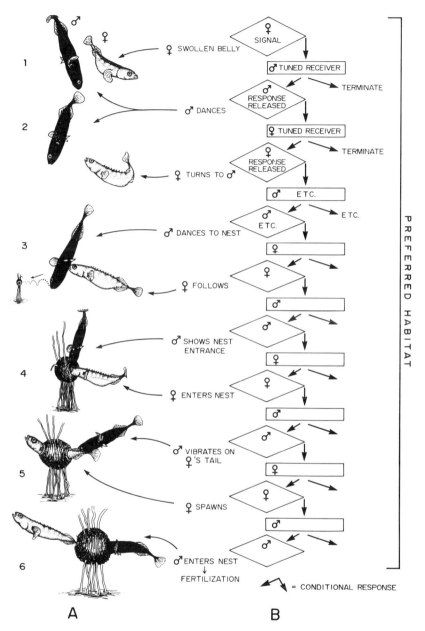

Fig. 1.2. *a*, The Specific-Mate Recognition System (SMRS) of *Pygosteus pungitius*, the ten-spined stickleback. *b*, Schematic illustration of the form of the SMRS as a coadapted signal-response reaction chain. (Adapted from Morris 1970[a] and Paterson 1985 [b].)

summation of the organismal behaviors (Fig. 1.1a, if one thinks of the larger system as *Oryx gazella,* the subsystems as oryx antelopes, the inputs as temperature, and the outputs as cooler brains, altered metabolism, behavior, etc.): The species *O. gazella* is not acting as a complex system with respect to thermoregulation. Many other activities, such as locomotion, feeding, and excretion, belong in this category, at least in many instances in many lineages.

Yet there are other cases of behaviors that depend on *among*-organism interactions. Where an entire lineage shares genetically based interactive behavior of this kind, with broadly predictable results in terms of population structure, variation-composition, and distribution, the lineage acts as a complex system, and those results are correctly termed heritable emergent properties of the lineage. (See Schull [1990], who explored species as massively parallel distributed processing systems, indeed as intelligent systems, with emergent properties based on mutation and recombination rates, on interdemic variation, and on the "species' norm of reaction" based on all the reaction norms of the component organisms.)

The most striking among-organismal organization is evident in biparental sexual reproduction [Fig. 1.1b,c and Fig. 1.2]. Because sexuality versus asexuality has featured prominently in species debates, let us briefly consider different reproductive modes.

Reproductive Modes

The distinction that is particularly relevant is that between (1) lineages that never have biparental sexual reproduction and (2) lineages that at least sometimes have biparental sexual reproduction.

A lineage of kind 1 (represented by Fig. 1.1a) is not a complex, among-organismal, reproductive system (although, of course, those organisms may interact in respects other than reproduction, such as by colony formation and by subdivision among them of feeding and defense roles and may constitute complex systems with respect to those interactions). Lineages with reproductive modes of kind 1 include the following (see glossary definitions from Bell [1982] and extensive discussions in Ghiselin 1974*a;* Williams 1975; Maynard Smith 1978; and Bell 1982): all lineages with obligate parthenogenetic **thelytoky** in either of its guises, **apomixis** and **automixis**, which are all uniparental modes (except biparental **gynogenesis**), and lineages that always use uniparental sexual modes such as obligate self-fertilizing hermaphrodites.

Lineages with reproductive modes of kind 2 follow the scheme in Fig. 1.1b,c at least some of the time. They comprise all those with obligate **amphimixis** (i.e., with syngamy between biparentally produced gametes); all lineages with parthenogenetic **arrhenotoky** (because although the haploid males are from unfertilized eggs, diploid females are produced from fertilized eggs); and all lineages with **heterogony** (with cyclical parthenogenetic life cycles in which one or more thelytokous generations are followed by an arrhenotokous or amphimictic generation).

Prevalence of Reproductive Modes

Many have objected to a sexual reproduction criterion for species on the grounds that several kinds of populations (those of kind 1) would be excluded from species status. (See review in Templeton [1989] who sums up his objection as the problem of "too little sex".) Thus, it is of interest what proportion of differentiated forms of life is in fact of kind 1, that is, that never engaged in biparental sex.

I have not found in the literature any estimated numbers of obligate self-fertilizing hermaphrodites. But Maynard Smith (1978) addressed such estimates for the other, larger category among reproductive modes of kind 1, namely that of obligate parthenogenetic thelytokes. He noted that "existing parthenogenetic populations must be descended from many hundreds of different ancestral lineages. Yet, with one important exception (the bdelloid rotifers), in no case does a major taxonomic group (subfamily or above) consist predominantly of thelytokous populations" (p. 53). On the same page he gave the example of the twenty-eight known thelytokous varieties of psocids (Insecta: Psocoptera) which "belong to thirteen different families, and in twelve of the twenty-eight cases there are sexual and parthenogenetic forms of the same nominal species." He concluded that a fundamentally similar picture is found in plants. It seems that thelytokous forms crop up in phylogenetically isolated positions, that these forms have hardly ever diversified after origin, and that *obligate* thelytoky may be rare. In fact, the major reviews (Ghiselin 1974*a*; Williams 1975; Maynard Smith 1978; and Bell 1982) suggest that the vast majority of distinct lineages among the highly diversified metazoans at least at some times engage in biparental sex.

Lineage as a Complex System due to the Fertilization System among Organisms

Figure 1.1b represents two kinds of subsystems, male and female, that share a fertilization system (the male-female interaction linkages in Fig. 1.1b) to constitute a larger system. The fertilization system of many taxa in which males and females remain in proximity is simple. In contrast, most mobile organisms have additional signal-response interactions that serve to bring sexual partners together for mating—the subsystem of the fertilization system that Paterson (1978) has termed the **Specific-Mate Recognition System** (SMRS). Among SMRSs, one can further note widely differing degrees of complexity. At one end of the spectrum are the minimal signal-response interactions of, for instance, the chemical recognition system found by Wiese and Wiese (1977) between mating types of the green alga *Chlamydomonas moewusii*. At the other extreme are some elaborate vertebrate SMRSs that include olfactory, auditory, tactile, and visual (color and behavioral) signals. In Figure 1.2 the form of the SMRS as a coadapted signal-response reaction chain is schematically illustrated (b) and exemplified by the moderately complex SMRS of *Pygosteus pungitius,* the ten-spined stickleback (a). (This and other courtship sequences are analyzed in Morris [1970]. Note that Fig. 1.2 shows the "classical" version of the courtship reaction chain in *P. pungitius,* while in reality most responses of either sex can be released by more than one action by the opposite sex. Yet Morris points out that the range of possible sequences that lead to fertilization is severely constrained.)

I suggest that a lineage of biparental organisms that share a unique fertilization system is itself a system, with both "vertical" linkages among organismal subsystems (by chains of ancestry and descent through time) and "horizontal" linkages (by interactions at particular times). To acknowledge the larger entity (here the species *P. pungitius*) as a system does not require that the linkages between male and female subsystems must be activated at all times. Inputs from the environment, together with the internal states of the subsystems, at some times result in a positive signal along the interaction linkages and at other times in a zero signal value, as is well known in nature for sexual systems. For instance, even in an obligate biparental sexual form like *P. pungitius*, reproduction is suppressed in environments that are unsuitable for SMRS function (as will be elaborated on below, after Morris 1970). In heterogonic lineages, such as *Aphis fabae* and many other aphids, many parthenogenetic generations can intervene between episodes of sexuality; and experimental manipulation of variables like temperature and photoperiod can elicit the switch from one reproductive mode to the other (Bell 1982). But the temporary absence of positive signals along the SMRS interaction linkages does not disqualify the lineage from constituting a sexual system.

Figure 1.1c represents the notion that the interactions among the subsystems can result in emergent systems properties. (I here mean the operational concept of emergence used in mathematical modeling of dynamic systems about which, for instance, Caswell et al. (1971:39) wrote: "It is an unfortunate fact that emergent properties have sometimes been given an almost mystical character in the literature. They *are* real phenomena, which do arise to confront anyone studying a complex system.") Fully elaborated sexual reproduction as a whole is a synapomorphy emergent at the lineage level that characterizes a huge living clade (if those who argue that it evolved only once are correct). Each unique SMRS, such as the stimulus-response reaction chain of the ten-spined stickleback (Fig. 1.2) and including all of its genetic and phenotypic components, is a complex emergent property, subdivisible into component emergent properties, of a particular lineage system, although particular SMRS components may be shared by clades over millions of years.

The sexual system has had enormous consequences for evolution, through its effects on the variation distributions within and among lineages and on rates of lineage diversification and extinction. While sexual lineages have massively diversified into the vast majority of distinct metazoan phenotypes on earth, asexual lineages characteristically have remained undiversified from their phylogenetically isolated origins (Williams 1975; Maynard Smith 1978; Arnold et al. 1989). As Maynard Smith (1978:54) puts it: "The facts fully support the . . . view that parthenogenetic varieties are doomed to early extinction." If evolutionary impact is considered a criterion for what should be encompassed by our favorite biological concepts, then the unique sexual systems among organisms are good candidates for species status.

Species Concepts in Relation to Complex Systems

The Phylogenetic Concept

The requirement under the Phylogenetic Concept that species be strictly monophyletic is based explicitly on nested homologies at the *levels of organisms and their constituents*. Whereas organismal characters that participate in sexual systems are used to deduce organismal genealogies, emergent characters of among-organismal systems are excluded. The set of phylogenetic species includes both asexual and sexual monophyletic groups of organisms. Thus, whereas this concept brings a strong hierarchical approach to deducing organismal genealogy, it ignores hierarchy in the systems sense by leaving out levels of organization above that of organisms.

An example within my family partly illustrates this species concept. My grandmother was born with the mutant phenotype of an abnormal left lower rib morphology. By chance assortment of alleles it happens that all her descendants so far bear this phenotype. Thus, we are a "smallest diagnosable cluster of individual organisms within which there is a parental pattern of ancestry and descent" (Cracraft 1983). Thus, I gather that under the Phylogenetic Concept we are a species. Yet, as we happen to live in different parts of the world without interactions (all our interactions, including sexual ones, are with "nonspecifics"), we are not a system of interacting subsystems. If we are a "species," we are likely to be an extremely ephemeral one.

The Recognition Concept

In contrast to the Phylogenetic Concept, the Recognition Concept focuses squarely on systems properties above the organismal level: the crux of the recognition species, and the phenomenon we can look at in nature to study it, is the system of "horizontal" linkages of interactions among sexual partners (Fig. 1.2). By virtue of strong stabilizing selection on the fertilization system (Paterson 1978, 1981, 1985), it also has potential stability through long time along the "vertical" linkages of ancestry and descent.

As mentioned above, the SMRS is an important subpart of the fertilization system in mobile organisms. Although entire clades of species may be characterized by individual components of the SMRS, such as a pheromonal, auditory, color, or behavioral phenotype of the male or female, a recognition species can be diagnosed only by the complete system based on a combination of such characters that is unique to a given cluster of related biparental organisms. (See example in Fig. 1.2. I emphasize that the choice of the fairly complex example in Fig. 1.2 should not be mistaken for "vertebrate-centric" bias against taxa with the rudimentary fertilization systems, which would not do justice to Paterson's [1978] intent. He has consistently argued that his concept applies equally to taxa, like many plants and the *Chlamydomonas* algae mentioned above, with simple systems.)

Paterson argued at length (convincingly, in my view) that divergence in the *post*fertilization system on its own (see Table 1.3) is highly unlikely to initiate

Table 1.3. Mayr's (1963:92) classification of "isolating mechanisms" augmented from Templeton (1989)
(The term *isolating mechanisms* is in quotes to indicate that it is inappropriate [Paterson 1978].)

Prefertilization isolation[a]

1. Premating isolation
    a. Ecological (habitat and temporal) isolation: potential mates do not meet because they mate in different habitats (or use different pollinators, etc.) or in different seasons
    b. Ethological isolation: potential mates meet but do not mate
2. Postmating but still prefertilization (prezygotic) isolation
    a. Mechanical isolation: copulation attempted but no sperm transfer takes place
    b. Gametic mortality or incompatibility: sperm transfer occurs but egg is not fertilized

Postfertilization (postzygotic) isolation

3. a. F1 inviability because zygote dies or produces a hybrid of reduced viability
   b. F1 sterility because hybrid is fully viable but partly or completely sterile
   c. F2 or backcross hybrid breakdown due to reduced viability or fertility
   d. Endoparasitic or cytoplasmic infection of a population results in fertility or viability breakdown after matings with uninfected individuals, although infected individuals are interfertile

[a]Under Paterson's (1978, 1985) Recognition Concept (1) prefertilization isolation (1a–2b) results as an incidental effect of divergence of the prefertilization system; and (2) in motile organisms the character subset of the fertilization system of especial importance is the Specific-Mate Recognition System (SMRS), underlying (1b) and, provided one accepts "lock-and-key recognition" during copulation and "gametic recognition" as part of the SMRS, also (2a) and part of (2b).

new separate branches on the "tree of life." Therefore, the species concept should focus on the causal network that matters: the *pre*fertilization—or simply fertilization—system. To my knowledge he was the first to argue this "to the hilt" of its consequences for branching in the sexual biota. In doing so he particularly had to address the logical and biological shortcomings of the Reinforcement Model of speciation according to which "isolation mechanisms" (Table 1.3) evolve by selection for isolating function (as espoused by Dobzhansky 1935, 1937; Ayala et al. 1974). Since the publication (Paterson 1978) many have quietly come to a closely similar conclusion on the improbability of reinforcement speciation (e.g. Mayr 1982; Lande 1979; Futuyma and Mayer 1980; Templeton 1981, 1989; Carson 1982; Butlin 1989).

The Isolation Concept (Biological Species Concept)

The elements of the Isolation Concept of species date from Dobzhansky's (1935, 1937) analyses and also Mayr's (1942, 1963; but see Eldredge, this volume). These were seminal in replacing the **typological species** concept with the neo-Darwinian synthesis of species as sexually reproductive units. This synthesis deserves credit for demolishing the idea of species as classes with defining essences (see Mayr 1970; the notion of species as logical individuals was later argued extensively by Ghiselin [1974*b*] and Hull [1976]).

It is curious that, while this tradition explored the implications of a reproductive system, especially for population variation, it focused strongly on its *negative* consequences—reproductive isolation—rather than on the system itself. Its central theme has been the importance of isolating *mechanisms* and how they could evolve by selection. Both prefertilization and postfertilization divergence to reproductive isolation are regarded as separately sufficient for speciation (Tables 1.1 and 1.3); thus, the idea of among-organismal systems is implicit in this concept (Fig. 1.1; Table 1.2), but in a negative sense. If species are to be construed in terms of sexual systems then, as Paterson has remarked (pers. comm.), it is as curious to construe them in terms of an "isolating function for hybridization avoidance" as it would be to view sex as being "for celibacy avoidance." It is hardly surprising that many speciationists (e.g. Templeton 1989) have come to agree with Paterson (1978) that this concept, although long known as the *Biological Species Concept,* is more appropriately termed the *Isolation Concept.*

The Cohesion Concept

The Cohesion Concept represents Templeton's (1989) effort to include all asexual and sexual forms (see the glossary for an outline of his ideas). He did not define cohesion more explicitly than genetically based "phenotypic cohesion." Among the "cohesion mechanisms" that promote phenotypic similarity by descent among organisms, he included very diverse phenomena: gene flow that promotes (while its absence preserves) genetic identity, the fertilization system *and* the postfertilization developmental system (and, thus, isolating mechanisms), genetic drift in monophyletic groups of organisms, natural selection for **niche** adaptation, and a variety of constraints on evolutionary change.

Table 1.4 gives my interpretation of how Templeton (1989) would apply his criteria of "genetic" and "demographic exchangeability" to decide on species status in different cases of two populations. *Case 2:* For two or more sexually reproducing populations to belong to the same species, "genetic exchangeability" between them is not sufficient on its own in this view. They have to be "demographically exchangeable" as well by virtue of a common "fundamental niche." For instance, the members of a **syngameon** (a syngameon is equivalent to a recognition species), because they are not demographically exchangeable, are separate cohesion species although they share genetic exchangeability. *Case 3a:* Demographic exchangeability can be sufficient on its own to decide species status without any reference to genetic exchangeability, as in the case of the "asexual species" recognized under this view. *Case 3b:* Yet, demographic exchangeability between two populations does not guarantee conspecificity. I gather that Templeton would recognize two sexual populations that still share a common habitat but have diverged postfertilization systems (i.e., that have lost genetic exchangeability) as two species.

In sum, the cohesion species clearly does not *require* an among-organismal interactive system; for instance, the consequences at the population level (genetic identity) of selection at the organism level or of drift are enough.

Table 1.4. Number of cohesion species recognized in different cases of two populations[a]

Demographic exchangeability between two populations	Genetic exchangeability between two populations	
	Yes	No
Yes	1. One Species	3a. Asexual: one species
		3b. Sexual: two species
No	2. Two Species	4. Two Species

[a]This table represents my interpretation of how Templeton [1989] would apply his criteria of "genetic and demographic exchangeability" to decide on species status.

Does a Species Need a Unique Component of Habitat Specificity?

The strong association in nature of distinct organisms with different combinations of climatic and other habitat variables (Darwin 1859) must surely be a prime candidate for the basis of a species concept. One can argue that species status should necessarily imply a unique relationship to habitat (note the subtle difference between a **habitat** and a **niche**). Before I compare the four species concepts in this respect, I raise five points regarding habitat specificity and the resources that organisms can use. In one way or another these points are relevant to all species concepts. In discussing these five points, when I refer to "species" I here have in mind not a particular bias with respect to problematic units (such as syngameons and sibling species), but those many units in the systematic literature the species status of which seems to enjoy widespread consensus.

Habitat Specificity as a Tolerance Range Based on Genetic and Ecophenotypic Variation

The habitat specificity of a species, with respect to a particular habitat variable (such as mean annual rainfall, temperature, or soil nutrient status) is always a *range* of tolerance within which life and reproduction can occur and outside which it is impossible for that species. The same is true for an organism's habitat specificity.

The habitat tolerance range of a species may be influenced at three levels: (1) genetic and phenotypic variation in resource use *among populations,* (2) *among organisms* within each population, and (3) the environmental tolerance range *within each organism,* resulting (3a) from the varied environmental tolerance of any particular phenotype or (3b) from the varied phenotypic expressions possible from a single organismal genotype in different environments. The latter (3b) is also known as the norm of reaction (see Stearns [1982] for numerous examples). Many lineages have broad reproductive norms of reaction, switching facultatively between different phenotypes, and from sexual to asexual reproduction, depending on environmental cues (Stearns 1982; Bell 1982). The nature and breadth of a norm of reaction itself are genetically based and can evolve (Stearns 1982). Thus, a species' habitat specificity is ultimately genetically based.

Estimation of Habitat-Specific Limits

The limits, with respect to variables such as temperature, rainfall, substrate, food, and vegetation cover, of a species' habitat specificity can in principle be quantified. For instance, Caithness (1990) used multivariate discriminant-

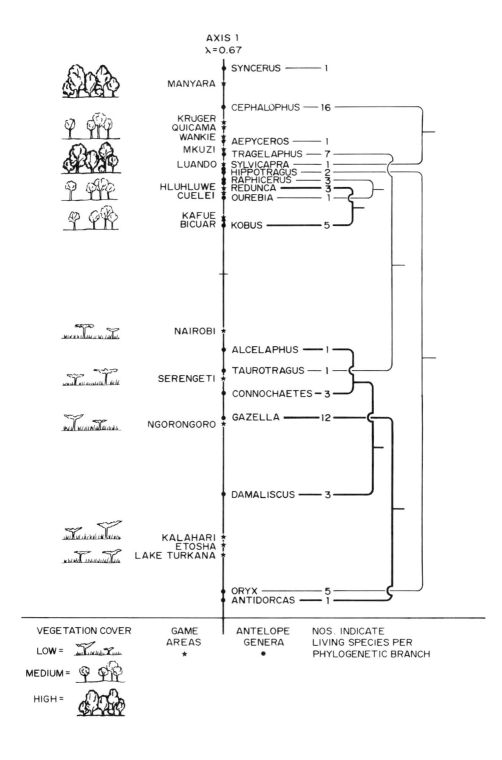

function methods and bioclimatic profile matching (Nix 1986) to estimate such limits for each of seventy-three African antelope species, in terms of twelve temperature- and rainfall-related variables.

Also, the relative importance of different habitat variables to a given lineage can be estimated. For instance, Greenacre and Vrba (1984) used correspondence analysis on antelope census data, from areas representing different ecosystems across subsaharan Africa, to investigate the relative causal influences of habitat components on the biogeographic patterns. This method is objective in not presuming any causal structure underlying the data. Instead, any nonrandom structure is revealed afterward by comparing the resulting distribution of taxa and areas along the axes with "supplementary" environmental variables for each game area. In our study these included mean annual rainfall and temperature, soil nutrient status, and vegetation cover coded from low to high proportions of woodcover compared to grasscover. We found (Fig. 1.3) that the vegetation cover codes plot in almost perfect ascending order against axis 1: among the variables we considered, *gross vegetational physiognomy* (i.e., not defined by plant species but by wood-to-grass proportions) was pinpointed as primary in the habitat specificities of the antelopes.

Long-term Heritability of Habitat Specificity

Within broad limits, components of habitat specificity can be heritable and characteristic for entire clades through millions of years. For instance, the cladistic relationships in Figure 1.3 (based on mitochondrial DNA [J. Gatesy pers. comm.] and skull characters [e.g. Vrba 1979; Vrba et al. 1994]) suggest this for vegetational habitat specificities in antelope taxa. Note that all eight species in the reduncine clade (*Kobus, Redunca*) are in the open woodland part, and all seven in the alcelaphine clade (*Alcelaphus*, etc.) are in the open grassland part of the vegetation spectrum. Each clade is known since at least five million years ago. The main point is that hypotheses of long-term constancy (due to heritability within and among species) of components of habitat specificity are testable cladistically: The distributions of species' habitat specificities among taxa are expected to form hierarchically nested sets and subsets just as those of the more conventional phenotypes do. There is already considerable support for the notion that species and clades can have long-term habitat fidelity (e.g. Vrba 1987, 1989, 1992).

Fig. 1.3. Axis 1 from a correspondence analysis (Greenacre and Vrba 1984) showing generic antelope frequencies in sixteen African game areas (*center*) to be strongly influenced by the habitat-variable vegetation cover (*left*). Cladograms (*right*) suggest that vegetation habitat specificities have been heritable characteristics (synapomorphies) for some clades through millions of years.

A Species' Habitat Specificity as a Unique Combination of Resource Ranges

It follows from the previous observation that a "unique habitat specificity" of a lineage or species is likely to involve only one or a few habitat components and often displacement of tolerance-range limits relative to those of related species while maintaining range overlap. Herein lies the reason, I believe, for the observations (e.g. Huntley and Webb [1989] for North American trees; Bush and Colinvaux [1990] for Central American forest communities; Coope [1979] and Coope and Brophy [1972] for beetles; Sutcliffe [1985] for European mammals) that, while species' habitats and geographic distributions during past climatic changes underwent large-scale shifts, they did not always shift in perfect "lockstep," so that the ecological associations of taxa were different to some extent in the past (even in the Quaternary) from those today. This is hypothetically depicted in Figure 1.4 in which two extant species, A and B, are compared with respect to three habitat variables. A and B differ but overlap in tolerance of temperature, rainfall, and substrate. A is more generalized with respect to mean annual rainfall and B with respect to substrate. As a result, A and B are allopatric today, but were partly sympatric during a past colder period. The main point here is that each species has a particular tolerance range for each habitat component that may or may not overlap or be identical to that of other species and a *combination* of such ranges for all its requirements that is unique.

Paleoclimatic Implications for Species' Habitats and Distributions

It is now generally accepted (e.g. Berger et al. 1984) that Earth's paleoclimate has cycled periodically between global cooling and warming, although not always accompanied by polar ice changes as during the Plio-Pleistocene. Three dominant cycles have been documented (the Milankovitch cycles): of roughly 100,000-, 40,000-, and 23,000-year periodicities. The best evidence for these astronomically caused cycles, from the Plio-Pleistocene, shows that the cycles were accompanied not only by large-scale expansion and retreat of ice at the poles (Denton 1985; Hays et al. 1976; Shackleton et al. 1984), but also by major climatic and vegetational changes in the land tropics (reviewed in Rind and Peteet 1985). These cycles, in some form, must have accompanied the entire history of life (see Olsen [1986] for the Triassic; Park and Herbert [1986] for the Cretaceous). So far we have data only for small parts of the fossil record. These show that over longer periods (one to several million years apart) the "background" Milankovitch cycles underwent major changes in mean and mode. Some have argued (review in Vrba 1992) that tectonic changes precipitated these major displacements of the Milankovitch cycle curve. (Note that these hypotheses suggest a causal role for tectonism only with respect to the *changes*

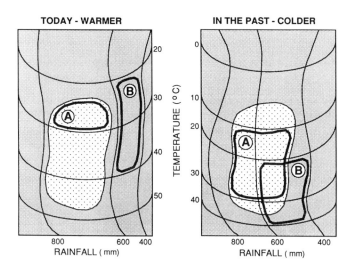

Fig. 1.4. Components of the habitat specificities of related species showing commonly overlapping ranges. Hypothetical species A and B, due to overlapping specificities for three habitat variables, have disjunct distributions today, yet had sympatric ones in the past.

in the nature of the Milankovitch cycles—the basic cycles are known to result from astronomical changes [e.g. Broecker and Denton 1990].)

The Milankovitch climatic amplitudes were large, at least for some periods and areas, relative to the habitat adaptations of most extant species (Vrba 1992b). And the less frequent changes in cyclic mean and mode must have been even more strongly felt by the biota. Thus, climatic and habitat changes of large scale swept back and forth over continents and in the oceans with periodicity about, for instance, 1/20th of the duration time of the average species of terrestrial mammals (about two million years, from observation of the African record). Much evidence shows (Vrba 1993) that most known species survived many climatic cycles by the passive response of geographic shifting and vicariance of their distributions, while maintaining habitat fidelity. This was by far the most common response of lineages (the rarer responses were extinction or evolution of novelty with or without lineage branching).

Most species occur today in a vicariated state, often with numerous allopatric populations. Many have cited this as a problem for species concepts (for instance, it is a barrier to species diagnosis under the Isolation Concept; see below). It has also been argued to negate system status for groups the organisms even if they share a common fertilization system, that is, for recognition species. Miller (1978:24) wrote: "The individual members (organisms) of a given species are commonly units of widely separated concrete systems. The reason the species is not a concrete system is that, though all its members *can* interbreed and interact, they do so only locally."

I suggest that the new paleoclimatic evidence necessitates a revision of our preconceptions based on the vicariated populations that we see today. Consider that each of all new species (under all concepts here considered) is expected to have originated as a single population, which spread initially in contiguity. Vicariance and allopatry followed only later with an extreme climatic change. But during regular background Milankovitch cycling, any one such climatic (for instance, cold) extreme is only one in a series of similar, regularly recurrent extremes, which is also true of the soon-to-follow (in geological terms) alternate (warm) extreme. Thus, in terms of "species time" or "lineage time," allopatry is an evanescent (yet recurring) phenomenon, with reconstitution of geographic distribution to sympatry in most cases soon to follow. The often-quoted metaphor "the gavotte of the chromosomes" in genetics applies analogously in biogeography as "the gavotte of the geographic distributions." The genetic metaphor describes the idea that the signal values of interaction linkages between different chromosomes are sometimes zero and at other times positive, as they alternate between quiescent separation and interactive contact on the cellular time scale. The analogous phenomenon occurs between males and females in populations and between populations sharing a fertilization system, on the organismal and species time scales, respectively.

Miller (1978) argued that a committee that meets once a year using the same committee rules or charter from year to year (with perhaps a total duration of that charter of twenty years with a turnover of members) is a good system. By analogy, a mammal species' population that resumes contact under the same "charter" (undiverged fertilization system) with other conspecific populations every 100,000 years or less must also be a good system (quite apart from the argument that the same system can exist in allopatry, such as Alaska, Hawaii, and the rest of the United States under a central governmental system; see Ghiselin [1974*a*] on individuality). This new perspective of a constant renewal of interactive linkages, a regularly recurring opportunity for mingling of sexual gene pools with potential obliteration of populationwide genetic divergence (given an intact fertilization system), casts a new light on more than one aspect of the species debate.

Species Concepts in Relation to Habitat Specificity

The Phylogenetic Concept. The Phylogenetic Concept does not require a species to have unique habitat specificity. Recall that under this view a species is "the smallest diagnosable cluster of individual organisms, within which cluster there is a parental pattern of ancestry and descent, beyond which there is not" (e.g. Nelson and Platnick 1981; Cracraft 1983). For example, any unique, small, selectively neutral modification to a protein that diagnoses a clade A of organisms, even if they exist in sympatry and with a habitat use identical to that of organisms in the sister-clade B, is sufficient to proclaim A as a species. Habitat relations do not feature at all under this concept.

The Cohesion Concept. The Cohesion Concept does directly address habitat specificity, in the guise of "demographic exchangeability": "complete demographic exchangeability occurs when all individuals in a population display exactly the same ranges and abilities of tolerance to all relevant ecological variables" (Templeton 1985:15). However, in this concept habitat specificity, or demographic exchangeability, is only one possible route [albeit a frequent one] to species status. Thus, the cohesion species does *not require* unique habitat specificity, because demographic exchangeability between two populations does not guarantee conspecificity. Recall Case 3*b* in Table 1.4: Templeton would recognize two sexual populations that still share a common habitat but have diverged postfertilization systems, that is, that have lost genetic exchangeability, as two species.

Templeton (1989) contributed significantly on the question of evolutionary forces that promote phenotypic uniformity—cohesion—in both asexual and sexual populations. His discussion of how drift and cohesion relate is especially interesting. He points out that (1) demographic exchangeability implies that the probability of a neutral or selectively favorable mutation going to fixation is exactly the same regardless of the individual in which it occurs; that (2) for the case of neutral alleles, "the rate at which genetic drift promotes identity-by-descent depends only on the neutral mutation rate and is therefore equally important in both small and large populations . . . predictions [which] are equally applicable to [sexual and] asexual organisms"; and that (3) in the neutral case, "making *only* the assumption of demographic exchangeability, it is inevitable that at some point in the future all the alleles will be descended from one allele that presently exists" (Templeton 1989:16).

Nevertheless, in spite of the strong stress on habitat relations, the cohesion species is so all-encompassing in its definition that it ends up without a consistent relation to this habitat criterion.

The Isolation Concept. The isolation species is also not necessarily specific for a unique habitat range. Numerous cases have been reported of postfertilization-isolated sister-"species" that mate freely (with heterozygote disadvantage) in hybrid zones in which both kinds of organisms use the same resources in the same habitat.

The Recognition Concept. To diagnose a recognition species *requires* only documentation of a unique fertilization system (Table 1.1). Yet, Paterson's (1978, 1981, 1985) specifications of a *system among organisms* and of conditions required for evolution of a new system (i.e., for speciation) have far-reaching *necessary* consequences, including that each species is uniquely habitat specific.

Miller (1978:1027) identified one of the quintessential characteristics of living systems as follows: Living systems "can exist only in a certain environment. Any change in their environment of such variables as temperature, air pressure, hydration, oxygen content of the atmosphere, or intensity of radiation . . . produces stresses . . . under [which] they cannot survive." For a recognition species this environment is the habitat-specific tolerance range in which the fertilization system can function. One can distinguish the tolerance-range limits for particular resources that this system requires—the system habitat specificity—from those limits needed by all life activities of organisms other than fertilization interaction—the economic habitat specificity (following Eldredge's distinction; 1985 and this volume). For example, Morris (1970:31) reports that the habitat conditions required for breeding of ten-spined sticklebacks (Fig. 1.2) "are profuse fresh vegetation of a type in which nests can be built, water which is not very fast moving (as this would cause too much movement of the weeds and endanger the nest), water which is well oxygenated and of a particular pH, and water which does not dry up in the possible high temperatures of late spring and early summer." He notes that breeding is absent or rare in other areas where the fish are nevertheless found, "for example the River Kennet in Wiltshire [is] in most places too swift-flowing for breeding purposes." Similarly, for many of the heterogonic lineages, such as the aphids mentioned earlier, the evidence indicates differences in habitat variables between the parthenogenetic and sexual phases (Bell 1982).

The crux of the recognition species' unique habitat specificity lies in the necessary conditions for evolution of a new system (speciation; Paterson 1985). Because the fertilization system comprises a coadapted signal-response reaction chain (Fig. 1.2), (1) it can only be changed by selection of small mutational steps involving male and female subparts in turn; and (2) an environmental change, acting on a population in allopatry from the parent sock, is *required* to set such divergence in motion. Unless the old habitat changes, strong stabilizing selection on the system obtains. Thus, one prediction for any two sister-populations is the equation: distinct fertilization systems = distinct habitat specificities. This can be tested in principle and without much difficulty in cases where habitat variables can be experimentally controlled (again, the stickleback research of Morris [1970] and others is a case in point).

This central implication is often missed by those who discuss the Recognition Concept. For instance, Templeton (1989:14) wrote: "the isolation and recognition concepts . . . have elevated a single microevolutionary force—gene flow—into the conclusive and exclusive criterion for species status." This is untrue for the Recognition Concept. It stresses that natural selection for a particular fertilization system can maintain single-species status across allopatric populations for long time periods, provided those populations are all subject to the same range of habitat conditions to which that species is adapted. Thus, gene flow here is irrelevant.

This new theoretical element on habitat specificity constitutes one of the fundamental differences between these two concepts: although both acknowledge prefertilization divergence as a cause of speciation, Paterson added a new focus on a recognition *system* that can change only in an environment different from the parental one. In contrast, the more traditional concept permits a wide variety of conditions under which "isolation mechanisms" can evolve, including divergence in the absence of habitat difference.

Species Concepts in Relation to Lineage Branching

What concerns me is not the extensive gray area that surrounds the actual process of lineage splitting before species divergence is unambiguously complete (particularly, of sexually reproducing lineages), but rather the expected long-term fate of incipient or newly arisen branches on the tree of life. For all incipient species branches there are favorable environments (their habitats) with respect to which environmental change represents habitat deterioration. Under all concepts, such onset of adversity may wipe out "young" species soon after origin. But the concepts differ sharply in their expectations of species longevity in the face of persistent favorable conditions.

One of the most highly corroborated and spectacular phenomena of life is its branching genealogy. In my view, a species concept should directly address that branching pattern. Any part of a given species concept that includes "species" that must, in terms of its own premises (including recently accumulated evidence on the implications of those premises), remain "stillborn buds" or tiny offshoots soon to be resorbed by the parent stem is irrelevant to the tree of life. I suggest that a reasonable requirement of any species concept is that it should be "economical" in being restricted to branches that *can* take off with "habitat luck" and that do so irreversibly.

The Phylogenetic Concept. Asexual phylogenetic species fulfill the "economical" requirement. But the sexual species do not. Recall the evidence that global paleoclimatic cycles resulted in large-scale changes in the geographic distributions of most taxa (Vrba 1992). For the moment, let us accept as species those taxa generally agreed to be species in the literature. Take, for example, a mammal species of two-million-years duration (average for mammals; average durations in many other groups were much longer [e.g. Stanley 1979]). If that mammal species had moderate tolerance to the cyclic changes, responding only to particularly severe extremes (which were about 100,000 years apart during part of the Pleistocene), then episodes of maximal vicariance and maximal recoalescence of its distribution might have recurred only with periodicity about 1/20th of the duration time. If that species was highly stenotopic and responsive to either the cold or the warm extremes in all cycles, then during its lifetime its distribution would have undergone nearly one hundred episodes of maximal vicariance and maximal recoalescence.

Thus, for most sexual taxa we may expect that phases of allopatric divergence (speciation under the Phylogenetic Concept) were more or less evanescent phenomena. "Diagnosable clusters" recurrently lost their genotypic distinctiveness as sympatric sexual interactions resumed: precisely during phases of maximal spread of their optimal habitats, their monophyletic identities ceased to exist. The Phylogenetic Concept recognizes even groups that are diagnosable by only minute genetic differences as species. Thus, one can expect from many to numerous phylogenetic species to exist (in actuality most will not be noticed, as argued below) within the average recognition species. Thus, the vast majority of all sexually reproducing species under the Phylogenetic Concept are not expected to be irreversible branches of the tree of life.

The Isolation Concept. Under the Isolation Concept some species, those with diverged fertilization systems, are expected to be irreversible branches. But species isolated only by postfertilization divergence are unlikely to lead to branches. As Paterson has pointed out since 1978 [1981, 1982, 1985, 1986], both theoretical and experimental results strongly suggest that postfertilization isolation is not an *initiating* cause of divergence of the fertilization system. Of course, in some cases of an allopatric population that is already postfertilization-isolated, the fertilization system may diverge independently, causally influenced not by reinforcement but by habitat change. But most such isolation species are expected to be the stillborn buds, mentioned above, that can never maintain branch identity when in sympatry with the parent species. Many evolutionists have come to an essentially similar conclusion (e.g. Mayr 1982; Lande 1979; Futuyma and Mayer 1980; Templeton 1981, 1989; Carson 1982; Butlin 1989). A category that is still cited by many as valid speciation via postfertilization isolation is speciation by polyploidy, particularly in plants. But, in my view, Paterson (especially 1981:117–118) has convincingly countered these arguments as well.

The Cohesion Concept. Asexual species under the Cohesion Concept are irreversible branches. So are sexual cohesion species evolved by divergence of the prefertilization system. But, among the several completely different kinds of species recognized under this concept, some fail the criterion of true and irreversible branching. Thus, Templeton (1989) includes two isolation species, distinguished only by postfertilization divergence, as two species because they lack "genetic exchangeability" (Table 1.4).

The Recognition Concept. All recognition species, once launched to the point where hybridization is negligible or has ceased, are necessarily implied to be irreversible branches with distinct durations that hold as long as appropriate habitats survive. If one focuses on the *system* as that which confers species identity, then the same species can continue even after giving off a branch with a distinct system and ends once its system ceases to exist.

Species Concepts in Relation to Monophyly

The Phylogenetic Concept. The Phylogenetic Concept is the only concept that claims that each species (from speciation to terminal extinction) must be strictly monophyletic based on nested homologies at the levels of organisms and their constituents. Bonde (1981:28), an adherent of this concept, reasoned the need for this criterion as follows: "If in the classification species have to be basic units (taxa) and phylogenetic relationships shall be precisely expressed in the system, then the ancestral species must cease to exist at speciations. Only in this way *all parts* (e.g. "chronosubspecies" or single fossil specimens) *within a species have exactly the same phylogenetic relationship to any* (part of) *other species in the system*" (italics in original). That is, even if the ancestral stem [A] is precisely the same as one of the descendent branches (A', the sister-branch being new species B) in every respect, this concept requires that the ancestor, A, become "extinct" at the speciation event, and that descendant branch, A', be recognized as a new species. In this view, the alternative of recognizing the persistent ancestral lineage (A-A') as the same species before and after giving off branch B, is unacceptable because it makes species A-A' paraphyletic. Recent arguments have gone further and suggested that there is no such thing as an ancestral species at all (e.g. Nelson 1989) or that all possibly ancestral lineages (for which only plesiomorphies are known) are metataxa or metaspecies and not species (Donoghue 1985; De Queiroz and Donoghue 1988, 1990). Thus, the species in this view is always a terminal taxon, as implied by its being a smallest diagnosable monophyletic group.

It is worth asking: precisely *what* is monophyletic in such a phylogenetic species?

The asexual case is clear: if a mutant gene X first appears in organism A and is transmitted to all its descendants, then A and those descendants are a monophyletic lineage characterized by X. Of course, as soon as an offspring appears with a new diagnostic gene Y, the previously recognized "species"—the X-organismal cluster—becomes paraphyletic and loses its species status.

The situation is more complex in the sexual case. A sexually out-crossing lineage can never be monophyletic at the organism level. Take the case of a dominant allele X which arises in a female (of genotype Xx, phenotype X') which mates with a non-X male (genotype xx, phenotype x'), in a population in which X eventually becomes fixed. Let us call the initial state, from mutation onward but previous to fixation, stage 1. During stage 1 there is indeed a smallest diagnosable (by gene X) cluster of organisms. Is this, under the Phylogenetic Concept, "species X"? If so, species X is of course nonmonophyletic at the level of organismal lineages. Even given the unlikely events that all F1 offspring of the mutant-X mother have X, and X-bearing individuals mate only with each other, and all further descendants of the mutant-X mother have X, one would still need an "immaculate conception" in the mutant-X mother to have a monophyletic X-cluster of organisms.

Let us call stage 2 the period that starts from fixation and lasts until a new mutant or immigrant destroys the XX-homozygosity of the population. Does "species X" start only with the population fixed for X at stage 2? I presume that, under the Phylogenetic Concept, this is what *has* to be claimed: the monophyly buck stops here in the sense that this first (or stem) unit of a new species, an "Eve-population" fixed for a diagnostic character, must be declared immune to investigation of its phyly. And this indeed seems to be what adherents do claim (e.g. Wheeler and Nixon 1990).

But how, when looking at variation in nature or in museum collections, does one recognize such an Eve-population? Wheeler and Nixon's (1990:77) stance on this is simply one of cladistic fiat: "species are the smallest terms analyzed by cladistic methods," and "in cases where cladistic analysis is possible . . . we are dealing with distinct species and not with infraspecific units." But I agree with De Queiroz and Donoghue's (1990:88) reply: "Wheeler and Nixon (1990) . . . wish to distinguish between inconstant 'traits' in populations [such as X in the stage 1 population in my example] and constant 'character states' of 'species' . . . present 'in all individuals of a terminal taxon.' But how is one to distinguish between variable traits and constant character states unless one recognizes beforehand the unit ('terminal taxon') within which the organisms bearing these attributes, whether variable or constant, occur?"

Similar problematic observations on monophyly apply to all cladistic analyses of sexually out-crossing lineages, and therefore also to the Cohesion Concept of biparental species and to the Isolation and Recognition concepts.

But these differ from the Phylogenetic Concept in claiming criteria for species other than strict monophyly.

The Recognition Concept. The Recognition Concept focuses on a *system* to diagnose a species. System status at a higher level does not depend on strict monophyly of subparts, but on the rules by which the subparts interact to constitute the system. For a new fertilization system to originate, at least one new male and one new female phenotypic and genotypic character pertaining to system interaction must evolve. As long as the system remains intact the species endures, even if it has in the interim given rise to one or more daughter species.

In sharp contrast, the Phylogenetic Concept insists that a sexual species cannot be allowed to persist as the same species after giving rise to a daughter species as it would then become paraphyletic. I would have reservations about this argument even if there were not the ambiguity on monophyly during the species' origin (mentioned above). Consider the same argument, but shifted to a lower level (in Vrba [1985] this is explored in relation to species as individuals, *sensu* Ghiselin [1974b]; the argument has more force if species are systems): instead of focusing on the species as the system with organismal subparts, focus on the metazoan organism as the system with subparts that are cells, and, within cells, genes. For instance, at the birth of my child some of the gene (and cell) lineages in my body became paraphyletic (quite apart from the fact that my genome was polyphyletic to start with); and yet I remain the same system. I repeat, system status at a higher level does not depend on strict monophyly of subparts, but on the rules by which the subparts interact to constitute the system. To say that a parent species must cease to exist once it gives off a branch, and be recognized as a new species if it persists without change after branching, is like saying that I ceased to exist at the birth of my daughter and since then must be named as a new individual.

Epistemology, with Special Reference to Extinct Species

I find the Cohesion Concept the most difficult species concept to apply. In attempting to encompass all groups of living organisms, it pays the heavy price of multiplied operational problems. First, the practical diagnosis of those cohesion species (including asexual clones and sexual syngameons) defined only by demographic exchangeability is unclear. How can one decide that too much or too little demographic exchangeability is present in a given case? Second, the diagnostic problems variously faced by the two sexual concepts of species, whether Recent or extinct, all apply to genetically exchangeable cohesion species as well.

The definition of the Isolation Concept is relational, as has often been pointed out (e.g. Ghiselin 1974b; Paterson 1981; Vrba 1985). Just as one can diagnose a "sister" only in relation to a sibling, so one can diagnose the boundaries of an isolation species only by evidence of reproductive isolation from a sister-species. Consequently, species status of allopatric populations presents a problem. If it is true that populations that are isolated only by postfertilization divergence are shortlived (Paterson 1978; Mayr 1982; Lande 1979; Futuyma and Mayer 1980; Templeton 1981, 1989; Carson 1982; Butlin 1989), then the predictions under this concept for systematic patterns in the recovered record of extinct species are expected to be quite similar to those of the Recognition Concept.

All the species concepts share the operational problem of norms of reaction, as all insist that speciation is a genetic event. The norm of reaction of most organisms (Stearns 1982) can result in different ecophenotypic expressions of the same genotype in response to different environments, and it is difficult to distinguish (especially in the fossil record) such ecophenotypic differences from genetically based ones. A large proportion of populations that would be judged as different phylogenetic species based on phenotypic diagnosis are likely to represent geologically ephemeral ecophenotypes. The same holds for many populations that appear to be separate species, of differing demographic exchangeability, under the Cohesion Concept. For instance, many or most cohesion species within syngameons may fall into this category. I suggest that the Recognition Concept is least vulnerable: the strong stabilizing selection on the fertilization system as a whole (Paterson 1978, 1981, 1985) is also expected to result in narrow norms of reaction for the components of this system. For the male *and* female SMRS characters (Fig. 1.2) both to change ecophenotypically in a changed environment, such that the SMRS still functions, would require the highly improbable situation that both change in a precisely covarying way. Selection is expected to remove reaction norms that result in any other kind of change. (The hypotheses that SMRS characters should have narrower variances and narrower norms of reaction are both testable. Lambert and Levey [1979], for instance, found support for the first hypothesis in *Drosophila melanogaster;* but I have not seen any tests of the second one.)

Next, I will consider some general issues in relation to paleontology. Then, because the problems and virtues of the Phylogenetic and Recognition concepts with respect to fossils jointly subsume those of the Cohesion and Isolation concepts, I will concentrate most of the remainder of my remarks in this section on the former two.

General Remarks on Operational Approaches to the Fossil Record

I think that we all agree on three issues relating to species, whether Recent or extinct:

1. We wish to study, name, and incorporate in our analyses of processes, those different kinds of entities, out there in nature both today and in the past, that are distinct and coherent entities.

2. As evolutionists we require that the biotic entities on which we focus our research energies should maintain coherence, manifested as stability of combinations of character states, at least through modest periods of time. (As a paleontologist I suggest that we should think in terms of at least 300,000 to one-half million years during which certain combinations of character states, including apomorphic and plesiomorphic states, must be consistently shared by a collection of organisms for that collection to qualify as a meaningful biotic entity.)

3. Such long-term stability implies a basis rooted in heritable characters and processes. The Phylogenetic Concept stresses the vertical genealogical linkages resulting from reproductive processes, while the other three concepts variously add emphases on horizontal linkages resulting from sexual and ecological processes; but all focus on processes. This agrees, for instance, with Michaux (this volume) who supports a species concept based on processes that lead to the coherence of the entities that we can see sufficiently clearly so as to be motivated to recognize them as biologically important.

Using these three criteria of meaningful coherence, stability through at least modest time intervals, and a basis in evolutionary processes, we can address two questions that have been often debated in relation to species in the fossil record:

Should we have a different concept for "species" in the fossil record? The answer has to be a resounding *No* if we consider that there is only one tree of life reflecting one history and one unified set of processes. While we can in principle recognize the presence of several different meaningful kinds of living entities (such as minimal clades, habitat clones, and species, in my usage), we should seek to comprehend them in both the modern and the extinct biota wherever the data allow us to do so. Turner (this volume) reviews the old acceptance that the "bread and butter species of the paleontologist" (morphospecies, or in more modern usage, a phylogenetic species, namely, "bread and butter" in the sense that the species concept is easy to apply) should differ from the biospecies (traditionally the isolation species). I agree with his conclusion that to argue for a different paleontological "species" is absurd. Further, as I point out below, *all* species concepts to varying extent lack "bread and butter" quality in application to the fossil record, as all require estimation of the truth and none is free of difficulties.

The second question is closely related to the first. *Should we hesitate to use a species (or other) concept just because it is difficult to apply to (and test in) most of the fossil record?* That is, should our articulation of concepts be subservient to an epistemological requirement that they should be testable in all or most of the fossil record? Again, the answer must be *No*, unless we want to argue that some of the great hypotheses of process in biology and geology were invalid at the time of their articulation simply because of then-current difficulties of testing them (or, for that matter, that they are invalid today because of persistent absence of relevant data from most of the geological record). For instance, if such subservience to the shortcomings of the geological record is taken seriously, then Milankovitch had no business to articulate his astronomical model of climatic cycles in the 1920s, as critical tests were only possible half a century later (review in Broecker and Denton 1990). And the Milankovitch hypothesis would still not be respectable, as it has only survived falsification in a minute fraction of the geological record. However, it has been supported in all cases in which testing is possible to date. (The same is true of many other major biological and geological hypotheses.) Thus, in my view, paleontologists use the same general concepts that have been found internally consistent and testable elsewhere in biology; and they may add some of their own that are first raised by, and tested in, the paleontological record. The intrinsic merit of a particular concept does not depend on the number of instances in which it can be tested, provided it is testable in principle and testable somewhere out there in the reality of obtainable data. If phylogenetic and recognition species, for instance, are judged to be meaningful entities, then let us recognize these distinctions theoretically and test for the existence of each where we can.

The Phylogenetic Concept. At first glance, the Phylogenetic Concept is easiest to apply to the fossil record, but upon closer examination there are conceptual problems. A concept can claim to have a good operational criterion for diagnosing species if that criterion succeeds in recovering units that preserve the theoretical meaning of species under that concept. I agree with De Queiroz and Donoghue (1988, 1990) that the ease of erecting some kind of cladogram or other does not save the Phylogenetic Concept from some problems in this respect.

Let us allow, giving the benefit of the doubt, that a unique character (discovered in some population or fossil collection) of a cluster of organisms—no matter how small—heralds the "clean beginning" of a new phylogenetic species. There still remain theoretical and operational difficulties with "quick and messy endings" in one of two ways. First, such a character-clade ceases to exist as a species as soon as a new character evolves within it. In epistemological terms, with every discovery of a new character, the status of species changes, that is, the concept is defined relationally and provisionally, pending exhaustive character analysis. Second, because we must expect repeated separations and reunions of populations over geologically short time intervals given cyclic climatic changes, hybridization must often result in a second kind of messy ending as recombination of character sets "fudges" previous species boundaries. Such "species genealogies" may be conceived to look more like an unusual piece of knitting than a diverging tree.

When one views life's genealogy in hindsight, there are relatively few species under this concept: only terminal autapomorphic branches are species, the vast remainder being metataxa (Donoghue 1985) banished from species status by their paraphyly.

The Recognition Concept. One problem in dealing with the extant biota is that the systematic character distributions of SMRSs in particular, and of fertilization systems in general, remain relatively little studied. It has also been held against this concept (and the Isolation Concept) that all lineages that never exhibit biparental sexual reproduction are excluded and that numerous cases exist in nature in which varying levels of hybridization blur species boundaries. On the positive side, all recognition species eventually do have definite beginnings of irreversible branches (unlike some isolation species) *and* "clean endings" (unlike most phylogenetic species). They also have a clear and testable diagnosis in principle (unlike some cohesion species), are not defined relationally, and, thus, are diagnosable in allopatry.

Because the Recognition Concept focuses on a particular and complex process among organisms, it entails a range of predictions for patterns including paleontological ones. Paterson's (1978, 1982) arguments predict not only particularly narrow variances in a species' SMRS characters in space and time, relative to other phenotypic characters, but also a pattern of punctuated equilibria, primarily in SMRS characters, and by pleiotropy also in other phenotypic characters (predicted independently, Paterson and James [1973] and Paterson [pers. comm.] of the articulation by Eldredge and Gould [1972]).

Insofar as the *skeletons* of extant survivors are found to include characters of the SMRS, this hypothesis is testable by examination of the fossil record of that monophyletic group. Consider that in cases where mating communication is primarily visual, a breakdown of communication between populations (speciation) should involve a shift in morphology (Vrba 1980, 1984). In such groups, rates of morphological change and speciation, under the punctuation model, should be positively correlated. Even relatively closely related species of visual "communicators" might be expected to be morphologically distinct. This explanation may be appropriate for the morphological differentiation between chimpanzee and man (King and Wilson 1975) and other higher primates, in speciose groups of *Haplochromis* (Fryer and Iles 1969) and the homoallozymic *Drosophila heteroneura* and *silvestris* (Carson 1976). Conversely, uniform species' flocks (e.g. minnows, Avise and Ayala [1976] and some fruitflies) may have an SMRS dominated by olfactory, auditory, or other nonvisual signals. The observation that frogs, with their bias toward auditory communication, include many sibling species would support this argument. Among mammals, particularly numerous sibling species have been recorded for rodents. Distinctive auditory and behavioral SMRSs have been reported for morphologically identical rodent species (Gordon and Dennet 1979).

The implication for the paleontologist is obvious (Vrba 1980): in monophyletic groups including extant species whose visual communication involves skeletal characters (e.g. Vrba 1984), the probability of successfully estimating species and species diversity patterns in the fossil record may be good. In contrast, it is hopeless in the case of groups of species that are sibling in terms of their hard parts. *The problem of extinct sibling species is shared by all species concepts.* In judging distinct fossil morphologies as species, the Recognition Concept of species requires the argument that the relevant morphologies are good symptoms of differing fertilization systems. I argue below that this argument is not as difficult to defend as is often claimed.

A Proposal for Paleontological Procedure

Two implications of the fossil record are important to the species issue. The first is a positive aspect, a strong and unique advantage relative to other data sets: the fossil record affords at least some direct evidence on the temporal persistence of any unique *combination* of morphological characters that we see in individual fossil organisms. Such a unique character combination might include at least one autapomorphy alongside a combination of plesiomorphies, or it might include only plesiomorphies relative to later cladistic branches, as expected of an ancestor. (Kimbel and Rak [1992] recently argued this and used the approach in conjunction with cladistic analysis to infer extinct species in the Plio-Pleistocene hominid record.) Long endurance of such a character combination of a sexual taxon allows certain inferences on process: the taxon did not interbreed with its sister-taxon of distinct character combination, either because allopatry alone (less likely, given what we know about paleoclimatic oscillations) or reproductive isolation in sympatry made it possible.

The second is the negative aspect of "gaps" in the fossil record—the low probability of preservation that excludes especially forms that are ephemeral in time and geographically rare. As already noted, most apparent novel combinations of phenotypes (which would be labeled phylogenetic species, if recovered) are small and evanescent in time, either because the population soon disappears altogether, or because the character combination dissolves upon each recurrent sympatry with the sister-taxon, or because of a changing norm of reaction. Included among these new ephemeral branches are those that constantly (in geological time) arise to rob ancestral taxa of their species status under the Phylogenetic Concept. We may infer that the gaps in the fossil record act mainly to hide all these ephemeral new taxa from our view. As a result cladists record far fewer branches overall than were really present; yet among these recorded branches, a high proportion of apparent phylogenetic *species* are really metaspecies. They have survived as species under the Phylogenetic Concept simply because their ephemeral subbranches that eclipsed them into metaspecies status are invisible to us.

The basic systematic procedures in paleontology are in principle those used in neontology; yet there are practical differences in how they can be applied. This topic deserves much more thorough and rigorous treatment than I can give it here. I offer only a few comments, including the following pragmatic conclusion: *The vast majority of sexual fossil taxa of sufficiently long duration to find their way into a cladogram at all, and then to be recognized by some as phylogenetic species, are recognition species* (with the caveat, under either concept, that separate species that are sibling in terms of their hard parts will not be detected). The phylogenetic metaspecies is also a potential recognition species—an ancestral recognition species. Recall that recognition species are predicted to remain discrete lineages that share certain combinations of character states, through long time in the face of the frequently recurring episodes of sympatry that paleoclimatologists insist have been a pervasive feature of the history of life. Under this conclusion the gaps in the fossil record act as a "great leveler." Whatever their semantic preferences, I suspect that most biologists would like to reserve a term for those *long*-lasting distinct units among sexually reproducing lineages. By an artifact of the fossil record, the different conceptual schools end up agreeing on most such fossil lineages that they should be called species!

I suggest that paleontologists, like neontologists, can in principle erect both cladistic hypotheses, with minimal clades and metataxa, *and* hypotheses of recognition species in relation to those cladograms. The latter hypotheses arise from the genealogy, together with patterns of character combinations in organisms, taxa, and through time and are testable by additional character information.

I will illustrate some relevant points by my cladistic analyses of skull characters in the Pliocene-Recent antelope tribe Alcelaphini (a cladistic revision of Vrba [1979] is now in preparation with J. Gatesy and R. DeSalle, incorporating additional fossil taxa to make a current total of about forty, as well as cladistics of the mitochondrial DNA sequences of the living taxa which number at least seven). I have argued that in Alcelaphini (and other bovid groups), minimal clades are unlikely to contain hidden sibling species. This is based on analogy with living alcelaphines in which all sister-taxa, which in sympatry behave like separate recognition species by not interbreeding, are distinguishable by skull characters.

First, the alcelaphine cladogram indicated metataxa that are hypothetical ancestral recognition species. An example is a *Damalops* sp., species 17 in Vrba (1979:Fig. 2), which is wholly plesiomorphic with respect to other cladistic branches from the same node as *Damalops* sp. or from higher nodes on the cladogram. *Damalops* sp. is known from numerous fossils, which consistently share a unique combination of character states (although they share no single autapomorphy) over half a million years. The hypothesis that this cluster of specimens represents an ancestor is falsifiable by future discovery of additional characters that disqualify it from such ancestry.

Second, there are extinct minimal clades that preserve a unique character combination (including at least one autapomorphy) for at least a few hundreds of thousands of years, which I hypothesize are terminal recognition species (such as *Parmularius angusticornis*).

Third, there are instances of clades, each containing at least one metataxon and one minimal clade, which I hypothesize to represent a single terminal recognition species. For instance, the taxon *Damaliscus niro* in Vrba (1979) has a combination of an autapomorphy (prominent, widely spaced transverse ridges on the horncores) with several plesiomorphies (one example is the wide spacing of the supraorbital foramina, an apomorphy of *Damaliscus*). This combination persists from 1.7 million years until the latest Pleistocene. *Within* this taxon, one set of specimens in an early stratum shares a subtly different horncore orientation from all others, making it a minimal clade within the larger clade; but this cluster is connected to other *D. niro* specimens in that stratum by intermediates, suggesting interbreeding, and is present in only that one assemblage, supporting the expectation based on interbreeding that it was an ephemeral manifestation. Hence, I suggest that the clade characterized by prominent, widely spaced transverse ridges on the horncores is the single recognition species *D. niro*.

Conclusion

I have examined the meaning of *species* under the Phylogenetic, Cohesion, Isolation, and Recognition concepts in terms of four phenomena: Does a concept require a species to have (1) *system status* through among-organismal interactions, (2) *unique habitat specificity*, (3) *irreversible lineage branching*, and (4) *monophyly*? These phenomena are fundamental to biotic organization and patterns at all levels of complexity. They should feature in distinctions that we make among different kinds of organismal groups.

One basic kind of organismal group is the "minimal clade" (minimally diagnostic character-clade; I suggest the possible abbreviation M-clade), termed *phylogenetic species* under the Phylogenetic Concept. Minimal clades, asexual or sexual, require study irrespective of the fact that most are very small and ephemeral.

A second important category is that of the obligately uniparental, habitat-specific clones—I suggest the term *habitat clones*—the evolution of which Templeton (1989) clarified, and which he regards as cohesion species held together by "demographic exchangeability." (*Habitat clone* here refers not only to obligate asexual but also to obligate uniparental sexual monophyletic lineages.)

A third fundamental unit is the "biparental sexual system" of the Recognition Concept, although most such systems are further subdivisible into minimal clades, and some are paraphyletic at the character level. Lineages that engage in biparental sexual reproduction at least sometimes differ crucially from those that never do so: the former have systemic organization *among* organisms while the reproductive system of the latter is confined to the organism level.

Between them, these three kinds of entities not only account for all of life but also refer to the basic kinds of living attributes.

I prefer to reserve the term *species* for the biparental system—the recognition species—because it requires three of the four criteria that I regard as fundamental biological phenomena (Table 1.2). The remaining criterion of character monophyly is violated once a recognition species gives rise to a daughter species. I argue that this requirement means as little when focusing on the species as a *complex system* as would the requirement that I became a new organismal system with a new name after incurring gene (and cell) paraphyly upon my daughter's birth. Eldredge (this volume) reviews questions by some systematists why species should differ from higher taxa (see Nelson 1989). The species under the Recognition Concept differs crucially in that it is a system among organisms, while clades of more than one species are not.

In addition to studying the boundaries in time and space of the species construed as a complex system, the relationship of the primary genealogical pattern of organismal character evolution—that is, of "minimal clades"—to the species must be addressed. Thus, I still agree with Avise et al. (1987:518) that "no longer will it be defensible to consider species as *phylogenetically* monolithic entities in scenarios of speciation and evolution" (see also Cracraft 1983, 1987; Nelson 1989). But I urge that we distinguish asexual from sexual lineages, and "minimal clades" from biparental sexual systems. The set of all groups of organisms that *look alike by descent* is divided by *differences in among-organism organization* too fundamental to squeeze them all into a single concept. Thus, I reject arguments that the term *species must* encompass all recognizable groups of organisms. As it is, among metazoans at least, the proportion of diagnosably different lineages that never use biparental sexual reproduction may be quite small (Maynard Smith 1978; Bell 1982).

Also, there is a common perception that to recognize only groups of a certain kind as species is to ignore other kinds—to deny the latter importance and reality (e.g. Templeton 1989). But that argument has no force. Take the concept of a metazoan organism. In most cases each such an organism is (1) a clone of cells *and* (2) an organization arising from rules of interaction among those cells. To confer the name *metazoan* in recognition of the importance of the additional factor 2 does not rob the unicellular clone of reality.

What we call the distinct kinds of organismal groups may largely be a matter of semantic taste. But the term *species* is so strongly imprinted on the minds of biologists, and so much interwoven with evolutionary thought, that it should belong to a kind of group with particularly significant properties. I have suggested four such properties: complex system status, unique habitat specificity, irreversible lineage branching, and monophyly at some included level. Even if one wishes to fall back on preevolutionary thinking, the properties of an among-organismal system, and of unique habitat specificity, remain compelling. I find the nonfulfillment of three out of these four criteria by the phylogenetic species a major flaw in a candidate species concept; and the only required property, monophyly, leaves something to be desired when examined closely. For instance, a sexually out-crossing lineage can never by monophyletic at the *organism* level.

On the one hand, there is something satisfactorily clear-cut about focusing only on one kind of criterion—genealogy—and ignoring all other kinds of phenomena. On the other hand, the price paid for designating these little clades as "species" seems too high in my view. The vast majority of phylogenetic species are *very* small, meaningless in terms of interactions among the subparts and with the environment, and ephemeral in time.

The Cohesion Concept seeks to encompass all organisms in terms of a unified set of evolutionary processes, the "cohesion mechanisms." In this the concept fails in my view. "Cohesion" is not a particular property and is not clearly defined. This results partly because its two guises, genetic and demographic exchangeability, are extremely different phenomena and partly because demographic exchangeability is poorly defined. In the bid to qualify as "everyone's species concept" the cohesion species ends up not fulfilling any one of the four investigated conditions consistently.

The Isolation and Recognition concepts share the requirement of a sexual system in a species (Table 1.2). But there are major ontological and predictive distinctions that are much more fundamental than, as widely misinterpreted, simply "two sides of the same coin" (e.g. Templeton 1989). Paterson's concept has added distinct new theoretical elements to the older one: First, he clearly articulated a hypothesis of process that negates the initiation of a new species branch by postfertilization divergence alone (as affirmed under the Isolation Concept, via "reinforcement of prefertilization isolation," e.g. Ayala et al. 1974). In this respect the ontological consequences of the two models are clearly different: Given persistence of the incipient species' habitat, recognition speciation results in irreversible lineage branching, while some isolation speciation does not. We are dealing here with more than a mere matter of opinion on what to call the new branch initiated solely by postfertilization divergence—under the Recognition Concept *such branches are not there*. Second, although both concepts acknowledge prefertilization divergence as a cause of speciation, Paterson's focus on a recognition *system* whose change is limited to occur only in a changed environment is new. In contrast, a wide variety of conditions under which "isolation mechanisms" can evolve are part of the more traditional concept, including divergence in allopatry in the absence of habitat difference. Thus, while the recognition species implies a unique habitat specificity, the isolation species cannot afford this requirement.

All these differences arise in one way or another from the stress by the one concept on *recognition mechanisms*, while the other focused on *isolating mechanisms*. First, the stress on reproductive *isolation* unhappily brought postfertilization divergence to share center-stage in the speciation process with prefertilization divergence. Second, the stress on *mechanisms* conjured up the unfortunate connotation of selection for isolation function and necessitated the model of selective reinforcement in secondary sympatry of postfertilization isolation in allopatry. I use the word *unfortunate* because it is now widely recognized that such reinforcement for speciation is highly unlikely to occur (as argued early on by Paterson [1978] and subsequently by many others) and that, as there is no other model around of selection for isolating function, both the isolating properties of postfertilization divergence alone and the uncomfortable notion of isolating mechanisms are irrelevant to speciation.

Perhaps the time has come to lay to rest the "ancestral" Isolation Concept of species and to acknowledge fairly that it has evolved into the Recognition Concept, via replacement by new theoretical "apomorphies" of some old misconceptions. We owe Paterson a debt for pointing to the internal dynamics of the biparental fertilization system itself and to their evolutionary consequences.

Summary

I have compared species concepts in terms of four criteria based on fundamental biological phenomena: Does a concept require a species to have (1) *complex system status*, (2) *unique habitat specificity*, (3) *irreversible lineage branching*, and (4) *monophyly*? At least three kinds of organismal groupings deserve theoretical and terminological distinction: the minimally diagnostic clade (= phylogenetic species; Cracraft 1983), the biparental sexual system (= recognition species; Paterson 1985), and the obligately uniparental, habitat-specific clone (= one kind of cohesion species; Templeton 1989). What these should be termed is largely a semantic matter. I prefer to use *species* only for the recognition species, because it fulfills criteria 1 through 3, and criterion 4 is irrelevant to the species viewed as a complex system. Under the Phylogenetic Concept, only criterion 4 is claimed, but it is applied inconsistently. The cohesion species (Templeton 1989), in encompassing all organisms, is too broad to fulfill any of the criteria consistently. The Isolation (= "Biological" [Mayr 1963]) Concept, which requires only criterion 1 in a way that is now widely recognized as misleading, should be laid to rest.

Acknowledgments

I thank Hugh Paterson for sharing with me some of his deep insight into natural history and particularly for persevering through many hectic (but fun) arguments in showing me new perspectives on the subject of "the unknown and the unknowable"! I am grateful to John Gatesy and J. A. Grant-Mackie for very useful comments.

Glossary

Species Concepts

agamospecies - A species that reproduces only nonsexually.
chronospecies - The successive species replacing each other in a phyletic lineage which are given ancestor-descendant status according to the geological time sequence (paleospecies, successional species).
Cohesion Concept of Species - "The cohesion concept species is the most inclusive population of individuals having the potential for phenotypic cohesion through intrinsic cohesion mechanisms" (Templeton 1989:12). Cohesion mechanisms may be classified as follows (after Templeton 1989:13):

  I. Genetic exchangeability—The factors that define the limits of spread of new genetic variants through *gene flow.*
     A. Mechanisms promoting genetic identity through *gene flow:*
        1. Fertilization system—organisms are capable of exchanging gametes leading to successful fertilization;
        2. Developmental system—the products of fertilization are capable of giving rise to viable and fertile adults.
     B. Isolating mechanisms: genetic identity is preserved by the lack of *gene flow* with other groups.
  II. Demographic exchangeability—The factors that define the fundamental niche and the limits of spread of new genetic variants through *genetic drift* and *natural selection.*
     A. Replaceability—*genetic drift* (descent from a common ancestor) promotes genetic identity.

B. Displaceability
   1. Selective fixation—*natural selection* promotes genetic identity by favoring the fixation of a genetic variant;
   2. Adaptive transitions—*natural selection* favors adaptations that alter demographic exchangeability, constrained by:
      a. Mutational constraints on the origin of heritable phenotypic variation,
      b. Constraints on the fate of heritable variation:
         i. Ecological constraints,
         ii. Developmental constraints,
         iii. Historical constraints,
         iv. Population genetic constraints.

Ghiselin's (1974a) species concept - Species are the most extensive units in nature such that reproductive competition occurs among their parts.

Isolation Concept of Species (also called Biological Species Concept) - Species are groups of interbreeding natural populations reproductively isolated from other such groups (Mayr 1940, 1963). "Isolating mechanisms" determine a species' limits. Either prefertilization or postfertilization isolating mechanisms (Table 1.3) are sufficient.

morphospecies - A group of individuals that are considered to belong to the same species on morphological grounds alone.

Phylogenetic Species Concept - A species is the smallest diagnosable cluster of individual organisms, within which cluster there is a parental pattern of ancestry and descent, beyond which there is not. (e.g. Nelson and Platnick 1981; Cracraft 1983). No sexual reproductive linkages are referred to. Thus, a species can be a cluster of asexual organisms.

Recognition Concept of Species - A species is that most inclusive population of individual biparental organisms that share a common fertilization system (Paterson 1978, 1981, 1985).

taxospecies - A species based on overall similarity determined by numerical taxonomic methods.

typological species (nomenspecies) - 1. A species defined on the characters of the type specimen(s); 2. A species is an *eidos* (type) with an unchanging *essence* (from Plato's [428–348 b.c.] concept of unchanging essences of, and discontinuities between, each eidos and each other).

Other Terms

amphimixis - Occurrence of syngamy between gametes produced by different individuals of different gender.

apomixis - The absence of both meiosis and syngamy among organisms that reproduce by eggs.

arrhenotoky - Production of haploid males from unfertilized eggs and diploid females from fertilized eggs.

automixis - Syngamy between meiotically reduced nuclei descending immediately from the same zygote (with or without formation of gametes).

gynogenesis - Thelytoky requiring pseudogamy, that is, penetration of ovum by sperm without sperm genome contributing genetic information to zygote; a form of parthenogenesis.

habitat - Of an organism or species, includes places plus the resources in those places that are necessary for life of that organism or species. The *fundamental habitat* of a species includes all the places plus necessary resources in which a species can live (although it may not be present there); and the *realized habitat* at a given time includes all those in which it does live.

habitat specificity - Of an organism or species, refers to the resource requirements of that organism or species.

heterogony - Cyclical parthenogenesis: a life cycle in which one or more thelytokous generations are followed by an arrhenotokous or amphimictic generation.

hierarchy - In the sense of Simon (1962:87): "Hierarchy has generally been used to refer to a complex system in which each of the subsystems is subordinated by an authority relation to the system it belongs to . . . I use hierarchy in the broader sense . . . [to] mean a system that is composed of interrelated subsystems, each of the latter being, in turn, hierarchic in structure until we reach some lowest level of elementary subsystem . . . A hierarchy can progressively form, or evolve, as stable subassemblies [i.e., the subsystems] form at increasingly higher levels."

niche, fundamental and realized - The "role or 'profession' of an organism in the environment; its activities and relationships in the community" (Krebs 1978:623). "The ecological role of a species in the community; conceptualized as the multidimensional space, of which the coordinates are the various parameters representing the condition of existence of the species, to which it is restricted by the presence of competitor species" (Lincoln et al. 1982:167).

resources - (My concept of resources is close to Lincoln et al.'s [1982].): Any components of the environment that can be utilized by an organism in its metabolism and activities, including temperature; relative humidity; pH; salinity; stream flow velocity; substrate characteristics; places for living, nesting, and sheltering; light; inorganic ions and molecules; all kinds of organic foods (such as prey); and mates and other mutualist organisms in the same or different species.

Specific-Mate Recognition System (SMRS) - A subpart of the fertilization system that is particularly important in mobile organisms for bringing sexual partners together for mating.

syngameon - Larger units, containing phenotypically and habitat-differentiated populations that hybridize naturally and have limited gene exchange. This concept is much used by botanists, e.g., by Grant (1981), who defines the syngameon as "the most inclusive unit of interbreeding in a hybridizing species group."

thelytoky - Parthenogenesis in which syngamy and meiosis from the same zygote occur, or in which meiosis and syngamy are absent.

References

Arnold, S., P. Alberch, V. Csanyi, R. Dawkins, S. Emerson, B. Fritzsch, T. Horder, J. Maynard Smith, M. Starck, E. Vrba, G. Wagner, and D. Wake. 1989. How do complex organisms evolve? Pp. 403–433 in D.B. Wake and G. Roth, eds. Complex organismal functions: integration and evolution in vertebrates. Dahlem Workshop Report. John Wiley, Chichester.

Avise, J.C., J. Arnold, R.M. Ball, E. Bermingham, T. Lamb, J.E. Neigel, C. Roeb, and N.C. Saunders. 1987. Intraspecific phylogeography: The mitochondrial DNA bridge between population genetics and systematics. Annu. Rev. Ecol. Syst. 18:489–522.

Avise, J.C., and F.J. Ayala. 1976. Genetic differentiation in speciose versus depauperate phylads: Evidence from the California minnows. Evolution 30:45–68.

Ayala, F.J., M.L. Tracey, D. Hedgecock, and R.C. Richmond. 1974. Genetic differentiation during the speciation process in *Drosophila*. Evolution 28:576–592.

Bell, G. 1982. The masterpiece of nature: The evolution and genetics of sexuality. California University Press, Berkeley.

Berger, A., J. Imbrie, J. Hays, G. Kukla, and B. Saltzman, eds. 1984. Milankovitch and Climate. Parts 1 and 2. Reidel, Dordrecht.

Bonde, N. 1981. Problems of species concepts in paleontology. Pp. 19–34 in International Symposium on Concepts and Methods in Paleobiology.

Broecker, W., and G.H. Denton. 1990. What drives glacial ages? Sci. Am. 262:48–56.

Bush, M.B., and P. Colinvaux. 1990. A pollen record of a complete glacial cycle from lowland Panama. J. Veg. Sci. 1:105–118.

Butlin, R. 1989. Reinforcement of premating isolation. Pp. 158–179 in D. Otte and J.A. Endler, eds. Speciation and its consequences. Sinauer, Sunderland, Mass.

Caithness, N. 1990. Patterns of speciation in the African antelope. Presented at International Congress of Systematic and Evolutionary Biology IV, Frostburg, Md., 1990.

Carson, H.L. 1976. Inference on the time of origin of some *Drosophila* species. Nature 259:395–396.

———. 1982. Speciation as a major reorganization of polygenic balances. Pp. 411–433 in C. Barigozzi, ed. Mechanisms of speciation. Alan R. Liss, New York.

Caswell, H., H.E. Koenig, J.A. Resh, and Q.E. Ross. 1972. An introduction to systems science. Pp. 3–78 in B.C. Patten, ed. Systems analysis and simulation in ecology. Vol 2. Academic Press, New York.

Coope, C.R. 1979. Late Cenozoic fossil Coleoptera. Evolution, biogeography, ecology, Annu. Rev. Ecol. Syst. 10:247–267.

Cooper, G.R., and J.A. Brophy. 1972. Late glacial environmental changes indicated by a coleopteran succession from North Wales. Boreas 1:97–142.

Cracraft, J. 1983. Species concepts and speciation analysis. Curr. Ornith. 1:159–187.

———. 1987. Species concepts and the ontology of evolution. Biol. Philos. 2:329–346.

Darwin, C. 1859. On the origin of species by means of natural selection; or, the preservation of favoured races in the struggle for life. John Murray, London.

Denton, G.H. 1985. Did the Antarctic ice sheet influence late Cainozoic climate and evolution in the southern hemisphere? S. Afr. J. Sci. 81:224–229.

De Queiroz, K., and M.J. Donoghue. 1988. Phylogenetic systematics and the species problem. Cladistics 4:317–338.

———. 1990. Phylogenetic systematics and species revisited. Cladistics 6:83–90.

Dobzhansky, T. 1935. A critique of the species concept in biology. Philos. Sci. 2:344–355.
———. 1937. Genetics and the origin of species. Columbia University Press, New York.
Donoghue, M.J. 1985. A critique of the biological species concept and recommendations for a phylogenetic alternative. Bryologist 88:172–181.
Eldredge, N. 1985. Unfinished synthesis. Oxford University Press, New York.
Eldredge, N., and S.J. Gould. 1972. Punctuated equilibria: An alternative to phyletic gradualism. Pp. 82–115 in T.J.M. Schopf, ed. Models in paleobiology. W.H. Freeman, San Francisco.
Fryer, G., and T.D. Iles. 1969. Alternative routes to evolutionary success as exhibited by African cichlid fishes of the genus *Tilapia* and the species flocks of the Great Lakes. Evolution 23:359–369.
Futuyma, D.J., and G.C. Mayer. 1980. Non-allopatric speciation in animals. Syst. Zool. 29:254–271.
Ghiselin, M.T. 1974a. The economy of nature and the evolution of sex. California University Press, Berkeley.
———. 1974b. A radical solution to the species problem. Syst. Zool. 23:536–544.
Gordon, D.H., and N. Dennet. 1979. Ultrasonic calls and courtship behavior of sibling species of multimate mice, *Praomys (Mastomys) natalensis,* and *Praomys (Mastomys) coucha*. Address to 1979 Zoological Society of Southern Africa Symposium on Animal Communication. Cape Town, South Africa.
Greenacre, M.J., and E.S. Vrba. 1984. A correspondence analysis of biological census data. Ecology 65:984–997.
Hays, J.D., J. Imbrie, and N.J. Shackleton. 1976. Variations in the earth's orbit: Pacemaker of the ice ages. Science 194:1121–1131.
Hull, D.L. 1976. Are species really individuals? Syst. Zool. 25:174–191.
Huntley, B., and T. Webb. 1989. Migration: Species' response to climatic variations caused by changes in the earth's orbit. J. Biogeogr. 16:5–19.
King, M.C., and A.C. Wilson. 1975. Evolution at two levels in humans and chimpanzees. Science 188:107–116.
Kimbel, W.H., and L.B. Martin, eds. 1993. Species, species concepts, and primate evolution. Plenum Press, New York.
Kimbel, W.H., and Y. Rak. 1993. The importance of species taxa in paleoanthropology and an argument for the phylogenetic concept of the species category. Pp. 461–484 in W.H. Kimbel and L.B. Martin, eds. Species, species concepts, and primate evolution. Plenum Press, New York.
Lambert, D.M., and B. Levey. 1979. The use of discriminant function analysis to investigate the design features of specific-mate recognition systems. Address to 1979 Zoological Society of Southern Africa Symposium on Animal Communication. Cape Town, South Africa.
Lande, R. 1979. Effective deme sizes during long term evolution estimated from rates of chromosomal rearrangement. Evolution 33:234–251.
Lincoln, R.I., G.A. Boxshall, and P.F. Clark. 1982. A dictionary of ecology, evolution and systematics. Cambridge University Press, Cambridge.
Maynard Smith, J. 1978. The evolution of sex. Cambridge University Press, Cambridge.
Mayr, E. 1942. Systematics and the origin of species. Columbia University Press, New York.
———. 1963. Animal species and evolution. Harvard University Press, Cambridge, Mass.
———. 1970. Populations, species and evolution. Harvard University Press, Cambridge, Mass.

———. 1982. Processes of speciation in animals. Pp. 1–19 *in* C. Barigozzi, ed. Mechanisms of speciation. Alan R. Liss, New York.
Miller, J.G. 1978. Living systems. McGraw-Hill, New York.
Morris, D. 1970. Patterns of reproductive behavior. Jonathan Cape, London.
Nelson, G. 1971. Paraphyly and polyphyly: Redefinitions. Syst. Zool. 20:471–472.
———. 1989. Species and taxa: Systematics and evolution. Pp. 60–81 *in* D. Otte and J.A. Endler, eds. Speciation and its consequences. Sinauer, Sunderland, Mass.
Nelson, G., and N. Platnick. 1981. Systematics and biogeography: Cladistics and vicariance. Columbia University Press, New York.
Nix, H. 1986. A biogeographic analysis of Australian elapid snakes. Pp. 4–15 *in* R. Longmore, ed. Atlas of Elapid snakes of Australia. Australian Government Printing Service, Canberra.
Olsen, P.E., 1986. A 40-million-year-old lake record of early Mesozoic orbital climate forcing. Science 234:842–848.
Otte, D., and J.A. Endler, eds. 1989. Speciation and its consequences. Sinauer, Sunderland, Mass.
Park, J., and T.D. Herbert. 1987. Hunting for paleoclimatic periodicities in a geologic time series with an uncertain time scale. J. Geophys. Res. 92:14027–14040.
Paterson, H.E.H. 1978. More evidence against speciation by reinforcement. S. Afr. J. Sci. 74:369–371.
———. 1981. The continuing search for the unknown and the unknowable: A critique of contemporary ideas on speciation. S. Afr. J. Sci. 77:113–119.
———. 1982. Perspective on speciation by reinforcement. S. Afr. J. Sci. 78:53–57.
———. 1985. The recognition concept of species. Pp. 21–34 *in* E.S. Vrba, ed. Species and speciation. Transvaal Museum Monograph No. 4. Pretoria.
———. 1986. Environment and species. S. Afr. J. Sci. 82:62–65.
Paterson, H., and S. James. 1973. Animal and plant speciation studies in Western Australia. J. R. Soc. West. Aust. 56:31–43.
Rind, D., and D. Peteet. 1985. Terrestrial conditions at the last glacial maximum and CLIMAP sea-surface temperature estimates: Are they consistent? Quat. Res. 24:1–22.
Schull, J. 1990. Are species intelligent? Behav. Brain Sci. 13:63–108.
Shackleton, N.J., J. Backman, H. Zimmerman, D.V. Kent, M. Hall, D.G. Roberts, D. Schnitker, J.G. Baldauf, A. Desprairies, R. Homrighausen, P. Huddlestun, J.B. Keene, A.J. Kaltenbach, K.A.O. Krumsieck, A.C. Morton, J.W. Murray, and J. Westberg Smith. 1984. Oxygen isotope calibration of the onset of icerafting and history of glaciation in the North Atlantic region. Nature 307:620–623.
Simon, H.A. 1962. The architecture of complexity. Proc. Am. Philos. Soc. 106:467–482.
Stanley, S.M. 1979. Macroevolution: Pattern and process. W.H. Freeman, San Francisco.
Stearns, S.C. 1982. The role of development in the evolution of life histories. Pp. 237–258 *in* J.T. Bonner, ed. Evolution and development. Springer-Verlag, Berlin.
Sutcliffe, A.J. 1985. On the track of ice age mammals. Harvard University Press, Cambridge, Mass.
Taylor, C.R. 1969. The eland and the oryx. Sci. Am. 220:88–95.
Templeton, A.R. 1981. Mechanisms of speciation—a population genetic approach. Annu. Rev. Ecol. Syst. 12:23–48.
———. 1989. The meaning of species and speciation: A genetic perspective. Pp. 3–27 *in* D. Otte and J.A. Endler, eds. Speciation and its consequences. Sinauer, Sunderland, Mass.

Vrba, E.S. 1979. Phylogenetic analysis and classification of fossil and recent Alcelaphini (Family Bovidae, Mammalia). Zool. J. Linn. Soc. 11:207–208.

———. 1980. Evolution, species and fossils: How does life evolve? S. Afr. J. Sci. 76:61–84.

———. 1984. Evolutionary pattern and process in the sister group Alcelaphini-Aepycerotini (Mammalia: Bovidae). Pp. 62–79 *in* N. Eldredge and S.M. Stanley, eds. Living fossils. Springer-Verlag, New York.

———. 1985. Introductory comments on species and speciation. Pp. ix–xviii *in* E.S.Vrba, ed. Species and speciation. Transvaal Museum Monograph No. 4. Pretoria.

———. 1987. A revision of the Bovini (Bovidae) and a preliminary revised checklist of Bovidae from Makapansgat. Paleont. Africana 26:33–46.

———. 1989. Levels of selection and sorting, with special reference to the species level. Oxford Surveys of Evolutionary Biology 6:111–168.

———. 1992. Mammals as a key to evolutionary theory. J. Mamm. 73:1–28.

———. 1993. Mammal evolution in the African Neogene and a new look at the Great American Interchange. Pp. 393–432 *in* P. Goldblatt, ed. Biological relationships between Africa and South America. Yale University Press, New York, in press.

Vrba, E.S., R. Vaisnys, J.E. Gatesy, R. DeSalle, and K.-Y. Wei. 1994. Analysis of paedomorphosis using allometric characters: The example of Reduncini antelopes (Bovidae: Mammalia). Syst. Biol. 43:92–116.

Wheeler, Q.D., and K.C. Nixon. 1990. Another way of looking at the species problem: A reply to de Queiroz and Donoghue. Cladistics 6:77–81.

Wiese, L., and W. Wiese. 1977. On speciation by evolution of gametic incompatibility: A model case in *Chlamydomonas*. Am. Nat. 111:733–742.

Wiley, E.O. 1978. The evolutionary species concept reconsidered. Syst. Zool. 27:17–26.

Williams, G.C. 1975. Sex and evolution. Princeton University Press, Princeton, N.J.

2

Species Concepts and the Interpretation of Fossil Data

BERNARD MICHAUX

Evolutionary Genetics Laboratory, Department of Zoology
University of Aucklund

PUNCTUATED EQUILIBRIUM (Eldredge and Gould 1972; Gould and Eldredge 1977) was not, despite what the subsequent debate seemed to suggest, exclusively about patterns of morphological change in the fossil record. A fundamental point was that these authors' views on morphological change were derived from their application of the Isolation Concept of species, combined with an allopatric model of speciation, to fossils. With one bold stroke Eldredge and Gould (1972) demonstrated to the paleontological community that fossil evidence does not "speak for itself." Fossil data, like any other, require interpretation, and the species concept the paleontologist employs is critical to this interpretation.

This crucial point raised in the punctuated equilibrium argument—that species concepts determine the way in which paleontological data are interpreted—was largely lost in the debate that followed. In 1979, Eldredge again argued that species concepts have logical priority in interpretation of data (Eldredge 1979). He outlined two approaches to the interpretation of fossil data based on distinct species concepts. These he termed a *taxic* approach, in which the biological reality of species is viewed as central, and a *transformational* approach, in which species are viewed as taxonomic conventions. He argued that instances of gradual transformation of morphology in the fossil record were routinely interpreted as examples of phyletic change in a species lineage and that this interpretation was dependent on viewing species as taxonomic conventions, that is, that species are defined solely in terms of gross morphology.

It was not that either he or Gould ever denied that examples of gradual character transformation existed, but rather that it was equally valid to interpret these transformations as a consequence of multiple speciation events when species were viewed as coherent natural groupings (Lambert et al. 1987), that is, defined (Michaux et al. 1990) by the process which leads to that coherence rather than by what they look like. In my view, the fundamental argument embodied in punctuated equilibrium is not about either pattern or process, but about interpretation and the logical priority of theory over data.

Morphological Difference and Biological Discreteness

> For most biological control programs to be effective, individuals that are released and carrying a "genetic load" must mate with the wild pest individuals. However, insects frequently have different species that appear to be the same to a human observer.
>
> Makela and Richardson, 1979

Paterson's (1980, 1985) general proposal—that we should base our concept of species on the basis by which individuals themselves recognize conspecific mates—has important implications in medical and pest management programs. Indeed, the very success of such programs is dependent on the accurate identification of these natural groupings which may not, initially at least, be identified on the basis of morphology (e.g. Paterson 1964; Richardson et al. 1982).

The results of biological control programs show clearly that one cannot assume a concurrence between our perception of natural groupings, based on taxonomic criteria, and those groupings that actually occur in nature. This is because, as Makela and Richardson (1979) point out, organisms have no difficulty in perceiving differences that our senses, particularly vision, cannot detect. The argument here is not one about taxonomic skill, or even about the relative frequency of misidentification of biological groupings. It is an argument about making unjustified assumptions.

While none would argue against the need to accurately identify the exact biological groupings involved in medical and pest management programs, such a consensus does not exist in paleontology where the majority of workers are content to employ a taxonomic species concept. Even when a "taxic" approach was adopted, most studies relied on taxonomic criteria to define species (but see Jackson and Cheetham [1990] as an exception). When it is important to correctly identify biological groupings—and I would argue that this is always so for theoretical reasons if not always, as with biological control programs, for financial or ecological reasons—then the concurrence has to be demonstrated.

How Can Biological Groupings Be Identified in the Fossil Record?

Vrba (1979, 1983, 1984) recognized the need to accurately identify biological groupings in the fossil record and made use of Paterson's Recognition Concept of species. She argued that in modern Alcelaphini the horns and facial bone structure are important elements in the recognition system of these animals and consequently used these structures to delimit fossil species.

I have earlier suggested another approach to the problem of identification of biological groupings in the fossil record. This method, used ideally in conjunction with the approach outlined above, is also suited to species for which elements of the recognition system are either unknown or have not been preserved and is outlined below (Michaux 1988, 1989a).

A Case Study: Amalda

In order to maximize the potential information available, the group chosen should not only have as complete a fossil record as possible, but also be common enough to collect in sufficient numbers from modern environments for adequate genetic analysis. They must also, of course, be sexually reproducing. There are many shallow marine groups in New Zealand that satisfy these criteria. One such group is the molluscan genus *Amalda* (Gastropoda: Olividae: Ancillinae), which has a fossil record in New Zealand dating back into the Eocene (Olson 1956; Michaux 1989b).

There are six extant *Amalda* spp. living off the coast of New Zealand. Four of these taxonomically defined species are common and live in shallow marine environments, the remaining two species being confined to deep water. Three of the four shallow water species are common in the fossil record of the Pliocene and Pleistocene (Michaux 1987).

Biological Status of the Taxonomic Species

Details of the fertilization system of *Amalda* species are unknown. However, the presence in this genus of a terminal appendage to the penis (Kilburn 1981) indicates that a tactile signal and response system is part of the overall fertilization system. Such characters are not, of course, preserved in the fossil record. In the absence of direct evidence about Specific Mate-Recognition Systems (SMRSs), one can follow a standard procedure of calculating whether gene frequencies based on collections from areas where taxonomically defined species occur in sympatry deviate from Hardy-Weinberg equilibrium. Such a procedure allows assessment of the biological reality (or not) of taxonomic species with a high level of confidence. The four taxonomic species of *Amalda* that form the basis of this example are biologically distinct in terms of the above criterion (Michaux 1987).

Morphometric Analysis

Having established the biological distinctness of the taxa under study, either by direct analysis of SMRS or indirectly through the establishment of genetic discontinuity, then one is in a position to establish the relationship between phenotypic variability of the extant specimens and phenotypic variability of fossil specimens through time.

Full details of the morphometric analysis of specimens are given in Michaux (1989a). Briefly, ten measurements, taken from each of 700 live-taken shells of the four biologically coherent study species, were used to define the appropriate phenotypes in multidimensional space. Because there were no external morphological characters that could be used to distinguish juveniles from adults, all size classes were included in proportion to their occurrence in each geographic sample. These samples were collected from throughout the species' geographical ranges. Canonical discriminant analysis was performed on the data, and a set of allocatory rules was derived. These allocatory rules were then applied to 644 fossil specimens of three of these biological species. Allocation results were similar for both extant and fossil specimens and are given in Table 2.1.

Figure 2.1 shows the mean canonical variate scores calculated from extant specimens of the four species, the circles representing 95 percent confidence surfaces. The first canonical axis accounts for 69.8 percent of the total variance, the second 25.8 percent. The cumulative variance accounted for by these two axes is therefore 95.6 percent. When the canonical axes scores for individual fossil specimens were calculated and plotted, these individual fossil specimens occupied the appropriate phenotypic space defined by their modern descendants (see Michaux 1989a: Figs. 3–5).

The post-Miocene history of the Wanganui basin, an important site of Tertiary and Quaternary sedimentation located on the west coast of the North Island, and in which numerous *Amalda* specimens are preserved in the sediments, is summarized in Table 2.2. The history of this region was one of complex and rapid changes in both climate and paleogeography (Fleming 1953). The major climatic fluctuations in this area, deduced by Fleming (1953) from floral and faunal evidence, are indicated in Figure 2.2. The Hautawan (formations 41–39 in Fig. 2.2) represented a time when temperatures were some 3° to 6°C lower than today, and glaciation extended as far north as 43° South. This period represented a thermal minimum for the Wanganui region. Temperatures subsequently increased (formations 38–33) to levels equivalent to those of the southern extremity of the South Island today. Further climatic deteriora-

Table 2.1. Allocation results from canonical discriminant functions
(Results are percentages of correct classification.)

Amalda species	N	Area[a]	Classified[b] as			
			A. australis	A. mucronata	A. novaezelandiae	A. depressa
Modern age						
A. australis	327		85	4		11
A. mucronata	163		2	98		
A. novaezelandiae	152		1	1	98	
A. depressa	58		5			95
Plio-Holocene						
A. australis	143	1	81	16	1.5	1.5
Plio-Pleistocene						
A. mucronata	135	1	3	95		2
Pleistocene						
A. novaezelandiae	117	1	16	6	78	
A. australis	50	2	84	14		2
A. mucronata	50	2	6	94		
A. novaezelandiae	50	2	12		88	

[a] Area 1 = Wanganui, Area 2 = east coast North Island.
[b] The rows are classified into the columns; thus 85 percent of modern A. *australis* are classified as *australis*, 4 percent, as *mucronata*, and 11 percent as *depressa*.

tion during formations 32–22 was followed by a progressive warming during the Putikian (formations 17–3) reaching a maximum during formations 16–14, when temperatures were estimated to have been 3° to 6°C higher than at present. Variable conditions existed during the Hawera stage that followed (formations 1 and 2).

The Wanganui basin had a complex geological history, with downwarping, margin tilting, variable sediment supply, and regional tectonics interplaying to give complex shoreline migration patterns. This resulted in the alternation of *australis/depressa* and *mucronata/novaezelandiae* collections as shallow coastal conditions alternated with deeper offshore environments. Tracking of a species' preferred habitat on a regional scale is a phenomenon reported by Coope (1979) for coleopteran taxa. Paterson (1985) has discussed the importance of this phenomenon with respect to a species' "equilibrium" phase, while Vrba (1985) has discussed its importance to vicariant events.

Plots of the canonical variate scores for the fossil samples of *A. australis* against formation number are shown in Figure 2.2. The pattern of variation of these scores through time is similar, and both patterns appear to track climatic change. During cold periods *A. australis* shells are larger and wider and have

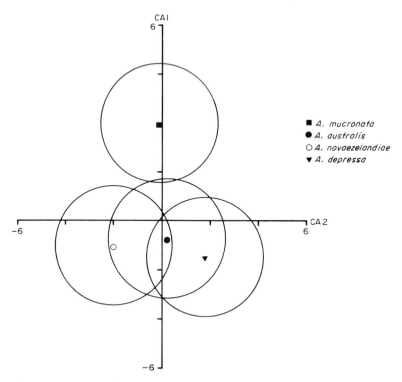

Fig. 2.1. Positions of the mean canonical variate axes scores for modern *Amalda australis*, *A. mucronata*, *A. depressa*, and *A. novaezelandiae*. Circles represent 95 percent confidence surfaces.

bigger apertures, with the converse during warmer conditions. Interestingly, the reverse appears true for *A. mucronata* (Fig. 2.3), cold climate being associated with smaller shells and warmer climate with larger shells. The phenotypic trajectories through time are thus climatically modulated for these species.

These results are also a clear confirmation of the oscillatory nature of a species' equilibrium phase predicted by Eldredge and Gould (1972). The term *stasis*, which implies a static, unchanging stability in form, is not appropriate for this phenomenon, which exhibits the characteristics of a dynamic equilibrium. It is quite probable that other examples of morphological stasis would, if viewed with a higher degree of resolution as suggested by Sheldon (1987), also exhibit this phenomenon.

On Inherent Limitations of Data: A Trilobite Example

Sheldon (1987, 1988a,b, 1990) reported the results of a comprehensive study of 15,000 Ordovician trilobite specimens. These specimens were sam-

Table 2.2. A synopsis of the stratigraphy of the Plio-Pleistocene sequence at Wanganui based on Fleming (1953)

Correlation	Stage	Symbol	Group	Formation[a]	Age[b] (Years B.P.)
Pleistocene	Hawera		Pouaki	1–2	
			Shakespeare	3–14	
					600,000
	Castlecliffian	Wc	Kai-iwi	15–21	
					850,000
			Okehu	22–26	
					1,100,000
	Nukumaruan	Wn	Maxwell	27–32	
			Nukumaru	33–38	
					1,790,000
Pliocene	Waitotaran	Ww	Okiwa	39–45	
			Paparangi	46–49	
					3,100,000

[a] Formation 1 = Brunswick formation; subsequent formations listed in Flemming (1953) are numbered sequentially.
[b] Ages are approximate.

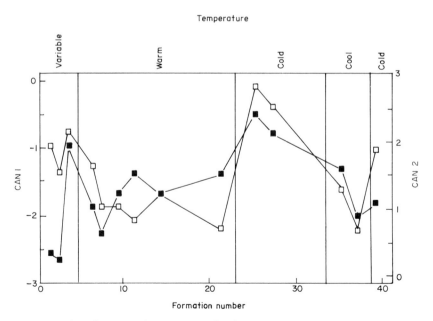

Fig. 2.2. Plot of CA1 and CA2 scores for fossil samples of A. australis versus formation number. Open squares represent CA1 scores; closed squares, CA2 scores. Note that age increases to the right. Refer to Table 2.2 for details of formation ages.

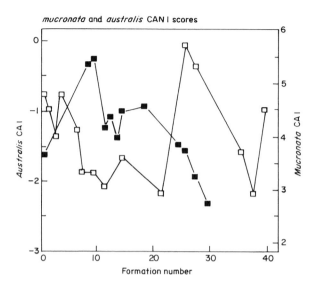

Fig. 2.3. Plot of CA1 scores for fossil samples of *A. australis* and *A. mucronata* versus formation number. Open squares represent CA1 scores for *A. australis;* closed squares, CA1 scores for *A. mucronata.* Note that age increases to the right. Refer to Table 2.2 for details of formation ages.

pled from seven sections of the *teretiusculus* Shales representing a time slice covering some three million years. The sections were divided into 400 sampling localities (with an average stratigraphic thickness of 23 cm). The *teretiusculus* Shales accumulated in a low-energy, poorly oxygenated marine basin of several hundred meters depth. The trilobites are thought to have scavenged on the basin floor feeding on small animals and the nutrient-rich mud. Sheldon's work demonstrated that both pygidial width (Sheldon 1988a:Fig. 11) and number of pygidial ribs (Sheldon 1987:Fig. 4) in these trilobite lineages show a net gradual (i.e., phyletic) increase. These net increases are achieved with frequent short-term reversals in six of the eight lineages studied.

Sheldon's (1987) study exemplifies the inherent limitation in using extinct taxa to understand species and speciation. This limitation has nothing to do with either the quality or quantity of data, which in this example are both beyond reproach, but stems from the circularity inherent in examining character change through time, using these character changes to delimit species, and then discussing speciation patterns through time (Levinton and Simon 1980). In such cases there is no possibility of using independent data to void tautological reasoning. As I have already detailed, these data can be obtained only from extant organisms. In addition, Sheldon's interpretation, that these character changes represent phyletic (i.e., within-lineage) change, is dependent on the assumption that biological groupings are accurately identified by changes in morphology.

Species Concepts and Extinct Higher Taxa

There is, of course, no possibility of determining what biological groupings existed within extinct higher taxa. Because of this inherent limitation, studies employing such groups can provide only weak and equivocal evidence concerning *speciation patterns* in the fossil record. However, even though workers with extinct groups are limited to employing a taxonomic definition of species, these groupings should reflect as accurately as possible the biological groupings that existed at the time.

Here species concepts can once again affect how the paleontologist views the problem. Sheldon (1987), utilizing a taxonomic species concept, has interpreted his data in a particular way. Consequently he states that "practical taxonomic subdivision of each lineage proved impossible. The apparent success of earlier Linnean nomenclature (with its implications of discrete species) could easily have been misinterpreted as evidence of punctuation and stasis, and it is probable that detection of many other gradualistic patterns has been hindered by ready application of binomial taxonomy to fossils" (Sheldon 1987:561).

If you view species as coherent natural entities your perspective is somewhat different. Assuming that trilobites reproduced sexually, it is highly probable that they used nonvisual modalities in their SMRSs (i.e., chemical or tactile signals) due to poor visibility in their habitat. As a general rule, species that use nonvisual modalities are more likely to be difficult to separate on the basis of external morphology. This would lead one to expect that quite subtle changes in morphology may signify the presence of discrete biological groupings. Analyzing the data using a multivariate ordination technique such as principal component analysis, which assumes no *a priori* structure, is an appropriate way of detecting such morphological groups. Once group structure has been established, a set of allocatory rules could be derived from canonical variate analysis to help assign specimens to the appropriate taxonomic species.

It would be an interesting exercise to reanalyze Sheldon's (1987) data in this way because I suspect that such an analysis would demonstrate the existence of morphological groupings within three of the lineages he studied. The nileids and the genera *Ogygiocarella* and *Cnemidopyge* appear to show group structures, and each would probably separate into two groups. In the nileids and the genus *Ogygiocarella*, these hypothesized groups correspond with a stratigraphic gap located between sections EP and PH (see Sheldon 1987:Fig. 4). In both cases the younger taxon is smaller (as indicated by pygidial width), with a greater average number of pygidial ribs, than the ancestral taxon. In the nileids these two groups have previously been described as separate genera, *Barrandia* and *Homalopteon.*

It seems to me that a legitimate general conclusion that could be drawn from this discussion is that the appropriate null hypothesis in studies where sufficient data are available (i.e., a series of specimens connecting taxonomically distinct end members), is that group structure *is* present in the data. Only when one rejected the null hypothesis would it be appropriate to hypothesize phyletic change. If the null hypothesis was rejected, then any conclusions drawn from the data concerning speciation patterns would be weakened because of the inherent limitation, discussed above, in defining species taxonomically.

Conclusions

For the past two decades there has been intense interest in what paleontological data might tell us about the "tempo and mode" of evolution. Sadly, this effort has been largely misdirected because workers have consistently attempted to demonstrate a pattern without first justifying the units by which that pattern is recognized. This chapter is an attempt to clarify the criteria by which evolutionary units (species) can be theoretically justified.

The major points of this clarification are as follows:

1. Species concepts are critical to the interpretation of fossil data.

2. Species definition has to be based on data that are independent of the fossil record to avoid tautological reasoning.

3. It must be demonstrated that taxonomically defined species accurately reflect any biological groupings present, as there are numerous examples in the biological literature which demonstrate that this assumption requires confirmation.

4. There are inherent limitations in fossil data. Any conclusions drawn from a study have to take account of these limitations. On the basis of the points discussed above, these limitations are overcome in decreasing degree by:

(a) Data from extant, sexually reproducing species with good fossil records, which demonstrate that groupings recognized by taxonomic criteria accurately reflect any biological groupings that exist. These data would ideally concern direct evidence of the fertilization system and genetic data from sympatric populations.

(b) Data from extinct species with close extant relatives. The type of evidence discussed in (*a*) could be obtained from modern species and applied, with varying levels of confidence, to closely related extinct species.

(c) Data from species of extinct higher taxa. In both (*b*) and (*c*), multivariate ordination techniques should be used to test the null hypothesis that the morphological data have structure to them.

Acknowledgements

I express my appreciation to Hugh Paterson, whose views on species have always given me a point of reference when thinking about biological phenomena, and to Roger Cooper and David Penny for their thoughtful comments.

References

Coope, C.R. 1979. Late Cenozoic fossil Coleoptera: Evolution, biogeography and ecology. Annu. Rev. Ecol. Syst. 10:247–267.

Eldredge, N. 1979. Alternative approaches to evolutionary theory. Bull. Carnegie Mus. Nat. Hist. 13:7–19.

Eldredge, N., and S.J. Gould. 1972. Punctuated equilibria: An alternative to phyletic gradualism. Pp. 82–115 *in* T.J.M. Schopf, ed. Models in paleobiology. W.H. Freeman, San Francisco.

Fleming, C.A. 1953. The geology of the Wanganui subdivision. N.Z. Geol. Surv. Bull., No. 52. DSIR publication, Wellington.

Gould, S.J., and N. Eldredge. 1977. Punctuated equilibria: Tempo and mode of evolution reconsidered. Paleobiol. 3:115–151.

Jackson, J.B.C., and A.H. Cheetham. 1990. Evolutionary significance of morphospecies: A test with Cheilostome bryozoa. Science 248:579–583.

Kilburn, R.N. 1981. Revision of the genus *Ancilla* Lamarck, 1799 (Mollusca: Olividae: Ancillinae). Ann. Natal Mus. 24:349–463.

Lambert, D.M., B. Michaux, and C.S. White. 1987. Are species self-defining? Syst. Zool. 36:196–205.

Levington, J.S., and C.M. Simon. 1980. A critique of the punctuated equilibrium model and implications for the detection of speciation in the fossil record. Syst. Zool. 29:130–142.

Makela, M.E., and R.H. Richardson. 1979. Hidden, reproductively isolated populations: One of nature's countermeasures to genetic pest control. Pp. 49–66 *in* R.H. Richardson, ed. The screw-worm problem. Texas University Press, Austin.

Michaux, B. 1987. An analysis of allozymic characters of four species of New Zealand *Amalda* (Gastropoda: Olividae: Ancillinae). N. Z. J. Zool. 14:359–366.

———. 1988. Organotaxism: An alternative "way of seeing" the fossil record. J. Theor. Biol. 133:397–408.

———. 1989*a*. Morphological variation of species through time. Biol. J. Linn. Soc. 38:239–255.

———. 1989*b*. Cladograms can reconstruct phylogenies: An example from the fossil record. Alcheringa 13:21–36.

Michaux, B., C.S. White, and D.M. Lambert. 1990. Organisms not species evolve: A reply to Ghiselin. Syst. Zool. 39:79–80.

Olson, O.P. 1956. The genus *Baryspira* in New Zealand. DSIR Bull., Wellington. No. 24.

Paterson, H.E.H. 1964. Direct evidence for the specific distinctness of forms A, B and C of the *Anopheles gambiae* complex Giles. Rev. Malar. 43:192–196.

———. 1980. A comment on "Mate Recognition Systems." Evolution 32:330–331.
———. 1985. The recognition concept of species. Pp. 21–29 *in* E.S. Vrba, ed. Species and speciation. Transvaal Museum Monograph No. 4. Pretoria.
Richardson, R.H., J.R. Ellison, and W.W. Averhoff. 1982. Autocidal control of screwworms in North America. Science 215:361–370.
Sheldon, P.R. 1987. Parallel gradualistic evolution of Ordovician trilobites. Nature 330:561–563.
———. 1988*a*. Trilobite size-frequency distributions recognition of instars, and phyletic size changes. Lethaia 21:293–306.
———. 1988*b*. Making the most of the evolutionary diaries. New Sci. 21:52–54.
———. 1990. Microevolution and the fossil record. Pp. 106–110 *in* D.E.G. Briggs and P.R. Crowther, eds. Paleobiology: A synthesis. Blackwell, Oxford.
Vrba, E.S. 1979. Phylogenetic analysis and classification of fossil and recent Alcelaphini (Bovidae: Mammalia). Zool. J. Linn. Soc. 11:207–228.
———. 1983. Evolutionary pattern and process in the sister-group Alcelaphini-Aepycerotini (Mammalia; Bovidae). Pp. 62–79 *in* N. Eldredge and S.M. Stanley, eds. Living fossils. Springer-Verlag, New York.
———. 1984. Patterns in the fossil record and evolutionary processes. Pp. 115–142 *in* M.-W. Ho and P.S. Saunders, eds. Beyond neo-Darwinism. Academic Press, London.
———. 1985. Environment and evolution: Alternative causes of the temporal distribution of evolutionary events. S. Afr. J. Sci. 81:229–236.

3

The Species in Paleontology

ALAN TURNER

Hominid Palaeontology Research Group
Department of Human Anatomy and Cell Biology, University of Liverpool

UNTIL THE EARLY 1970s there was broad agreement that the fossil record could be simply and directly married to natural selection, Mendelian inheritance, and molecular biology, to consolidate and extend the "modern synthesis" of Huxley (1942). The leading figure in efforts to integrate paleontology with evolutionary biology was undoubtedly Simpson (1944, 1953), but others such as Mayr, Dobzhansky, and Stebbins all described the construction of that greater synthesis, and their own roles as its architects, with infectious optimism. The clear implication was that only details need be added for a full understanding to emerge.

In a sense that was so, although it has always been evident that many of the details involve complex problems, as Mayr (1978) has stressed. But some of the details, far from being the decorations on the finished product envisaged by the architects, have proved to be part of the foundations. What was missing has turned out to be a clear understanding of what species are and how speciation actually occurs.

This gap in understanding, by its very nature, has been prevalent in both neontology and paleontology. Species are after all a common currency in both disciplines, even if speciation is something that the neontologist need generally consider only in theory. But its effect on ideas about evolution has been more insidious in paleontology, because it is there that we look for evidence of evolution as a fact of history. In contrast, it is from neontology that we derive our ideas about what species are, but if we are wrong it does not, at first sight, necessarily affect us too badly. We may, of course, fail to understand what separates an apparently homogeneous group into true species, each with different behavior patterns, and we may fail in our attempts to control pests, as Huxley (1942:156) clearly recognized in the case of malaria-transmitting mosquitoes (see Michaux, this volume). But if our idea of species is awry, one could argue, we will still see the difference between a sheep and a sheepdog, or an alpaca and an aardvark, a point also essentially made by Huxley. We are unlikely to waste our time trying to persuade males and females from such different animals to mate, and in that sense it may seem to be of little overall importance to get the theory of what species actually are precisely right.

But if we apply an incorrect concept of species to the fossil record, and to our interpretation of evolution based on the deployment of speciation over time, other problems arise. We risk being seriously misled by a cumulative error in our understanding of the development of life. However much we may seek to avoid it, faulty ideas about evolution over geological time will feed back into our notions of microevolutionary mechanisms and complete the circle of arguments about tempo and mode.

In truth, the species has long been acknowledged as a problem in paleontology. For this contribution I shall chart what I consider to be some significant points in the development of ideas about it over the past several decades, without any pretense at dealing with everything ever written on the subject. In particular, I want to underline the fact that until the development of Paterson's Recognition Concept we really were stumbling in the dark, equipped with inadequate ideas of what constitutes a species, how speciation might actually take place, and how to integrate the theoretical process with the historical pattern of speciation in the fossil record.

Species and Paleontology

Some paleontologists have simply ignored the problems posed by the species, as Romer (1933, 1945, 1966) effectively did. By the time he wrote the third edition of his book on the vertebrates he was able to push the whole problem to one side with a passing reference to the "synthetic theory" available in works that the reader "would be well advised to consult" (1966:3). Others, in the spirit of the modern synthesis, have incorporated ideas from neontology into their own, at times extensive, discussions of the whole issue (see, e.g. Beerbower 1968; Kurtén 1968; Carroll 1988; Eldredge 1985a, 1989). But an indication of the persistent seriousness of the problem, and of the disparate views held by those active in the field, may be seen in the decision by the Systematics Association to hold a meeting in 1954 entirely devoted to the species concept in paleontology.

In his introduction to the resultant publication, Sylvester-Bradley (1956:1) argued that most paleontologists of the day recognized "two distinctly different sorts of species." The "biospecies" was the equivalent of the neontologist's species, which implied adherence to the definition given by Mayr (1942:120) as "groups of actually or potentially interbreeding populations reproductively isolated from other such groups." The second, what Sylvester-Bradley called the "bread and butter" species of the paleontologist, was the "morphospecies." It is apparent from the other contributions that the confounding element in efforts to achieve a single "sort" of species fell into two, interrelated parts. The first was the time element and the resultant changes seen in the fossil record. The second was the fundamental problem of an inadequate definition of species that could apply to fossils, a problem simply exacerbated by the introduction of time into the discussion. But the overall view seemed to be that change within and between species in the fossil record was gradual in nature, just as it was held to be by classic population biology.

This view, of course, raised the problem of how to distinguish between species. While authors such as Rhodes (1956:40) argued that paleontological classification should strive to indicate genetic relationships, others disagreed fundamentally. Arkell (1956:99), in a wonderfully reactionary essay, saw only "a danger to palaeontology inherent in too much concern with the theoretical aspects of taxonomy," and felt that the units employed by the paleontologist must be arbitrary. He even argued (p. 97) that the only criterion for a taxon in paleontology was its "usefulness" in providing manageable groups and that different groups require different species concepts, conveying the forthright impression that he would be almost as happy classifying pieces of concrete as he would fossil organisms. Others again, such as Joysey (1956:86), took the view that chronological classification was essential. Indeed, Joysey was still championing such an approach nearly twenty years later, when he said, "I advocate that a continuous evolutionary sequence should be arbitrarily subdivided on chronological grounds" (1972:269). In his concluding remarks on the symposium, George (1956:124) strove to amalgamate genetic ("biological"), morphological, and chronological species into a single overall approach on the grounds that "there is no inherently 'right' way of achieving taxonomic integration."

All that was in 1954, and the debate might seem to be long behind us. But uncertainties over the nature of the species taxon in paleontology have continued to be reflected in the literature (Wiley 1978, 1979; Eldredge and Cracraft 1980; Vrba 1980, 1984*b*, 1985*a;* Tattersall 1986; Bock 1986). Moreover, the idea that fossil lineages can be divided somewhat arbitrarily into species at convenient points in the stratigraphy in the absence of other criteria has been repeated by others. Simpson (1961:165) and Mayr (1969:35–36) both lent their considerable authority to that view, and it was strongly implied by Kurtén (1968:238) in his discussion of mammalian evolution in the Pleistocene of Europe. Perhaps the most extreme statement of this position was made by Campbell (1979:569) in his survey of the problems of classification and nomenclature confronting physical anthropologists. As he put it then:

> It is clear that the boundaries of sequent taxa, be they genera, species or subspecies, should be conveniently agreed time-lines, rather than diagnostic mor-

phological features such as the famed cerebral rubicon of Keith. This means that *both anatomy and dating are necessary to create the taxonomy of fossil lineages* [italics in original]. It follows that the development of a reliable chronology is one of the most important characteristics of the recent period of research, and as this paper will show, *new dates give us new taxonomy* [italics added].

I consider Campbell's statement extreme because I doubt if the suggestion that new dates should indeed give us a new taxonomy would find favor with many paleontologists. But the overall stance adopted is not so unusual. The result, at least until the end of the 1960s, was a reasonably coherent view of evolution held by paleontologists and neontologists alike, albeit with a few areas of obscurity about just what biological units the paleontologist actually employs. Although the details varied, the essence of most standard treatments of species in the fossil record was one of pragmatics and, at the risk of distortion, may be summarized as follows. It is impossible to know whether the organisms in question actually interbred on a regular basis (and therefore constituted what we would think of as a species), but some means of establishing taxa in the fossil record must be devised, and morphology (or rather the presence of morphological gaps) is the usual basis for decision making. The term *species* is used for the resultant groupings, but we should, of course, understand that this is done simply for convenience and that no precise parallel with the term *species* used for groups of living animals should *necessarily* be assumed. Evolution in the fossil record is therefore population biology plus time plus an inevitable (and apparently acceptable) measure of uncertainty.

Interpretations of the horse lineage offer perhaps the clearest example of the perspective. A trend toward monodactyly, greater size, and increased hypsodonty over several tens of millions of years are still frequently cited as a textbook example of the gradual nature of evolution at work, in line with the arguments put forward by Matthew and Chubb (1921), Romer (1933), and a host of other authors including, of course, Simpson (1944, 1951, 1953). A note of contradiction is introduced because change, in the conventional view, is expected to be gradual and cumulative, while absolute differences are expected to occur between taxa when speciation results from the accumulation of change over time, the usual paleontological dilemma. The various horse taxa are usually distinguished on the basis of detailed differences in tooth enamel patterning despite the overall assumption of a gradual and linear trend in evolutionary development (see, by way of example, the discussion by Churcher and Richardson [1978]). The mechanism for the actual process of speciation is never really made clear in such treatments, although seemingly correlated events such as chromosomal variations have increasingly been invoked as causal factors by some population biologists (Bush et al. 1977; Bush 1981) eager to incorporate newer ideas about an episodic pattern of change as discussed in the next section.

New Ideas, New Problems

The relatively coherent picture of evolution held until the end of the 1960s was seriously thrown out of focus when first Eldredge (1971) and then Eldredge and Gould (1972) argued that morphological gaps in the fossil record might point to the episodic nature of evolution and not to a series of preservational accidents. They suggested that phyletic gradualism, the slow and cumulative transformation of entire populations marred by an occasional break in the record, owed its dominance to intellectual inertia and an initial muddling of concepts by Darwin himself. In its place, they proposed that successive episodes of rapid change (punctuations) are followed by long periods of stable structure (equilibria).

These authors were led toward the idea of episodic change by the theory of allopatric speciation in small, peripherally isolated populations. Their emphasis on speciation in allopatry came about "because it is the allopatric, rather than the sympatric, theory that is preferred by biologists" (Eldredge and Gould 1972:94). Such a process, they argued, was incompatible with the idea of gradual change between species. But, although they highlighted the futility of much paleontological debate about how to reconcile the differences between biospecies and paleospecies, they stopped short of addressing the problem of the nature of species and the process of speciation from first principles. Unfortunately, the allopatric theory of speciation does not actually deal with the nature of species and offers a poor understanding of the process of speciation. Indeed, one of the major difficulties posed by speciation in allopatry was expressed by Mayr himself (1963:548) when he noted that isolating mechanisms, an essential part of the Biological Species Concept that underpins his theory of speciation in allopatry, "are *ad hoc* mechanisms. It is therefore somewhat difficult to comprehend how isolating mechanisms can evolve in isolated populations."

It is self-evident that any proposed character arising in allopatry cannot be an isolating mechanism, since it has not evolved in any sense to protect the "integrity of the species," and must be considered an *effect* in the sense employed by Williams (1966). But although Eldredge and Gould sought support from speciation in allopatry, they took no account of the difficulties posed by the requirement for isolating mechanisms to arise in such circumstances and restricted their concern to pattern rather than process. They were content simply to take what they needed from the idea of allopatric speciation and to argue (1972:94) that "a peripheral isolate develops into a new species if *isolating mechanisms* evolve that will prevent re-initiation of gene flow if the new form re-encounters its ancestors at some future time" (italics in original). As a result they left the way open for critics of punctuated equilibrium to argue that evolutionary biology could equally well predict a gradual pattern of change, whatever theory of speciation was "preferred by biologists," as they put it.

Not surprisingly, the idea of punctuated equilibrium met with considerable opposition from members of a paleontological community firmly committed to the modern synthesis (see the discussion by Gould and Eldredge [1977] and Eldredge [1985c] and arguments put forward by Gingerich [1984, 1985] and by Hecht and Hoffman [1986]). Some of the criticisms even led Gould (1985:3) to characterize them as "distortions so obtuse that I can only regard them as willful." But a model of evolution rooted in the synthesis also continued to receive strong support from evolutionary biologists outside the paleontological community, in both specialist and popular treatments (Mayr 1978; Ayala 1978, 1983; Stebbins and Ayala 1981; Barton and Charlesworth 1984; Carson and Templeton 1984; Butlin 1987; Maynard Smith 1981, 1987; Coyne et al. 1988; Chandler and Gromko 1989).

The irony inherent in the proposal for punctuated equilibrium put forward by Eldredge and Gould is that it was rooted firmly in the ideas of the modern synthesis, which saw evolution by slow, progressive change within the species as a result of directional selection. Speciation was then expected to occur largely as a function of the time over which such within-species changes had accumulated and was itself seen as the product of the same slow, progressive change that from time to time spilled over into the production of a new species. Mayr's proposal for allopatric speciation did little to challenge that notion, since it simply argued for the divergence of geographically separated populations and the development of isolating mechanisms as part of the same overall process. When Eldredge and Gould challenged the intellectual inertia of phyletic gradualism they struck directly at the heart of the consensus view of evolution held by adherents of the modern synthesis, which may do much to explain the vehemence of the reaction. But they did so with a proposal that was not sufficiently radical in its departure from the accepted wisdom of the synthesis and that simply made the heretical suggestion that speciation might indeed be rapid and the gaps in the morphological continuity of the record therefore real. In so doing, of course, they also challenged one of the implicit beliefs of orthodox paleontology, the ludicrous notion that we are lucky to have such gaps; otherwise we should have no means of recognizing taxa in a world where gradual change prevails, as Eldredge and Cracraft (1980:116) have pointed out.

At the heart of the subsequent debate over punctuated equilibrium has lain a failure to separate within-species change from speciation. This stems in part from use of the term *stasis* to describe the periods between speciation events and from Eldredge and Gould's failure to deal with the nature of species and the precise mechanism of speciation. Critics of punctuated equilibrium have since felt that a demonstration of gradual change over time in a given species, expected under the terms of the synthesis, is sufficient evidence to reject the punctuationist interpretation in favor of phyletic gradualism. But the rate of change within a species actually offers no answer to questions about the speed at which it became that species or eventually gave rise to another. The error that leads to a conflation of the two stems from an inadequate concept of the species itself, and once that is overcome the problem resolves itself. I shall return to this point in the next section.

I have suggested that the theory of allopatric speciation employed by Eldredge and Gould (1972) did not deal with the nature of species and gave at best a partial understanding of speciation itself. Gould (1980a:123) has since rejected the primacy of allopatric speciation and suggested the incorporation of sympatric models involving the fixation of chromosomal variants as isolating mechanisms. Eldredge (Eldredge and Cracraft 1980:126) has argued that, to the systematist, both parapatric and sympatric speciation might mimic the effect of allopatric speciation in peripherally isolated populations. Such proposals imply episodic change and can be fitted into the punctuated equilibrium picture of evolutionary tempo, although the causal connection between chromosomal rearrangements and speciation is open to very serious question (Paterson 1981), and speciation in sympatry has little theoretical or empirical support (Mayr 1963; Paterson 1981, 1985). But such shoring up of the edifice calls in question the strength of the original argument for punctuated equilibrium if the concept of the species and the model of speciation can be so readily altered or extended. In these circumstances both the extent and the inconclusive nature of the debate since 1972 have, with only modest hindsight, been entirely predictable.

More New Ideas, Fewer (or Different) Problems

A solution to the problems of understanding evolutionary tempo and mode in the fossil record has been provided by Paterson in various publications, most notably in 1985 and 1986. The essential point as made by him (Paterson 1985) when he stressed that "*any* view of species must be cast in genetic terms if it is to be useful in understanding the process of evolution" (italics added). That statement echoes those of the more perceptive participants in the 1954 meeting on species in paleontology. Arbitrary paleontological units, as advocated by Arkell (1956), may have their uses in some or other scheme for pigeonholing samples, but to argue that paleontologists deal with different categories of "species" is to remove any point of contact with neontology or, for that matter, with reality. The debate over tempo and mode in the fossil record concerns the deployment of speciation, not the appearance of arbitrarily defined units, as Wiley (1979:215) has stressed, and unless speciation is clearly understood to refer to genetic species the whole debate is pointless. Of course, the paleontologist may choose to operate outside the constraints of a genetic view of species, but then the whole business of studying fossils does indeed risk becoming a form of stamp collection as implied by Gould (1980b).

Under the Recognition Concept, a species is defined as "that most inclusive population of individual, biparental organisms which share a common fertilization system" (Paterson 1985). The details of the Recognition Concept are adequately presented elsewhere. The point to stress here is that the fertilization system of a species will be stable because of the coadapted nature of its male and female components, the effect of stabilizing selection in the organism's normal habitat, and the large population size in normal circumstances. Stability is likely in components of the fertilization system that serve to bring mating partners together, what Paterson has termed the Specific-Mate Recognition System (SMRS). Speciation occurs when a new SMRS is produced by the action of directional selection on the original system in a small, isolated subset of the population in a different habitat, ensuring an inability to recognize and mate with members of the original population. In contrast, changes resulting from directional selection operating on characters unrelated to the fertilization system can occur at any time within the lifetime of a species and are effectively decoupled from those at speciation. Characters related to the fertilization system will therefore remain stable while the species persists, while others may change in response to selective pressures as the population becomes finely tuned to its new circumstances. But gradual changes resulting from such tracking of the environment say nothing about the deployment of speciation. The importance of all this for paleontology cannot be overemphasized. Speciation will be episodic, and the fossil record will show evidence of punctuated equilibrium rather than gradual change between species.

This view of speciation differs fundamentally from that incorporated in Mayr's Biological Species Concept, which underlay Eldredge and Gould's original argument for punctuated equilibrium. Species within the framework of the Recognition Concept cannot be arbitrary divisions of the morphological pattern over time in a lineage but are an effect of processes with a clear genetic basis. Because the Recognition Concept provides a robust explanation for stasis in species-specific characters, it also offers a better understanding of the equilibrium phases between the punctuations than that first proposed by Eldredge and Gould (1972:114). They suggested that the coherence of a species existing in nearly independent local populations might result from the species' origin "as a peripherally isolated population that acquired its own powerful homeostatic system." They argued for the reinforcing effect of such a homeostatic mechanism, but were unable to offer a concrete model for its action and could only suggest that "the answer probably lies in a view of species and individuals as homeostatic systems." We may now see that the effect of stabilizing selection on the signal and response chain of the SMRS is the key point in the argument for stasis and that it answers the call by Maynard Smith (1981) for an explanation of the phenomenon.

The extent to which the argument over phyletic gradualism versus punctuated equilibrium as the best description of evolutionary tempo and mode has been misplaced is now clear. Confusion has stemmed from an unfortunate use of the term *gradualism,* a misleading equation of the term *equilibrium* with stasis, and a failure to separate within-species changes from speciation events. As originally coined by Eldredge and Gould (1972:89), *phyletic gradualism* was clearly meant to mean gradual change between species in contrast with punctuated equilibrium. But because those authors themselves failed to distinguish within- from between-species changes clearly enough, confusion was set in train from the outset. *Gradualism* is also employed in the literature to describe the nature of accumulated intergenerational morphological change within a species, which simply must be gradual. *Equilibrium* as used by Eldredge and Gould referred to the continued existence of a species between the speciation event that brought it into being and any subsequent extinction. The term *stasis* has come to be regarded as synonymous with *equilibrium,* but it refers to the absence of any evidence for the operation of directional selection upon the species during its lifetime and not to the lifetime itself (see also Michaux, this volume). However, Eldredge and Gould chose to equate equilibrium with stasis and with an absence of change, and left the way clear for those who reject punctuated equilibrium based on a demonstration of gradual intraspecific change and the assumption that this implies phyletic gradualism. *Gradualism* and *punctuated equilibrium* are therefore not useful terms to employ for contrasting versions of the same thing, but should be used to describe totally separate phenomena, unless the distinction between speciation and within-species changes are to be blurred (as of course they usually are). Indeed, *phyletic gradualism,* seen from the perspective of the Recognition Concept of species, is a rather meaningless term. Geologically, a speciation event will always appear short, perhaps to the point of virtual invisibility. Efforts to demonstrate that evolution is gradual by producing evidence for a slow change in one or other character of a species in the fossil record simply say nothing about the deployment of speciation. Confusion over this point underlies the discussion by Carroll (1988) of evolutionary tempo and is evident in his conclusion (p. 575) that evolutionary rates are "certainly neither *gradual* nor *punctuational,* but *irregular* or *opportunistic*" [italics in original].

Of course, the Recognition Concept is not a panacea for the problems of identification that face the paleontologist. We are still left with the simple fact that fossils preserve only parts of an organism and that features of possible taxonomic value are often missing. The identification of species in the fossil record is therefore always going to present problems. Moreover, it follows from an acceptance of the Recognition Concept that the nature of the differences between species is unlikely to be absolute, since only those characters related to the fertilization system will *necessarily* change at speciation. Morphological continuities will be enforced through phylogenetic inertia, and all change must be gradual from generation to generation unless we are to accept the notion of "hopeful monsters." Put most simply, paleontologists will detect speciations with greatest reliability in the fossil record when the fossils in question represent biparental organisms that bear characters of importance in the fertilization system of the species, such as the genitalia of insects, horn cores of the Bovidae, antlers of the Cervidae, or cranial ornaments in dinosaurs. They will have greater problems when they deal with families like the Equidae and Suidae, where many taxa are established on the basis of dental morphology (Turner and Chamberlain 1989). Continuities and partial preservation of tissues aside, sibling species are simply not amenable to recognition in the fossil record, and there may very well be an overall tendency to underestimate the number of true species in certain families or even orders (see also Vrba, this volume). This problem has been discussed in relation to the Hominidae by Tattersall (1986), although there are grounds for thinking that we may be able to approach the problem in a sensible manner if we adopt the perspective of the Recognition Concept (Turner 1986; Turner and Chamberlain 1989).

To date, acceptance of the value of the Recognition Concept of species in both neontology and paleontology has been slow. In the field of paleontology there has been some initial appreciation of the benefits (Vrba 1980, 1984a,b, 1985a,b, this volume; Turner 1985, 1986; Turner and Chamberlain 1989), and there is now evidence of interest among a wider circle (Crompton 1989; Tattersall 1989) beyond those former colleagues and students of Paterson who have applied his ideas to their own work. In a most significant step, Eldredge (1985a,b) has devoted increasing attention to the topic and presents (1989) a lengthy discussion of the SMRS in his latest book on the subject of macroevolution, which includes a clear statement of the difference between speciation and within-species changes. But although Eldredge's incorporation of Paterson's ideas affords considerable insight and addresses many of the issues discussed above, his analysis of developments appears somewhat flawed. One gains from it no sense that the Recognition Concept offers a radical alternative to the concept of species and the mechanism of speciation employed in the original formulation of the punctuated equilibrium model. Instead, the idea of the SMRS is simply incorporated into the argument for periods of stasis being interrupted by episodes of speciation, with the claim (1989:119) that disruption to the SMRS was covered in the original discussion of rapid adaptive changes in small, peripherally isolated populations presented by Eldredge and Gould (1972) and indeed went beyond the conclusions reached by Paterson. I find that part of the argument difficult to sustain.

In contrast, most if not all of the criticism of the Recognition Concept has been from the neontological perspective and appears to have been based largely on an unshaken acceptance of the continued validity of the modern synthesis and the Biological Species Concept (e.g. Coyne et al. 1988) and, in some cases (for instance, Raubenheimer and Crowe 1987), a seeming inability to grasp the point at issue. Paterson (1981, 1982, 1985, 1988) and others (e.g. Masters and Spencer 1989) have demonstrated the illogicalities inherent in many of the criticisms, but a revolution in thought is still seemingly required on the issue. The clear value of the concept in the paleontological context may be the best hope of producing that revolution.

Acknowledgments

I thank David Lambert for inviting me to contribute to this volume in honor of Hugh Paterson, whose ideas have done so much to shape my own views on species in the fossil record. I am grateful to Ian Tattersall and Lars Werdelin for constructive comments and to The Leverhulme Trust for research funding.

References

Arkell, W.J. 1956. Species and species. Pp. 97–99 *in* P.C. Sylvester-Bradley, ed. The species concept in paleontology. The Systematics Association, London.
Ayala, F.J. 1978. The mechanisms of evolution. Sci. Am. 239:48–61.
———. 1983. Microevolution and macroevolution. Pp. 387–402 *in* D.S. Bendall, ed. Evolution from molecules to men. Cambridge University Press, Cambridge.
Barton, N.H., and B. Charlesworth. 1984. Genetic revolutions, founder effects and speciation. Annu. Rev. Ecol. Syst. 15:133–164.
Beerbower, J.R. 1968. Search for the past. 2d ed. Prentice-Hall, Englewood Cliffs, N.J.
Bock, W.J. 1986. Species concepts, speciation and macroevolution. Pp. 31–57 *in* K. Iwatsuki, P.H. Raven, and W.J. Bock, eds. Modern aspects of species. Tokyo University Press, Tokyo.
Bush, G.L. 1981. Stasipatric speciation and rapid evolution in animals. Pp. 201–218 *in* W.R. Atchley and S. Woodruff, eds. Evolution and speciation. Cambridge University Press, Cambridge.
Bush, G.L., S.M. Case, A.C. Wilson, and J.L. Patton. 1977. Rapid speciation and chromosomal evolution in mammals. Proc. Natl. Acad. Sci. 74:3942–3946.
Butlin, R.K. 1987. Species, speciation and reinforcement. Am. Nat. 130:461–464.
Campbell, B.G. 1979. Some problems in hominid classification and nomenclature. Pp. 567–581 *in* C. Jolly, ed. Early hominids of Africa. St. Martin's Press, New York.
Carroll, R.L. 1988. Vertebrate paleontology and evolution. Freeman, New York.
Carson, H.L., and A. R. Templeton. 1984. Genetic revolutions in relation to speciation phenomena: The founding of new populations. Annu. Rev. Ecol. Syst. 15:97–131.
Chandler, C.R., and M.H. Gromko. 1989. On the relationship between species concepts and speciation processes. Syst. Zool. 37:190–200.
Churcher, C.S., and M.L. Richardson. 1978. Equidae. Pp. 379–422 *in* V.J. Maglio and H.B.S. Cooke, eds. Evolution of African mammals. Harvard University Press, Cambridge, Mass.

Coyne, J.A., H.A. Orr, and D. J. Futuyma. 1988. Do we need new species concept? Syst. Zool. 37:190–200.
Crompton, R.H. 1989. Mechanisms for speciation in Galago and Tarsius. Human Evol. 4:105–116.
Eldredge, N. 1971. The allopatric model and phylogeny in Paleozoic invertebrates. Evolution 25:156–167.
———. 1985a. The ontology of species. Pp. 17–20 in E.S. Vrba, ed. Species and speciation. Transvaal Museum Monograph No. 4. Pretoria.
———. 1985b. Unfinished synthesis: Biological hierarchies and modern evolutionary thought. Oxford University Press, Oxford.
———. 1985c. Time frames: The evolution of punctuated equilibria. Princeton University Press, Princeton, N.J.
———. 1989. Macroevolutionary dynamics. McGraw-Hill, New York.
Eldredge, N., and J. Cracraft. 1980. Phylogenetic patterns and the evolutionary process. Columbia University Press, New York.
Eldredge, N., and S. J. Gould. 1972. Punctuated equilibria: An alternative to phyletic gradualism. Pp. 82–115 in T.J.M. Schopf, ed. Models in paleobiology. W.H. Freeman, San Francisco.
Gingerich, P.D. 1984. Punctuated equilibria: Where is the evidence? Syst. Zool. 33:335–338.
———. 1985. Species in the fossil record: Concepts, trends and transitions. Paleobiol. 11:27–41.
George, T.N. 1956. Biospecies, chronospecies and morphospecies. Pp. 123–137 in P.C. Sylvester-Bradley, ed. The species concept in paleontology. The Systematics Association, London.
Gould, S.J. 1980a. Is a new and general theory of evolution emerging? Paleobiol. 6:119–130.
———. 1980b. The promise of paleobiology as a nomothetic evolutionary discipline. Paleobiol. 6:96–118.
———. 1985. The paradox of the first tier: An agenda for paleobiology. Paleobiol. 11:2–12.
Gould, S.J., and N. Eldredge. 1977. Punctuated equilibria: The tempo and mode of evolution reconsidered. Paleobiol. 3:115–151.
Hecht, M.K., and A. Hoffman. 1986. Why not neo-Darwinism? A critique of paleobiological challenges. Oxford Survey of Evolutionary Biology 3:1–47.
Huxley, J. 1942. Evolution: The modern synthesis. George Allen and Unwin, London.
Joysey, K. A. 1956. The nomenclature and comparison of fossil communities. Pp. 83–94 in P.C. Sylvester-Bradley, ed. The species concept in paleontology. The Systematics Association, London.
———. 1972. The fossil species in space and time: Some problems of evolutionary interpretation among Pleistocene mammals. Pp. 267–280 in K.A. Joysey and T.S. Kemp, eds. Studies in vertebrate evolution. Oliver and Boyd, Edinburgh.
Kurtén, B. 1968. Pleistocene mammals of Europe. Weidenfeld and Nicolson, London.
Masters, J.C., and H.G. Spencer. 1989. Why we need a new genetic species concept. Syst. Zool. 38:270–279.
Matthew, W.D., and S.H. Chubb. 1921. Evolution of the horse. American Museum of Natural History, New York.
Maynard Smith, J. 1981. Macroevolution. Nature 289:13.
———. 1987. Darwinism stays unpunctured. Nature 330:516.
Mayr, E. 1942. Systematics and the origin of species. Columbia University Press, New York.

———. 1963. Animal species and evolution. Harvard University Press, Cambridge, Mass.
———. 1969. Principles of systematic zoology. McGraw-Hill, New York.
———. 1978. Evolution. Sci. Am. 239:39–47.
Paterson, H.E.H. 1981. The continuing search for the unknown and unknowable: A critique of contemporary ideas on speciation. S. Afr. J. Sci. 77:113–119.
———. 1982. Perspective on speciation by reinforcement. S. Afr. J. Sci. 78:53–57.
———. 1985. The recognition concept of species. Pp. 21–29 in E.S. Vrba, ed. Species and speciation. Transvaal Museum Monograph No. 4. Pretoria.
———. 1986. Environment and species. S. Afr. J. Sci. 82:62–65.
———. 1988. On defining species in terms of sterility: Problems and alternatives. Pacific Sci. 42:65–71.
Raubenheimer, D., and T.M. Crowe. 1987. The Recognition species concept: Is it really an alternative? S. Afr. J. Sci. 83:530–534.
Rhodes, F.H.T. 1956. The time factor in taxonomy. Pp. 33–52 in P.C. Sylvester-Bradley, ed. The species concept in paleontology. The Systematics Association, London.
Romer, A.S. 1933. Vertebrate paleontology. Chicago University Press, Chicago.
———. 1945. Vertebrate paleontology. 2d ed. Chicago University Press, Chicago.
———. 1966. Vertebrate paleontology. 3d ed. Chicago University Press, Chicago.
Simpson, G.G. 1944. Tempo and mode in evolution. Columbia University Press, New York.
———. 1951. Horses. Oxford University Press, New York.
———. 1953. The major features of evolution. Columbia University Press, New York.
———. 1961. Principles of animal taxonomy. Columbia University Press, New York.
Stebbins, G.L., and F.J. Ayala. 1981. Is a new evolutionary synthesis necessary? Science 213:967–971.
Sylvester-Bradley, P.C. 1956. The new paleontology. Pp. 1–8 in P.C. Sylvester-Bradley, ed. The species concept in paleontology. The Systematics Association, London.
Tattersall, I. 1986. Species recognition in human paleontology. J. Human Evol. 15:165–175.
———. 1989. The roles of ecological and behavioral observation in species recognition among primates. Human Evol. 4:117–124.
Turner, A. 1985. The recognition concept of species in paleontology, with special consideration of some issues in hominid evolution. Pp. 153–158 in E.S. Vrba, ed. Species and speciation. Transvaal Museum Monograph No. 4. Pretoria.
———. 1986. Species, speciation and human evolution. Human Evol. 1:419–430.
Turner, A., and A. Chamberlain. 1989. Speciation, morphological change and the status of African *Homo erectus*. J. Human Evol. 18:115–130.
Vrba, E.S. 1980. Evolution, species and fossils: How does life evolve? S. Afr. J. Sci. 76:61–84.
———. 1984a. Evolutionary pattern and process in the sister group Alcelaphini-Aepycerotini (Mammalia: Bovidae). Pp. 62–79 in N. Eldredge and S.M. Stanley, eds. Living fossils. Springer-Verlag, New York.
———. 1984b. Patterns in the fossil record and the evolutionary process. Pp. 115–142 in M.-W. Ho and P.S. Saunders, eds. Beyond neo-Darwinism. Academic Press, London.
———. 1985a. Introductory comment on species and speciation. Pp. ix–xviii in E.S. Vrba, ed. Species and speciation. Transvaal Museum Monograph No. 4. Pretoria.
———. 1985b. Environment and evolution: Alternative causes of the temporal distribution of evolutionary events. S. Afr. J. Sci. 81:229–236.
Wiley, E.O. 1978. The evolutionary species concept reconsidered. Syst. Zool. 27:17–26.

———. 1979. Ancestors, species and cladograms: Remarks on the symposium. Pp. 211–225 *in* J. Cracraft and N. Eldredge, eds. Phylogenetic analysis and paleontology. Columbia University Press, New York.

Williams, G.C. 1966. Adaptation and natural selection. Princeton University Press, Princeton, N.J.

4

Kinji Imanishi's Biological Thought

KIYOHIKO IKEDA

Biological Laboratory, Faculty of Education
Yamanashi University

ATUHIRO SIBATANI

Wissenschaftskolleg zu Berlin and Biological Laboratory,
Faculty of Humanities, Kyoto Seika University

P<small>ROF.</small> K<small>INJI</small> I<small>MANISHI</small> (1902–1992), since the 1940s a father figure in Japan for many different fields related to biology and anthropology, is now known to the West through articles by Sibatani (1983a,b), Imanishi (1984), and Halstead (1985, 1988). These articles are all related to his evolutionary theory on which he published many books and articles during the last period of his academic activity (e.g. Imanishi 1977, 1980, 1984, 1987). This was long after his retirement around 1966 from the chair of anthropology at the Institute for Human Sciences, Kyoto University, and from his duties as President of Gifu University around 1971. Because Imanishi took a strong anti-Darwinian or antiselectionist stance when articulating his thought on evolution, and because this coincided with the explosive growth of debate about evolution in the West since 1972, his name has been associated, mainly in the West but also in Japan in recent years, with the image of an unorthodox, and even nonscientific, theorist on evolution.

However, before he returned, toward the end of his academic activities, to the initial point in biology where he started his career as research scientist, he had worked in various fields of scientific pursuit, each time within a minority circle (Ogushi 1985). Indeed, through his work in such "marginal" areas he founded a number of new schools of research, some of which have by now risen to internationally acclaimed positions. These aspects of Imanishi's contribution to Japanese science are less well known than his position on evolution, but without acknowledging these it would be unjustified to generally criticize Imanishi's work and his insights into biology.

Thus, from the beginning of the 1940s Imanishi was at the center of a group of competent quantitative or mathematical ecologists who were then newly rising in Kyoto, and although he never took up ecology in his own work, his influence extended well beyond Kyoto (Ito 1990), largely owing to his perceptivity to what was then most essential in ecology, or more generally, in whole biology. Then he turned to biosociology and led the primate study group in Japan. This group's approach of identifying individual animals in the field by directly recognizing them and naming them in the natural language, rather than labeling them, made, through its uniqueness and high productivity of novel findings, a great impact on the then-rising primatology in the world at large (Asquith 1986). He also assisted the development of cultural anthropology and ethnology in the Kansai area, which has since become the center of these disciplines in Japan (Umesao 1989). All these activities are associated, in one way or another, with his approach to nature as a leader of expeditions to Asia, Oceania, and Africa. The last of these resulted in the establishment of a substantial center of African studies in Kyoto University. It can thus be seen that Imanishi's activities followed a certain pattern (Ogushi 1985). Once he had started scientific research in a marginal area, inspiring and gathering around him a group of young and able scientists, and had then seen the group establish a reputable field of research, he then left seeking yet another unrecognized area of research.

Finally, Imanishi chose to revisit, as another marginal area (at least in Japan until the end of the 1970s), the field where he had begun his work in biology but had left before fully exploring it. That field was evolutionary biology. It was here that he eventually declared he could no longer tolerate the current trend in scientific practice, so much so that he finally declined to be identified as a scientist (Imanishi 1984). Surely what he tried to preach at the end sounded quite unpalatable to scientists, but if one wanted to reconstruct biological science, what he was trying to say might have made some sense.

During this time one of the present authors, Sibatani, worked as a Lepidoptera taxonomist/morphologist turned molecular biologist turned developmental biologist and was looking askance at Imanishi's activity. Finally he decided to write a critique of Imanishi's latest books (Sibatani 1981) after Imanishi (1977, 1980) had begun to write quite a few books on evolutionary biology. This was at least partly to fathom the matter with which Imanishi was so concerned. Sibatani (1985) then began to advocate structuralism in biology via a critique of molecular biology. Through this process Sibatani had gradually developed an interpretation of Imanishi's theory with a structuralist slant (Sibatani 1982, 1984, 1985).

This chapter presents a summary of the position taken by one of the authors, Sibatani, concerning Imanishi, to be followed by a further analysis of Imanishi's biology being advanced by the other author, Ikeda (1989).

Ikeda (1988) started his structuralist biology in the wake of Sibatani's (1985) declaration of structuralism in biology. Like Sibatani, Ikeda (1987) was initially severely critical of Imanishi's evolutionary theory. However, after working on theories in structuralist biology, scientific epistemology, and evolutionary biology, it became apparent to Ikeda that the young Imanishi (1941, 1949) had sensed something which was quite akin to Ikeda's own structuralism as well as some of the forerunners of structuralist biology in the history of European science, such as Spinoza and Goethe (Ikeda 1989).

Problems of Lifestyle Partitioning and Congener Recognition

Imanishi started his biological work on the taxonomy and ecology of mayflies in Japan, especially those living in mountain streams. It was through his studies on the ecdyonurid taxa (Insecta: Ephemeroptera) that he became committed to investigating the social aspects of animals. He found that species of related genera had similar lifestyles; in the case of *Ecdyonurus* and *Epeorus,* the flat-shaped nymphs of this group move by sliding on stone surfaces in the river bed. However, the morphology and habitat of each species differed characteristically according to the geographic (flow course) and physical (rapid or slow stream) factors of the river. Of these factors, the latter may involve, besides the flow rate, temperature and oxygen concentration of the water, the quantity and type of food material, and surface conditions of stones. What Imanishi found was that different species lived in their respective habitats without much overlap or intervening void spaces, the overall result being the partitioning of habitat space (and time) by individual species or, more generally, the partitioning of lifestyle. Thus, throughout the likely habitats of ecdyonurids, except for a downstream littoral zone in winter, each of the various areas or zones along and across the stream were occupied by one and only one species; the various zones tended to converge into one zone toward cool torrential upstreams of the Alpine and northern districts of Japan.

These observations led Imanishi to define the existence of animals as societies of species, where individuals of a species have a particular relationship with each other, which is distinct from that with the members of other species societies. Hence, allied species having similar lifestyles occupy the contiguous habitats partitioned according to their particular ways of living, in relation to geographic, temporal, and other physical characteristics of the habitat. The juxtaposed or other equivalent societies of different species, which occupy no overlapping habitats,* are called the *dooi-syakai,* meaning equivalent societies or societies of the same rank. Beyond this, higher and more complex hier-

---

*This is so, at least theoretically; as in physics, Imanishi apparently spoke of the ideal state, the reality being always contaminated by noise.

archies of animal societies were envisaged, encompassing more widely differing species and higher taxa.

In relation to these species societies Imanishi seems to have held the firm conviction that each species is governed by its own laws that bind its individuals. He apparently saw the existence of a general law, which underlies such laws for individual species, as invariant in its essence from the very beginning of life. This he described as as *syutaisei* or subjective existence and later as protoidentity (more on this concept later) (Ikeda 1989).

In working out such a scheme for the organization of animal societies, Imanishi was naturally aware that the choice of preferred habitats by respective species societies largely depended on the physical variables of the habitat. However, he assumed from examples of fish habitat in mountain streams that the habitat range of a species also depended on the presence of other species, which in mutually exclusive manner occupied contiguous but nonoverlapping habitats. Hence, the actual boundary of the partitioned habitats was determined biologically rather than physically. This part of his idea was misunderstood by his British critics (Halstead 1985, 1988; Rossiter 1986), who stressed the significance of physical factors in habitat choice. Such factors were only the prerequisite of habitat partitioning, which was overridden by Imanishi, who wanted to stress the biological aspect within this constraint (Sibatani 1987, 1989*a*).

Tenets Characteristic of Imanishi's Thoughts

In later years Imanishi (1984) said that he had based his thoughts on evolution upon those early experiences with mayflies. However, what he wrote on evolution contained several enigmatic formulas (Imanishi 1977, 1980) that made him rather popular among his lay admirers, but had the reverse effect on scientists and led to his becoming a victim of predatory journalists as well. Such formulas included:

1. The species and the individual are two separate entities, but they nevertheless together represent a unique entity to be conceived as the specia or species society.

2. All individuals of a specia are equivalent, in the sense that they follow the law of species or specia, so any individual will be able to play a potentially equivalent role in evolution. This would mean that if a new species were being formed, all the members of the old species, or the relevant part thereof, would simultaneously proceed to change in the same way. In other words, the entire membership of the species (or its significant portion) changes when it is time for it to change.

3. Individuals of a species society are able to identify themselves as mem-

bers of the species they belong to and, furthermore, as members of the particular group to which they belong (as in a macaque society); this "awareness" of membership in a society (both in a group such as a swarm, school, or flock *and* in specia) is called the protoidentity, not to be confused with the notion of human individuals finding their own "identities" in modern society.

4. Phylogeny repeats ontogeny, in the sense that every member of the entire biological world must act as a part of the unified and organized whole, cooperating rather than competing, as cells do in embryogenesis. The law that makes this possible must have been continuously held by all the members of the living world up to the present time.

5. The hierarchy of processes as well as entities involving organisms may be, from the top downward, the geocosmos, integral organismic society or *holospecia,* specia, and individual.

6. "Dogma or the legend of the creation of life" dictates that, as in chemical reactions, the same processes must have taken place in different entities, thus generating in parallel the first organisms as a group. Those first organisms must have been still quite comparable to each other and thus equivalent. Hence, the origin of life must have been the origin of species (be it sexually or asexually reproduced), involving, from its very beginning, quite a few individuals which proceeded and behaved in a unique and similar way governed by the law of the specia or the law of discrete, individual biological existence.

A few comments on these tenets may be relevant here. First, tenet 2 is the obvious thesis from which the rejection of natural selection as the motive force of organic evolution derives as a corollary. Quite remarkably, this tenet seems to be supported by recent experiments, especially those reported by Hall (1988, 1990) using nongrowing bacteria under stressed conditions, at least at the phenomenological level. Cells in a developing embryo and during postnatal ontogeny die in rather predetermined, regulated ways (to be found, for instance, in the nematode *Caenorhabditis elegans;* the separation of digits in the tetrapod limb; the 120-day-long life span of human erythrocytes; senescence; and the periodical renewal of plant bodies in deciduous trees or perennial herbs). Hence, the view of ontogeny as subject to cellular competition may not be adequate (tenet 4). Chemical reactions cannot be viewed as uniform at the level of individual molecules when complex entities such as polymers are involved. Here, the argument of tenet 6 will break down as long as one sticks to the concept of entities or components of a system. One must envisage, as something invariant, systems or rather structures, but not the entities or substances. This point will be elaborated upon below. It will suffice to say here that the idea of a nested hierarchy expressed in tenet 5 should be approached with care, to avoid mixing up entities and structures (Ikeda 1988). Moreover, even as entities, the actual organic world does not necessarily entail nested hierarchies, as evidenced by the use of the same nucleotides in various permutations for different triplet codes as well as genes, or the repeated occurrence of twenty amino acids in various proteins.

That said, there are a few interesting points in the recent statements made by Imanishi. First, according to Imanishi (1987), the awareness of membership should extend to the recognition of a particular spot of the habitat, where conspecifics come together to assemble, or begin to assemble starting from a single individual. Moreover, according to Takemon (1989), there is some evidence that the occurrence of nymphs of *Ephemera strigata* is restricted to those parts of a stream in proximity to air spaces suitable for mating and oviposition by imagines. This is despite the presence of other sections of the stream that would allow nymphal life to be safely completed. In other words, the nymphs choose a submerged aqueous habitat according to the physical conditions prevailing in the air space above the habitat, a quality they cannot directly survey. As the nymphs do not spread beyond these places, which are delimited only by certain conditions above the water, the association of the two habitats above and below the water surface must be sensed by nymphs while still in the local stream. The means of determining such an association must be extremely subtle and might even involve remote sensing. Such competence on the part of mayfly nymphs may have been conceived by Imanishi (1984) as their being part of the "geocosmos."

Second, Imanishi (1987) wanted to classify his "holospecia" into two systematically irrelevant groups: those which live in groups and those which do not. Apparently he saw the group-formers as more "advanced" (or specialized for higher functions) in evolutionary terms than the non-group-former. This may be likened to cells which aggregate and those which do not, roughly corresponding to unicellular and multicellular organisms.

Third, Imanishi's holospecia is reminiscent of the latest development in artificial life research (*neobiology?*). According to a report by Waldrop (1990), physicists and computer scientists, including Doyne Farmer, Stewart A. Kauffman, and Christopher Langton, are demonstrating that simulation of complex systems with a large number of units having highly diverse characters indicates that biodiversity may be generated by a law governing such a complex system, thus casting doubt on the neo-Darwinian assumption of the evolution of biodiversity primarily through mutation and natural selection. If that view is correct, it would imply that the emergence of diverse taxa is the logical consequence of a structure, still remaining largely unknown, which underlies the ecosystem, or the living system at large, and is hence equivalent to what Imanishi conceived as holospecia from which individual specia should have been generated. This is in contrast to the conventional view that the latter emerged individually during the course of history, guided by random mutation and selection, and have thus generated the biocoenosis.

Congener Recognition and Specific-Mate Recognition

At an early stage in his scientific life Imanishi recognized that lifestyle partitioning or *sumiwake* occurred in mayfly nymphs in freshwater streams. He took up this problem again during his retired life when he dedicated himself to exploring his antipathy toward the functionalist aspects of the Darwinian theory of evolution. Incidentally, he totally accepted the Darwinian view on the existence of organic evolution and even concurred with the idea that organic evolution tends to increase the total biological productivity through increasing the density of specia per unit area. As for the mechanism of such protoidentity, Imanishi has never put forward his ideas clearly. He has been fluctuating between the notion that organisms' recognition of suitable habitats is shared by all the members of a specia and the idea that they recognize their congeners through discrimination of all the other organisms. Imanishi perhaps tended to ascribe the mayfly nymph's remarkable behavior of mutual specific discrimination or *sumiwake* to the former, the birds' and mammals' specific aggregation to the latter, but he was never explicit in this respect.

By contrast, he always held the recent rise of sociobiology as superficial, probably because he sensed that sociobiology lacked an awareness of deep structure extant in the organic world (see below). This deep structure he himself has never succeeded in articulating. Instead, he has taken resort in generalized, and even quite enigmatic, expressions which, on the one hand, have led to misunderstandings, and indeed complete dismissal, and, on the other hand, have been a source of inspiration and poststructuralist text-reading in the sense that there is neither a real author nor a unique meaning in any discourse, scientific or otherwise (Foucault 1969), though Imanishi himself finally declined to be identified as natural scientist. This probably reflected the frustrations of a scientist in search of his own identity, unable to express his intuitions in the explicit terms with which he was formerly familiar.

In the meantime, Sibatani (1981, 1982, 1983a) decided to tackle the problem raised by Imanishi of the organism's perception of conspecific identity. Then, thanks to Michael Soulé's suggestion, his attention was drawn to Paterson's (1978) Specific-Mate Recognition System which was later generalized into the specific fertilization system (Paterson 1985). However, it was very clear from the outset that Imanishi and Sibatani, as well as Hendrichs (1988) and Kortmulder and Sprey (1990), who have more recently dealt with similar subjects, were all primarily interested in the recognition, by the members of the species, of other members of the same species, regardless of sex. It was quite obvious that in the case of "habitat segregation" by mayfly nymphs, which are of course sexually immature, sexual identity does not come to the fore. Thus, if these insects recognize fellow members, rather than sensing the microhabitat directly as suggested by Imanishi (1987) for lifestyle partitioning, they must do so regardless of the sex of the congeners. The term *congeners* is an expression advocated by Kortmulder (1986). It is broader than the term *conspecifics*, including occasional misidentifications on the part of the organism, and hence obviously preferable. In this connection, Sibatani (1989*b*) was interested in the conspecific male-to-male association of flying butterflies, especially in prolonged co-rotating flights of two conspecific males facing each other. The behavior is evidence for a positive and direct recognition of congeners. Clearly this was not the indirect place-mediated association as in the case of lekking, or rejection of strangers by territory keepers, but an active choice to stay associated with a particular individual while moving in the air. Sibatani (1989*b*) thus holds the view that Paterson's mate recognition system, which probably underlies the origin and maintenance of the species in eukaryotes, must be secondary to a more general system which involves cells' and organisms' recognition and association with their likes (i.e., in relevant ways, including the lateral inhibition to keep congenors sufficiently apart). Such recognition and association among congeners, but not the mate recognition system or the recognition of microhabitat as suggested by Imanishi, must underlie the formation of multicellular organisms (if not eukaryotes only). It must also underlie the particular ways of animal life for foraging, defense and so on, other than those directly related to sexual behaviors.

Probably, to an average biologist, the foregoing argument may sound confusing. However, we need to envisage the mechanisms involved in cell-to-cell adhesion and those in individuals' recognition through chemical entities as being rather similar. Although these similarities are superficial, they are not unfounded on structuralist thinking, a matter which we will explore in later sections. One point should be noted here. Paterson was led to the concept of specific-mate recognition through Darwinist thinking, and quite logically via species' self-sustainability through sexual reproduction. Paterson thus constructed his theory focused on speciation, which needs a novel specific recognition system. However, for Imanishi, and also for the present authors (see below), what is primary is the law specifying a species, or the logical structure allowing its existence; neither Imanishi nor the present authors have particularly been bothered by the mechanism allowing the emergence of a new species or new recognition system. For them perhaps the recognition system is not completely independent of the rest of the species character, in the network of relations which make up the law governing a species. In this sense, it may be appropriate to express the view held by the authors that, in spite of the superficial resemblance and its intrinsic interest, Paterson's specific-mate recognition or fertilization system is distinct from Imanishi's concept of protoidentity.

Each touches the other only tangentially, although this aspect in itself is not without interest.

Introduction to a Structuralist Viewpoint

The notion of the primacy of congener recognition convinced Sibatani (1984, 1985) of the fact that organisms or species are governed by a law binding their individuals. Individuals are clearly visible, but laws are hidden. Thus the relationship between the individual and specia pertaining to a given species would be like that between the *parole* and the *langue* of a given language as illustrated in Saussure's theory of linguistics. The former may be described as something personally characterizable which is generated from the law embodied in the latter. Thus the enigmatic concept of twoness being equal to oneness for the individual/specia relationship as emphasized by Imanishi can be given an interpretation that should be palatable to a wider audience. It was at this point that Sibatani became convinced of the prospect that structuralism would be useful in theoretical biology and that there is something worthwhile in pursuing Imanishi's thoughts not only in evolutionary but also in general biology.

The Origin of Lifestyle Partitioning

By extending his search of evidence for the reality of congener recognition as the basis of lifestyle partitioning, Sibatani (1983a,b, 1984) noted that nymphs of ecdyonurid mayflies must choose their habitat actively while being carried downstream by the water after hatching at the oviposition site upstream. The choice must naturally be guided by the physical conditions they prefer, but as already mentioned, there is evidence that the nymph's choices are further limited by the conditions above the water to be required only later by the imago which will emerge into the air, mate, and then oviposit. Thus many potential habitats, where nymphal stages could be completed, are not chosen (Takemon 1989). It would thus be plausible to expect the young nymph to positively choose habitats not only for the preferred physical variables but also to effect a minimum overlap with nymphs of other species, giving rise to the surface phenomenon called the lifestyle partitioning. For such goals, the nymph should be able to sense much more than just physical variables of the habitats, in order to decide on what is an optimum habitat. There is evidence from the field and from experimentation that this habitat partitioning represents a once and for all choice of habitat, rather than being something gradually developed during the nymphal period, starting with initially mixed habitation and leading to the eventual segregation seen in mayflies and in newly hatched aphid nymphs (see Sibatani 1987, 1989a). With the caddisfly genus *Hydropsyche* (Trichoptera), the microhabitat segregation seems not to be competition, since a high living density of one species is not reduced by the temporary disappearance of a second species occupying a contiguous habitat (Tanida 1984). The last result differs from the situation referred to by Imanishi (1980) with freshwater fishes in macrohabitat segregation and the first instar aphid nymphs (Sibatani 1989a) referred to above. It may be that lifestyle partitioning is a pathway-invariant phenomenon like some processes in ontogeny and regeneration. If so, this would certainly point to the presence of some deep-seated structure underlying the observed partitioning phenomenology.

There are serious difficulties in developing an alternative model for the evolution of lifestyle partitioning using differential increase of gene frequency through gradual processes helped by natural selection. These difficulties consist mainly of two points. First, we have to explain lifestyle partitioning as a general rather than specific phenomenon to be observed in many taxa and habitats. Second, the model must not make use of congener recognition and interspecific distinction in any form, either chemical suppression, specific avoidance, or other mechanisms that would work differentially against the conspecific and other individuals. This is because the model would then no longer serve as the alternative to the hypothesis for the primacy of congener recognition (Sibatani 1987), but become an argument that must accept the preexistence of some ability to differentiate separate species on the part of organisms.

Saussure Revisited

Before Ikeda (1988) set out to reinterpret Saussure's theory of general linguistics for his work on theoretical biology in the vein of structuralism, Ikeda (1987) had found Imanishi's thought unacceptable as a natural science. Later Ikeda (1988) was able to point out that Saussure gave two inverse definitions for the *parole/langue* dichotomy in his first and combined second and third lectures. To quote from the original material of Saussure's lectures (Engler 1967), "Par la parole on désigne l'acte de l'individu réalisant sa faculté au moyen de la convention sociale qui est la langue ⟨définition⟩. Dans la parole, il y a une idée ⟨de⟩ réalisation de ce qui est permis par la convention sociale" (p. 32, fragment 160, II R 6).

This version, found in the second and third lectures, corresponds to the general understanding of the terms. However, the definition given in the first lecture apparently asserts the opposite (Engler 1967): "Tout ce qui est amené sur les lèvres par les besoins de discours et par une opération particulière, c'est la *parole*. Tout ce qui est contenu dans le cerveau de l'individu, le dépôt des formes ⟨entendues et⟩ pratiquées et de leur sens, ⟨c'est⟩ la *langue*. De ces deux sphères, la sphère ⟨parole⟩ est la plus sociale, l'autre est la plus complètement individuelle. La langue est le réservoir individuel; tout se qui entre dans la langue, c'est-a-dire dans la tête, est individuel" (pp. 383–84, fragment 2560, IR 2.23).

Meanwhile, Chomsky (1986) demonstrated that the innate ability of a person to speak a language need not involve accurate learning; that the correct use of grammar of a language by a child is innate once it begins to speak in that language. Thus, the deep-seated structure which enables every child to make correct use of a given language must be innate. Hence, people's ability to be bound by a *langue* of any language should not be ascribed to the socially binding power or consensus of the *langue*, but to the structure and its particular configuration innately held by the people for this *langue*, which is activated to materialize by appropriate triggers.

Ikeda (1988) assumed that these triggers must be the *parole* spoken by the people, to which the child is exposed. In other words, the *langue*, which lies as a deep-seated structure to generate the *parole*, exists innately in every person's brain as a structure activated in a material basis.

This structure is isomorphic to the ones held by other people because this structure is characteristic of the human species; it can be activated through entraining by the actual *parole*. Instead of assuming the Chomskyian universal grammar with its transformation rules, Ikeda (1988), helped by the concept of pluralism of structure as derived from the axiom of its arbitrariness (Sibatani 1989c, 1991), preferred to argue as follows (Appendix A). The *langues* corresponding to all the existing languages must exist as potential, and usually exclusive, alternative structures in each person's brain. This means that the *langue* does not exist as an abstract rule floating between persons in society, but is rather firmly rooted in the brain of each human individual. The *langue* is thus encapsulated in the human brain. By extrapolation, not only human speech but also all forms of human cognition, scientific and otherwise, must also be based on the activation of the preexisting structure, implying that only the structures out there in the world that are isomorphic with those existing in the human brain can be recognized and understood by human beings; and that those not isomorphic are, after all, beyond the reach of human understanding (Ikeda 1988) (Appendix B).

Imanishi as a Structuralist

Now the same idea can be applied to the species concept. It is obvious from reading Imanishi's works from his first book (Imanishi 1941) down to the latest ones that for him the species was a real entity rather than a conceptual tool for scientific treatment of the biological world. This concept of the real existence of species is, as Ikeda argues (1989), basically structuralist and quite akin to Ikeda's (1988) own theory of language. In this theory the present authors emphasize their view that the structure for language was acquired by humans once and for all, without producing any actual language used among people. However, once the faculty of language existed, the underlying basic structure for that faculty made possible all the structures and configurations for different languages. The only thing needed for the realization of any one possible structure had to wait until some human group eventually reached the stage where they activated that particular structure which preexisted the actual "discovery" in the human brain of the cognate language. The authors would likewise argue that the species is the configuration of a preexisting structure and is encapsulated in individual organisms (like the structure for a particular language or the *langue* is encapsulated in human brain), rather than being as an abstract relationship floating between individuals of the same species. Individual organisms are vehicles of a certain configuration of the structure (shared by many species), which has materialized in the actual world. Hence, species and individuals may be understood as two aspects of one structure, like two sides of a coin, and this structure with its specific configuration should correspond to Imanishi's specia.

According to the view of the authors, Imanishi had foresight into structuralism as early as 1941, but this was obviously too early for any biologist to have grasped its real meaning, let alone the position of structuralist thinking in the controversy of reductionism and holism (or mechanistic view as against vitalism). Thus it was not possible even for Imanishi himself to distinguish structuralism from holism. Reductionism believes that the whole can be understood from the properties of the parts, the relation between the parts also being determined by the properties inherent in them. Holism categorically denies this and stresses that the properties of the whole are determined from the top down and cannot be reduced to its parts. However, the structuralist view (Ikeda 1989) asserts that the properties of the structure, as a binding principle of the parts, cannot be reduced to the properties of the parts even if the components of the structure are firmly based on the properties of the parts. This is because structure is groundless and arbitrary (Sibatani 1989c, 1990, 1991; Elder 1989) in the sense that an alternative plausible structure could have been generated at the expense of the existing one (see Appendix A). Hence it cannot be explained from its parts, but its nature is inherent in this structure alone. Hence components of the structure are bound by this arbitrariness, and hence the assertion of holism may only apply to each separate structure and its relationship to its parts. However, holism is, like reductionism, strongly monistic, asserting that the principle or law of the whole applies throughout the hierarchical organization of all the existences. It is here that the structuralist is decidedly divorced from the holist, because of the axiom of pluralism in arbitrary structures.

Imanishi could not realize the distinction between structuralism and holism and was unable to give proper expression to his earlier intuitions. He was trapped by holism and in his intellectual isolation ineffectually scattered obscure utterances on evolution during the last period of his work. This view helps, for example, to see through the inconsistency in his interesting thoughts about the "dogma or legend of the origin of life." As discussed above, Imanishi compared the presumed uniformity of individuals in the process of species formation with the uniformity of a chemical reaction at the level of individual molecules. However, the chemical complexity in biopolymers make this comparison difficult, unless helped by the template which does not exist in the inanimate world. Here, Imanishi confused substance or component with structure.

According to the authors' latest view, it is the structure and not the substance which is invariant, and this provides the framework of reference for our scientific understanding of all the variable existences and processes (Ikeda 1989). This is because the structure is an imaginary construct in our brain and is, by definition, *invariant* (like the rule of chess), regardless of whether it can or does actually exist in the world out there, independent of our brain's operation (Appendix B).

The authors' argument may confuse materialist readers, but actually it is based on materialist (and definitely *not* idealist) thinking, in so far as we admit that all human imaginations and formalisms have a firm materialist basis in what we call human brain. This stance may be called cerebrist (Yoro 1989) rather than naive materialist. What we argue is that humans understand the outside world using the imaginary structures that we construct in our brains. These imaginary structures, because of the very nature of the definition of structure as a formalized description of relationships, are all invariant and may or may not correspond to the outside world. If these structures (such as a certain concept or formulation of species) can make us feel that we have understood the outside world (such as organisms), it may mean that a structure isomorphic with the one in our brain exists there outside of our brain. However, the latest theoretical stance taken by Ikeda (1990) has gone beyond this statement. We would now argue that our science tries to describe what is variant (and hence difficult to grasp) in terms of the invariant, such as names, rules, formalisms, etc., that are all unambiguous, which *we* construct in our brain. In this sense, structures we ordinarily assume to have existed prior to the actual emergence of many real biological species, or any entities whatsoever, need not have existed in that way in the outside world, as far as *our* science is concerned (see Appendix A).

In his early period, Imanishi (1941) was already convinced that the morphological sameness or similarity—or generation of similarity by the organisms—was based on and generated through the "structure" held by organisms, but he extended his thought to the identification concept—or recognition of similarity by individual members of organisms. This is obviously related to various degrees of relatedness and thus akin to the concept of human culture with all its subtle differences and obvious multiplicity. Thus, Imanishi found that the concept of culture was useful and essential in understanding animals and their evolution. As in the case of language and species, culture is also firmly rooted, according to our view, in the existing structure in the brains of individuals in both animals and humans. Only those social institutions that correspond to the preexisting structure in the brain can be accepted as culture. However, the arbitrariness factor and the vast number of possible structures or their configurations for various forms of culture render the actual choice of one configuration among many alternative ones as a creative act (Castoriadis 1975).

This type of approach to biology via human sciences, we believe, is complementary to reductionism, and may be expressed as reverse-reduction (Sibatani 1990). It will give us a new way of seeing biological phenomena, which may uncover some unknown principle or principles underlying a number of unsolved problems in biology.

Summary

Kinji Imanishi is known in the West as the founder of a particular antiselectionist theory of organic evolution with minimal scientific value. In this chapter we give evidence for his unmistakable contributions to various fields such as taxonomy and ecology of mayflies, biosociology, primate research, anthropology, and African studies. Imanishi was a pioneer who consecutively left newly established fields for the quest of new ones. The authors analyze the starting point of his evolutionary thought, the lifestyle partitioning that led Imanishi to the idea of species society or specia (with a related concept of protoidentity or congener recognition) and compare the latter with the idea of the Specific-Mate Recognition System developed by Hugh Paterson. The result is that the two concepts touch only tangentially.

Through this operation the authors are led to a structuralist theory of biology, largely based on Saussure's theory of linguistics, and in this context examine Imanishi's thought on evolution and his earlier works on broader aspects of biological systems. They thus come to the conclusion that the young Imanishi already had insights that are very close to the aspects of biological structuralism that the authors are now developing. These probably came too early to be understood properly by Imanishi's contemporaries and followers. Imanishi himself has never realized a true structuralist theory, having been unable to find a proper distinction between holism and structuralism, and has thus failed to persuade other scientists. Here the authors revisit his earlier works and elevate his thoughts to the position they deserve in the history of biology. Above all, they analyze his concept of species as a real entity rather than a conceptual tool and give, on the basis of a reinterpretation of Saussure's theory of linguistics, a novel framework in which the species is conceived as a really existing system corresponding to a structure encapsulated in individual organisms.

Appendix A: Apparent Immanence and Necessity of Physical Laws

Some (or most) readers will be perplexed by the authors' statement that all the structures, including the most basic ones determining elementary physical processes, are arbitrary. Here they are presenting an alternative framework for considering science, which axiomatically is top-down in its approach with the aim of generating different conceptual landscapes about everything (Sibatani 1990). Seen from structures pertaining to humans, the axiom of arbitrariness of structures may be expanded down to the most basic level of the structural series, where only one unique law exists, where alternative proposals have proved to be wrong (Ikeda 1988, 1990). This situation may appear to indicate that physical laws are all immanent and necessary. However, the uniqueness of physical laws does not contradict the axiom of arbitrariness of structures. Simply it means that there is only one arbitrary structure; that is, it could have been otherwise, but it has just happened to be this way. Modern theoretical physics has approached the problem more or less similarly. Moreover, once an arbitrary structure establishes itself, it acts as the constraint which binds all the elements and units pertaining to it with a sweeping necessity. At least by looking into objects this way, basic distinctive lines drawn between the living and nonliving worlds can be erased. The possibility is also not excluded that some structures in the living world happen to be unique. The implications of this view have been examined against the background of Greek philosophy as well as modern physics with its latest development, in the context of structuralist philosophy of science as recently formulated by Ikeda (1990). So far our view has remained uncontested by reviewers and hence remains unscathed at the moment.

References

Ikeda, K. 1988. Koozoosyugi seibutgaku towa nani ka (What is structuralist biology?). Kaimeisya, Tokyo. In Japanese.
———. 1990a. Koozoosyugi kagakuron no booken (An adventure of structuralist philosophy of science). Mainiti Sinbun Sya, Tokyo. In Japanese.
Sibatani, A. 1990. Stability of arbitrary structures and its implications for heredity and evolution. Wissenschaftskolleg zu Berlin Jahrbuch 1988/1990:206–217.

Appendix B: Summary of Structuralist Philosophy of Science

Contrary to what has been unanimously believed, science need not assume the existence of an external world. "I," as a scientist, must at least assume the reality of the existence of myself; otherwise I could not engage myself in science; hence the science in question would not exist and the prerequisite of science would become trivial. If the existence of "myself" is certain, then the existence of my thoughts and experiences cannot be doubted. To put it simply, the former is the idea and the latter the phenomenon. Of these, what matters in relation to the external world is the phenomenon (Ikeda 1990*a*).

In our everyday life we inescapably experience what has been unknown to us, through which we presume that there must be something which is independent of ourselves. It is certainly possible to suppose that all the phenomena are nothing but illusions that take place inside "us." Unfortunately, it is impossible to falsify this kind of thinking through mere logic. Hence the reality of the external world is at least non-self-evident.

Science is a system of theories to explain phenomena. In order for us to share common theories, that is, to generate science, we have to accept the existence of something that is common between the phenomena taking place within us. If the external world exists, this must be it. Thus, scientists would generally presume that the reality of the external world is a prerequisite of science, where the objectivity of science is guaranteed by the reality of the external world.

By contrast, structuralist philosophy of science asserts that science can dispense with such a spurious assumption as the reality of the external world. Occam's razor would favor the structuralist philosophy of science over the one depending on the reality of the external world.

What actually guarantees the objectivity of science is the isomorphism of a rule that transforms a phenomenon into words as well as a formalism in the rule relating different words. For instance, we use the same word *dog* for different animals although we know how not to apply this word to cats. However, once we call them *dogs*, universality of the word *dog* makes it possible for us all to share common understanding. Universality of dog does not reside in the external world but in the word, which, however, is arbitrary as to which phenomena should be called "dog." Were it not for the words *wolf* or *coyote*, these animals would be called dogs as well. Such thinking (Ikeda 1990*b*) is akin to nominalism or structuralist linguistics originating from Saussure; hence the use of the name of structuralist philosophy of science employed to express these views.

As regards the formalism of the rule relating words, a scientific description "dog is larger than rat" is obtained in accordance with a general statement "A is larger than B," where A denotes dog and B rat. The formalism itself is quite objective. We believe that this logic embodies scientific objectivity.

Which is more objective, then, the isomorphism in phenomenon-word transformation or the formalism relating words? Of course the latter. None of the phenomena that we experience is invariant. Materialists who believe in the reality of the external world have tried to find invariant entities existing in the external world, so far in vain. Everything that appeared to have an invariant existence has turned out to be noninvariant. Examples include atom, atomic nucleus, proton, and absolute space. This is because our observation and experience entail some shift of time, and hence we are not able to observe invariance. In other words, all phenomena are in fact variant.

According to the definition that science is an attempt to explain phenomenon with words and their relationship, the phenomenon is variant, but the word and relationship are invariant. Hence we can say that science is an attempt to decipher the variant in terms of the invariant, or alternatively that science is a game to decipher what is visible in terms of what is invisible.

Now, if the word and formalism that construct scientific theories do not really exist in the external world, they must reflect patterns of human cognition and thinking. Then, if these patterns did change, the same phenomena could be explained by entirely different scientific theories. In scientific theories that assume reality of the external world as the basis of objectivity, at least one of the two different scientific theories proposed for the same phenomenon should be wrong. However, in a structuralist philosophy of science, it is trivial that scientific theory changes with a change in patterns of cognition and thinking; such a statement is simply nonsensical. Naturally, what looks like a better-organized theory for the phenomenon in question would be more readily accepted, but this is simply due to the better fitting of such a theory to the pattern of cognition and thinking of the time rather than the ultimate truthfulness or correctness of the theory.

References

Ikeda, K. 1990*a*. Koozoosyugi kagakuron no booken (Adventure of structuralist philosophy of science). Mainiti Sinbun Sya, Tokyo. In Japanese.
———. 1990*b*. Is a self-recursive biology possible? Riv. Biol.–Biol. Forum 83:93–106.

Acknowledgements

We express our hearty thanks to David Elder and Louisa Macmillan for linguistic correction of this chapter.

References

Asquith, P.J. 1986. Imanishi's impact in Japan. Nature 323:675.
Castoriadis, C. 1975. L'institution imaginaire de la société. Edition du Seuil, Paris.

Chomsky, N. 1986. Knowledge of language: Its nature, origin and use. Praeger, New York.
Engler, R. 1967. Ferdinand de Saussure: Cours de linguistique générale. Édition critique. Otto Harrassowitz, Wiesbaden.
Foucault, M. 1969. Qu'est-ce qu'un auteur? Bull. Soc. Fr. Philos. 63:73–104.
Hall, B.G. 1988. Adaptive evolution that requires multiple spontaneous mutations. I. Mutations involving an insertion sequence. Genetics 120:887–897.
———. 1990. Spontaneous point mutations that occur more often when advantageous than when neutral. Genetics 126:5–16.
Halstead, B. 1985. Anti-Darwinian theory in Japan. Nature 317:587–589.
———. 1988. Imanishi Sinkaron Hihan no Tabi (A journey for critique of the evolutionary theory of Imanishi). (Nakayama, T., and Sakuramati, S., trans.), Tukizi Syokan, Tokyo. In Japanese.
Hendrichs, H. 1988. Lebensprozesse und wissenschaftliches Denken. Zur Logik der Lebendigkeit und ihrer Erstarrung in den Wissenschaften. Karl Alber, Freiburg (Reviewed by Meyer, P. 1990. J. Soc. Biol. Struct. 13:175–177.)
Ikeda, K. 1987. Sinkaron ni okeru sinpo-syugi (Progressionism in evolutionary theories). Yamanasi Daigaku Kyooikugakubu Kenkyuu-hookoku (Mem. Faculty Lib. Arts and Educ., Yamanashi University) 37:48–58. In Japanese.
———. 1988. Koozoosyugi seibutgaku towa nani ka (What is structuralist biology?). Kaimeisya, Tokyo. In Japanese.
———. 1989. Koozoosyugi to sinkaron (Structuralism and evolutionary biology). Kaimeisya, Tokyo. In Japanese.
Imanishi, K. 1941. Seibutu no Sekai (The world of living things). Koobundoo, Tokyo. Reprinted 1972, Koodansya, Tokyo. In Japanese.
———. 1949. Sesibutu syakai no ronri (The logic of biosocieties). Mainiti Sinbun Sya, Tokyo. Reprinted 1979, Mainiti Sinbun Sya, Tokyo. In Japanese.
———. 1977. Daauin-ron: Dotyaku sisoo karano rezisutansu (On Darwin: Resistance from the indigenous thoughts). Tyuukoo Sinsyo, Tyuuoo Kooron Sya, Tokyo. In Japanese.
———. 1980. Syutaisei no sinkaron (An evolutionary theory based on subjectivity/autonomy). Tyuukoo Sinsyo, Tyuuoo Kooron Sya, Tokyo. In Japanese.
———. 1984. Conclusion to my study of evolutionary theory. J. Soc. Biol. Struct. 7:357–368.
———. 1987. Mure-seikatusya-tati (Those animals living together). Kikan Zinruigaku (Anthropology Quarterly) 18:82–96.
Ito, Y. 1990. Nippon no seitaigaku—tokuni Imanisi Kinji no hyooka to kannrennsite (Ecology in Japan, especially in relation to the evaluation of Kinji Imanishi). Seibutu Kagaku 42:176–191.
Kortmulder, K. 1986. The congener: A neglected area in the study of behavior. Act Biotheor. 35:39–67.
Kortmulder, K., and T.E. Sprey. 1990. The connectedness of all that is alive and the grounds of congenership: Beyond a mechanistic interpretation of life. Riv. Biol.–Biol. Forum 83:107–127.
Ogushi, R. 1985. Imanishi-gakuha no keihu (The lineage of the Imanishi school). Kanazawa Daigaku Kyooiku Kaihoo Sentaa Kiyoo (Bulletin of the Open Education Center of Kanazawa University) 5:55–67. In Japanese.
Paterson, H.E.H. 1978. More evidence against speciation by reinforcement. S. Afr. J. Sci. 74:369–371.
———. 1985. The recognition concept of species. Pp. 21–29 in E.S. Vrba, ed. Species and speciation. Transvaal Museum Monograph No. 4. Pretoria.

Rossiter, A. 1986. Evolutionary "classics" may self-destruct. Nature 332:315–316.
Saussure, F. de. 1972. Cours de linguistique générale. Édition critique (T. de Mauro, ed.). Payot, Paris.
Sibatani, A. 1981. Imanishi sinkaron hihan siron (Critique of K. Imanishi's theory of evolution). Asahi Press, Tokyo. In Japanese.
———. 1982. Watasi ni totte kagaku towa nani ka (What is science for me?) Asahi Sinbun Sya, Tokyo. In Japanese.
———. 1983a. Kinji Imanishi and species identity. Riv. Biol. Perugia 76:25–42.
———. 1983b. Anti-selectionism of Kinji Imanishi and social anti-Darwinism in Japan. J. Soc. Biol. Struct. 6:335–343.
———. 1984. Atogaki (Postscript). Pp. 254–259 in K. Imanishi, A. Sibatani, and S. Yonemoto, eds. Sinkaron mo sinka-suru (Evolutionary theory is also evolving). Riburopooto (Libroport), Tokyo. In Japanese.
———. 1985. Molecular biology: A structuralist revolution. Riv. Biol.–Biol. Forum 78:373–397.
———. 1987. Sumiwake o doo kangaeru ka (How to interpret *sumiwake* or lifestyle partitioning?). Seibutu Kagaku (Biological Sciences) 39:179–191. In Japanese.
———. 1989a. Sumiwake-ron no tenkai to sinka-ronsoo (Exploration of the lifestyle partitioning concept and debate in evolutionary biology). Pp. 164–172 in A. Sibatani and K. Tanida, eds. Nippon no suisei kontyuu (Aquatic insects of Japan). Tookai Daigaku Syuppankai, Tokyo. In Japanese.
———. 1989b. Conspecific recognition in male butterflies. Riv. Biol.–Biol. Forum 82:15–38.
———. 1989c. Stability of arbitrary structures: Its implications for heredity and evolution. Riv. Biol.–Biol. Forum 82:348–349.
———. 1990. Stability of arbitrary structures and its implications for heredity and evolution. Wissenschaftskolleg zu Berlin Jahrbuch 1988/1989:206–217.
———. 1991. Structuralist biology in Japan. Bull. Kyoto Seika Uni. 1: 309–324.
Takemon, Y. 1989. Monkageroo zoku no uka, hansyoku yoosiki to ryuutei bunpu (The mode of emergence and reproduction and the distribution along the course of flow in the genus *Ephemera*). Pp. 9–41 in A. Sibatani and K. Tanida, eds. Nippon no Suisei-kontyuu (Aquatic insects of Japan). Tookai Daigaku Syuppankai, Tokyo. In Japanese.
Tanida, K. 1984. Larval microlocation on stone faces of three *Hydropsyche* species (Insecta; Trichoptera), with a general consideration on the relation of systematic groupings to the ecological and geographical distribution among the Japanes *Hydropsyche* species. Physiol. Ecol. Japan. 21:115–120.
Umesao, T. 1989. Kenkyuu keiei-ron (Essays on research enterprise). Iwanami Syoten, Tokyo. In Japanese.
Waldrop, M.M. 1990. Spontaneous order, evolution, and life. Science 247:1543–1545.
Yoro, T. 1990. Yuinooron (Cerebrism). Seidosya, Tokyo.

5

Mosquito Species Concepts: Their Effect on the Understanding of Malaria Transmission in Africa

RICHARD H. HUNT AND MAUREEN COETZEE

Medical Entomology, Department of Tropical Diseases
School of Pathology of the South African Institute for
Medical Research and the University of the Witwatersrand

Mosquitoes have long been known to transmit disease pathogens to man. Ronald Ross's discovery (1897) that the *Anopheles* mosquitoes were the vectors of human malaria parasites was in fact almost an accident. He spent from 1895 to 1897 dissecting and examining culicine mosquitoes. The discovery of parasites in two large mosquitoes, which Ross described simply as having "dapple wings," focused his attention on the genus *Anopheles*. It is of interest to note the marked change in the malariologist's morphological techniques since that time. Compare the simplistic descriptions of mosquitoes by Ross (1897) to the wealth of detail in later morphological descriptions of the group (e.g. Belkin 1962; Harrison and Scanlon 1975; Gillies and De Meillon 1968; Gillies and Coetzee 1987). This change was brought about not only because of the incrimination of anophelines as vectors of malaria, but because of the realization that not all species in the genus are vectors (Watson 1915, 1921; Swellengrebel and De Buck 1938). When active mosquito control became possible with the advent of insecticides, it was essential for the malariologist to know if a particular species was a potential vector or not. This logic lead Swellengrebel and De Buck (1938) to coin the phrase "species sanitation" to describe Sir Malcolm Watson's method of dealing only with vector species and ignoring the nonvectors.

Ever since the days of Ross, research in malaria epidemiology has frequently revealed the importance of an understanding of the nature of species and speciation. Initially, of course, species were described purely from a morphological viewpoint, and even biological characteristics such as biting and resting behavior (Giles 1902; Theobald 1903) were largely ignored. Although Mayr and Dobzhansky were actively publishing their biological species concept in the 1930s and 40s, scant attention was paid to the implications of the concept by entomologists in the field of mosquito taxonomy (Edwards 1941; Evans 1938; Hopkins 1952).

The first indication that morphologically similar species might exist in the anopheline mosquitoes was the realization that *Anopheles maculipennis* consisted of a group of species. As early as 1901 Italian workers had been aware that the incriminated vector of malaria in Europe, *An. maculipennis*, quite frequently occurred in large numbers in areas free of malaria (Harrison 1978), giving rise to the term "anophelism without malaria." By 1937 it had become obvious that the taxon *An. maculipennis* consisted of more than one species (Hackett 1937). Hackett and co-workers were able to demonstrate that the taxon comprised five species, morphologically similar as adults, some of which were vectors of malaria parasites and some not. All five species, although superficially very similar in appearance, can be separated by small morphological differences. In fact, major and obvious differences were found between the eggs of two of them (Falleroni 1926).

The European situation is well described by Swellengrebel and De Buck (1938) in their book *Malaria in the Netherlands*. They describe the difficulty facing those charged with the control of malaria in a situation where an apparently homogeneous vector population occurred in both malarious and nonmalarious areas. They contrasted this with the apparently much simpler situation in Asia where the vectors could be identified morphologically. Swellengrebel and De Buck conducted elegant mating and cross-mating studies on their populations, and while obviously favoring the idea that they were dealing with two biologically distinct entities, they shied away from calling them species. They contented themselves with calling them "elementary species" (*sensu* De Vries 1905) to differentiate between species in the taxonomic tradition of Linnaeus and "really existing entities" in nature.

This group of species was resolved formally by using mainly classical morphological methods that rely on clear structural differences with no intergrades. This enables a museum taxonomist to classify individual specimens and allocate them to the appropriate museum drawer. Apart from Swellengrebel and De Buck (1938), entomologists working on mosquito systematics at that time followed this methodology, and biology and behavioral characteristics generally were not considered to be of any great significance in the classification of species.

Essentially, the *An. maculipennis* complex was first resolved using morphological characteristics that helped explain behavioral anomalies. Later, the application of techniques such as chromosome cytology and electrophoresis, to test genetical concepts of species, revealed that species complexes are indeed rather common in anophelines.

The History of the *Anopheles gambiae* Complex

It is now more than ninety years since Ross and co-workers (1900) discovered that the mosquitoes which today are known as members of the *Anopheles gambiae* Giles (1902) complex were highly efficient vectors of human plasmodia and filarial parasites. Originally known as *Anopheles costalis*, the name *gambiae* came into general use only after 1932 (Christophers 1924; Edwards 1932). Dönitz in 1902 rejected the name *costalis* on the grounds that the common species known as *costalis* did not correspond with the description of this species given by Loew in 1866. However, Theobald (1903) defended the name of *costalis* because "the species has been so long known as *costalis* by all the important medical men in Africa that endless confusion would ensue" should the name be changed. Such was Theobald's authority that the name *gambiae* did not finally replace *costalis* until the publication of Edwards' monograph in 1932.

For the next thirty years publications on the differences noticed in the biology of *An. gambiae* were numerous. It was noted that the larval habitats varied from open, sunlit, freshwater pools (e.g. De Meillon 1937, 1941; Evans 1938; Haddow et al. 1947) to underground, cement-lined, water tanks (De Meillon 1938), shaded pools (Causey et al. 1943), marshes (Vincke and Parent 1944), flooded, well-vegetated islands (Parent and Demoulin 1945), and pools with high salinity (e.g. Evans 1931; Ribbands 1944; Muspratt in De Meillon 1947; Muirhead-Thomson 1951). Similarly, the adult biology also proved to be variable. Although *An. gambiae* in many areas was largely endophilic (indoor-resting) and anthropophilic (man-biting) (Gordon et al. 1932; Barber et al. 1932; Symes 1932; Gibbins 1933; De Meillon 1941), as more data were collected it became evident that the extent of zoophily (animal-biting) was often surprisingly high. For example, in Ethiopia only 57 percent of the mosquitoes collected had fed on man (Corradetti 1938) and in Zimbabwe 37–70 percent (Bruce-Chwatt and Gockel 1960).

With the advent of residual insecticide spraying of houses around 1947, a large number of studies were concerned with the resting behavior of the species. It was demonstrated in West Africa by Muirhead-Thomson (1947*b,c*) that DDT, because it irritated *An. gambiae* adults, caused avoidance behavior. This resulted in a dramatic change in house-resting behavior. Muirhead-Thomson (1950) demonstrated that BHC (benzine hexachloride) did not affect the mosquitoes in this way. Mastbaum (1954, 1957*a,b*) working in Swaziland found that BHC eliminated the house-resting population of *An. gambiae*. However, he found that large numbers of larvae persisted in areas where malaria transmission had been interrupted effectively by six years of residual spraying. This was interpreted as natural selection of zoophilic (animal-biting) and exophilic (outdoor-resting) behavior and was called behavioral resistance. This would, according to the argument, have serious implications for malaria control. Such outdoor-resting populations were considered to be out of reach of the insecticide so that spraying with residual insecticides would be of little value. Analysis of mosquito bloodmeals using the precipitin test revealed that in many areas this exophilic population of *An. gambiae* fed predominantly on cattle (Ramsdale in Mattingly 1963, unpublished document World Health Organization WHO/MAL 389.63). This was reassuring, but as the apparent change was not understood, there was always the specter of the outdoor population returning to its old man-biting ways.

In retrospect it is surprising that the truth about the taxon *An. gambiae* was not arrived at sooner. Ribbands (1944) and Muirhead-Thomson (1945, 1947a) had presented clear evidence that the two salt-water "varieties," *An. melas* on the West coast and *An. merus* in the East, were in fact distinct species. In the case of *An. melas,* Muirhead-Thomson (1947a) presented data showing hybrid sterility when salt-water tolerant animals were crossed with freshwater individuals. These results were ignored, mainly because of the work of Bruce-Chwatt (1950), who reported that he had repeated Muirhead-Thomson's experiments and found no sterility. Subsequent work has shown that this cross always gives rise to completely sterile male offspring (Davidson et al. 1967). However, Bruce-Chwatt's influence was so great at that time that the possibility of more than one species in the taxon *An. gambiae* was largely ignored.

Davidson (1958), while investigating the genetics of insecticide resistance in some West African populations of "*An. gambiae,*" found that in some cases when he crossed susceptible with resistant strains, the resulting male progeny were sterile. He explained this phenomenon as being associated with the resistance gene (presumably a pleiotropic genetic effect). In further studies, Davidson and Jackson (1962) called their strains forms A and B and suggested that the sterility indicated "a divergence within the species *An. gambiae.*" Later, however, Davidson (1964) admitted that all could be better explained if each group or form was considered a separate species.

Repeating the work of Muirhead-Thomson (1947a) but using the East African salt-water breeder *An. merus,* Paterson (1962) and Kuhlow (1962) separately showed sterility between the crosses of *An. merus* and the freshwater breeding *An. gambiae.* They concluded on this evidence that *An. merus* was a separate species. Paterson et al. (1963) reported the existence in southern Africa of a third freshwater member of the complex, which they called form C. The cross-mating characteristics of form C showed male sterility between it and forms A and B, *melas* and *merus.*

In 1964 Paterson (1964a) presented evidence for the coexistence of the three freshwater forms A, B, and C in Zambia without hybridization. He did this by collecting wild females, inducing them to lay eggs and then examining the male progeny for sterility. If random mating was taking place then one could expect some egg batches to contain males with atrophied testes. However, all the broods examined had normal, fertile males, indicating that the wild females had mated conspecifically. Furthermore, examination of the giant polytene chromosomes found in the salivary glands of fourth instar larvae showed no evidence of asynapsis. He concluded that, according to the Biological Species Concept and population genetic theory, they were all good species. He discussed the implications of this conclusion several times (Paterson 1963, 1964a,b, 1968).

It was not until Paterson (1962, 1963, 1964a) applied the Biological Species Concept of Mayr (1942, 1963) to the problem that it became clear that the taxon *An. gambiae* was in fact a group of five morphologically similar species. It is now known that the group consists of at least six different species (see Gillies and Coetzee 1987). They vary considerably in their biology and behavior. These differences are not only of academic interest because they profoundly influence the vectorial capability of the various species and their resting behavior. However, it has often proved difficult to persuade local health authorities to adopt different attack methods depending on which species is present and responsible for malaria transmission in a particular area. In any case, malaria control by the large-scale use of insecticides for house spraying in tropical Africa is largely ruled out on the grounds of cost. Indications are that control in the future will probably be more dependent on community-based programs using methods such as insecticide-impregnated bed nets. Such changes in the approach to control are unlikely to eliminate the need for correct species identification. In fact, the use of insecticides with strong repellent effects is likely to exacerbate the problems experienced in areas where *An. arabiensis* is the vector (Sharp et al. 1990). Situations where avoidance in response to pyrethrum and DDT leads to survival of *An. gambiae* s. l. have long been known (Muirhead-Thomson 1947b).

Paterson's ideas were not immediately accepted by the scientific community, and many workers continued to refer to the species as "forms" or "mating types" and questioned their specific status mainly because of the lack of morphological differences between the species (Hamon 1963; Davidson 1964; Coz and Hamon 1964). However, after 1964 most workers accepted that they were good species (Davidson et al. 1967) although they were only officially named much later (Mattingly 1977).

Discussion

The consequences of the recognition that *An. gambiae* was a complex of several species were numerous. The hypothesis that *An. gambiae* changed its habits from a man-biting, house-resting, vector species to a nonvector, animal-biting, outdoor-resting species was shown to be wrong. What, in fact, people were looking at was two different species with very different behavior patterns. This had a marked effect on the antimalarial spraying campaigns in various parts of Africa. The idea advanced by Davidson (1958) that insecticide resistance genes were in some way associated with the sterility found in hybrids produced by crossing some resistant and susceptible populations of *An. gambiae* was also shown to be incorrect.

Subsequent work on anopheline mosquitoes in other parts of the world has revealed the existence of many species complexes and in doing so have resolved some of the problems in exactly the same way. For example, in the Oriental region *An. maculatus* consists of six species (Rattanarithikul and Green 1986); *An. culicifacies,* three species (Green and Miles 1980; Subbarao et al. 1983); and *An. dirus,* five species (Baimai et al. 1988). In Australia, *An. farauti* is known to be a complex (Bryan 1970; Mahon et al. 1981), as is *An. annulipes* (Green 1972). In the Afrotropical Region the taxon *An. marshallii* comprises four species (Lambert 1979, 1981); *An. pharoensis,* two species (Miles et al. 1983); *An. coustani,* two species (Coetzee 1983); and *An. ziemanni,* two species (Coetzee 1984). Of them all, the *An. gambiae* complex is arguably the most significant because of its medical importance and because it was the first to be resolved by applying a definite genetical concept of species (that of Mayr [1942]) using genetical approaches. Thus, Paterson's contribution to the understanding of the *An. gambiae* complex in Africa has had worldwide impact on the understanding of vector-borne diseases in general.

At the time of his work on the *An. gambiae* complex, Paterson subscribed to the Biological Species Concept of Mayr (1942) and Dobzhansky (1951). He firmly believed then, as now, that "only the genetical species definition has any meaning in nature" (Paterson 1963), and he used population genetical theory in his arguments for the specific status of members of the complex. However, it was probably the insights gained while working on the practical problems of malaria transmission by members of the *An. gambiae* complex that led Paterson to formulate his Recognition Concept of species (Paterson 1985).

By so doing, Paterson has provided the means for further insights into species complexes. The *An. gambiae* complex is again an excellent example. Recent work in West Africa indicates that *Anopheles gambiae* s. str. may consist of more than one species (Bryan et al. 1982; Toure et al. 1984; Coluzzi et al. 1985; Coluzzi 1988). These workers have applied the Isolation Concept of species (Mayr 1963) in an effort to interpret chromosomal polymorphisms found to be correlated with various populations of *An. gambiae* s. str. For example, in Mali three chromosomal "forms" exist, designated "Bamako," "Savanna," and "Mopti" (Toure et al. 1984; Coluzzi et al. 1985). No heterozygotes between the Bamako and Mopti forms are found in nature. At localities where all three are sympatric, the various polymorphic inversions are not in Hardy-Weinberg equilibrium, perhaps indicating positive assortative mating.

However, unlike crosses between recognized species of the *An. gambiae* group, there is no sterility in hybrid offspring, nor are there any morphological differences. Coluzzi and co-workers refer to these forms as "incipient species." Similar situations exist in the Gambia (Bryan et al. 1982) and Nigeria (Coluzzi et al. 1985), although the chromosome inversions involved at these localities are different. Coluzzi et al. (1985) conclude that "ethological isolation" appears to be the only explanation for the genetic discontinuities. Their reluctance to regard the taxa as separate species is probably because, as adherents to the Isolation Concept, they expect "other isolating mechanisms" to be present in "good" species (e.g. seasonal isolation, habitat difference, or sterility). If the Recognition Concept (Paterson 1985) is applied, the chromosomal inversions are seen as markers of positive assortative mating, and as individuals of the various populations are not mating in nature, they are separate species.

Coluzzi et al. (1985) further contend that the speciation in the *An. gambiae* complex is a recent event and give five reasons in support of this: (1) postmating isolating mechanisms are incomplete in that only male hybrids of the recognized species are sterile, and the "precopulatory, ethological barriers" break down under laboratory conditions; (2) morphological differentiation is at a very low level; (3) Nei's genetic distances are also very low; (4) differences in the polytene chromosomes involve only changes in the banding sequences due to paracentric inversions; (5) the malaria vectors are closely associated with man—"it would be difficult to hypothesize its evolution and its wide diffusion in Tropical Africa in the absence of man." None of these, in fact, supports the contention that the group is recently evolved. If one applies the Recognition Concept it can be seen that:

1. Partial sterility and the degree of completeness of "ethological" precopulatory barriers are effects accumulated during the speciation process. It would be difficult to postulate how they could change much with time after speciation because of stabilizing selection in a panmictic population (Lambert and Paterson 1982). The fact that the mate recognition system breaks down in the laboratory is not relevant to the natural situation.

2. Morphological differentiation is probably not indicative of the length of time since speciation in groups of cryptic species. It can be argued that, as anopheline mosquitoes are mostly nocturnal, vision probably plays a limited role in their mating behavior. Therefore, there would be no selection for the acquisition of subtle morphological characters to assist in specific-mate recognition. Subsequent to speciation, morphological change would be very limited in groups that have virtually no morphological polymorphisms on which natural selection can act.

3. Coluzzi (1988) gives a table showing Nei's genetic distance for several different groups of sibling species within *Anopheles*. It is interesting that he includes the *An. marshallii* group, as this undermines his argument that a

small genetic distance indicates recent speciation and that morphological differentiation indicates a longer period. The genetic distance of 0.03 between *An. marshallii* species A and C (described by Lambert and Coetzee in 1982 and named *An. letabensis* and *An. hughi*, respectively) is of the same order as that for the taxa *An. gambiae* Mopti and Bamako. A major point here is that there are marked morphological differences between the two *marshallii* species in all life stages. In the *An. gambiae* group, morphologically very similar species have much larger genetic distances. We must therefore conclude that neither Nei's genetic distance nor morphological differences are reliable indicators of how long ago these taxa speciated. Lambert and Paterson (1982) argue that "at the level of closely related species ... there is no consistent correlation between morphological resemblance and genetic distance, because no set amount of genetic divergence can be found to accompany speciation events."

4. With regard to the level of rearrangement in the polytene chromosomes, there is no particular reason why any changes to the chromosomes should occur at speciation. In all anophelines where complete chromosome homologies have been worked out, the only rearrangements found are paracentric inversions and, occasionally, whole arm translocations. *Anopheles funestus–An. vaneedeni* and *An. atroparvus–An. labranchiae* are examples of pairs of species with homosequential chromosomes. In both cases the species can be separated using morphological characters. Any changes to the chromosomes that do occur should therefore be seen as coincidental, chance events and not causal in relation to the speciation process.

5. Finally, if the association of the malaria vectors with humans indicates that the species evolved in conjunction with humans, then the powerful evidence that humans evolved in Africa (Cann et al. 1987) suggests the possibility of an extremely long association.

It can be seen from the above arguments that the use of the Recognition Concept of species can provide different insights, in this case into malaria vectors. These insights are clearly not purely academic because accurate identification of species and a clear understanding of the delimitation of their gene pools are vital in medical and veterinary entomology.

Summary

A historical review of the malaria vector *Anopheles gambiae* and its related species is given. Changes in the interpretation of data can be seen to occur when different species concepts are applied. Historically, the application of the "Biological Species Concept" (or Isolation Concept) resulted in the recognition of the taxon *An. gambiae* comprising at least six morphologically similar species. These species all exhibit some degree of hybrid sterility when cross-mated and possess different polytene chromosome arrangements. Today, work in West Africa indicates that more species exist within the taxon *An. gambiae* s. str. These show no fixed differences in chromosome banding sequences, morphology, or hybrid viability, the only differences being polymorphic inversion frequencies of different populations. Application of the Isolation Concept has resulted in these populations being referred to as "incipient" species on the grounds that only ethological isolating mechanisms are operating. Application of the Recognition Concept of species results in the populations being regarded as "good" species because positive assortative mating is taking place in nature. The usefulness of biological and epidemiological data depends on the accuracy of species identification.

References

Baimai, V., R.E. Harbach, and U. Kijchalao. 1988. Cytogenetic evidence for a fifth species within the taxon *Anopheles dirus* in Thailand. J. Am. Mosq. Control Assoc. 4:333–338.
Barber, M.A., J.B. Rice, and J.Y. Brown. 1932. Malaria studies on the Firestone Rubber Plantation in Liberia, West Africa. Am. J. Hyg. 15:601.
Belkin, J.A. 1962. The mosquitoes of the South Pacific. University of California Press, Berkeley.
Bruce-Chwatt, L.J. 1950. Recent studies on insect vectors of yellow fever and malaria in British West Africa. J. Trop. Med. Hyg. 53:71–79.
Bruce-Chwatt, L.J., and C.W. Gockel. 1960. A study of the blood-feeding patterns of *Anopheles* mosquitoes through precipitin tests. Bull. W. H. O. 22:685.
Bryan, J.H. 1970. A new species of the *Anopheles punctulatus* complex. Trans. R. Soc. Trop. Med. Hyg. 64:28.
Bryan, J.H., M.A. Di Deco, V. Petrarca, and M. Coluzzi. 1982. Inversion polymorphism and incipient speciation in *Anopheles gambiae* s. str. in the Gambia, West Africa. Genetica 59:167–176.
Cann, R.L., M. Stoneking, and A.C. Wilson. 1987. Mitochondrial DNA and human evolution. Nature 325:31–36.
Causey, O.R., L.M. Deane, and M.P. Deane. 1943. Ecology of *Anopheles gambiae* in Brazil. Am. J. Trop. Med. 23:73.
Christophers, S.R. 1924. Provisional list and reference catalogue of the Anophelini. Indian Med. Res. Mem. 3.
Coetzee, M. 1983. Chromosomal and cross-mating evidence for two species within *Anopheles (A) coustani* (Diptera: Culicidae). Syst. Entomol. 8:137–141.
———. 1984. A new species of *Anopheles* (*Anopheles*) from Namibia (Diptera: Culicidae). Syst. Entomol. 9:1–8.
Coluzzi, M. 1988. Anopheline mosquitoes: Genetic methods for species differentiation. Pp. 411–430 *in* W.H. Wernsdorfer and I. McGregor, eds. Malaria. Principles and practice of malariology. Vol. 1. Churchill Livingstone, Edinburgh.
Coluzzi, M., V. Petrarca, and M.A. Di Deco. 1985. Chromosomal inversion intergradation and incipient speciation in *Anopheles gambiae*. Boll. Zool. 52:45–63.
Corradetti, A. 1938. La malaria nella regione Uolla-Jegiu nel periodo Luglio-Ottobre 1937. Boll. Soc. Ital. Biol. Sper. 13:115.

Coz, J., and J. Hamon. 1964. Le complexe *Anopheles gambiae* en Afrique Occidentale. Riv. Malar. 43:233–244.
Davidson, G. 1958. Studies on insecticide resistance in anopheline mosquitoes. Bull. W. H. O. 18:579–621.
———. 1962. *Anopheles gambiae* complex. Nature 196:907.
———. 1964. *Anopheles gambiae*, a complex of species. Bull. W. H. O. 31:625–634.
Davidson, G., and E. Jackson. 1962. Incipient speciation in *Anopheles gambiae* Giles. Bull. W. H. O. 27:303–305.
Davidson, G., H.E.H. Paterson, M. Coluzzi, G.F. Mason, and D.W. Micks. 1967. The *Anopheles gambiae* complex. Pp. 211–250 *in* J.W. Wright and R. Pal, eds. Genetics of insect vectors of disease. Elsevier Publ. Co., Amsterdam.
De Meillon, B. 1937. A note on *Anopheles gambiae* and *Anopheles funestus* in northern Rhodesia. Publ. S. Afr. Inst. Med. Res. 7:306–312.
———. 1938. A note on *Anopheles gambiae* Giles and *Anopheles coustani var. tenebrosus* Dönitz from South Africa. S. Afr. Med. J. 12:648–650.
———. 1941. Estudos entomologicos da colonia de Mozambique I. (Lourenco Marques, Impresna Nacional de Mozambique).
———. 1947. The Anophelini of the Ethiopian geographical region. Publ. S. Afr. Inst. Med. Res. No. 49, Johannesburg.
De Vries, H. 1905. Species and varieties. The Open Court, Chicago.
Dobzhansky, T. 1951. Genetics and the origin of species. 3d ed. Columbia University Press, New York.
Dönitz, W. 1902. Beitrage zur Kentniss der *Anopheles*. Z. Hyg. Infektkr. 41:15–88.
Edwards, F.W. 1932. Genera insectorum. Diptera Fam. Culicidae. Brussels 258 pp.
———. 1941. Mosquitoes of the Ethiopian region. III. Culicine adults and pupae. Br. Mus. (Nat. Hist.), London.
Evans, A.M. 1931. Observations made by Dr. M.A. Barber on a melanic, coastal race of *Anopheles costalis*, Giles (*gambiae*) in southern Nigeria. Ann. Trop. Med. Parasit. 25:443–453.
———. 1938. Mosquitoes of the Ethiopian region. II. Anophelini adults and early stages. Br. Mus. (Nat. Hist.), London.
Falleroni, D. 1926. Fauna anofelica italiana e suo "habitat" (paludi, risaie, canali), metodo di lotta contro la malaria. Riv. Malar. 5:553–593.
Gibbins, E.G. 1933. The domestic *Anopheles* mosquitoes of Uganda. Ann. Trop. Med. Parasit. 27:15–25.
Giles, G.M. 1902. A handbook of the gnats or mosquitoes. John Bale, Sons and Danielsson, London.
Gillies, M.T., and M. Coetzee. 1987. A supplement to the Anophelinae of Africa south of the Sahara. Publ. S. Afr. Inst. Med. Res. No. 55.
Gillies, M.T., and B. De Meillon. 1968. The Anophelinae of Africa south of the Sahara. Publ. S. Afr. Inst. Med. Res. No. 54.
Gordon, R.M., E.P. Hicks, T.H. Davey, and M. Watson. 1932. A study of the house haunting Culicidae occurring in Freetown, Sierra Leone. Ann. Trop. Med. Parasit. 26:273–345.
Green, C.A. 1972. The *Anopheles annulipes* complex of species. Proc. Fourteenth Int. Congr. Entomol. Canberra (Abstracts), p. 286.
Green, C.A., and S.J. Miles. 1980. Chromosomal evidence for sibling species of the malaria vector *Anopheles (Cellia) culicifacies* Giles. J. Trop. Med. Hyg. 83:75–78.
Hackett, L.W. 1937. Malaria in Europe: An ecological study. Oxford University Press, Oxford.

Haddow, A.J., J.D. Gillett, and R.B. Highton. 1947. The mosquitoes of Bwamba County, Uganda. V. The vertical distribution and biting cycle of mosquitoes in rain-forest, with further observations on microclimate. Bull. Entomol. Res. 37:301–330.
Hamon, J. 1963. Comment on Paterson (1963b) at Seventh Int. Congr. Trop. Med. Malaria, Rio de Janeiro. 5:225–226.
Harrison, B.A., and J.E. Scanlon. 1975. Medical entomology studies. II. The subgenus *Anopheles* in Thailand (Diptera: Culicidae). Contrib. Am. Entomol. Inst. 12, No. 1.
Harrison, G. 1978. Mosquitoes, malaria and man: A history of the hostilities since 1880. John Murray, London.
Hopkins, G.H.E. 1952. Mosquitoes of the Ethiopian region. I. Larval bionomics of mosquitoes and taxonomy of culicine larvae. Br. Mus. (Nat. Hist.), London.
Kuhlow, F. 1962. Beobachtungen und experimente uber den *Anopheles gambiae* komplex, Abtrennung von *Anopheles tangensis* n.sp. Zeit. Tropenmed. Parasit. 13:442–449.
Lambert, D.M. 1979. *Anopheles marshallii* (Theobald) is a complex of species. Mosq. Syst. 11:173–178.
———. 1981. Cytogenetic evidence of a possible fourth cryptic species within the taxon *Anopheles marshallii* (Theobald) (Diptera: Culicidae) from northern Natal. Mosq. Syst. 13:168–175.
Lambert, D.M., and M. Coetzee. 1982. A dual genetical and taxonomic approach to the resolution of the mosquito taxon, *Anopheles (Cellia) marshallii* (Culicidae). Syst. Entomol. 7:321–332.
Lambert, D.M., and H.E.H. Paterson. 1982. Morphological resemblance and its relationship to genetic distance measures. Evol. Theory 5:291–300.
Loew, H. 1866. Beischreibung einiger Afrikanischen Diptera Nemocera. Berl. Ent. Z. 10:55–62.
Mahon, R.J., P. Miethke, and P.I. Whelan. 1981. The medically important taxon *Anopheles farauti* (Diptera: Culicidae) in Australia. Aust. J. Zool. 29:225–232.
Mastbaum, O. 1954. Observations of two epidemic malaria seasons (1946 and 1953) before and after malaria control in Swaziland. Trans. R. Soc. Trop. Med. Hyg. 48:325–331.
———. 1957a. Past and present position of malaria in Swaziland. J. Trop. Med. Hyg. May 1957:3–11.
———. 1957b. Malaria control in Swaziland: Some observations during the first year of partial discontinuation of insecticides. J. Trop. Med. Hyg. August 1957:190–192.
Mattingly, P.F. 1977. Names for the *Anopheles gambiae* complex. Mosq. Syst. 9:323–328.
Mayr, E. 1942. Systematics and the origin of species. Columbia University Press, New York.
———. 1963. Animal species and evolution. Harvard University Press, Cambridge, Mass.
Miles, S.J., C.A. Green, and R.H. Hunt. 1983. Genetic observations on the taxon *Anopheles (Cellia) pharoensis* Theobald (Diptera: Culicidae). J. Trop. Med. Hyg. 86:153–157.
Muirhead-Thomson, R.C. 1945. Studies on the breeding places and control of *Anopheles gambiae* and *A. gambiae* var. *melas* in coastal districts of Sierra Leone. Bull. Entomol. Res. 36:185–252.
———. 1947a. Studies on *Anopheles gambiae* and *A. melas* in and around Lagos. Bull. Entomol. Res. 38:527–558.

———. 1947b. The effects of house spraying with pyrethrum and with DDT on *Anopheles gambiae* and *A. melas* in West Africa. Bull. Entomol. Res. 38:449–464.

———. 1947c. Recent knowledge about malaria vectors in West Africa and their control. Trans. R. Soc. Trop. Med. Hyg. 40:511–536.

———. 1950. DDT and gammexane as residual insecticides. Trans. R. Soc. Trop. Med. Hyg. 43:401–412.

———. 1951. Studies on salt-water and fresh-water *Anopheles gambiae* on the East African coast. Bull. Entomol. Res. 41:487–502.

Parent, M., and M.L. Demoulin. 1945. La faune anopheline a Yangambi (Ineac). Biologie *A. moucheti* Evans specialement. Applications statistiques. Recl. Trav. Sci. Med. Congo Belge No. 3:159.

Paterson, H.E.H. 1962. Status of the East African saltwater-breeding variant of *Anopheles gambiae* Giles. Nature 195:469–470.

———. 1963. The species, species control and antimalarial spraying campaigns. S. Afr. J. Med. Sci. 28:33–44.

———. 1964a. Direct evidence for the specific distinctness of forms A, B and C of the *Anopheles gambiae* complex. Riv. Malar. 43:191–196.

———. 1964b. "Saltwater *Anopheles gambiae*" on Mauritius. Bull. W. H. O. 31:635–644.

———. 1968. Evolutionary and population genetical studies of certain Diptera. Ph.D. dissertation, University of Witwatersrand, Johannesburg.

———. 1985. The recognition concept of species. Pp. 21–30 *in* E.S. Vrba, ed. Species and speciation. Transvaal Museum Monograph No. 4. Pretoria.

Paterson, H.E.H, J.S. Paterson, and G.J. Van Eeden. 1963. A new member of the *Anopheles gambiae* complex. A preliminary report. Med. Proc. 9:414–418.

Rattanarithikul, R., and C.A. Green. 1986. Formal recognition of the species of the *Anopheles maculatus* group (Diptera: Culicidae) occurring in Thailand, including the descriptions of two new species and a preliminary key to females. Mosq. Syst. 18:246–278.

Raubenheimer, D., and T.M. Crowe. 1987. The recognition species concept: Is it really an alternative? S. Afr. J. Sci. 83:530–534.

Ribbands, C.R. 1944. Differences between *Anopheles melas* (*A. gambiae* var. *melas*) and *Anopheles gambiae*. Ann. Trop. Med. Parasit. 38:85–99.

Ross, R. 1897. On some peculiar pigmented cells found in two mosquitoes fed on malarial blood. Br. Med. J. 1786–1788.

Ross, R., H.E. Annett, and E.E. Austen. 1900. Report on the malaria expedition of the Liverpool School of Tropical Medicine and Medical Parasitology. Mem. Liverpool Sch. Trop. Med. 2:22.

Sharp, B.L., D. Le Sueur, and P. Bekker. 1990. Effect of DDT on survival and blood feeding success of *Anopheles arabiensis* in northern KwaZulu, Republic of South Africa. J. Am. Mosq. Contr. Assoc. 6:197–202.

Subbarao, S.K., K. Vasantha, T. Adak, and V.P. Sharma. 1983. *Anopheles culicifacies* complex: Evidence for a new sibling species, species C. Ann. Entomol. Soc. Am. 76:985–988.

Swellengrebel, N.H., and A. De Buck. 1938. Malaria in the Netherlands. Scheltema and Holkens, Amsterdam.

Symes, C.B. 1932. Note on the infectivity, food and breeding waters of anophelines in Kenya. Rec. Med. Res. Lab., Nairobi No. 4:1.

Theobald, F.V. 1903. A monograph of the Culicidae or mosquitoes. Br. Mus. (Nat. Hist.), London.

Toure, Y.T., V. Petrarca, and M. Coluzzi. 1983. Nuove entita del complesso *Anopheles gambiae* in Mali. Parassitologia 25:367–370.

Vincke, I.H., and M. Parent. 1944. Un essai de lutte antimalarienne specifique à Stanleyville. Bull. Assoc. Inq. Faculte Tech. Hainaut. Mons 8:1.

Watson, M. 1915. Rural sanitation in the tropics: Being notes and observations in the Malay Archipelago, Panama and other lands. John Murray, London.

———. 1921. The prevention of malaria in the federated Malay states. John Murray, London.

6

Toward Operationality of a Species Concept

JACK P. HAILMAN

Department of Zoology
University of Wisconsin

A VAST AMOUNT has been and continues to be written about species concepts. I draw from this fact two conclusions: evolutionary scientists agree that the notion of a species is important yet continue to disagree on how a species should be defined. This chapter addresses the discrete issue of whether it is possible to devise an operational approach to defining the "interbreeding" or "mate recognition" criterion of the species concept for sexually reproducing organisms.

On Defining the Sexually Reproducing Species

Most biologists probably believe the sexually reproducing species to be well defined, perhaps even operationally defined. This section presents species definitions, explains what operationalism in biology means, shows why the widely accepted definition of the sexually reproducing species is neither clear nor operational, and reviews some possible approaches to making it operational.

Species Concepts and Evolution

The Latin root of the word *species* refers to outward appearance as in shape and form. Therefore, it is not surprising that early naturalists called organisms that looked alike "a species." In the task of pigeonholing living things, biologists assigned names to such species and began assembling similar species into larger groups. Our ability to "see" similarities and differences has been extended to traits ranging from molecular composition to animal behavior, but the similar/different criterion remains at the heart of most practical sorting of organisms into species. It is difficult to conceive of a meaningful alternative to similarity/difference in classifying asexually reproducing organisms and those known only from the fossil record.

The advent of Darwinism altered the conception of higher taxonomic categories without fundamentally affecting the species notion. Groupings of species had to be based on similarities that reflected common ancestry rather than just any similarities. Although it would be perfectly logical to group together bats with those species of insects and birds that fly (the majority, although not all, insect and bird species fly), such a grouping would not reflect phylogeny. Bats have a more recent common ancestor with other mammals than with either birds or insects, and both insects and birds have more recent common ancestors with nonflying animals than they do with one another. Although the contemporary school of taxonomy called phenetics essentially rejects evolutionary criteria for defining higher categories, most modern systematics is explicitly phylogenetic in purpose. Despite this change in the nature of classification, no similar revolution in how we define species occurred until the middle of the twentieth century.

It is almost always possible to find precursors to any major conceptual breakthrough in the history of biology, but the recognition that a different sort of species concept is possible for sexually reproducing organisms is rightly attributed to Ernst Mayr. In one of the earliest, if not the very first, explicit statements on the subject, Mayr (1940) stated that species are "groups of actually or potentially interbreeding natural populations which are reproductively isolated from other such groups" (see also Mayr 1942). Variants of this definition are still echoed in virtually all modern textbooks of biology (e.g. Campbell 1990; Starr and Taggart 1992; Purves et al. 1992).

There are a few semantic problems with the quoted definition that have occasionally been circumvented by restatements. For example, breeding is most clearly viewed as an activity of individuals rather than populations. Another minor problem is the pleonasm created by the unnecessary specification of reproductive isolation from other such groups, which are obviously of the same species if there is interbreeding. Therefore, with no injustice to the original intent, one could say that species of sexually reproducing organisms are *naturally occurring groups of actually or potentially interbreeding individuals.*

This species definition for sexually reproducing organisms has stood the test of a half-century so well that apparently only Paterson (e.g. 1980, 1981) has tried seriously to reorient it. Paterson has emphasized the criteria used by individuals to recognize a mate: a Specific-Mate Recognition System (SMRS). He sees this SMRS as the fundamental property of a species, established at the time of speciation and essentially unaltered thereafter. Lest too narrow a view be taken of this notion, it is easily extended to recognition among gametes for plants and those animals that shed gametes into the environment without coming together for spawning or copulation.

Paterson's species notion rests upon prezygotic mechanisms and therefore denies any possible role of reinforcing selection. It is possible under Mayr's definition that breeding between individuals of different species is inhibited by postzygotic mechanisms, such as hybrid embryos that die or hybrid offspring that are sterile. It is therefore possible to reason that would-be parents engaging in such ill-fated matches are wasting valuable gametes, and hence such mating attempts would be selected against. Such selection would promote refinement of mate recognition, and it is specifically this possibility that Paterson argues against.

It is not a purpose of this essay to compare the two foregoing views of sexually reproducing species or to answer the question as to whether reinforcing selection for mate recognition occurs. I have argued elsewhere against reinforcing selection in a more general context (Hailman 1982). Instead, the present task is first to point out that neither species concept is operationally defined (for the same reason) and then to see if an operational approach is feasible. In fact, from the viewpoint as to where the concepts founder, there is no fundamental distinction between Mayr's and Paterson's sexually reproducing species; they differ only in ways that are not the concern of this essay, so will be treated for present purposes as variants of the same concept.

Operationalism in Evolutionary Biology

Operationalism as a method for assuring communicability of scientific concepts grew naturally out of physics early in the twentieth century (Bridgman 1927, 1938, 1945). Physicists could not see and handle entities such as electrons and so had to define them in terms of measurements made by instruments. Furthermore, physicists quickly recognized that even the measured value of a variable had no precise meaning unless the operations by which it was measured were made explicit. For example, the speed of an airplane as measured by an instrument on board that senses how fast air rushes by the plane delivers one value whereas the distance between departure and arrival points divided by the time taken to make the flight may give a quite different answer. In order to communicate a variable such as "speed" one must stipulate the operations by which it is measured. Physicists had to define entities such as "electron" by criterion values of measured variables. And causal relationships were defined as logical functions relating measured variables. These three things—variables, criterion values, and functions—are the stuff of which science is made, and they will be communicable and hence "objective" only if based on explicit procedures of measurement.

There is a philosophical concern connected with operationalism that is not at issue here. Logical positivists—some principal figures being Bridgman, Alfred North Whitehead, and Bertrand Russell—tended to consider the "real" world as nothing more than our perception of it: things exist only if we can verify them empirically. I am not concerned here with philosophical issues of reality, but only with communicability among scientists. I take the view that our models of the external world are communicable, and therefore "objective," only if they are based in principle on measurable variables.

Biologists, being able to see and handle many of their plants and animals, were slower than were physicists to build their component sciences on operational principles. After all, it was (in most cases) disarmingly obvious what a biological entity such as an individual animal was, so why waste time trying to define it in some way based on measurements? Troublesome cases were decided more or less by fiat. For example, every introductory biology textbook until recent times simply announced to students that siphonophores such as the Portuguese man-of-war (*P. physalia*) were colonies rather than individual organisms—without making explicit how this counter-intuitive judgment was reached. In point of fact, many different operational definitions of an individual are possible, and by some a man-of-war is a colony and by others it is an individual organism. Just as an airplane can travel at two different speeds simultaneously, so is a Portuguese man-of-war simultaneously an individual and a colony—depending upon the definitions used.

Explicit operationalism is being increasingly adopted in biology today, although evolutionary studies have been slow to adopt. The tardiness may be due to the complexity of evolutionary concepts and lack of recognition of the power of operational thinking. Still, many attempts to make evolutionary biology operational are in evidence, as in algorithms for determining phylogenetic trees; concepts such as homology, analogy, and their kin (Hailman 1976); and the notion of adaptation (Hailman 1988). Whenever it is vague as to how one should determine whether a given case meets a stipulated criterion, then the criterion is likely to be improved by more explicit operationalism. The criterion of "interbreeding" or "mate recognition" in species definitions for sexually reproducing organisms is a case in point.

Problems with "Interbreeding" and "Mate Recognition"

The problem with "interbreeding" or "mate recognition" criteria is remarkably easy to state, yet I have never encountered such a statement in the literature. Put as a query, one may ask how to judge whether or not two groups of natural populations are interbreeding (or potentially interbreeding). Or, how is one to judge whether the individuals of a naturally occurring group share a mate recognition system?

Pursuing these questions perhaps a bit further than necessary, it is obvious that all members of one population could not possibly interbreed with all members of another. This is one of the reasons that Mayr's definition contains the phrase "potentially interbreeding." Even that caveat simply compounds the problem because the definition does not reveal how to assess potentiality. Furthermore, except in hermaphroditic organisms and those (like some wrasses and parrotfishes) that may change sex, sexually reproducing organisms cannot even interbreed with all members of their own population. This latter problem could be circumvented by adding the stipulation that interbreeding connotes "with the opposite sex," a point that is more clearly implied by Paterson's "mate recognition" criterion. Both formations of species, however, suffer from the same vagueness: as not every male could possibly interbreed with or recognize as a potential mate every female in a population (unless it were quite small), and similarly not every female could mate with or recognize as a potential mate every male, the criteria of interbreeding and recognition are simply undefined.

If stipulated criteria of a definition are unspecified, then there is no assurance that two equally competent persons will arrive at the same end point when attempting to apply the definition. Suppose there exists a population of, say, a hundred small flycatchers equally divided between the sexes, males of which sing a distinctive song. If one person adopts the criterion "flying toward a playback speaker broadcasting a recording of the song" as indicating potential interbreeding or mate recognition on the part of females, some percentage of females will qualify. Another person might decide to inject all the females with estrogen and then see if they respond to the playback. Suppose, then, these two procedures identify different collections of females, which have some but not all individuals in common (a likely result of such procedures). Which collection is the species? If both procedures are deemed reasonable interpretations of "potentially interbreeding" or "mate recognition," then both collections define different and overlapping species. That conclusion violates the notion of reproductive isolation and the implication (of all species concepts) that every individual in nature belongs to one and only one species.

Definitions do not necessarily have to be explicit about measuring operations in order to be considered operational, so long as those operations are clearly implied. However, in the case of definitions of species in sexually reproducing organisms, no operations for assessing "interbreeding" or "mate recognition" are implied. As the critical criterion is ambiguous in both cases, no definitions logically exist. The road out of this tangle is toward operationalism, and it forks immediately into two paths.

Operational Approaches to a Species Concept

The alternative paths toward operationality are to abandon notions such as interbreeding and mate recognition on the one hand, or to make them operational on the other. As an example of pursuing the first path, one might consider defining a sexually reproducing species as the smallest group of organisms that was consistently distinct in morphology from all other groups (Cronquist 1978:15). This definition, if taken seriously, would elevate males and females of sexually dimorphic forms to species status as well as create other ludicrous conundrums. Apologists for cladistic analyses of phylogeny get into similar problems when defining species "objectively," and Mayr (1988:326–328) has laid out some telling objections.

To return to any morphologically based, similarity/difference approach to species, even if based explicitly on measurable variables, would be throwing the baby out with the bath. There *is* something special about sexually reproducing organisms, which Mayr and Paterson have attempted to finger. In order to preserve that something special, one needs to try to define it in some more operationally oriented manner. Notions of "interbreeding" or "mate recognition" seem to imply that members of the same species are connected with one another in some way that involves the reproductive process. It is the nature of this connection that we seek to clarify and assess, and there are various types of candidate connections.

For example, many species have apparently species-specific pathogens or parasites. Those already existing are of no help, for they merely constitute an ordinary taxonomic character like some part of the anatomy. However, one could in principle create genetically new pathogens and parasites, introduce these to an individual, and chart their spread. The collection of organisms infected at asymptotic spread would then define the species. Some problems with this approach are obvious: some individuals may be sufficiently resistant that the infecting agent cannot be detected in them, and in many cases pathogens spread across what we consider to be species boundaries (e.g. flu viruses of man come from birds as well as other mammals).

As a more relevant attempt to assess reproductive networks, it might be possible to introduce a new and beneficial gene into one individual and chart its spread to new individuals of succeeding generations. Not many years ago this suggestion would have been dismissed as fantasy, but with modern genetic engineering it is today merely futuristic rather than unthinkable. The spread of the beneficial gene in succeeding generations would sooner or later halt, and all those individuals possessing it could be defined as conspecific. Some drawbacks of this approach are immediately obvious: one does not want to muck around with the gene pools of wild species (we have already altered their environments immensely), and the time required to reach asymptote in the geographic spread of the beneficial gene might render the method impractical for all but very small populations. Worse yet is the high probability, due to genetic factors such as random mutation and laws of Mendelian inheritance, that the beneficial gene never would reach all members of what we consider to be the same species. Of course, if the beneficial gene endowed the possessors with such selective superiority that all those lacking it would eventually disappear from the species, the method might work. But the wait for the results of selection would be even longer than the wait for genetic spreading, so the whole notion must be abandoned as manifestly impractical.

It would appear that if reproductive connectivity is to be made operational, the only feasible route involves assessment of actual breeding history. Even in highly polygynous and the few polyandrous species, no individual mates with more than a relatively few individuals of the opposite sex. Although sib-sib matings can occur, brothers and sisters are connected reproductively by virtue of the reproductive processes of their parents.

Suppose, then, one attempted to define species in terms of genealogical connectivity. If I had a brother (which I do not), and he married into the Smith family while I married a Brown girl, the Hailmans, Smiths, and Browns could all be connected through the offspring of my brother and me, and our common parents. Could one not pursue such connectivity to create a network that would define the species? Disregarding temporarily the obvious impracticality of working out genealogical relationships for most animal populations, the connectivity approach presents other problems. How far back must records go to create the genealogical network? The answer clearly depends on the degree of outbreeding within local areas of a species' range, and for widely distributed species the number of generations required might be enormous. Nonetheless, this approach may not present unsurmountable problems. Those of us who are genealogical buffs frequently find common ancestors within fifteen or so generations back: thus I discovered quite fortuitously that I am related to one of my former doctoral students and to the former wife of a colleague in my department.

A more troublesome problem relates to the required multigenerational data. Consider two species of chickadees (*Parus*) that are believed to be closely related (and in fact occasionally hybridize in certain areas of contact): the Carolina (*P. carolinensis*) and black-capped (*P. atricapillus*) chickadees. The latter, in particular, has such a large geographic range and such a low dispersal distance that it is conceivable that one might have to go back genealogically past the point of their separation from a common ancestor in order to connect all the individuals within one of the two species. We might therefore end up concluding that all members of both species were in fact members of the same species. The connectivity approach might succeed only if connectivity is such as to "network" currently living individuals without going far enough back in genealogical history to the last speciation event.

Indeed, the universality of the genetic code of base-pair triplets determining amino acids suggests that all life sprang from a common source, so that all individuals of every species show connectivity if one goes back enough generations. So, for the connectivity approach to succeed logically it requires that all individuals of a species be connected by links occurring since the last speciation event. It might be that this required property exists universally in biological organisms—there are in fact ancillary reasons to believe that it does—but insofar as I can tell there is no proof. Furthermore, it is necessary to identify when in the genealogical history the last speciation event occurred, for one has to have a criterion for deciding when to stop making the retrograde links.

An Example: Florida Scrub Jay Connectivity

With the kind permission of Glen E. Woolfenden and John W. Fitzpatrick, I have analyzed their genealogical records on the Florida scrub jay (*Aphelocoma c. coerulescens*) to assess the feasibility of a connectivity approach to defining a sexually reproducing species. The goal of the analysis was to see how connectivity among individuals increases as a function of increasingly distant connections.

Biology of the Florida Scrub Jay

The Florida scrub jay is a relatively large passerine bird that is restricted to oak scrub habitat of the Florida peninsula. In western United States and Mexico scrub jays classified in various other subspecies live in similar, but more diverse, habitats—separated from the Florida peninsula by about 1500 miles in which no similar habitat and no scrub jays occur. Apparently, a once-continuous range was bifurcated by Pleistocene glaciations in mid-continent. The timing of the last speciation event is unknown, but the only extant sympatric congener is the gray-breasted, formerly Mexican, jay (*A. ultramarina*) of Mexico, which occurs as far north as southern Arizona, where its range overlaps that of the scrub jay, and apparently no hybrids have ever been reported. A third member of the genus, the unicolored jay (*A. unicolor*) occurs in humid montane forests and oak-pine habitat of Central America.

The Florida scrub jay is basically a nonmigratory, permanently territorial, permanently paired, permanently monogamous resident of the Florida oak scrub, which is completely saturated by conterminous jay territories (Woolfenden and Fitzpatrick 1984). Unlike western scrub jays, where young birds survive in peripheral habitats that apparently will not support successful nesting, Florida scrub jays stay on their natal territories or become adopted by breeding pairs other than their genetic parents. Those few young Florida scrub jays that attempt to establish territories or home ranges outside of the scrub soon disappear, apparently mainly being taken by hawks. It thus appears that Florida does not have suitable peripheral habitat for survival, so nonbreeding jays must remain as "helpers" on the territory of a breeding pair until they can become breeders by pairing with a widow or widower on a nearby territory (or by a few other means). Dispersal is low, with males often becoming breeders in a territory adjacent to their natal territory and females averaging only one territory removed. There are cases, however, of banded individuals moving as far as about 15 km. Prebreeding mortality is high, as in almost all animals, but once a Florida scrub jay becomes a breeding bird it may survive for a decade or longer, and lifetime reproductive output of a few individuals can thus be impressively high.

A few more empirical findings about reproduction are necessary background. As stated above, Florida scrub jays are "basically" monogamous and permanently paired; rare exceptions do occur. There are known to me two documented cases of bigamy, one (not on the study tract) that has persisted for several years. There are also very rare cases of divorce and repairing. A breeding bird commonly, however, loses a mate (presumably to predation) and always attempts to re-pair, remating being apparently more successful among males than females. Therefore, a given bird may have several mates over its lifetime. Finally, there is strong incest avoidance among the jays: fathers never pair with their daughters nor mothers with their sons. Brother and sister never pair if growing up together on their parents' territory. If, however, one sibling leaves the territory to become a breeder before the other hatches, and the first subsequently loses its mate, it could pair with the younger sibling. Such pairings have happened a few times in the nearly quarter century of the population study but were not encountered in my analysis.

Operations of the Connectivity Analysis

The heart of my methodology was to connect individual jays by "links." In the present analysis there is only one type of link: parent-offspring. Thus the two parents of an individual are not linked because they paired, but rather because they are each linked to the offspring. The connectivity of parents is thus two units, that of parent and offspring one unit. Graphically, linkage may be represented by an arrow connecting parent and offspring, with the arrowhead at the offspring end. (This is merely one type of diagram used commonly in genealogic and genetic studies, the explanation a bit labored here to make clear the operations behind my analysis.) In short, linkage requires successful reproduction; even pairs that built a nest, laid eggs, and had hatchlings were not considered to be linked in my analysis—unless at least one of the hatchlings subsequently fledged and lived long enough to acquire its full complement of color bands from Woolfenden (about three months of age). My linkage criterion was therefore stringent.

The question naturally arises as to whether the reputed parents are actually the genetic parents of offspring that come from eggs in their nest. At least two potentially confounding factors could exist: the female may have received extra-pair copulation(s) with a male or males other than her mate, or eggs in the nest might have been laid by some other female. Neither factor appears to be of concern in the Florida scrub jay. Woolfenden and coworkers (pers. comm.) have begun DNA fingerprinting studies of the jays and, to the surprise of no jay researcher, have found so far no evidence of any young being of parentage other than the breeding pair. Although a final statement on this issue must await more extensive samples and analysis, it appears now that if extra-pair copulations do occur in Florida scrub jays they are so rare as to be of no consequence for connectivity studies.

The other possibility—eggs being laid by a female other than the breeding female—seems similarly remote. Egg dumping is known for a few species of birds (mainly nonpasserines), where it constitutes a sort of intraspecific nest parasitism, but not even the suspicion of this phenomenon has been raised with Florida scrub jays. Clutch sizes show a symmetrical, unimodal distribution, with no indication of supernumerary eggs. Furthermore, territorial intrusions are immediately challenged, with no individuals other than the breeding pair and their helpers allowed to remain once detected. Helper females are driven away from the nest by the breeding female during nest-building, egg-laying, and incubation, being allowed at the nest only after the eggs have hatched (at which point helpers commonly bring food to the nestlings). And male breeders actively interfere with any courtship display, manipulation of nesting material, or other signs of breeding in helpers on their territory (pers. obs.). The one type of exception to all this evidence is the very rare occurrence of polygyny (only bigamy) where two females act as if they were breeders and both might contribute eggs to the clutch. There are no such cases of this in the sample I analyzed, and the sole, well-documented case (R.L. Mumme, pers. comm.) is not in jays of the main study tract of Woolfenden and Fitzpatrick. DNA fingerprinting would be as sensitive to eggs from other than the breeding female as it is to extra-pair copulations, and as noted above, results to date show no evidence of any complicating factors.

Assuming, then, that the breeding pair of a given territory are the parents of the offspring in the nest of that territory, connectivity based on census data is genetic connectivity. There are many possible starting points and rules for developing connectivity, the following being those I used for this analysis. The geographic area of study was defined as that area of Archbold Biological Station (Highlands County, Florida) censused monthly by Woolfenden, Fitzpatrick, and their co-workers. This has since become known as the "demography tract" to distinguish it from other areas on the large station, where Florida scrub jays are also under study by a variety of workers. I took one point in time (January 1989, when I began work on this chapter) and listed all birds on the tract, as shown by the census for that month, as the starting point.

It turned out that this choice of study birds from the beginning of 1989 represented a kind of worst-case scenario because of dispersal into the study tract of birds whose ancestry is completely unknown. In retrospect, perhaps I should have chosen birds from about 1980, thus providing a decade of ancestry and a decade of descent. My analysis is thus restricted to a pedigree network connecting ancestors and is thereby robbed of some portion of its power to connect birds. In order to alleviate (but unfortunately not eliminate) the dispersal problem, I discarded from analysis 28 first- or second-year birds that had dispersed into the study tract from elsewhere and were too young to have bred. The remaining 109 individuals thus became the study population.

The 109 birds were designated by their color-band combinations and written down (in computer graphics). Parent-offspring arrows were then drawn among appropriate birds from past census data. (That statement makes it seem easier than it actually was. There existed only hardcopy monthly census sheets for about twenty years. From these, as needed during the progression of the analysis, I created a database of genealogical relationships in Hypercard for the Apple Macintosh computer. The task proved to be an undertaking of stagger-

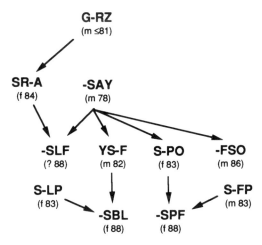

Fig. 6.1. An example of a connectivity "group" of eleven Florida scrub jays defined by one unit of parent-offspring links (arrow from parent to offspring). Color-band combinations always contain three letters and one hyphen (the "dash"), one of the letters being S or Z (silver and zilber, respectively, the U.S. Fish and Wildlife Service–numbered band). Other colors are: A, azure; B, blue; F, flesh; G, green; L, lime; O, orange; P, purple; R, red; Y, yellow; and W (not shown), white. Below each band combination is the sex and year of hatching of the bird; the designation ≤81 (G-RZ at top) means hatched in 1981 or an earlier year. Some sexes of young birds were not known when the analysis commenced in 1989. This is a particularly large group defined by first-order connectivity, primarily due to -SAY, a male that was 11 years old at the starting date and is still alive (as of 1993).

ing proportions, considering the repeated manual searches required in hundreds of raw census sheets. The Hypercard stack has more than 1600 cards, each representing an individual jay, with a summary of its territorial home by month, its parents, its breeding status each year, its mate each year, and the offspring produced each year. Search time to locate a particular card can exceed ten minutes.) Each set of birds that was connected by arrows formed a "group" as exemplified by Figure 6.1.

Note that the methodology is not purely one of common ancestry. In the figure, S-FP (lower right) is connected with all the other birds because he is the father of -SPF, whose mother (S-PO) provides the critical link to other birds through her father (-SAY). It might be, for example, that S-FP shares an ancestor with G-RZ (top of the diagram), but no evidence in the records attests to this possibility.

I then began tracing every parent and every offspring of the 109 birds, a process that began linking previously separate groups through birds that had died before the starting date for the study. (In order to do this I had to create a second database for each group membership so that included birds could be searched for by computer.) Figure 6.2 shows how an intermediary bird (WLS-)

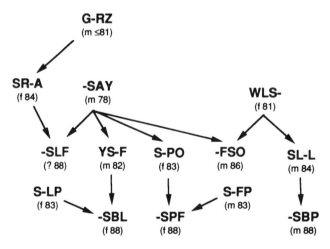

Fig. 6.2. An example of enlarging a group by finding a link (in this case, a common ancestor) in genealogical searches. The birds SL-L and -SBP are added because WLS- was the mother of both -FSO and SL-L. The original group shown in Fig. 6.1 was actually enlarged by many such second-order connections (not shown), so as to encompass 26 of the 109 birds alive in 1989 (the largest such group resulting from second-order linkages). Full connectivity diagrams are far more complicated because of the number of nonstudy birds that are ancestors providing linkages and because two birds can be connected by more than one path of arrows.

connected two previously separate groups, this bird being the mother of both SL-L (hatched in 1984) and -FSO (hatched in 1986).

The process of searching and connecting continued until no further connections could be made. One limitation encountered was that the study tract increased in area over the two decades of data, so as one moved backward in time the genealogical connections became increasingly sparse. Woolfenden began study of the jays in 1969, banding adults of unknown age and parentage, but because the jays are long lived, it was well into the 1980s before the entire population consisted of birds hatched on the study tract (parentage and age known) or young birds dispersing into it (age only known, sometimes not precisely). Again, in retrospect these problems would be less serious if one analyzed, say, birds from 1980 instead of 1989, and worked forward as well as backward in developing connectivity.

Results of the Analysis

The results can be summarized graphically by plotting the number of groups created by parent-offspring linkages as a function of the number of links required to form the groups, as shown in Figure 6.3. Because of the operations I used to find linkages—essentially stopping a line of pursuit when a linkage between two individuals was found—the results have no assurance of parsimony. I also did not attempt to devise an algorithm for path analysis to find the minimum path among those graphed. Therefore, Figure 6.3 really represents the maximum number of groups: the real number of groups is the value plotted or some smaller number.

Figure 6.3 shows that as the number of links increases, the number of groups decreases to a minimum, which in this case was ten groups, reached after seven linkage units. Nearly two-thirds of the birds ($68/109 = 62.4$ percent) formed the largest group, with the next largest being 14 birds, and the remaining eight groups having 5 or fewer birds each (Table 6.1). At a linkage of just three units, nearly half ($50/109 = 45.9$ percent) of the birds formed a single group.

The results provide a satisfactory but qualified answer to the posed question: How far back in time does one need to go in order to show connectivity of birds alive at a given point in time? The oldest bird required for asymptotic linkage (which was within the largest group) was one banded as an adult when Woolfenden began the study in 1969. All of the small groups remain unlinked to the large one because pedigree analysis stops with an immigrant from outside the study tract. No small group depends for its internal linkage upon birds banded more than about a decade ago, and most have internal linkages of only a few years back.

Therefore, despite the multiple limitations encountered in the analysis, about two previous decades of reproductive connectivity is sufficient to link all

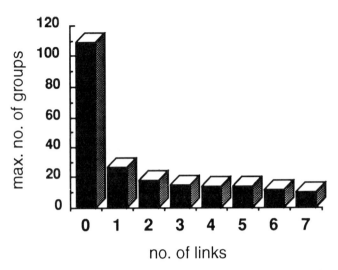

Fig. 6.3. The number of groups defined by increasing numbers of links. The function falls to a minimum of ten groups at seventh-order linkage, where genealogical data were exhausted. The largest group at seventh-order linkage contains two-thirds of the individuals (Table 6.1). As discussed in the text, the number of groups at any linkage between 1 and 6, inclusive, could be smaller than that shown because no formal algorithm was used to find the most efficient paths of linkages; hence the vertical axis is labeled as the "maximum" number of groups.

Table 6.1. Distributions of group sizes as linkage increases

Group size	Number of links							
	0	1	2	3	4	5	6	7
1	109	6	5	4	3	3	2	2
2		3						
3		7	4	4	4	4	4	4
4		4	3	2	2	2	2	2
5		3	2	2	2	2	2	1
7		2						
10			1					
11				1	1	1		
14			1	1	1	1	1	1
15		1						
16		1						
20			1					
26			1					
50					1			
51						1	1	
63							1	
68								1

individuals whose pedigree does not terminate with immigrant dispersers. Because the study tract has arbitrary geographic boundaries, there is no reason to posit any fundamental difference between birds that disperse within the study tract (known ancestry) and those that disperse into it from outside (unknown ancestry). Therefore, two decades is a reasonable general figure for the time required to connect, by pedigree network, all the Florida scrub jays alive at a given point in time and space. If this figure were misestimated by an order of magnitude, 200 years would still fall well before the last speciation event. Indeed, even 2000 years would provide sufficient assurance that Florida scrub jays can be connected reproductively without nearly approaching the timing of the last speciation event.

Conclusions

By way of concluding this essay, it is useful to point out that details of the connectivity operations could be changed and to comment on some things that an operational species definition does not accomplish. Then a summary formulation of the species definition is offered.

Connectivity Operations

Details of the operations used to measure connectivity in the Florida scrub jay example are not the only possible ones, and there is no guarantee the choice was optimal. For example, one could use different criteria for linkages. The one used required the production of independent young that lived at least approximately three months, but one could easily set a looser criterion at the production of eggs or a tighter one at offspring survival to a year of age.

The analysis made is basically that of a pedigree network, but variants and alternatives are possible. A pedigree *sensu stricto* is the ancestry of a given individual, but the connectivity method used is not based solely upon common ancestry. Parents, for example, are connected by two links through their common offspring, even when no common ancestor of the two parents was found by the analysis (e.g. Fig. 6.2). The known pedigree of each focal bird was traced, but these pedigrees were connected with one another by collateral relatives as well as common ancestors. Hence the method may be called that of a "pedigree network."

Furthermore, there might be merit in working forward in time in addition to the pedigree-network approach used. Working forward is what genealogists refer to as family-history analysis: beginning with an individual (or couple) and listing all their descendants. A "family-history network" would be conceptually parallel with the pedigree network, in which each focal individual's descendants are traced and those lineages are interrelated by links to collateral relatives. In order to assess quickly the potential of this family-history approach, I looked at raw census data for the breeding season of 1989 (the one immediately following the January 1989 census from which the 109 study birds were taken). The 1989 breedings connect at least one of the smaller final groups of Figure 6.3 with the large group. The breedings also connect some of the 28 young immigrants excluded from analysis with existing groups, and make links among other of the immigrants. Thus, as might be expected, a single forward year improves connectivity quickly and reinforces the rationale for doing such analyses on birds that were alive about a decade ago so that links could be forged forward in time—either instead of, or in conjunction with, pedigree-network analysis.

Finally, the example of the Florida scrub jay connectivity suggests the possibility of using different criterion values for accepting the generality that the individuals investigated are all connected by breeding. Perhaps it would rarely or never happen that every focal animal could be connected to every other focal animal because of a variety of practical problems such as those encountered in the jay example. Therefore, one could develop criteria which, if met, justify the conclusion that all focal animals are of the same species—and if not met, leave that conclusion unsupported. The parallel with hypothesis testing in statistics is obvious, where there is general agreement on an arbitrary criterion of $\leq .05$ for a probability value to be considered "significant." For example, Figure 6.3 shows that a linkage of one unit makes a dramatic effect upon creating groups. In that case the first link reduces the number of groups (i.e., the number of individuals at the outset) to about a quarter of the total ($27/109 = 24.8$ percent to be precise). One could establish a rule of thumb by which first-order analysis had to collapse the number of unconnected groups to some criterion value, such as 25 percent or less.

Table 6.1 suggests at least two other approaches to judging the completeness of connectivity. A criterion could be established requiring that a certain percentage of the individuals analyzed fall into one interconnected group. In the case of the scrub jays, nearly two-thirds of the birds (actually $68/109 = 62.4$ percent) were in a single group when the analysis reached the limits of available data at seventh-order linkage. Another approach would be to require that a certain large percentage of the individuals be connected with at least one other individual. Here, the usual statistical alpha level might be a convenient value of choice. If one asked of the jay data that 95 percent of the birds be connected with at least one other bird of the focal animals, this criterion requires that fewer than $0.05 \times 109 = 5.45$ birds be in group sizes of one. Table 6.1 shows that this criterion is met by second-order linkage.

What Connectivity Cannot Accomplish

The connectivity approach is intended to rescue the definition of the sexually reproducing species from its inherent ambiguity of stipulations such as "potentially interbreeding" and "mate recognition." In order to understand that it is intended as no more than this, consider some specific tasks that connectivity does not accomplish.

First, connectivity is not proposed as a practical tool of taxonomy. Decisions as to how many species should be recognized and how individual organisms should be assigned to each always have been and will remain fundamentally based on similar/different criteria. This situation will not be changed regardless of whether classical morphological traits are used or these are augmented by characters derived from behavioral data, nucleotide sequences, or other kinds of studies. If for no other reason, the millions and millions of species of living organisms present a task so great that no alternative to similarity/difference is conceivable. What purveyors of the species concept for sexually reproducing organisms can hope to accomplish is a taxonomic emphasis on traits that reflect breeding connectivity.

Second, the reasonableness of connectivity as an operational statement of what is really intended by vague terms such as *interbreeding* and *mate recognition* cannot be assessed in just any species. The Florida scrub jay was chosen for analysis because relevant data were available. Such a test of the connectivity approach would not be feasible in most sexually reproducing species for a whole host of reasons. Consider two other well-studied avian species. Even though the black-capped chickadee (*Parus atricapillus*), like the Florida scrub jay, is nonmigratory, banding studies suggest that the young typically disperse about 5 km in a random direction as soon as they are weaned. In our study population of color-banded birds in Madison, Wisconsin, we have located subsequently only a few individuals that were hatched in the study area. Furthermore, chickadees are not nearly as long lived as scrub jays, making linkages difficult. Migratory animals present further problems in determining linkages. Some young birds return to the vicinity of their natal territories, but others seem not to do this. The snow goose (*Chen hyperborea*) in North America pairs on the wintering grounds in areas bordering the Gulf of Mexico, and the pair tends to return to the natal colony of the female in subarctic Canada. Therefore, connectivity analysis is practical only through matrilineal links. Oceanic animals, including many coral-reef and pelagic fishes as well as various invertebrates, have planktonic young. Even if the use of genetic markers were feasible, finding offspring of specific parents would clearly be impossible.

Third, connectivity cannot solve the problem of allopatric populations. The "potentially" part of the interbreeding notion in the species definition for sexually reproducing organisms is a double-edged sword. This essay has addressed the edge concerning what the notion of interbreeding (or mate recognition) really means in a population where no animal could possibly mate with or even court all members of the opposite sex. The other aspect of "potentially interbreeding" refers to individuals that, due to their geographic separation, will never meet and hence physically could not possibly pair, much less produce offspring. Whether or not to consider allopatric forms as the same species because (in the judgment of researchers) they would/will interbreed if/when coming into contact is an old problem that has been dealt with in many ways. Perhaps the most sensible solution, in terms of practical taxonomy, is to recognize a category of allospecies—closely related and highly similar, but geographically separate, populations (see Mayr [1980] for a discussion concerning avian allospecies). It seems unlikely that connectivity or any possible alternative operational approach to defining "potentially interbreeding" or "mate recognition" could solve the problem as to whether similar but allopatric populations are of one or more species.

Last, connectivity would not seem to portend any fundamental changes to the way in which hybridization is treated. Matings between individuals considered to be of different species ordinarily fall into two general categories. There may be occasional matings anywhere within a broad sympatric area of two species, or the matings may occur principally in contact areas between two largely allopatric species. In either case the decision as to whether one or two species are really involved is made arbitrarily according to the frequency of hybridization. Connectivity networks might provide new ways of viewing the degree of hybridization and its potential effect on the gene pools involved, but is unlikely to affect the way in which hybridization rates are actually assessed in the field or the taxonomic decisions based upon such information.

The Species of Sexually Reproducing Organisms

This essay has sought to eliminate the inherent ambiguity in the definition of the sexually reproducing species embedded in such undefined notions as "interbreeding" and "mate recognition." Several operational alternatives were considered, and from them "breeding connectivity" was chosen as the most promising and most attune with the underlying views of Mayr and Paterson. Therefore, it is possible to propose that a species of sexually reproducing organisms may be considered a *naturally occurring group of individuals that are linked by a breeding connectivity network*. "Breeding connectivity network" is operational in that it implies a specific family of measurement procedures based on genealogical data. The uncovered linkages *result from* individuals of the opposite sex, or their gametes, within a network being preferentially attracted to one another (mate recognition). The linkages also *result in* networks that are discontinuous with one another (interbreeding and reproductive isolation).

Summary

Definitions of the species in sexually reproducing organisms focus on "potential interbreeding" (Mayr) or "mate recognition" (Paterson). As not every individual of one sex could possibly even attempt to mate with every individual of the other sex, no clear meaning can be attached to the "interbreeding" or "recognition" stipulation. Nevertheless, the notions underlying such terms seem to imply an important property that would be discarded by reverting to any procedure that relies solely upon similar/different criteria for defining species in sexually reproducing organisms. Several candidate approaches to operationalizing "interbreeding" and "recognition" concepts founder either in failing to capture the essence of the concepts or in being so impractical that empirical verification would be impossible. Breeding connectivity based on genealogical information provides the most promising approach, which was evaluated for the monogamous Florida scrub jay (*Aphelocoma c. coerulescens*) from data gathered by Woolfenden and Fitzpatrick over twenty years. Analysis revealed numerous practical constraints that render such data sets less than ideal, but connectivity based on pedigree did link most of the birds living on the study tract at one specific point in time (January 1989). Suggestions for improving connectivity analyses were offered, and tasks not accomplished by operationalizing the species concept were made explicit. In conclusion, one might offer for consideration the definition of a species in sexually reproducing organisms as a "naturally occurring group of individuals that are linked by a breeding connectivity network."

Acknowledgments

I am grateful above all to Hugh Paterson, who in 1979 convinced me (not without considerable effort) that many of my ideas concerning evolutionary biology were strikingly convergent on his own. I want also to mention that Ernst Mayr first interested me (as an undergraduate) in species concepts. Glen Woolfenden and John Fitzpatrick kindly provided the data on which my analysis of connectivity was based, and my wife, Liz, spent many evenings and weekends helping me create the database and do the genealogical searches underlying the analysis. For their accommodating patience through two years of my reluctant involvement in academic administration, I also thank editors Lambert and Spencer. Comments on the first draft of the manuscript were provided by T. Garland, M. Lambrechts, E. Mayr, A.G. Rodrigo, and G.E. Woolfenden; I hope to have met their helpful criticisms satisfactorily.

References

Bridgman, P.W. 1927. The logic of modern physics. Macmillan, New York.
———. 1938. The intelligent individual and society. Macmillan, New York.
———. 1945. Some general principles of operational analysis. Psychol. Rev. 52:246.
Campbell, N.A. 1990. Biology. 2d ed. Benjamin/Cummings, Redwood City, Calif.
Cronquist, A. 1978. Once again, what is a species? In: Biosystematics in agriculture. Beltsville symposia in agricultural research. Vol. 2. John Wiley, New York.
Hailman, J.P. 1976. Homology: Logic, information and efficiency. Pp. 181–198 *in* R.B. Masterton, W. Hodos, and H. Jerison, eds. Evolution, brain, and behavior: Persistent problems. Lawrence Erlbaum, Hillsdale, N.J.
———. 1982. Evolution and behavior: An iconoclastic view. Pp. 205–254 *in* H.C. Plotkin, ed. Learning, development, and culture. John Wiley, London.
———. 1988. Operationalism, optimality and optimism: Suitabilities versus adaptations of organisms. Pp. 85–116 *in* M.-W. Ho and S.W. Fox, eds. Evolutionary processes and metaphors. John Wiley, New York.
Mayr, E. 1940. Speciation phenomena in birds. Am. Nat. 74:249–278.
———. 1942. Systematics and the origin of species. Columbia University Press, New York.
———. 1980. Problems of the classification of birds: A progress report. Pp. 95–112 *in* Acta XVII Congr. Int. Ornithol., Deutsche Ornithologen Gellschaft, Berlin.
———. 1988. Toward a new philosophy of biology: Observations of an evolutionist. Harvard University Press, Cambridge, Mass.
Paterson, H.E.H. 1980. A comment on mate recognition systems. Evolution 34:330–331.
———. 1981. The continuing search for the unknown and unknowable: A critique of contemporary ideas on speciation. S. Afr. J. Sci. 77:113–119.
Purves, W.K., G.H. Orians, and H.C. Heller. 1992. Life: The science of biology. 3d ed. Sinauer, Sunderland, Mass.
Starr, C., and R. Taggart. 1992. Biology: The unity and diversity of life. 6th ed. Wadsworth, Belmont, Calif.
Woolfenden, G.W., and J.W. Fitzpatrick. 1984. The Florida scrub jay: Demography of a cooperative-breeding bird. Princeton University Press, Princeton, N.J.

7

Fitness and the Sexual Environment

HAMPTON L. CARSON

Department of Genetics and Molecular Biology
University of Hawaii

D<small>ARWINIAN FITNESS</small> is a property of the sexually produced individual organism, such as a single human person, *Drosophila* fly, or plant. The concept expresses the capacity of that individual, relative to his or her fellow members of a local population, to make a genetic contribution to the ensuing generation. As so defined, fitness is a fundamental tenet of population genetics and evolution. This property relates to *genetic* components rather than direct effects of the environment on the soma of the individual. In most cases, fitness clearly has a polygenic basis. In gathering empirical data on the genetic basis of fitness, the population geneticist has relied mostly on measurements of survival, fertility, or fecundity, properties of an individual that are relatively easy to measure.

More important, however, and more difficult to measure, are crucial physiological attributes such as sexual behavior and sexual activity that are directly proximal to the sexual union prior to reproduction. These acquire exaggerated importance for the individual during the brief time in its life that attempts are made to engage in sexual reproduction. Although the term *fitness* has generally been used as an all-inclusive term, I choose to emphasize the strong influence of components proximate to the sex act by invoking the concept of the *sexual environment* as contrasted with the *ambient environment*. Data on the intrademic genetic variability of such attributes are needed. Here I discuss this subject in the light of some of Hugh Paterson's ideas.

Mate Recognition

In 1976, Hugh Paterson first advanced some provocative ideas on the complex syndrome of characters manifested by potential mates in animal populations. These considerations arose largely from his view that efficient sexual reproduction is of paramount importance for the species. He argued that reproductive success is served particularly well within each species by "a specific mate recognition system comprising a number of coadapted stages" (1978). The coadaptation referred to is intersexual: a signal from one potential sex partner evokes a particular response from the other; these events may then culminate in a chain of alternating signals and responses between the individuals, resulting in the success or failure of copulation.

Ethologists studying courtship routines have shown that behavioral interactions between male and female are often exceedingly complex and detailed, involving various combinations of visual, auditory, tactile, or olfactory signals that are usually deployed in precise temporal and spacial patterns (e.g. Blum and Blum 1979; Alcock 1979; Thornhill and Alcock 1983). Some morphological characters that strengthen and reinforce such signals appear to have evolved exclusively to fulfil a strong supporting role in the behavioral exchanges. Stimulated by Paterson's general approach, but differing with him on some emphases, I develop here some notions about the biological significance of these systems.

Paterson hypothesizes that such a behavioral interaction system between the sexes has evolved as a process whereby the participants are able to efficiently recognize one another's *species status*. In this chapter, I will adduce data and arguments contending that these systems serve an additional and even more fundamental function. I subscribe to the view that they consist primarily of a process that conjoins in sexual reproduction single individuals of high fitness (see Carson 1987*a*). To a great extent, courtship appears to involve the process whereby mating pairs are chosen by selection from recombinational fields of genetically variable males and females. Depending on the species, these fields may be reduced by differential survival, then sharply narrowed by intermale struggle for access to females. The field of males is further narrowed by female choice. In a broad sense, therefore, Paterson's emphasis on the idea that selection favors intraspecific reproductive efficiency through relevant behavior may be retained as a guiding principle. Paterson's recognition process serves to aggregate a conspecific field of individuals whereas the final mate choice is a finer-grained process that determines which individual or individuals among those aggregated actually differentially succeed in reproduction.

Each mate recognition system was considered by Paterson to define a common fertilization system for the group. This concept forms the basis of Paterson's definition of the species (1985). The genetic basis of sexual coadaptation is thought by him to have evolved in allopatry as a syndrome of characters promoting efficient conjunction of the sexes. According to this theory, the recognition system evolves largely intraspecifically and has not been significantly influenced by competitive sexual challenges from individuals belonging to related sexual systems. If it happens that the system serves at some later time to reproductively isolate individuals of the group from others having similar systems, this is viewed as a fortuitous "effect" of the system rather than a "function" of it.

I have found these formulations both fascinating and revolutionary. For nearly ten years after they were proposed, little attention was given to them in the literature. Recently, however, extensive discussion has ensued (e.g. in the recent book of Otte and Endler 1989). I am in general agreement with Paterson's arguments about the fortuitous, secondary nature of intrinsic reproductive isolation. As he has pointed out, this concept suggests the Biological Species Concept is inadequate, since a key feature of the definition views the species in terms of its intrinsic isolation from other such groups (Mayr 1963). Like Paterson, I cannot agree with Coyne and Orr (1989) that the Isolation Concept is especially useful since it narrows "the problem of the origin of species to the origin of reproductive isolating mechanisms." I prefer the Paterson view that the mating system is an intraspecifically developed syndrome of characters that serves the efficiency of sexual reproduction. In this chapter, I am not concerned with pursuit of any *interspecific* selective effects on the mating system. Rather, I argue that it is first and foremost a system for intrapopulational screening for relative fitness of potentially reproducing individuals. This is accomplished largely through a dynamic system of sexual selection, in which recombinational fields of genetically variable mates vie to populate the next generation at the expense of other coexisting conspecific organisms. The latter are viewed as making, relative to the sexually selected reproductive elite, a minor contribution, if any, to the next generation. Below I will elaborate on this view, which is also supported by West-Eberhard (1984).

Genetic Variation in Courtship Behavior

The Specific-Mate Recognition System (SMRS), as Paterson conceives it, serves the efficient conjunction of the sexes, a fundamental act of reproduction in the species. Although not explicitly discussed by Paterson, the implication is strong that over time the chain of signals and responses of the SMRS has become rigid, being largely fixed genetically within the species. In 1978, he stated that "Any mutation influencing a link in this chain to any significant extent will be selected against." Further, Paterson admits the possibility of the existence of only a small genetic variance on each of the components of the system. Lambert (1982) and Millar and Lambert (1984) have also stressed what they perceive as the stability, and therefore the species-diagnostic nature, of the mating systems of several species of *Drosophila*. Their discussion implies that they consider such a system to be fixed in all of its essentials throughout the distribution of the species. The methods used for their work, however, are inadequate to reveal the presence of genetic variation in relative mating success of individual specimens within populations, a topic that will be reviewed later in this chapter.

In my view, a strong element of typological thinking about species' characteristics exists in this argument. Thus, it does not take into account the evidence for genetically based intraspecific variability in such systems. The use of the word *specific* as descriptive of the mate recognition system carries the implication that it consists of a rigid set of characters, invariably and universally manifested by all members of the species. Now, this may be true in a broad sense if one confines one's attention simply to the recognition aspect. Nevertheless, I believe that recognition of the species status of the potential mate is only a small part of the significant biological function of the behavioral syndromes manifested by courting organisms. The contrary view, which I have stated briefly above, adheres closely to Darwinian sexual selection and holds that the interaction systems between the sexes serve principally as a program whereby the Darwinian fitnesses of the individuals actually participating in reproduction may be kept high. In short, the system serves the efficiency of sexual reproduction through sexual selection.

Recognition of the Individual: Its Role in Sexual Selection

Sexual selection can proceed only in the presence of individual variability. In a sexually reproducing, cross-fertilizing organism, a massive amount of data have accumulated showing that variation in individual capacity for successful reproduction in such a population is the rule (see the special emphasis on the assay of individual performances in Clutton-Brock 1988). Like all variability, behavioral traits display both genetic and environmental components. As is well known, Darwin was the first to suggest that the courtship procedure serves individual fitness through the process he referred to as sexual selection. The courtship or territorial display of one male individual of a species may vary slightly from that of another. Insofar as this slight difference has a genetic component, no matter how small, the opportunity exists for mate sexual display and choice to function as a selective device, effectively producing and embellishing complex syndromes of slightly varying sexually oriented characters displayed by the competitors.

Female Choice

As Darwin suggested, it is generally acknowledged that sexual selection may operate in two ways, either through intermale struggle for territorial position favorable to reproduction or through an active choice of mate exerted by one sex, principally the female. Female choice has been proposed in view of the evidence that many male characters appear to serve as "female persuaders" rather than weapons of aggression directed at other males. The female choice idea, regardless of the precise physiological mechanisms whereby it may operate, is relevant to the present discussion, since female choice in particular would amount to a very delicate and cryptic form of the sexual selection process. Much recent attention has been given to the subject of sexual selection and specifically female choice (e.g. O'Donald 1980; Clutton-Brock et al. 1982; Bateson 1983; Eberhard 1985). Empirical data are badly needed.

The widespread existence of a "rare-male" effect (see review in Ehrman and Parsons 1981) deserves attention in any discussion of female choice. Although alternative interpretations are possible, these experiments suggest that females of diverse species are indeed able to react in delicate and precise ways to differing fields of competing males. Further, there appears to be a role for pheromones in recognition of the individual (Averhoff and Richardson 1976; Ehrman and Probber 1978).

Genetics of Sexual Selection

A major correlate of the sexual selection theory holds that secondary sexual characters are usually subject to "runaway" selection, affecting such characters in the male sex (e.g. Fisher 1958). This means that any genetic variant contributing to the fitness of the individual that reaches the final consummation of the sexual act will tend to become fixed in the population, being opposed principally by natural selection affecting the survival of the variant phenotype. This emphasis on fixation is often taken to mean that, under the system, a population will not be able to achieve balanced polymorphism for such characters and thus will not be able to carry cryptic genetic variability for them. In my view, this concept errs, particularly in its assumption that large mutational changes must be involved in the genetics of behavior. I will return to this point later.

Were the runaway process to be generally true, then genetic individuality of the separate members of the population might indeed be eroded to the point that fitness differentials between individuals in the population are no longer maintained. There is, in fact, no compelling evidence for such uniformity of individuals. Perhaps it is the assumption that sexual selection must always run away to fixation that has led Paterson, Lambert, and others to view courtship and display elements as if they were rigidly fixed.

Relative homozygosity at allozyme loci has, I think erroneously, been taken to indicate a lack of genetic variability in all other aspects of the gene pool. Recently, however, McCommas and Bryant (1990) have obtained interesting data on this point, using a series of controlled bottlenecks in housefly populations. In their experiments, they show that allozyme variability is lost at a rate that follows neutral expectations. Quantitative traits, however, did not show a comparable decreased variance; in fact, an *increase* in variance in some quantitative traits followed the bottleneck (see the section on episodic release of genetic variability below).

This finding suggests some new interpretations of the possible fate of populations that show low allozyme variability. For example, relative allozyme homozygosity in the northern elephant seal has been taken to mean that the gene pool of this subspecies is depauperate in variability due to past population bottlenecks (Bonnell and Selander 1974). On the other hand, the data of Le Boeuf and Reiter (1988) suggest that as few as 3 percent of males of this population are responsible for 48 to 92 percent of all matings during a breeding season. Accordingly, fitness differentials may exist. Although genetic parameters of sexual behavior have not been studied, there is no proof of the lack of balanced polymorphism as a genetic basis for behavioral variability in this and in other such cases. The methods used to assay genetic variability (e.g. allozymes) have usually not included parallel analyses of genetic variance of quantitative characters. I believe the evidence indicates that sexual selection is usually an ongoing functional process, with an underlying genetic system based on balanced polymorphism. In this view, the system is considered capable of undergoing future episodic change, in which fitness parameters may be refined or acquire novel properties through novel genetic recombination.

Polygenic Inheritance of Secondary Sexual Characters

There are data suggesting that, like other important characters of the organism, the genetic basis of secondary sexual characters is polygenic in nature (Lande 1980a,b, 1981; Carson and Lande 1984). Under such a system, intraspecific variability pertaining to courtship can be carried in complex systems of balanced polygenic heterozygosity (Carson 1987a,b). Such blocks of genes will be subject to recombination at each meiotic event, and the variable products will be exposed to individual selection, especially sexual selection. Although there is some evidence for the participation of a major gene the effects of which are modified by polygenes (Templeton 1981), very simple scenarios, wherein a major gene arises by mutation and moves directly to fixation, find little support in the data. Thus, sexual selection is more likely to proceed over the main span of its evolutionary history by small increments of change in each generation. Periods of equilibrium, without directional movement, may nevertheless continue to serve sexual selection efficiently, since recombinant individuals deviating from the mean may have high survival fitness but will be less successful in the ultimate reproductive act.

Episodic Release of Genetic Variance

As in evolutionary changes that support increased adaptation to the ambient environment, sexual selection may temporarily emerge from an essentially balanced state into an episode of rapid change that extends over a number of generations. What might precipitate such an episode is the subject of much speculation and, unfortunately, little empirical evidence. Environmental factors that subject the population to stress may elicit changes in the genetic system that temporarily favor the release of cryptic genetic variability (Parsons 1987). For example, there is suggestive evidence that events that stress the genetic system, such as population bottlenecks, may be followed by significant increases in genetic variance for quantitative characters (e.g. Lints and Bourgois 1984; Bryant et al. 1986; Bryant and Meffert 1988; Goodnight 1987, 1988; Carson and Wisotzkey 1989). An unexpected result is that changes observed following population bottlenecks do not consist simply of a decrease in variance through loss of alleles at some loci by random drift. The workers cited above have documented an *increase* in genetic variance for quantitative characters following abrupt restriction of population size. Thus, new variance appears to be released or activated following a situation of gamete disequilibrium or other novel opportunity for recombinational episodes (see review in Carson 1990).

Episodic hybridization events have been frequently associated with increases in genetic variance; indeed, this is a property of most hybrid zones (Hewitt 1989). In this latter case, as well as in population bottlenecks, the release of variance appears to be largely recombinational in nature. Although point mutations may increase in hybrids, one is not forced to look to such a source for increased ability of the population to respond to selection. Increase of mutant variability, however, may indeed play a role in fueling episodes of directional selection. The large literature on transposable elements (see Engels 1983) stresses the abrupt activation of the mutagenic effects of these elements under certain conditions. Although the proximate cause of these powerful genetic effects is not yet clear, their very existence underscores the importance of episodes in the release of genetic variability on which selection may operate.

McDonald (1989) has discussed the episodic nature of the effects of retroviral elements in particular, stressing that such effects may be realized in unstable, peripheral, or bottlenecked populations. Like other transposable elements, retroviruses not only can act as mutagens but may also participate in regulatory changes depending on their point of integration into the DNA of the host.

Genetic Variability in Courtship-related Characters

What is the evidence that genetic variability for courtship-related characters actually exists in natural populations? This question is important since all variants of sexual selection theory require that such genetic variability be present. This is true whether sexual selection operates by a runaway process or by balanced polygenic polymorphism that undergoes shifts, releasing latent variability. Surprisingly, very little empirical evidence exists on this point. The classical experiments of Bateman (1948) focused on the differential reproductive ability of males of *Drosophila melanogaster*. He found that individual males carrying different large-effect marker mutants differ strongly in reproductive success and that the contribution of males to the next generation is more variable than that of females. Although many of the effects might be attributable to pleiotropy of the mutant markers, the specimens used came from a number of genetic backgrounds. Thus, the males differed not only in the mutant marker but in the residual polygenic genotype as well.

Genetic balance of the type proposed here is not an interpopulational system: characteristically, it serves the choice system that exists *within* a breeding population, or deme. In this regard, the experiments of Bateman were unnatural, as it was necessary to disturb deme balance in the strains used when large mutant markers were introduced. Some experiments will now be mentioned that avoid this artificial experimental design.

Petit et al. (1980) used wild strains carrying much natural genetic variability to study male mating success in *D. melanogaster* under competitive conditions. They found that one-quarter of the males did not mate at all, whereas another 25 percent mated two or three times. Half of the matings were performed by half of the males, each of which mated only once. Similar disproportionate success of certain individual males from wild strains, reproducing as demes in the laboratory, has been reported for the species *Drosophila silvestris* (Spiess and Carson 1981; Carson 1986). In these experiments, wild-type males from wild populations or large laboratory populations were marked with spots of paint and placed in cages in intraspecific competition with one another. Results from within each of two different wild strains were similar. In each case, 30 percent of healthy and actively courting males do not successfully copulate at all, whereas 33 percent yield nearly 70 percent of the observed matings (Carson 1986). A disproportionately high frequency of heterokaryotypes is represented among the successful males (Carson 1987a,b). These results recall the strongly differential mating success reported by Bosiger (1974) in *D. melanogaster* and ascribed by him to heterozygote superiority.

Making use of a Mexican population of *Drosophila pseudoobscura* with a very high level of natural inversion polymorphism, Anderson et al. (1979) were able to demonstrate differential reproductive success by certain wild males in the population. Like the case of *D. silvestris,* this finding is important since the genetic markers used to identify individual male genotypes were intrinsic to the natural wild population. Thus, the mating system was not perturbed by the necessity of engineering the insertion of genetic markers.

The elaborate courtship of male *D. silvestris* includes deployment of several rows of long tibial cilia against the abdomen of the female at a crucial final stage of courtship. In a study that emphasized the conditions existing in a single wild population of this species, the phenotypic variation within this population was found to have moderate to high heritability (Carson and Lande 1984; Carson 1985). The cilia character also responds to artificial selection for both high and low number, confirming the existence of genetic variability in the wild population (Carson and Teramoto 1984). This character is also variable geographically. Crosses between phenotypically different strains show that both sex-linked and autosomal segregating genetic units underlie the differences, but no major gene appears to be involved (Carson and Lande 1984).

Considerable genetic analysis has been carried out on the acoustic behavior of crickets. Bentley (1971), Bentley and Hoy (1972), and Hoy and Paul (1973) have used hybrids and backcrosses between two species of *Teleogryllus* to analyze the inheritance of male mating calls. Although there is no simple dominance, sex-linked factors are present. Like in *Drosophila silvestris,* analysis of the data support the conclusion that there are polygenic autosomal modifiers as well. That sex-linked genes are involved in both *Drosophila* and crickets suggests that sex linkage may be widespread in nature as a basis for secondary sexual characters of males. In *Colias* butterflies (Taylor 1972), the wing patterns important in courtship also show sex linkage, although of course in these latter insects, the female is the digametic sex.

Interspecific crosses have been used for genetic analyses of courtship sounds produced by the males of various species of the *virilis* group of *Drosophila* (Hoikkala and Lumme 1984, 1987). Using diallel reciprocal crosses of several interfertile species, it was found that in some cases there is a decisive role for X-linked genes whereas in others the differences are based on polygenes on each of the autosomes. In general, the pattern of polygenic inheritance and sex linkage is similar among all these cases from diverse insect forms.

Most of the studies mentioned above reveal genetic variability that exists between closely related species that can hybridize naturally or are capable of being hybridized artificially. In most of these cases, little attention has been given to intrapopulational variability of the kind that could serve as raw material for sexual selection. In the case of *Drosophila silvestris,* however, intrapopulation polymorphism appears to be maintained by blocks of heterotically balanced polygenes. Such a genetic state is compatible with the data on interspecific crosses in the *virilis* group, in *Colias,* and in crickets. In fact, crosses between *D. silvestris* and its very close relative, *Drosophila heteroneura,* also yield fertile hybrids (Val 1977). Inheritance of the tibial cilia character follows a pattern close to that found in crosses between phenotypically different strains of *D. silvestris* (Bryant and Carson 1979). This suggests the relevance of the data cited above on interspecific courtship sounds made by crickets to what might be expected within populations.

The above cases appear to provide to sexual selection a polygenically variable field of males, differing only slightly from one another. Cases are known, however, wherein the males fall into a small number of distinct classes that are segregating in the population. For example, among courting male crickets (*Gryllus integer*), Cade (1981) discovered that some males call and others do not. The latter successfully mate by silently intercepting females attracted to calling males. Selection experiments were used to demonstrate that this polymorphism is indeed genetically based, a result that extends the interspecific analysis of Hoy and Paul (1973) to include intraspecific variability of acoustic behavior in crickets. In the Coho salmon, males are also of two types. The large "hook-noses" are adept at fighting, whereas the smaller, early-maturing "jacks" sneak in to breed from the perimeter of the breeding ground. The two breeding strategies appear to be genetically balanced in the population (Gross 1985).

In males of a number of insect species, the width of the head has evolved an extraordinary breadth, placing the eyes in a stalklike lateral position. In the well-known Hawaiian species *Drosophila heteroneura,* for example, hypercephaly appears to play a role in sexual selection through male-to-male combat (see Fig. 1 in Kaneshiro and Boake 1987). A relatively broadened head is found in a number of species of the neotropical drosophiloid genus *Zygothrica* (Grimaldi 1987; Grimaldi and Fenster 1989). In some species, males are dimorphic or polymorphic for a hypercephalic condition; in one case, conspecific males manifest a bimodal distribution of head widths. Variable groups of males of the same *Zygothrica* species are found in nature on fungal lek areas: that this polymorphism has an underlying genetic balance is strongly indicated. Dimorphism is also known in males of the Afrotropical species *Drosophila pugionata,* in which some individuals have a normal anterior orbital bristle whereas in others it has been transformed into a strong spine (Tsacas et al. 1981).

Behavioral Mutants in *Drosophila melanogaster*

Among the very large number of mutants that have been screened in *D. melanogaster* are a number that affect courtship behavior at different stages of the process (Hall 1981). For example, there are mutants that produce an altered or cacaphonous male song, and others in which the male reaches the stage at which copulation is normally attempted but fails to do so. Even after copulation is achieved, mutants are known that interfere with completion of the sexual act, such as "coitus interruptus" or "stuck." Pains have been taken to ascertain that the principal effects are behavioral, as judged from the absence of obvious morphological alterations and the presence of general vigor.

The above discussion of major behavioral mutants should not be interpreted as supporting a model whereby either morphological or behavioral evolution is thought to proceed by such large single changes. Indeed, the opposite appears to be the case. Lande (1980b) states the case strongly: "most (if not all) cases of morphological evolution during allopatric speciation involve polygenic characters." This appears to also be true of genetic variability affecting secondary sexual characters that are related to sexual selection (Lande 1981). This author has employed quantitative genetic models of the joint evolution of male secondary sexual characters and different types of female mating preferences. He concludes that polygenic mutation, recombination, and assortative mating can maintain the additive genetic variance and covariance nearly constant in spite of selection that tends to deplete the variability. Lande further suggests that, under his model, random drift in female mating preferences in small populations might initiate an evolutionary episode producing novel differentiation in the secondary sexual characters of males. This is in accord with the observations of Kaneshiro (1987) on asymmetrical mating preferences.

Sexual Selection in Organisms in Which Males Provide More Than Gametes

Most of the discussion and examples used in this chapter have come from a consideration of basically polygamous systems of mating where males are promiscuous and invest little or nothing except gametes in their offspring. This is because this chapter has emphasized cases from organisms used because of their convenience for genetic studies: most of these make only gamete investments. The elaborations of the sexual process that involve mate guarding or feeding of females by males are also interpretable as manifestations of individual fitness and should, as has been maintained by Darwin and others, come under the broad sweep of sexual selection.

Summary

Mate recognition at the individual level, as distinct from the recognition of species status, serves the important function of fitness assessment of a potential mate. I have proposed that the males in each local population of a species exist as a field of genetically variable individuals that differ from one another in fitness. Evidence for the existence of genetic variability in mating performance in males has been reviewed. The variance in mating success, like other types of genetic variance, usually exists, I contend, in a balanced state. Accordingly, variation among the males is traceable to segregation from complex balanced polygenic polymorphisms with variance due to genetic recombination. Under certain demographic conditions, such as bottlenecks and hybridizations, there may be an episodic release of genetic variation that serves as raw material for rapid morphological advance under sexual selection. Most of this variation is of a polygenic nature, and a process of "runaway" selection that fixes mutants of large effect appears not to be involved. In this chapter I have argued that mating is rarely random. Crucial environmental selection in many organisms is thus concentrated in the *sexual environment,* rather than the *ambient environment.* Selection from an intraspecific field of males appears to favor relatively few individuals; this results in effective population sizes far below census. Reproduction is thus embodied in an intrapopulational reproductive elite.

Dedication

This chapter is affectionately dedicated to Hugh Paterson, whose bold thinking has done so much to invigorate evolutionary biology.

References

Alcock, J. 1979. Animal behavior: An evolutionary approach. Sinauer, Sunderland, Mass.
Anderson, W.W., L. Levine, O. Olivera, J.R. Powell, M.E. de la Rosa, V.M. Salceda, M.I. Gaso, and J. Guzman. 1979. Evidence of selection by male mating success in natural populations of *Drosophila pseudoobscura.* Proc. Natl. Acad. Sci. 76:1519–1523.
Averhoff, W.W., and R.H. Richardson. 1976. Multiple pheromone system controlling mating in *Drosophila melanogaster.* Proc. Natl. Acad. Sci. 73:591–593.
Bateman, A.J. 1948. Intra-sexual selection in *Drosophila.* Heredity 2:349–368.
Bateson, P., ed. 1983. Mate choice. Cambridge University Press, London.
Bentley, D.R. 1971. Genetic control of an insect neuronal network. Science 174:1139–1141.
Bentley, D.R., and R.R. Hoy. 1972. Genetic control of the neuronal network generating cricket (*Teleogryllus gryllus*) song patterns. Anim. Behav. 20:478–492.
Blum, M., and N. Blum, eds. 1979. Sexual selection and reproductive competition in insects. Academic Press, New York.
Bonnell, M.L., and R.K. Selander. 1974. Elephant seals: Genetic variation and near extinction. Science 184:908–909.
Bosiger, E. 1974. The role of sexual selection in the maintenance of the genetical heterogeneity of *Drosophila* populations and its genetic basis. Frontiers Biol. 38:167–184.
Bryant, E.H., S.A. McCommas, and L.M. Combs. 1986. The effect of an experimental

bottleneck upon quantitative genetic variation in the housefly. Genetics 114:1191–1211.
Bryant, E.H., and L.M. Meffert. 1988. Effect of an experimental bottleneck on morphological integration in the housefly. Evolution 42:698–707.
Bryant, P.J., and H.L. Carson. 1979. Genetics of an interspecific difference in a secondary sexual character in Hawaiian *Drosophila*. Genetics 91:s15–s16.
Cade, W.H. 1981. Alternative male strategies: Genetic differences in crickets. Science 212:563–564.
Carson, H.L. 1985. Genetic variation in a courtship-related male character in *Drosophila silvestris* from a single Hawaiian locality. Evolution 39:678–686.
―――. 1986. Sexual selection and speciation. Pp. 391–409 *in* S. Karlin and E. Nevo, eds. Evolutionary processes and theory. Academic Press, London.
―――. 1987*a*. The contribution of sexual behavior to Darwinian fitness. Behav. Genet. 17:597–611.
―――. 1987*b*. High fitness of heterokaryotypic individuals segregating naturally within a long-standing laboratory population of *Drosophila silvestris*. Genetics 116:415–422.
―――. 1990. Increased genetic variance after a population bottleneck. Trends Ecol. Evol. 5:228–230.
Carson, H.L., and R. Lande. 1984. Inheritance of a secondary sexual character in *Drosophila silvestris*. Proc. Natl. Acad. Sci. 81:6904–6907.
Carson, H.L., and L.T. Teramoto. 1984. Artificial selection for a secondary sexual character in males of *Drosophila silvestris* from Hawaii. Proc. Natl. Acad. Sci. 81:3915–3917.
Carson, H.L., and R.G. Wisotzkey. 1989. Increase in genetic variance following a population bottleneck. Am. Nat. 134:668–673.
Clutton-Brock, T.H., F.E. Guiness, and S.D. Albon. 1982. Red deer: Behavior and ecology of two sexes. University of Chicago Press, Chicago.
Clutton-Brock, T.H., ed. 1988. Reproductive success: Studies of individual variation in contrasting breeding systems. University of Chicago Press, Chicago.
Coyne, J.A., and H.A. Orr. 1989. Two rules of speciation. Pp. 180–207 *in* D. Otte and J.A. Endler, eds. Speciation and its consequences. Sinauer, Sunderland, Mass.
Eberhard, W.G. 1985. Sexual selection and animal genitalia. Harvard University Press, Cambridge, Mass.
Ehrman, L., and P.A. Parsons. 1981. Behavior genetics and evolution. McGraw-Hill, New York.
Ehrman, L., and J. Probber. 1978. Rare *Drosophila* males: The mysterious matter of choice. Am. Sci. 66:216–222.
Engels, W.R. 1983. The P family of transposable elements in Drosophila. Annu. Rev. Genet. 17:315–344.
Fisher, R.A. 1958. The genetical theory of natural selection. Dover, New York.
Goodnight, C.J. 1987. On the effect of founder events on epistatic genetic variance. Evolution 41:80–91.
―――. 1988. Epistasis and the effect of founder events on the additive genetic variance. Evolution 42:441–454.
Grimaldi, D.A. 1987. Phylogenetics and taxonomy of *Zygothrica* (Diptera: Drosophilidae). Bull. Am. Mus. Nat. Hist. 186:103–268.
Grimaldi, D.A., and G. Fenster. 1989. Evolution of extreme sexual dimorphisms: Structural and behavioral convergence among broad-headed Drosophilidae and other Diptera. Am. Mus. Novitates 2939:1–28.

Gross, M.R. 1985. Disruptive selection for alternative life histories in salmon. Nature 313:47–48.

Hall, J.C. 1981. Sex behavior mutants in *Drosophila*. BioSci. 31:125–130.

Hewitt, G.M. 1989. The subdivision of species by hybrid zones. Pp. 85–110 *in* D. Otte and J.A. Endler, eds. Speciation and its consequences. Sinauer, Sunderland, Mass.

Hoikkala, A., and J. Lumme. 1984. Genetic control of the difference in male courtship sound between *Drosophila virilis* and *D. lummei*. Behav. Genet. 14:257–268.

———. 1987. The genetic basis of evolution of the male courtship sounds in the *Drosophila virilis* group. Evolution 41:827–845.

Hoy, R.R., and R.L. Paul. 1973. Genetic control of song specificity in crickets. Science 180:82–83.

Kaneshiro, K.Y. 1987. The dynamics of sexual selection and its pleiotropic effects. Behav. Genet. 17:559–569.

Kaneshiro, K.Y., and C.R.B. Boake. 1987. Sexual selection and speciation: issues raised by Hawaiian *Drosophila*. Trends Ecol. Evol. 2:207–211.

Lande, R. 1980a. Sexual dimorphism, sexual selection and adaptation in polygenic characters. Evolution 34:292–305.

———. 1980b. Genetic variation and phenotypic evolution during allopatric speciation. Am. Nat. 116:463–479.

———. 1981. Models of speciation by sexual selection on polygenic traits. Proc. Natl. Acad. Sci. 78:3721–3725.

Lambert, D.M. 1982. Mate recognition in members of the *Drosophila nasuta* complex. Anim. Behav. 30:438–443.

Le Boeuf, B.J., and J. Reiter. 1988. Lifetime reproductive success in northern elephant seals. Pp. 344–362 *in* T.H. Clutton-Brock, ed. Reproductive success. University of Chicago Press, Chicago.

Lints, F.A., and M. Bourgois. 1984. Population crash, population flush and genetic variabillity in cage populations of *Drosophila melanogaster*. Genet. Select. Evol. 16:45–56.

Mayr, E. 1963. Animal species and evolution. Harvard University Press, Cambridge, Mass.

McCommas, S.A., and E.H. Bryant. 1990. Loss of electrophoretic variation in serially bottlenecked populations. Heredity 64:315–321.

McDonald, J.F. 1989. The potential evolutionary significance of retroviral-like transposable elements in peripheral populations. Pp. 190–205 *in* A. Fontdevila, ed. Evolutionary biology of transient unstable populations. Springer-Verlag, New York.

Millar, C.D., and D.M. Lambert. 1984. The mating behavior of individuals of *Drosophila pseudoobscura* from New Zealand. Experientia, Basel 41:950–952.

O'Donald, P. 1980. Genetic models of sexual selection. Cambridge University Press, London.

Otte, D., and J.A. Endler, eds. 1989. Speciation and its consequences. Sinauer, Sunderland, Mass.

Parsons, P.A. 1987. Evolutionary rates under environmental stress. Evol. Biol. 21:311–347.

Paterson, H.E.H. 1976. The role of postmating isolation in evolution. Unpublished resume of a paper read at the symposium Application of genetics to insect systematics and analyses of species differences, Fifteenth International Congress of Entomology, Washington, D.C. August 23, 1976.

———. 1978. More evidence against speciation by reinforcement. S. Afr. J. Sci. 74:369–371.

———. 1985. The Recognition concept of species. Pp. 21–29 *in* E.S. Vrba, ed. Species and speciation. Transvaal Museum Monograph No. 4. Pretoria.
Petit, C.P., P. Bourgeron, and H. Mercot. 1980. Multiple matings, effective population size and sexual selection in *Drosophila melanogaster*. Heredity 45:281–292.
Spiess, E.B., and H.L. Carson. 1981. Evidence for sexual selection in *Drosophila silvestris* of Hawaii. Proc. Natl. Acad. Sci. 18:3088–3092.
Taylor, O.R. 1972. Random vs. non-random mating in the sulfur butterflies, *Colias eurytheme* and *Colias philodice*. Evolution 26:344–356.
Templeton, A.R. 1981. Mechanisms of speciation—a population genetic approach. Annu. Rev. Ecol. Syst. 12:23–48.
Thornhill, R., and J. Alcock. 1983. The evolution of insect mating systems. Harvard University Press, Cambridge, Mass.
Tsacas, L., D. Lachaise, and J.R. David. 1981. Composition and biogeography of the Afrotropical Drosophilid fauna. Pp. 197–259 *in* M. Ashburner, H.L. Carson and J.N. Thompson, eds. The genetics and biology of *Drosophila*. Vol. 3a. Academic Press, London.
Val, F.C. 1977. Genetic analysis of the morphological differences between two interfertile species of Hawaiian *Drosophila*. Evolution 31:611–629.
West-Eberhard, M.J. 1984. Sexual selection, competitive communication, and species-specific signals in insects. Pp. 283–324 *in* T. Lewis, ed. Insect communication. Academic Press, London.

Consequences of the
Recognition Concept for
Speciation, Ecology, and Evolution

8

Models of Speciation by Founder Effect: A Review

HAMISH G. SPENCER

Department of Zoology, University of Otago
and
Museum of Comparative Zoology, Harvard University

> The origin of new species, signifying the origin of essentially irreversible discontinuities with entirely new potentialities, is the most important single event in evolution.
>
> Ernst Mayr, 1963

A THOROUGH UNDERSTANDING of speciation is of vital importance to the study of evolution because it is through speciation that a permanent increase in the amount of biological diversity is established (Dobzhansky 1937; cf. Futuyma 1987). In discussing various theories, several authors have suggested that small population size plays an important role in the genesis of new species. Hugh Paterson, for example, has grouped the various speciation models into two classes: those for which there is evidence they model events that have actually occurred (Class I) and those for which such evidence is lacking, although such events remain theoretical possibilities (Class II) (Paterson 1981). He goes on to argue that the only Class I models are those involving founder events. It seems appropriate, therefore, to more closely examine the theory underlying these models, in the hope that distinctions (and similarities) among them may be more easily identified.

In addition, application of the Paterson's Recognition Concept means that our attention is focused on the features of a species that make it a coherent whole. It is when these features are altered that speciation occurs. Different models predict different effects of the founder event on the Specific-Mate Recognition System (SMRS) and its genetics, and these may also be used to clarify the application of the models to particular real cases. For example, Kaneshiro (1989) has suggested that sexual selection may play an important role after a founder event in Carson's founder-flush model of speciation. The

Recognition Concept has fundamental implications for how we think any sexual selection might act (Spencer and Masters 1992). Moreover, Vrba (this volume) and Carson (1989 and this volume) have argued that the Isolation Concept is an inadequate concept for the study of speciation. Paterson's view reveals that the Isolation Concept forces an examination of the side effects of speciation (for a recent example see Coyne and Orr 1989).

There is both field and experimental evidence for the importance of founder effects in speciation. The explosion of species of Hawaiian *Drosophila* is the most widely quoted example of field evidence (see Throckmorton [1966] and Carson and Kaneshiro [1976] for reviews). These species are often endemic to one island (or one group of islands which in the geological past was but one) and have closely related species (as determined by numerous chromosomal, molecular, behavioral, and morphological features) on neighboring islands. Since the Hawaiian archipelago is a chain of geologically recent volcanoes, each island must have been colonized by flies from a neighbor. The high level of endemism suggests that such colonization would have been by very few individuals and that speciation was a not unlikely outcome. The time scale of the volcanic eruptions also suggests that no ecological climax communities would become established on young islands. Hence, colonizing species could easily have been subject to repeated bottlenecks.

Further field evidence comes from populations at the periphery of a species' distribution, which tend to be morphologically distinct from those near the center of the range. Several examples are summarized in Mayr (1954): New Guinean *Tanysiptera* kingfishers, Australasian *Halcyon* kingfishers, and Mediterranean *Lacerta* lizards (see also Mayr 1963). These populations are usually small, existing at the limit of the species' ecological and physiological tolerance and hence subject to population bottlenecks.

Experimental evidence is provided by Powell (1978, 1989), who worked with *Drosophila pseudoobscura* and found a significant tendency towards intra-strain mating in previously randomly mating lines that had passed through repeated founder-flush cycles (i.e., small populations that rapidly increase in size after being formed by the random sampling of a large population). Arita and Kaneshiro (1979) and Ahearn (1980), using Hawaiian *Drosophila* species, also noted increased homogamy in the laboratory. Templeton and his colleagues (reviewed in Templeton 1989) used parthenogenetic lines of *Drosophila mercatorum* to show that even a single female may harbor sufficient genetic variation to produce significant differences in reproductive behavior. Bryant et al. (1986a,b) subjected laboratory populations of the housefly *Musca domestica* to bottlenecks of various sizes and noted significant amounts of among-population differentiation within a given bottleneck size. They later provided experimental evidence that founder-flush cycles need not lead to a long-term decrease in fitness caused by the increased level of inbreeding during the bottleneck (Bryant et al. 1990).

Carson and Templeton (1984) have reviewed the founder-event-mediated speciation theories and proposed a taxonomy of three theories, namely, Mayr's genetic revolution, Templeton's genetic transilience, and Carson's founder-flush. In their subsequent discussion, however, they add a fourth theory, Carson's flush-crash. I will follow (for now) this augmented taxonomy, because it makes useful distinctions between the theories in terms of population genetics.

The first of these theories is Mayr's (1954) peripatric or genetic revolution theory. Under this theory, the founder event leads to a strong increase in homozygosity (p. 202). This increase exposes the surviving alleles to selection more often (as homozygotes) and so changes the genetic environment. Such a change occurring at many loci adds up to the genetic revolution. Moreover, changes at one locus may well affect those at others. Because the founder population is geographically isolated, the changes are protected from homeostatic effects of gene flow. The change in genetic environment leads to the building of a new coadapted gene pool and the formation of a new species.

The second theory in Carson and Templeton's schema is what Templeton (1980) called genetic transilience. In contrast to Mayr's theory, the founder event does not decrease genetic variation; it merely disorganizes the genetic system so that a proportion of the protected variation becomes exposed to selection. For example, polymorphisms maintained at an equilibrium by some form of balancing selection may have the frequencies of their constituent elements (alleles or haplotypes) changed so that selection pushes the polymorphism to a different equilibrium. Under genetic transilience, it is hypothesized that the population grows rapidly after the founder event. As in genetic revolution, the population is subject to strong selective pressures at all times. The selection acts primarily on modifier loci and is caused by the alteration of the frequencies of genes with major pleiotropic and epistatic effects.

The third theory, Carson's (1970, 1971, 1982) founder-flush, is similar to genetic transilience except that there is no selection immediately after the founder event. While the population is small it is subject to little or no selection, and, as it grows, selection is gradually reimposed. The chief source of selection is the external environment. The founder-flush cycle can be repeated several times until a new genetic balance is reached. Carson emphasized that both the fluctuating population size and the relaxation of selection are vital components of the model in that they are both needed to disorganize the balanced parental gene complex.

Finally there is Carson's (1968) flush-crash theory. Under this theory, the population first flushes as a result of relaxed selection, revealing a large amount of genetic variability and allowing recombination among previously unfit types. Dispersal or a sudden reduction in population size (a crash) due to the reimposition of selection samples this pool, and selection then eliminates most of the daughter populations. Some populations, however, may possess new fit genetic combinations, inaccessible under the original selection regime, and these populations will survive. Several of these flush-crash cycles then lead to a new species being formed.

Although the theories proposed derive many of their ideas from Wright's shifting balance theory (see Wright [1977] for a complete discussion), they often have little mathematical underpinning. They are described in vague terms such as "polygenic balanced system" (Carson 1982) and "genetic revolution" (Mayr 1954). When these authors have become more precise they have been sharply criticized by theoretical population geneticists: see, for example, Lewontin's (1965) comment on, and Lande's (1980) critique of, Mayr's genetic revolution theory, and Barton and Charlesworth's (1984) reply to Carson and Templeton (1984). Some of these arguments are investigated below.

A Critique of the Theories

Genetic Revolution

The fact that an event has occurred does not necessarily tell us how it happened. Mayr (e.g. 1982c) was clearly incorrect, therefore, to equate good evidence for the occurrence of speciation with founder events with evidence for his genetic revolution theory. Indeed, ever since the genetic revolution theory was propounded by Mayr (1954) as an extension of his ideas on the founder principle (Mayr 1942), its proposed mechanisms have been subject to frequent and adverse criticism (e.g. Lewontin 1965; Lande 1980; Templeton 1982; Barton and Charlesworth 1984). This criticism has had curiously little effect on Mayr's view. (As recently as 1978 Mayr wrote that it was startling "how little population genetics has contributed to our understanding of speciation.") In his 1982 (Mayr 1982a) rechristening of genetic revolution as "peripatric speciation," he was insistent on some of the most often criticized features of his theory (in particular the effect of the founder event on homozygosity) (see also Mayr 1982b,c).

Most, if not all, of Mayr's critics have pointed out that a founder event does not necessarily lead to a great increase in homozygosity (Lewontin 1965; Lande 1980; Templeton 1982). Nei et al. (1975) modeled the effects of a founder event on neutral heterozygosity for various bottleneck sizes, N_C, and population growth rates, r, after the crash. Although they were modeling neutral alleles, Nei et al. believed that these results would also apply to most selected alleles since, in small populations, drift is more often important than natural selection. They found that, even with the maximally severe bottleneck size of 2 (i.e., one fertilized female), most of the decrease in heterozygosity occurred within ten generations. If the population recovered sufficiently rapidly (i.e., had an r greater than about 0.5) then the heterozygosity level would be reduced by no more than about 50 percent of its original value. With a larger bottleneck of 10, r was less important, and the reduction was much smaller. With a larger r of 1.0 (such as would be quite possible on a newly colonized island with large areas of suitable habitat and few predators) the reduction was negligible, and even with an N_C of 2, no more than an absolute decrease of 4 percent was likely.

When heterozygosity was reduced, however, Nei et al. (1975) reported a more disturbing result: the extremely long time that heterozygosity levels take to increase again under mutation pressure. In all cases the increase started around 10^5 generations, and original levels were not reached until 10^{10} generations. As Nei et al. pointed out, this time is often longer than the evolutionary life of a species. Yet the *Drosophila* of the island of Hawaii have quite normal levels of enzyme polymorphism (Carson and Kaneshiro 1976).

Restoration of genetic variation by mutation is less of a problem if quantitative characters are considered: typical levels of heritability can be rebuilt in the order of a few thousand generations (Lande 1980). However, if the theory is recast so that the founder event is hypothesized to lead to a large decrease in quantitative variation, then the maximum effect of selection will be tiny, contrary to that required under Mayr's theory.

Thus Mayr's claim that the founder event leads to a dramatic increase in homozygosity is true only if the bottleneck is very small and the subsequent population growth rate is very low ($N_C = 2$, $r = 0.1$ in Nei et al.'s model). If there is such a reduction in heterozygosity, however, then it is unlikely that selection will have much effect since it requires genetic variation on which to act. Moreover, mutation will not restore the level of polymorphism required for selection (or for that matter observed in populations) for a very long time, in stark contradiction to the assertions of Carson (1971) and White (1978). This line of argument suggests that any population subject to a heterozygosity-depleting bottleneck will become extinct.

When pressed on this point Mayr responded (e.g. Mayr 1965) that what is really important is the number of alleles at a locus. The model of Nei et al. (1975) also investigated this variable and found it extremely sensitive to N_C and less so to r, because of the extreme loss in the first generation at the bottleneck. Mayr suggested that the reduction in allele number leads to selection for these alleles which have higher homozygote fitnesses. These alleles, he argued, are likely to be different alleles from the common ones in the ancestral population, which have been selected to be "good mixers" (i.e., good heterozygotes) in the face of previous gene flow from other populations.

This last argument appears to be wrong on several points. First, with a large ancestral population, it is not the native alleles that are selected by the immigrants to have high heterozygote fitnesses, but rather the reverse. Spencer and Marks (1988) have shown that in large populations, governed primarily by deterministic natural selection, the only immigrants to invade a population successfully are those with high heterozygote fitnesses. Second, the alleles most common in large populations subject to gene flow are likely to have high heterozygote and homozygote fitnesses (Marks and Spencer 1991). Furthermore, those with high homozygote fitnesses are likely to be common (Marks and Spencer 1991). Hence selection for those alleles that do well as homozygotes is more likely to reconstruct a very similar gene pool to the ancestral one, albeit a somewhat depauperate version in terms of allele number.

In conclusion, therefore, Mayr's theory requires several contradictory assumptions—either a reduction in heterozygosity whereby any subsequent recovery is almost forever prolonged, or the prevalence of alleles which are rare to begin with and likely to be lost through selection.

Genetic Transilience

Genetic transilience was the name given by Templeton (1980) to his modification of Mayr's (1954) genetic revolution theory. Templeton's reasons for the new terminology were (1) that the words *genetic revolution* implied that many or most loci were involved (which he believed was not true) and (2) that Mayr's population genetics were wrong (see above) and the correct formulation said very different things about the changes in the population. Mayr has objected to this change, claiming that (1) the speciation event is not a transilience, which to Mayr has saltationist connotations (Mayr 1982b), and (2) we should not coin new terms for small changes in interpretation lest we "drown in terminology" (Mayr 1982c:885). Ironically, in the same year Mayr renamed genetic revolution *peripatric speciation* merely to emphasize its difference from the traditional dumbbell theory of allopatric speciation (Mayr 1982b).

Templeton did indeed use a number of arguments from population genetics in his reformulation. He recognized that the founder event must not reduce variation to a level at which selection is impotent. Consequently, he altered the emphasis of the model from an increase in homozygosity per se to the radical alteration of allele frequencies at a few loci, by sampling, and their further alteration by natural selection because of strong epistasis between the loci. The essential feature of the theory, however, remains the profound alteration of allele frequencies brought about by the bottleneck sampling and the continued action of natural selection, resulting from the changed genetic environment in building a new coadapted gene pool. It seems to me, therefore, that a new name is scarcely needed and genetic revolution should suffice (particularly since Mayr's original hypothesis now seems untenable).

Templeton (1980) made a number of predictions on what would be the most favorable conditions for a genetic revolution. Traits influenced by a large number of additive loci, each with a small effect, are not likely to be important, because suitable founder events (i.e., those that leave sufficient variability for selection) are not likely to have a large enough effect on the sum of the alleles. Conversely, traits influenced by a few major genes with strong epistasis among them could well be profoundly changed. This argument has been rejected by Barton (1989) since within-population variance is reduced by the same factor regardless of the number of alleles. The importance of epistatic effects depends on the specific model, and we clearly need to know much more about the genetics of SMRSs.

Templeton framed the dual requirements of a sufficient level of variability (for natural selection) and a major change in allele frequencies at major loci in terms of two sorts of effective population size. Variance effective population size, N_{ev}, controls for the sampling variance of the allele frequency in the offspring generations. If the frequency of the allele in the parental generation is p, then the offspring generation's frequency will come from a (binomial) distribution with variance $p(1-p)/2N_{ev}$. In contrast, the inbreeding effective population size, N_{ef}, controls for the level of inbreeding in a population and is defined as the reciprocal of the probability that two randomly chosen gametes from the parental generation come from the same individual (Crow and Kimura 1970). It is naturally related to the decrease in heterozygosity: the change in heterozygosity, $\Delta H^t = H^t - H^{t-1} = -H^{t-1}/2N_{ef}$, where H^t is the heterozygosity in the tth generation. Note that N_{ef} is affected mostly by the numbers of parents, whereas N_{ev} is affected by the numbers of offspring. In an expanding population, therefore, N_{ef} is usually less than N_{ev} (Crow and Kimura 1970). For a "large" randomly mating monoecious population at a fixed size, with the number of offspring per parent distribution as a Poisson random variable and no selection, $N_{ev} = N_{ef} = N$ (Crow and Kimura 1970).

Templeton (1980) argued that for a genetic revolution to succeed, the bottleneck must be such that N_{ev} is not too small (to preserve variation for natural selection) and yet the decrease in N_{ef} is large (so as to change the genetic environment sufficiently by increasing the level of homozygosity).

Barton and Charlesworth (1984) severely criticized this formulation, pointing out that any genetic revolution would be the cumulative effect of the sequence of generations starting with the crash. Hence, they argued, the two measures are effectively the same. The loss of variation and the increase in the level of homozygosity are inseparable, even though, as Carson and Templeton (1984) pointed out, the population is growing rapidly after the founder event: N_{ef} is less than N_{ev} for every generation. If the population sizes from the founder event onward are $N^0, N^1, N^2, \ldots, N^n$, then the heterozygosity will be reduced by a factor of $H_n/H_0 = \Pi_{i=0}^{n-1} \{1 - [1/(2N^i)]\}$ relative to that in the ancestral population (Barton 1989). Concomitantly, the variance of an allele frequency p^n, around the ancestral frequency, p, among different populations is given by $p(1-p)(1-H_n/H_0)$ (Barton 1989). Thus, Barton (1989) continued, any founder event that causes a significant level of population differentiation necessarily leads to a significant decrease in heterozygosity. This reasoning ignores the effects of selection, but Rouhani and Barton (1987) have argued that it is a good approximation unless selection were strong. Moreover, if selection were strong, it would more likely act to return the population to its previous state (less any genetic variation lost).

Templeton (1980) went on to use the respective formulas for N_{ef} and N_{ev} to predict what ancestral and founder population characteristics (e.g. mating structure, mean number of offspring per individual, population density) give a greater probability of genetic revolution. The ancestral population should be large (and so also will be N_{ef} and N_{ev}) and panmictic (so that much of the variability will be within individuals, and N_{ev} of the founder population will not be too small). The founder event's sampling process should also be important. A sample of genetically correlated individuals will decrease or increase the probability of genetic revolution, depending, respectively, on the relative importance of N_{ev} and the decrease in N_{ef}. If the level of within-individual variation is sufficient to allow natural selection (as Templeton expected it to be in a large, panmictic population), then a genetically correlated sample should increase the probability of a genetic revolution. In a small inbred population, however, a genetically correlated sample would have the opposite effect.

The population subsequent to the founder event should grow rapidly (see previous discussion). Not only does this flush increase the chance for a genetic revolution, but it clearly increases the probability of survival. Templeton also claimed that a hierarchical population structure (e.g. Wright's [1932] Island model of several small demes with but occasional gene flow among them [see also Wright 1977]) enhanced the probability of a genetic revolution, since it increases inbreeding, yet preserves the total amount of genetic variation. It seems to me, however, that since most of the selection is operating within demes, it would be necessary to maintain within-deme variability, not merely that within populations.

Assortative mating, in the sense of Lewontin et al. (1968), does not alter allele frequencies (and hence maintains variation) but does increase inbreeding. Templeton thus argued that assortative mating changes the probability of a genetic revolution. This concept of assortative mating is usually, however, quite unrealistic. Many apparently assortative mating schemes lead to changes in allele frequencies, and, certainly, in the face of selection, an alteration of genotype frequencies (a natural result of assortative mating) would be expected to result in allele frequency changes (Spencer 1993). Thus, homogamic mating could well decrease the probability of a genetic revolution.

Finally, Templeton suggested that looser linkage (manifested as a larger chromosome number, a larger total genomic map length, or a system with few crossover suppressors) would enhance the chance of speciation. This increase arises, he argued, because with tighter linkage a substantial proportion of the population may not be inbred (even if the mean level of inbreeding is the same) (Franklin 1977) and would resemble the ancestral population. Selection for these individuals could reconstruct essentially the same genome and prevent a genetic revolution. Certainly, looser linkage permits the generation of more novel types by recombination, and in the founder population these may reach a reasonable frequency.

Many of these arguments may be resolved with better information about the genetics of SMRSs. For example, under genetic transilience, different levels of genetic variation in SMRS characters will alter the effect of a genetically correlated founder population. Most of the time it is not sufficient to merely have information about the genetics of characters unimportant in mate recognition. Unfortunately, we often do not know the importance of the characters whose genetics we do know.

In summary, then, genetic transilience is genetic revolution with better population genetics. The founder event and recovery must permit sufficient variation for selection. Selection acts throughout the bottleneck and flush, building a new coadapted gene complex. The selection pressures arise from the altered genetic background.

Founder-Flush

Carson has long been interested in the genetic consequences of population bottlenecks and has proposed two speciation theories by his count: flush-crash (Carson 1968) and founder-flush (Carson 1970). They have had a somewhat confused history: the original description of the founder-flush theory (Carson 1970) was not identified as such, and in a longer essay on founder-flush the following year Carson (1971) drew no difference between the two models, with numerous references to the older flush-crash model. It is clear that more recently Carson considered them very different (Carson and Templeton 1984), taking Charlesworth and Smith (1982) to task for simulating flush-crash while discussing founder-flush. The major difference between the two is that in flush-crash the population flush, during which selection is absent or relaxed, occurs before the founder event, whereas in founder-flush it occurs after. The difference was considered extremely important because of Carson's view of speciation as an initial disorganization of the coadapted gene complex and its subsequent reorganization into something different in the new species (Carson 1982). Under flush-crash, therefore, the disorganization occurs in the initial flush with its permissive selection allowing all sorts of previously unfit types to survive and the crash merely samples the modified gene pool (Carson

1968, 1973, 1975). In contrast, under founder-flush the crash first samples the ancestral population, before the relaxed selection allows new types to form (Carson 1970, 1971; Carson and Templeton 1984).

Founder-flush differs from genetic revolution in that it has no selection operating in the bottleneck: selection is reintroduced only as the population regains its maximum size. This difference has a number of important consequences. First, the disorganization of the genome can take place over a longer time—there is not just the sampling effect of the founder event, common to all the theories, but also the period of relaxed selection during which recombination can generate types that previously would not have survived (or else would have remained at very low frequencies). Second, in contrast to the claim of Barton and Charlesworth (1984), under selection fixation of an allele is more likely for many selection schemes. In a small population even heterozygote advantage can increase fixation rate if the deterministic equilibrium frequency is outside the range of 0.2 to 0.8 (Robertson 1962). Certainly, with a sharp change in the genetic environment such equilibria are quite possible. Barton (1989) remained unconvinced of the importance of the relaxation of selection immediately after the founder event. For the same reasons outlined in the discussion of genetic transilience, he argued that any founder event causing substantial changes in the gene pool will concomitantly reduce the variability.

Barton and Charlesworth's (1984) attack on the founder-flush theory made extensive use of Wright's (1932) metaphor of an adaptive landscape, likening a speciation event to an adaptive peak shift. Charlesworth and Smith's (1982) model, they argued, showed that a reduction in variability reduces the frequency of a stochastic peak shift. Carson and Templeton (1984) replied that Charlesworth and Smith's (1982) result was a consequence of (1) having selection act immediately after the crash and (2) having only one flush-crash cycle, which eliminated the effect of the bottleneck sampling. Indeed, Barton (1989) agreed that a series of bottlenecks is more likely to lead to a peak shift.

Barton and Charlesworth (1984) also claimed that adaptive valleys shallow enough to permit a reasonable frequency of peak shifts fail to provide enough postmating isolation to keep the populations at the two peaks differentiated in anything but the selected loci. The acquisition of postmating isolation is, however, neither a sufficient nor a necessary component of speciation (Paterson 1981; Spencer et al. 1987). The Hawaiian *Drosophila* are clearly defined by their mating behavior (Carson and Kaneshiro 1976; Spieth 1984; Hoy et al. 1988), and, indeed, such phenomena are often a major part of a species' mate recognition system (Paterson 1980). It seems to me, therefore, that we should not be looking at occasional incidental by-products of speciation, but rather at the origin of positive assortative mating (in the wide, nonrandom sense). Nevertheless, Barton and Charlesworth's (1984) argument can be generalized, for example, as in Rouhani and Barton (1987), to say that selection strong enough to provide stable peaks leads to a low probability of a peak shift (but see below).

Again, the wider application of the Recognition Concept would be an enormously helpful procedure. Paterson's ideas are also illuminating when considering hypotheses involving sexual selection (Spencer and Masters 1992). Kaneshiro (1989) argued that while the population is small after the founder event, discriminating females previously maintained in the population by a genetic correlation with successful males will not be able to find a mate. Hence, there is a shift toward less discriminating females (and otherwise less successful males) which is manifested in the asymmetric mate preferences reported by Kaneshiro and others (see Kaneshiro [1989] for a review). Kaneshiro's model depends, however, on a genetic correlation between female discrimination and male mating ability. It seems to me that a view of mate recognition systems whose function is the achievement of syngamy suggests no reason for the existence of such a correlation.

One of Barton and Charlesworth's conclusions is that the probability of speciation by founder event is low (e.g. Barton and Charlesworth 1984:141, 157). Here they are in agreement with Mayr (1982*a*), Templeton (1981, 1982), and Carson (Carson and Templeton 1984), and it is not clear to me, either, why this low probability should be a problem. Most founder events will not lead to speciation: probably most cause no change at all, and the next most likely outcome is extinction because of the severity of the founder event. Moreover, if Wright's adaptive-landscape metaphor is a good one (and critics of founder speciation theories seem to believe that it is), then surely restricted population size (such as a founder event) is a prerequisite for a reasonable frequency of peak shifts (Wright 1931, 1932, 1988), if the landscape itself is not changing. The high level of single-island endemism in the Hawaiian drosophilids, however, does suggest that the founder-flush model does not completely describe speciation in these flies. If the most likely outcome of a founder event is no change, then why do we not find more widespread species?

Flush-Crash

The flush-crash hypothesis is very similar to founder-flush, but differs in that the flush period precedes the founder event. This increase from the normally prevailing size is permitted by the relaxation (or absence) of selection, whereas under founder-flush, the flush represents a return to normal population size. The flush culminates in a crash that samples the large population, although it is not clear to me how this sampling occurs. For example, Carson (1968) wrote that the population at the crest of the flush would become subject to many density-dependent selective factors (such as shortages of food and space), which cause the subsequent crash. Thus the sample (say, of emigrants) will not be random with respect to these selective pressures, in contradiction to Carson's claim. It is even harder to see how flush-crash could lead to random sampling of those individuals who do not emigrate. Yet Carson's (1975) figure and text were unequivocal about the absence of selection throughout the flush and crash. Of course, the crash may be produced by selection that does not select for the previously fittest types, but this scenario would seem to me to be a form of the traditional dumbbell theory (Mayr 1963). Moreover, a selection-induced crash may easily deplete the genetic variation.

After the crash the population flushes again, either to a stable level where strong selection is reintroduced, or else to a larger value which leads to another crash. Carson and Templeton (1984) suggested that several flush-crash cycles would be necessary to produce speciation.

Flush-crash, therefore, is rather like founder-flush, but with a prior, disorganizing flush. If the crash is caused by strong selection, however, the effect of this flush may be nullified (or even reversed) since genetic variation may be minimal. Founder-flush seems to me to be more likely to have random sampling. There is thus a greater probability of retaining less fit individuals from the Wrightian selective valley, which can then generate the genotypes of the alternative peak.

Charlesworth and Smith (1982) developed a two-diallelic-loci model of flush-crash in which there were two polymorphic fitness peaks. An essentially infinite population was initiated at one of these peaks and not subject to selection for several generations (corresponding to the flush). A random sample was then made (corresponding to the crash), and this finite population iterated under selection for several more generations (the bottleneck). Finally, the population was deterministically iterated to a fitness peak (the "nearest"), and which peak this was determined whether or not a peak shift had occurred. They found that the probability of a shift was low, but greatest with a long flush (≥ 100 generations) and a short (two generations), medium-sized (six individuals) bottleneck. With long flushes, the level of linkage disequilibrium was much reduced—with their parameters it was 0.006 of its original value by 100 generations—essentially negligible. In other words, to obtain a high probability of a peak shift, the population had to be almost exactly at the lowest point on the adaptive landscape (which, of course, was not operating at the time). These optimal conditions are ones that intuitively would be expected to lead to a high level of disorganization, while at the same time maintaining variation.

Carson and Templeton (1984) were highly critical of this model, pointing out that selection is relaxed or absent during the bottleneck of both flush-crash and founder-flush, not strong as in Charlesworth and Smith's model. They noted that relaxed selection could accentuate the initial disorganization of the founder event. Certainly, this greater disorganization would reduce the length of the flush necessary to result in a higher probability of a peak shift.

In summary then, flush-crash is founder-flush, with a preceding flush period of relaxed selection. The effect of this period may enhance or reduce the probability of speciation, depending on the dynamics of the sampling process that initiates the crash.

Summary

Of the four models of founder-event-induced speciation recognized by Carson and Templeton (1984), it seems to me that only two are sufficiently distinct to warrant separate names. I recognize only genetic revolution (under which I subsume peripatric speciation and genetic transilience) and founder-flush (which includes as a variant flush-crash). The continued interest of evolutionary biologists in such models requires that they be given a firmer theoretical base. Many questions about if and how often founder events are involved in speciation and what population parameters are important clearly remain unanswered. Charlesworth and Smith (1982), for example, suggested that speciation was an unlikely outcome of flush-crash cycles and moreover, that genetic change was most easily accomplished with relaxed selection and moderate crash-population sizes—hardly the conditions envisaged by Carson. More widespread application of Paterson's ideas on species (e.g. in deciding what changes in speciation and in viewing sexual selection) would go a long way toward answering many of these questions.

Acknowledgments

I am grateful to N. Barton, K. Kaneshiro, D. Lambert, R. Lewontin, and J. Masters for providing numerous helpful suggestions which greatly improved previous versions of this essay (with which, of course, they do not necessarily agree). The work was supported, in part, by National Institutes of Health grant GM-21179 to R.C. Lewontin and an An Wang Fellowship (Harvard University), a Fulbright Travel grant, and University of Otago Division of Sciences grant DFZ 845 to H.G. Spencer.

References

Ahearn, J.N. 1980. Evolution of behavioral reproductive isolation in a laboratory stock of *Drosophila silvestris*. Experientia, Basel 36:63–64.
Arita, L.H., and K.Y. Kaneshiro. 1979. Ethological isolation between two stocks of *Drosophila adiastola* Hardy. Proc. Hawaii Entomol. Soc. 23:31–34.
Barton, N.H. 1989. Founder effect speciation. Pp. 229–256 *in* D. Otte and J.A. Endler, eds. Speciation and its consequences. Sinauer, Sunderland, Mass.

Barton, N.H., and B. Charlesworth. 1984. Genetic revolutions, founder effects and speciation. Annu. Rev. Ecol. Syst. 15:133–164.
Bryant, E.H., L.M. Combs, and S.A. McCommas. 1986a. Morphometric differentiation among experimental lines of the housefly in relation to a bottleneck. Genetics 114:1213–1223.
Bryant, E.H., S.A. McCommas, and L.M. Combs. 1986b. The effect of an experimental bottleneck upon quantitative genetic variation in the housefly. Genetics 114:1191–1211.
Bryant, E.H., L.M. Meffert, and S.A. McCommas. 1990. Fitness rebound in serially bottlenecked populations of the house fly. Am. Nat. 136:542–549.
Carson, H.L. 1968. The population flush and its genetic consequences. Pp. 123–137 in R.C. Lewontin, ed. Population biology and evolution. Syracuse University Press, Syracuse, New York.
———. 1970. Chromosome tracers of the origin of species. Science 168:1414–1418.
———. 1971. Speciation and the founder principle. University of Missouri Stadler Genetics Symposia 3:51–70.
———. 1973. Reorganization of the gene pool during speciation. Pp. 274–280 in N.E. Morton, ed. Genetic structure of populations. Hawaii University Press, Honolulu.
———. 1975. The genetics of speciation at the diploid level. Am. Nat. 109:83–92.
———. 1982. Speciation as a major reorganization of polygenic balances. Pp. 411–433 in C. Barigozzi, ed. Mechanisms of speciation. Liss, New York.
———. 1983. Chromosomal sequences and inter-island colonizations in the Hawaiian Drosophila. Genetics 103:465–482.
———. 1989. Genetic imbalance, realigned selection, and the origin of species. Pp. 345–362 in L.V. Giddings, K.Y. Kaneshiro, and W.W. Anderson, eds. Genetics, speciation and the founder principle. Oxford University Press, New York.
Carson, H.L., and K.Y. Kaneshiro. 1976. *Drosophila* of Hawaii: Systematics and ecological genetics. Annu. Rev. Ecol. Syst. 7:311–345.
Carson, H.L., and A.R. Templeton. 1984. Genetic revolutions in relation to speciation phenomena: The founding of new populations. Annu. Rev. Ecol. Syst. 15:97–131.
Carson, H.L., and J.S. Yoon. 1982. Genetics and evolution of Hawaiian *Drosophila*. Pp. 297–344 in M. Ashburner, H.L. Carson, and J.N. Thompson, eds. The genetics and biology of Drosophila. Vol. 3b. Academic Press, London.
Charlesworth, B., and D.B. Smith. 1982. A computer model of speciation by founder effects. Genet. Res. 39:227–236.
Coyne, J.A., and H.A. Orr. 1989. Two rules of speciation. Pp. 180–207 in D. Otte and J.A. Endler, eds. Speciation and its consequences. Sinauer, Sunderland, Mass.
Crow, J.F., and M. Kimura. 1970. An introduction to population genetics theory. Harper and Row, New York.
Dobzhansky, T. 1937. Genetics and the origin of species. Columbia University Press, New York.
Franklin, I.R. 1977. The distribution of the proportion of the genome which is homozygous by descent in inbred individuals. Theor. Pop. Biol. 11:60–80.
Futuyma, D.J. 1987. On the role of species in anagenesis. Am. Nat. 130:465–473.
Hoy, R.R., A. Hoikkala, and K. Kaneshiro. 1988. Hawaiian courtship songs: Evolutionary innovation in communication signals of *Drosophila*. Science 240:217–219.
Kaneshiro, K.Y. 1989. The dynamics of sexual selection and founder effects in species formation. Pp. 279–296 in L.V. Giddings, K.Y. Kaneshiro and W.W. Anderson, eds. Genetics, speciation and the founder principle. Oxford University Press, New York.

Lande, R. 1980. Genetic variation and phenotypic evolution during allopatric speciation. Am. Nat. 116:463–479.

———. 1985. Expected time for random genetic drift of a population between stable phenotypic states. Proc. Natl. Acad. Sci. 82:7641–7645.

———. 1986. The dynamics of peak shifts and the pattern of morphological evolution. Paleobiol. 12:343–354.

Lewontin, R.C. 1965. Comment. Pp. 481–484 in H.G. Baker and G.L. Stebbins, eds. The genetics of colonizing species. Academic Press, New York.

Marks, R.W., and H.G. Spencer. 1991. The maintenance of single-locus polymorphism. II. The evolution of fitnesses and allele frequencies. Am. Nat. 138:1354–1371.

Mayr, E. 1942. Systematics and the origin of species. Columbia University Press, New York.

———. 1954. Change of genetic environment and evolution. Pp. 188–213 in J. Huxley, A.C. Hardy and E.B. Ford, eds. Evolution as a process. Allen and Unwin, London.

———. 1963. Animal species and evolution. Harvard University Press, Cambridge, Mass.

———. 1965. Comment. P. 481 in H.G. Baker and G.L. Stebbins, eds. The genetics of colonizing species. Academic Press, New York.

———. 1970. Populations, species, and evolution. Harvard University Press, Cambridge, Mass.

———. 1978. Review of *Modes of Speciation*. Syst. Zool. 27:478–482.

———. 1982a. Speciation and macroevolution. Evolution 36:1119–1132.

———. 1982b. Processes of speciation in animals. Pp. 1–19 in C. Barigozzi, ed. Mechanisms of speciation. Liss, New York.

———. 1982c. The growth of biological thought. Harvard University Press, Cambridge, Mass.

Nei, M., T. Maruyama, and R. Chakraborty. 1975. The bottleneck effect and genetic variability in populations. Evolution 29:1–10.

Paterson, H.E.H. 1980. A comment on "mate recognition systems." Evolution 34:330–331.

———. 1981. The continuing search for the unknown and unknowable: A critique of contemporary ideas on speciation. S. Afr. J. Sci. 77:113–119.

Powell, J.R. 1978. The founder-flush speciation theory: An experimental approach. Evolution 32:465–474.

———. 1989. The effects of founder-flush cycles on ethological isolation in laboratory populations of *Drosophila*. Pp. 239–251 in L.V. Giddings, K.Y. Kaneshiro, and W.W. Anderson, eds. Genetics, speciation and the founder principle. Oxford University Press, New York.

Robertson, A. 1961. Inbreeding in artificial selection programmes. Genet. Res. 2:189–194.

Spencer, H.G. 1993. Assortative mating versus selective mating: Is the distinction worthwhile? Social Biology 39:310–315.

Spencer, H.G., D.M. Lambert, and B.H. McArdle. 1987. Reinforcement, species, and speciation: A reply to Butlin. Am. Nat. 130:958–962.

Spencer, H.G., and R.W. Marks. 1988. The maintenance of single-locus polymorphism. I. Numerical studies of a viability selection model. Genetics 120:605–613.

Spencer, H.G., and J.C. Masters. 1992. Sexual selection: Contemporary debates. Pp. 294–301 in E.F. Keller and E.A. Lloyd, eds. Keywords in evolutionary biology. Harvard University Press, Cambridge, Mass.

Spieth, H.T. 1984. Courtship behaviors of the Hawaiian picture-winged *Drosophila*. Univ. Calif. Publ. Entom. 103.

Templeton, A.R. 1980. The theory of speciation via the founder principle. Genetics 94:1011–1038.

———. 1981. Mechanisms of speciation: A population genetic approach. Annu. Rev. Ecol. Syst. 12:23–48.

———. 1982. Genetic architectures of speciation. Pp. 105–121 *in* C. Barigozzi, ed. Mechanisms of speciation. Liss, New York.

———. 1989. Founder effects and the evolution of reproductive isolation. Pp. 329–344 *in* L.V. Giddings, K.Y. Kaneshiro, and W.W. Anderson, eds. Genetics, speciation and the founder principle. Oxford University Press, New York.

Throckmorton, L.H. 1966. The relationships of the endemic Hawaiian Drosophilidae. Stud. Genet. III. Univ. Tex. Publ. 6615:335–396.

White, M.J.D. 1978. Modes of speciation. W.H. Freeman, San Francisco.

Wright, S. 1931. Evolution in Mendelian populations. Genetics 16:97–159.

———. 1932. The roles of mutation, inbreeding, crossbreeding and selection in evolution. Proc. Sixth Int. Cong. Genet. 1:356–366.

———. 1977. Evolution and the genetics of populations. Vol. 3. University of Chicago Press, Chicago.

———. 1988. Surfaces of selective value revisited. Am. Nat. 131:115–123.

9

Outcomes of Negative Heterosis

MICHAEL G. RITCHIE AND GODFREY M. HEWITT

School of Biological Sciences
University of East Anglia
Norwich, Norfolk, NR4 7TJ U.K.*

The Significance of Negative Heterosis to the Recognition Concept of Species

A COMMON CRITICISM of Hugh Paterson's Recognition Concept of species is that the main factor which distinguishes it from the biological definition, the change in emphasis from isolation to recognition, is merely semantic because the central importance of gene flow (more correctly, of potential recombination) remains unaltered (e.g. Raubenheimer and Crowe 1987). The main difference between the two concepts becomes apparent when considering the evolution of the intrinsic barriers that prevent genetic intermingling between species. If these barriers are due to mating behavior (or any other factor) that has evolved for the purpose of preventing gene flow, then the species involved are being defined by the action of true reproductive isolating mechanisms (Dobzhansky 1951; Williams 1966). In contrast, if the boundaries are incidental side effects of intraspecific processes such as adaptation to environmental conditions, then a definition such as Paterson's Recognition Concept would seem to reflect more accurately how species actually evolved.

When two taxa meet and produce unfit hybrids there may be selection for the evolution of mechanisms preventing hybridization. Such "reinforcement" is widely regarded as a cause of speciation and supposes that isolation evolves as a direct response to deleterious hybridization or "negative heterosis," to follow the terminology of Lambert et al. (1984). It therefore seems to us that a critical approach to assessing the Recognition Concept must involve estimating the frequency with which the process of reinforcement has been responsi-

---

*Current address for Michael G. Ritchie: Biological and Medical Sciences, Bute Medical Building, University of St. Andrews, St. Andrews, Fife, Scotland KY16 9TS

ble for the evolution of reproductive isolation. Paterson (1985) states that the Recognition Concept is a testable hypothesis but gives no precise details of what he considers a decisive test would involve. Masters et al. (1988) list four predictions which they consider come directly from the Recognition Concept: that characters affecting reproductive isolation (i.e., those of the Specific-Mate Recognition System, or SMRS) will show little geographic or historical variation; that the complexity of fertilization systems will be independent of the proximity of other species; that animal taxa with nonvisual sexual signals will contain many sibling species; and that speciation is almost always allopatric. However, we feel that these predictions are less decisive tests of the Recognition Concept. The stability of the SMRS is more of a test of the efficacy of stabilizing selection than of the Recognition Concept and probably lacks empirical evidence (Butlin, this volume; but see White et al., this volume). The predictions of complexity and speciation rates are yet to be tested, but are indirect and not conclusive. Sympatric species could have more "complex" SMRSs because greater diversity allows sympatry, or they could have diverged in order to prevent competition or interference, not to improve isolation. Also, although many models of sympatric speciation assume reinforcement following the evolution of hybrid inviability due to habitat specialization, this is not necessarily the case. If reproductive isolation was simply a pleiotropic consequence of specialization to the new habitat, independent of the other species, we cannot see how the Recognition Concept is necessarily violated. Coyne et al. (1988) have further criticized the resolving power of the predictions of the Recognition Concept proposed by Masters et al. (1988).

Given the above, it is not surprising to find that Paterson has been a leading critic of reinforcement. His criticisms are many, but most significant are models suggesting that the extinction of either one of the hybridizing taxa or one set of the genes that contribute to the negative heterosis is a much more likely outcome than reinforcement. It has been pointed out repeatedly that other models and selection experiments have had to maintain numbers artificially in order to prevent this (see Paterson 1978; Lambert et al. 1984; Spencer et al. 1986).

However, reinforcement is typically expected to take place following the secondary contact of two differentiated taxa. This is likely to occur at a narrow zone of parapatry, with the massive bulk of the two taxa not being directly exposed to the deleterious hybridization but always being present to flow into the contact region. Paterson (1978) has argued that this will counteract reinforcement, and this is undoubtedly true (see also Sanderson 1989). However, this gene flow will also counteract extinction by replacing alleles that are being selected against in the zone of parapatry (one could make an analogy with First World War trench warfare). Therefore total extinction of one form or set of alleles would seem to be unlikely or at best would be a localized and temporary effect. Presumably one reason why Paterson (1978), Lambert et al. (1984), and Spencer et al. (1986) favor extinction as an alternative to reinforcement is that their models are specifically of sympatric rather than parapatric interactions. Given the importance they place on the allopatric model of speciation (e.g. Paterson 1981), it is surprising that they do not state how they expect this sympatry to be achieved. Parapatry must be more likely.

There are several further possible outcomes of secondary contact at a zone of parapatry in addition to reinforcement or even extinction (see Endler [1977] or Barton and Hewitt [1983] for reviews). The simplest is neutral introgression where genes of the two taxa flow into each other unheeded. If dispersal is extremely long range relative to the distribution of the taxa, fusion could theoretically follow. Another possibility is an equilibrium situation where dispersal is balanced by selection against hybrids—a "tension zone" (Key 1974; Barton 1983). If this is the case we would expect to find unfit individuals within the hybrid zone, and there are clear cases of this in frogs, mice, and grasshoppers, for example (Hewitt 1988). Such zones can be very stable, and Barton and Hewitt (1985) listed over 170 examples, many of which were probably of ancient origin. This fact alone constitutes a considerable challenge to the reinforcement hypothesis (Murray 1972). Another outcome that is rarely discussed is amelioration of the hybrid dysfunction. We define this as selection for alleles and genotypes that reduce the negative heterosis in the zone by directly counteracting the unfitness of heterozygotes (or deleterious recombinant genotypes).

Largely in response to Paterson's work, we have been examining a hybrid zone in the acridid grasshopper *Chorthippus parallelus* for evidence of reinforcement in a parapatric situation, along with our colleague Roger Butlin. Here we would like to review some of this work as a "case history" of the outcome of secondary contact, and we shall then discuss the possibility of amelioration in further detail. We emphasize amelioration in our discussion only because it seems to us to be a relatively neglected possibility. Our available data from the study of *Chorthippus parallelus* certainly do not provide conclusive evidence for amelioration, but we hope this essay may stimulate others to investigate all the possible outcomes of secondary contact.

The Hybrid Zone in *Chorthippus parallelus*

During the last ice age, the Pyrenean mountain range was fairly extensively covered by ice, obviously greatly increasing this area's potential to act as a barrier to gene flow. The ice finally retreated around 10500–8000 B.P., creating a classic setting for secondary contact between many forms that had been confined to refugia during the ice age. The grasshopper *Chorthippus parallelus* shows evidence of this process with two subspecies *C. p. parallelus* and *C.*

p. erythropus forming a hybrid zone which runs along the main ridge of the mountain range (Butlin and Hewitt 1985a; Hewitt 1989). The typical habitat of both subspecies is moist meadow below an altitude of around 2100 meters; therefore they have come into contact only around the lower east and west ends of the mountain range and through the lower cols of the main mass of the mountains. Where the transition between the subspecies has been mapped, it is usually found to be coincident with the geographic border between France and Spain. Our work has concentrated on two transitions which occur through the Col de la Quillane and the Col du Pourtalet (Fig. 9.1).

Considerable selection against heterozygotes will have occurred when the subspecies made contact, because, as Hewitt et al. (1987) have shown, laboratory-produced hybrid males are sterile (see their Fig. 1). Several further crosses between pure *C. p. parallelus* and *C. p. erythropus* populations have been made since then, including populations from either side of the Col de la Quillane. These regularly produce sterile F1 males with severe testes dysfunction. The F1 females are apparently normal, and backcross males have intermediate testes structure and disturbed sperm bundle formation. Crosses of populations within each subspecies on either side of the Pyrenees produce normal male and female progeny, and a more recent set of crosses suggests that epistatic interactions between the subspecific genomes is involved (Virdee and Hewitt 1992).

It is difficult to imagine extinction within the hybrid zone of one set of the alleles that cause this sterility. The cols where the hybrid zones lie act as channels through which dispersal occurs. Grazing by sheep and cattle is common and ancient in this region, and the cols can provide an excellent habitat for this grasshopper, which is often extremely numerous. Continual inflow of alleles from the bulk of the subspecies range is likely to counteract any local extinction that selection (or drift) may achieve. It is conceivable, however, that local extinction of one set of alleles could occur within a very isolated patch in a heterogeneous distribution. At least one of the transects studied (through the Col de la Quillane) seems to display a patchy population structure, with some of the contact region still being forested and grasshoppers abundant only in clearings, especially by the roadside. The frequency of gene flow into such patches is clearly a critical factor that only further study can reveal.

Patches within a zone may be more persistent when the subspecies involved are adapted to different environments which are themselves patchily distributed within the zone. This scenario has been suggested for the hybrid zone between *Gryllus firmus* and *G. pennsylvanicus* (Harrison and Rand 1989). Harrison and Rand consider a multiplicity of such patches will make reinforcement more likely than in a traditional broadfronted contact zone. This is for two main reasons: the existence of relatively stable patches will counteract

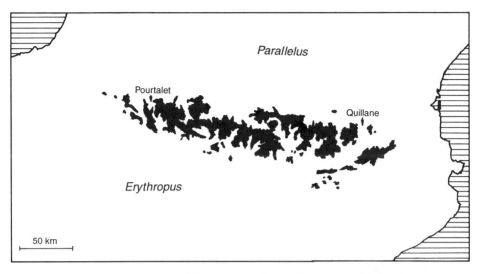

Fig. 9.1. The mountain range of the Pyrenees is a major geographic barrier separating *Chorthippus parallelus parallelus* and *C. p. erythropus*. The hybrid zone between them lies in low cols traversing the mountains, and the Col du Pourtalet and the Col de la Quillane are indicated. Land above 2000 meters in altitude is shaded.

extinction, therefore leading to a more prolonged period of hybridization; furthermore, the existence of many small contact zones around patches will provide multiple opportunities for local reinforcement. We do not consider these obvious consequences of patchiness. The age of such a patchy zone is no more than that of a broadfronted zone. Habitat choice would reduce the interaction between the two hybridizing forms, thus reducing the production of hybrids and further reducing the proportion of each population subject to reinforcing selection. Patches above a certain critical size persist because they are protectively ringed by a tension zone; there is an absence of hybridization within each patch. It may even be the case that fine patchiness could increase the likelihood of extinction through genetic drift.

Nichols et al. (1990) have described a patchy region of a hybrid zone that does not seem to involve habitat choice. Patches can form and persist where population density is low (see also Nichols 1989), allowing greater introgression of the subspecies (or at least overlapping of their ranges) than we would otherwise expect. These patches seem to persist because hybridization is reduced through population structure, which would hence reduce the selection for reinforcement.

Fitness and the Hybrid Zone

In order both to understand the current structure of any hybrid zone and to infer processes that may have occurred within it in the past, it is clearly important to ask how selection is operating within contemporary hybrid populations. One indirect demonstration that selection may be occurring is the observation of clines of varying width through the hybrid zone. For example, different estimates of cline width have been obtained for allozymes, morphology, song characteristics, and chromosomal characters for *C. parallelus* through the Col de la Quillane (see Hewitt 1989). One would expect narrower clines for characters under selection or those linked with such genes (Barton 1983). If selection is still operating within hybrid populations, we may expect to see males with reduced follicle size or disturbed sperm bundle formation within the hybrid zone. Consider the results of an analysis of male testes taken from a transect through the Col du Pourtalet, which has a more continuous distribution of grasshoppers than the transect at the Col de la Quillane. In the summer of 1985 males were collected from pure populations well away from the zone and at approximately 200-m intervals during a walk through the hybrid zone, in the center of the transition in morphology. Usually ten males were collected, their testes were dissected out and fixed, and the follicle length was measured. Any parasitized males (i.e., containing a nematode worm or dipteran maggot) were omitted from the analysis—these were not more common within the hybrid zone (cf. Sage et al. 1987).

Figure 9.2 shows how follicle length varied through the col. There is no significant variation among populations ($F = 1.535$, $df = 13,98$, NS), and no population has significantly shorter follicles than either of the parental populations. Hewitt et al. (1987) show mean follicle length of F1 males to be around 100 units and those of backcrosses to be around 150 units. All populations examined here had a mean follicle length of around 200 units or more (Fig. 9.2). An examination of sperm organization within the follicles also failed to indicate any poorly functioning testes, and no individual with testes typical of those of laboratory F1s or backcrosses was ever found.

Reinforcement and the Hybrid Zone

The lack of observable dysfunction within any contemporary hybrid population sampled so far has several possible explanations. The first of these that we would like to consider, which is the most important in the context of reinforcement and the Recognition Concept, is that assortative mating may have evolved within the contact zone, thereby preventing the production of deleterious heterozygous gene combinations. This assortment need only be with regard to genes contributing to the unfitness, and therefore is not incom-

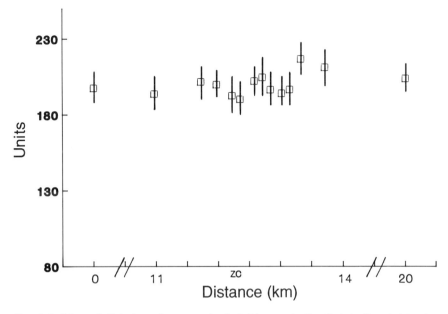

Fig. 9.2. Mean follicle length across the hybrid zone in the Col du Pourtalet, plus 95 percent confidence intervals. The populations at the extreme left and right are pure, and the approximate position of the center of the transition in morphology is indicated by ZC. F1 males have follicle lengths around 100 units; backcross males, around 150 units (80 units = 1 mm).

patible with the observation of considerable gene flow for some characters. We have adopted a variety of approaches to seek evidence of the operation of the process within the hybrid zone. Several involved what could be termed the "classical" approach to this, that is, seeking evidence of inverse clines in characters affecting the SMRS, as these are presumably indicative of greater premating isolation near the hybrid zone. Full accounts are available elsewhere, so we only summarize the major results here.

The SMRS of *Chorthippus parallelus* would appear to be typical of many acridid orthoptera, consisting largely of male stridulations and appropriate female preferences. However, additional factors such as pheromones are almost certainly involved (Ritchie 1990). There are several differences in the songs of the two subspecies, but these change through the Col de la Quillane in apparently monotonic curves, with no evidence of any increased divergence near the zone (Butlin 1989; Ritchie et al. 1990).

Variation in female preferences for pure males of the subspecies has been examined by allowing females from a transect through the Col de la Quillane a choice between two reference types of male, one of each subspecies. Variation among populations should reflect variation in female preference, and this is quite pronounced, suggesting the presence of substantial variation among populations, even within subspecies. This variation is not obvious in male song characters. The cline in female preference across the zone is extremely interesting. It is unusually narrow, which suggests selection of some form is operating on the preferences. There is even a suggestion that the most consistent preference for *C. p. parallelus* males occurs on the north side of the zone, as would be predicted if preferences there had been reinforced. However, this pattern is not a statistically better explanation of the variation than a sharply stepped cline model, and even the maximum level of preference shown is not strong (Butlin and Ritchie 1991).

Assortative mating between populations spanning the hybrid zone has been examined (Ritchie et al. 1989). The pattern is partly confounded by differences between the assortment shown by virgin and remating females, and between females of different wing morphs within one of the populations used. With virgin females, the strength of assortative mating decreases between populations from close to or within the hybrid zone, the opposite result from that predicted by reinforcement. If rematings are included, females from pure populations do not remate assortatively, producing a pattern with increased assortment near the zone. Though this superficially resembles the reinforcement pattern, once more it is not a better explanation than a step in assortment centered around the morphological transition.

Taken together, these experiments do not provide strong evidence suggesting that reinforcement is occurring within this hybrid zone, though they certainly do not exclude the possibility. This is partly because the approach of looking for inverse clines is not necessarily the best test of the reinforcement hypothesis, at least not when applied to hybrid zones where a hybrid population might be composed of mixed assorting genotypes. A more central prediction of the hypothesis is that assortative mating within a hybrid zone should result in offspring of high fitness relative to matings that are set up at random. This is because, if assortative mating has evolved in response to deleterious hybridization, preventing the expression of genes causing assortative mating should result in the reappearance of the hybrid unfitness.

We have therefore examined the relative fitness of offspring resulting from forced random matings versus free, potentially nonrandom matings within two hybrid populations taken from the center of the morphological transition at the Col du Pourtalet (full details are in Ritchie et al. 1992). We know that F1 hybrid males are sterile but have not measured other fitness parameters of them, such as hatch rates or viability. Therefore, a wide range of measures were made on the offspring produced during this experiment. These measures have also been carried out on pure populations of each subspecies, allowing a broad comparison of measures of fitness. Fitness parameters noted were: the proportion of families which hatched; number of nymphs hatching per egg pod; the time till hatching; the survival until adult per family; and the developmental rate per family. For one hybrid population, the testes of male offspring were dissected out, and testis follicle length was measured. Giving individuals the potential to mate nonrandomly did not significantly improve the fitness of the offspring for any of these parameters. Perhaps most significant is that the testes of males from one of the hybrid populations whose mothers were denied the opportunity to assort did not suffer any unfitness like that so obvious in the F1 and backcross laboratory-produced males (Fig. 9.3).

Even though there is no fitness difference due to assortment, all the fitness measures varied significantly among populations. However, this is not in the pattern one would expect were hybrid dysfunction occurring within the zone center. Figure 9.4 gives a typical result, that for survival from hatching till adult. Both hybrid populations hatched and survived as well as the "best" pure population. They hatched earlier and developed more quickly than one of the pure populations. In fact, a gradation across the populations for these last two parameters suggested a genetic difference between the subspecies with the hybrids simply having an intermediate value. Once more, we can find no evidence of strong fitness reductions associated with these contemporary hybrid populations.

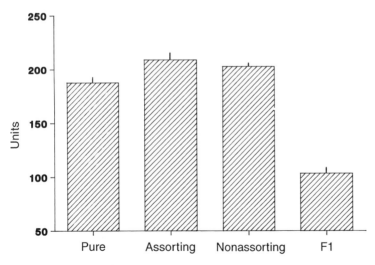

Fig. 9.3. Testes follicle lengths (+1 s.e.) of hybrid males whose mothers were either allowed or not allowed the opportunity to mate assortatively, plus those of pure and F1 males for comparison (80 units = 1 mm). There is no significant difference between the two mating groups.

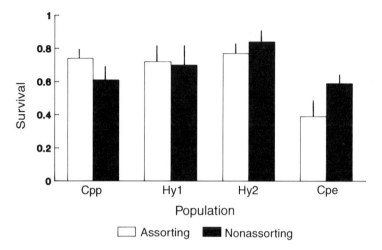

Fig. 9.4. Mean nymphal survival rate (i.e., from hatching until adult) per family (+1 s.e.), classified by whether the mother was given the opportunity to mate assortatively. Results are shown for a pure population of each subspecies and two hybrid populations. There is significant variation among populations, but mating group has no effect, nor is there a significant population by mating group interaction, such as would occur following reinforcement.

Problems Detecting Reinforcement, and Other Possible Outcomes

A prime reason for using the *C. parallelus* hybrid zone as a model system with which to test the reinforcement hypothesis was the observation that laboratory-produced hybrid males are sterile, and backcross males also have severely disrupted spermiogenesis. Thus, hybrid dysfunction should have exerted an influence on the fate of the contact between these taxa. However, we have not found a contemporary hybrid population displaying any indication of testicular dysfunction or any other sign that might indicate that the populations were less fit than parentals. The assortment and fitness experiment demonstrates that this is not due to the evolution of assortative mating within the populations sampled. This is consistent with most of the previous studies we have carried out looking for evidence of reinforcement at this hybrid zone. If the apparent increase in female preference for *C. p. parallelus* males north of the zone at the Col de la Quillane is due to selection originating within the zone, then such a process has apparently not occurred within the zone at the Col du Pourtalet (where the hybrid populations used in the fitness study were collected). Clearly it is desirable that both types of study can be repeated at similar sites.

However, it is still theoretically possible that reinforcement could be occurring undetected in this hybrid zone. Without sampling every few meters one cannot be certain that one has simply failed to detect a narrow region of dysfunction within which assortative mating has evolved. Models predict that the width of a region containing heterozygotes will be inversely proportional to the degree of selection against them (Barton 1983). Dispersal estimates appropriate to *C. parallelus* (of around 30 m per generation, Virdee and Hewitt 1992) and a selection coefficient of 0.5 (complete sterility of one sex) imply that a region of around 100 m width is theoretically possible were the sterility caused by a single gene ($w = 8\,d/s$ approximates the expected width, where w = width, d = dispersal, and s = selection [Barton and Hewitt 1981]). As the number of genes contributing to dysfunction increases, and recombination reduces the effective selection on any single individual, the region is likely to be wider. However, the extent of dysfunction seen under these conditions will be lessened, since few, if any, individuals will be heterozygous for most of the loci contributing to unfitness. We can therefore imagine two extremes: a very narrow region with sterile males, or a wide region containing near-normal males. Where between these two extremes the zone in *C. parallelus* lies will depend on the genetic architecture of the dysfunction and in particular whether the simple heterozygous dysfunction model is appropriate or if epistasis is primarily responsible (Virdee and Hewitt 1992).

There is a further point that should be remembered when considering the possibility of failing to detect a narrow zone of dysfunction. All our sampling was done within the morphological transition, because we expected the clines in most characters to be coincident (this is the case in most hybrid zones studied in detail [Barton and Hewitt 1985]). However, if a region of sterility (and possibly assortative mating) is displaced some distance from the zone center we may be sampling in the wrong place. Subsequent study has shown that the clines for different morphological characters do vary in both position and width, especially through the Col du Pourtalet (Bultin et al. 1991) and a marker on the X chromosome may be substantially displaced (Rubio, pers. comm.). Therefore, even in this well-studied system, decisive information concerning the possibility of reinforcement is still lacking. It is difficult to know how a truly decisive experiment could be carried out without identifying the genes responsible for dysfunction and the alleles carried by mating partners in the field.

As we stated in the beginning of this chapter, there are other potential outcomes of secondary contact with negative heterosis besides reinforcement. One is a narrow band of dysfunction at a stable equilibrium (with reinforcement probably being prevented by the swamping effects of gene flow into the region [Barton and Hewitt 1981; Sanderson 1989]). Another potential outcome is that recombination breaks up the many gene associations contributing to the unfitness, causing a wide zone composed of many coincident gene clines each with low selection coefficients. Reinforcement would seem to be unlikely under these conditions. Both these possibilities fit with conventional tension zone theory (e.g. Barton 1983). A further possibility is amelioration of the hybrid unfitness through the evolution of modifier genes or gene combinations which counteract or prevent its expression, rather than preventing heterozygotes from being formed.

It is not yet possible to distinguish any of these possibilities for our study of the hybrid zone in *C. parallelus*. The lack of observed dysfunction might favor the broad and recombined or ameliorated hypotheses, but as stated above, a narrow zone of dysfunction with or without reinforcement could still be present. Further work will have to be carried out before we can decide which is more likely. In particular, the genes causing the F1 sterility will have to be identified, and their distribution in the field mapped.

Amelioration and Other Hybrid Zones

The possible outcome that has received least attention is amelioration, yet this could in some ways be a more likely outcome than reinforcement. We would therefore finally like to briefly discuss this hypothesis further. Two main questions will be addressed: Is there evidence of such a process and is it more or less likely than reinforcement?

First, can we find evidence of such a process? A review of the best studied hybrid zones (Barton and Hewitt 1985) found ten examples where hybrids were demonstrably less fit than parentals, against six examples where no dysfunction had been found. However, measures of fitness from the laboratory (usually of F1 hybrids) and contemporary hybrids found in the field are usually not comparable. Distinguishing these two types of studies possibly implies that fit hybrids are more likely to be found in the field (three field studies versus one laboratory study). Inviable hybrids are found in three laboratory studies to four field ones. Such small numbers of comparisons do not allow any generalizations to be made. In particular, we need more studies that assess the fitness of both F1 and F2 hybrids and contemporary field hybrids before any estimate can be made of the likely frequency of amelioration. An example of this is the work of Barton and Hewitt (1981) and Nichols and Hewitt (1988) with *Podisma pedestris* where clear hybrid dysfunction exists in the laboratory and there are several indications of weaker nymphal unfitness also occurring in the field. Even here the two types of study are not directly comparable because no laboratory fitness estimates have been made on backcross and F2 generations.

Plainly one of the greatest problems facing any attempt to identify amelioration is the difficulty in distinguishing a broad, recombined zone from one created following selection against dysfunction. A detailed knowledge of the genetic architecture of the hybrid dysfunction would be essential to allow an assessment of the likelihood of the broad recombined option. A simple polygenic determination would obviously favor this. Also, one could cross between contacts in a subdivided hybrid zone such as occurs in *C. parallelus*. Dysfunction might reappear if different forms of amelioration had occurred in different contacts. Other possible approaches could be to cross laboratory-produced into field hybrids and compare the dysfunction found, and to examine the hybrid zone to assess the relative frequencies of different recombinant products in conjunction with laboratory measures of their fitness.

Perhaps the only clear example of amelioration to date is the pattern of chromosomal rearrangements found in the hybrid zone in central England between two races of the shrew *Sorex araneus*. The parental karyotypes contain different metacentrics following centric fusion. Hybrids would face severe meiotic disturbance due to monobrachial homology. However, in the hybrid zone the four basic chromosome arms involved are very frequently found as acrocentrics. These would be favored as they cannot produce the monobrachial hybrids whoever they mate with (Searle 1987).

Finally, what would be genetically more likely, reinforcement or amelioration? Sanderson (1989) considered the effect of a single locus modifier of either dysfunction or assortative mating on a tension zone. He found that either type of modifier could increase in frequency only if there was little selection against the modifier away from the zone, and that the gene had a great effect. Otherwise gene flow into the zone swamped the evolution of the modifier. If established, either modifier would increase zone width, but with a greater production of heterozygotes in the case of amelioration. Relatively little is known as yet concerning how genes act to cause hybrid dysfunction and mating behavior; therefore it is difficult to evaluate a model assuming such simple genetic control. However, we believe some factors might favor amelioration. The first is a lack of the need for strong linkage. The evolution of prezygotic isolation requires that genes causing assortative mating are maintained in strong linkage disequilibrium with one set of "dysfunction" genes. Recombination between the gene associations is a major difficulty with models of the process (e.g. Felsenstein 1981). There may be less need for such linkage with modifier genes, which could act on heterozygotes rather than in concert with specific parental allelic combinations. The genes could therefore be favored with both parental races, which would greatly increase the probability of incorporation (Sanderson 1989). A second factor that might favor the process is the possible role of single genes. It has been recognized that a simple genetic control of behavior and dysfunction would aid reinforcement (e.g. Butlin 1987). However, it seems likely that characters affecting behavioral isolation will be polygenic (Charlesworth et al. 1987; Butlin and Ritchie 1989). Though some single genes may have major effects (Kyriacou and Hall 1980; Beiles et al. 1984;

Stratton and Uetz 1986), this does not seem to be common. The genetics of characters affecting hybrid dysfunction are relatively difficult to analyze, and available studies might constitute a biased sample, but there is clear evidence for the sex chromosomes being involved (Charlesworth et al. 1987; Coyne and Orr 1989). This would presumably make the dysfunction similar to a character under relatively simple genetic control because the genes responsible would be strongly linked and, again, single genes can be involved (Coyne and Charlesworth 1986). The nature of modifier genes is virtually unanalyzed, though studies in other contexts show this analysis to be possible (e.g. Clarke and Sheppard 1960; Clarke and McKenzie 1987). We would expect there to be a greater scope for single genes having a major effect on countering dysfunction in the case of amelioration.

An example potentially illustrating several of these genetic features is the species pair *Drosophila mojavensis* and *D. arizonensis*. Their ranges overlap in the sonoran desert, where they show greater premating isolation than in allopatry. Male offspring of a *D. mojavensis* male and a *D. arizonensis* female are sterile, though the other three reciprocal offspring classes are fertile. Following an elegant series of studies, E. Zouros and co-workers have elucidated some features of the genetics of the system. Female preference for homogametic males is determined by major genes on at least two chromosomes, with all other chromosomes also contributing a little to preference. Furthermore, the male characters involved are controlled by major genes on two chromosomes, neither of which also carries either of the two factors having greatest effect on female preference (Zouros 1981). Whether any of these major genes are in fact single genes or clusters of genes on the same chromosome is unknown. Male sterility has also been the subject of genetic analysis. One cause of this is an interaction between the Y chromosome of *D. arizonensis* with at least two *D. mojavensis* autosomes (Vigneault and Zouros 1986). A male containing only a *D. arizonensis* Y chromosome with all other chromosomes originating from *D. mojavensis* will be fully sterile (Pantazidis and Zouros 1988). However, there is a single gene on the fourth chromosome of *D. arizonensis* that can "rescue" this hybrid sterility. Only a single copy of this gene is necessary (Pantazidis and Zouros 1988). Therefore, in this case, it seems that a gene with the potential to act as an "amelioration" gene was already within the gene pool of one of the hybridizing taxa. We give this example only to illustrate elements of a potential system of amelioration ("lethal hybrid rescue," a single gene which makes the normally inviable male hybrids between *Drosophila simulans* and *D. melanogaster* viable, might be another example, and other *Drosophila* genes with similar effects have been found [Hutter et al. 1990]). It seems hybrids between *Drosophila mojavensis* and *D. arizonensis* are virtually never found in nature (Zouros, pers. comm.).

Zouros and d'Entremont (1980) argued that reinforcement was a likely explanation for the behavioral isolation found between sympatric races, and many subsequent workers have agreed with this interpretation. However, we (and no doubt Hugh Paterson) would disagree with this. Had the species mated with each other when making secondary contact, the relatively weak postmating isolation suggests that substantial genetic introgression would have occurred and been detectable within the sympatric races. We think this is more likely to represent a case where sympatry was possible due to the prior evolution of ethological barriers in those subraces which are now sympatric. To conclude, Paterson (1978) argued that: "When two formerly allopatric populations become sympatric after diverging genetically there are two possible nontrivial outcomes, not just one [i.e., extinction as well as reinforcement]. Few authors appear to have noted this. It would appear certain that natural selection will in all cases act to increase mean fitness by eliminating the genetic basis of the heterozygote disadvantage."

He further argued that reinforcement had received something akin to positive discrimination due to a (perhaps subliminal) wish to see natural selection "producing good things" (Paterson 1985). Objective genetic considerations, largely resulting from Paterson's work, have certainly revealed difficulties with the process (see Butlin 1987; Spencer et al. 1986). In the quotation above Paterson clearly saw extinction as a more likely outcome. We would say here that there are at least four possible outcomes: stable tension zones, ameliorated hybrid zones, the extinction of one taxon, or reinforcement. We suspect that amelioration may be a more likely means of eliminating heterozygote disadvantage than either reinforcement or extinction, but tension zones probably remain the most likely outcome.

Summary

Critical to an evaluation of Paterson's Recognition Concept of species is an understanding of the role of the reinforcement model of speciation in creating premating isolation. Paterson criticized this theory by arguing that extinction was a more likely outcome of negative heterosis. We argue that this is not the case where secondary contact occurs at a parapatric boundary, which is the most likely aftermath of genetic divergence in natural settings. Studies of hybrid zones which form at such contacts do not provide strong evidence of reinforcement. It is as yet unclear what is the most common outcome of the negative heterosis, however. Some hybrid zones show evidence of narrow regions of unfitness persisting in the field. If gene arrangements are responsible for the hybrid unfitness, recombination might break them up. A further alternative response is the amelioration of hybrid unfitness through the evolution of modifier genes diminishing the strength of the deleterious heterozygous effects.

Acknowledgments

It is a pleasure to acknowledge Roger Butlin, who contributed greatly to all the work and ideas discussed in this chapter. All other members and allies of the University of East Anglia grasshopper group have contributed in many ways, particularly Marise East and Neil Sanderson. Nick Barton, Alexandre Piexoto, and one anonymous referee gave constructive comments on the manuscript, and Hilary Ritchie was a great help in the field. Grasshoppers were collected in the Parc National de Pyrenees Occidentales with the kind permission of the authorities. Most important, we are grateful to the Science and Engineering Research Council for funding.

References

Barton, N.H. 1983. Multilocus clines. Evolution 37:454–471.
Barton, N.H., and G.M. Hewitt. 1981. Hybrid zones and speciation. Pp. 109–145 in W.R. Atchley and D.S. Woodruff, eds. Evolution and speciation: Essays in honour of M.J.D. White. Cambridge University Press, Cambridge.
———. 1985. Analysis of hybrid zones. Annu. Rev. Ecol. Syst. 16:113–148.
Beiles, A., G. Heth, and E. Nevo. 1984. Origin and evolution of assortative mating in actively speciating mole rats. Theor. Pop. Biol. 26:265–270.
Butlin, R.K. 1987. Speciation by reinforcement. Trends Ecol. Evol. 2:8–13.
———. 1989. Reinforcement of premating isolation. Pp. 158–179 in D. Otte and J.A. Endler, eds. Speciation and its consequences. Sinauer, Sunderland, Mass.
Butlin, R.K., and G.M. Hewitt. 1985a. A hybrid zone between *Chorthippus parallelus parallelus* and *Chorthippus parallelus erythropus* (Orthoptera: Acrididae): Morphological and electrophoretic characters. Biol. J. Linn. Soc. 26:269–285.
———. 1985b. A hybrid zone between *Chorthippus parallelus parallelus* and *Chorthippus parallelus erythropus* (Orthoptera: Acrididae): Behavioral characters. Biol. J. Linn. Soc. 26:287–299.
Butlin, R.K., and M.G. Ritchie. 1989. Genetic coupling in mate recognition systems: What is the evidence? Biol. J. Linn. Soc. 37:237–246.
———. 1991. Variation in female mate preference through a grasshopper hybrid zone. J. Evol. Biol. 4:227–240.
Butlin, R.K., M.G. Ritchie, and G.M. Hewitt. 1991. Comparisons among morphological characters and between localities in the *Chorthippus parallelus* hybrid zone (Orthoptera: Acrididae). Philos. Trans. R. Soc. Lond. B 334:297–308.
Charlesworth, B., J.A. Coyne, and N.H. Barton. 1987. The relative rates of evolution of sex chromosomes and autosomes. Am. Nat. 130:113–146.
Clarke, C.A., and P.M. Sheppard. 1960. The evolution of mimicry in the butterfly *Papilio dardanus*. Heredity 14:163–173.
Clarke, F.M., and J.A. McKenzie. 1987. Developmental stability of insecticide resistant phenotypes in blowfly: A result of canalysing natural selection. Nature 325:345–346.
Coyne, J.A., and B. Charlesworth. 1986. Location of an X-linked factor causing sterility in male hybrids of *Drosophila simulans* and *D. mauritiana*. Heredity 57:243–246.
Coyne, J.A., and H.A. Orr. 1989. Two rules of speciation. Pp. 180–207 in D. Otte and J.A. Endler, eds. Speciation and its consequences. Sinauer, Sunderland, Mass.

Coyne, J.A., H.A. Orr, and D.J. Futuyma. 1988. Do we need a new species concept? Syst. Zool. 37:190–200.
Dobzhansky, T. 1951. Genetics and the origin of species. 3d ed. Columbia University Press, New York.
Endler, J.A. 1977. Geographic variation, speciation and clines. Princeton University Press, Princeton, N.J.
Felsenstein, J. 1981. Skepticism towards Santa Rosalia; or, why are there so few kinds of animals? Evolution 35:124–138.
Gosalvez, J., C. Lopez-Fernandez, J.L. Bella, R.K. Butlin, and G.M. Hewitt. 1988. A hybrid zone between *Chorthippus parallelus parallelus* and *Chorthippus parallelus erythropus* (Orthoptera: Acrididae): Chromosomal differentiation. Genome 30:656–663.
Harrison, R.G., and D.M. Rand. 1989. Mosaic hybrid zones and the nature of species boundaries. Pp. 111–133 *in* D. Otte and J.A. Endler, eds. Speciation and its consequences. Sinauer, Sunderland, Mass.
Hewitt, G.M. 1988. Hybrid zones—natural laboratories for evolutionary studies. Trends Ecol. Evol. 3:158–167.
———. 1989. The subdivision of species by hybrid zones. Pp. 85–110 *in* D. Otte and J.A. Endler, eds. Speciation and its consequences. Sinauer, Sunderland, Mass.
Hewitt, G.M., R.K. Butlin, and T.M. East. 1987. Testicular dysfunction in hybrids between parapatric subspecies of the grasshopper *Chorthippus parallelus*. Biol. J. Linn. Soc. 31:25–34.
Hutter, P., J. Roote, and M. Ashburner. 1990. A genetic basis for inviability of hybrids between sibling species of *Drosophila*. Genetics 124:909–920.
Key, K.H.L. 1974. Speciation in the Australian morabine grasshoppers: Taxonomy and ecology. Pp. 43–56 *in* M.J.D. White, ed. Genetic mechanisms of speciation in insects. Australia and New Zealand Book Co., Sydney.
Kyriacou, C.P., and J.C. Hall. 1980. Circadian rhythm mutations in *Drosophila melanogaster* affect short-term fluctuations in the male's song. Proc. Natl. Acad. Sci. 77:6729–6733.
Lambert, D.M., M.R. Centner, and H.E.H. Paterson. 1984. A simulation of the conditions necessary for the evolution of species by reinforcement. S. Afr. J. Sci. 80:308–311.
Masters, J.C., R.J. Rayner, I.J. McKay, A.D. Potts, D. Nails, J.W. Ferguson, B.K. Weissenbacher, M. Allsop, and M.L. Anderson. 1987. The concept of species: Recognition versus isolation. S. Afr. J. Sci. 83:534–537.
Murray, J. 1972. Genetic diversity and natural selection. Oliver and Boyd, Edinburgh, Scotland.
Nichols, R.A. 1988. Genetical and ecological differentiation across a hybrid zone. Ecol. Entomol. 13:39–49.
———. 1989. The fragmentation of tension zones in sparsely populated areas. Am. Nat. 134:969–977.
Nichols, R.A., E.A. Humpage, and G.M. Hewitt. 1990. Gene flow and the distribution of karyotypes in the alpine grasshopper *Podisma pedestris* (Orthoptera: Acrididae). Boll. San. Veg. Plagas 20:373–379.
Pantazidis, A.C., and E. Zouros. 1988. Location of an autosomal factor causing sterility in *Drosophila mojavensis* males carrying the *D. arizonensis* Y chromosome. Heredity 60:299–304.
Paterson, H.E.H. 1978. More evidence against speciation by reinforcement. S. Afr. J. Sci. 74:369–371.

———. 1981. The continuing search for the unknown and unknowable: A critique of contemporary ideas on speciation. S. Afr. J. Sci. 77:113–119.

———. 1985. The recognition concept of species. Pp. 21–29 *in* E.S. Vrba, ed. Species and speciation. Transvaal Museum Monograph No. 4. Pretoria.

Raubenheimer, D., and T.M. Crowe. 1987. The recognition concept of species: Is it really an alternative? S. Afr. J. Sci. 83:530–534.

Ritchie, M.G. 1988. A Pyrenean hybrid zone in the grasshopper *Chorthippus parallelus* (Orthoptera: Acrididae): Descriptive and evolutionary studies. Ph.D. dissertation, University of East Anglia, Norwich, England.

———. 1990. Are differences in song responsible for assortative mating between subspecies of the grasshopper *Chorthippus parallelus* (Orthoptera: Acrididae)? Anim. Behav. 39:685–691.

Ritchie, M.G., R.K. Butlin, and G.M. Hewitt. 1989. Assortative mating across a hybrid zone in *Chorthippus parallelus* (Orthoptera: Acrididae). J. Evol. Biol. 2:339–352.

———. 1990. Mating behavior across a hybrid zone in *Chorthippus parallelus* (Zetterstedt). Boll. San. Veg. Plagas 20:269–275.

———. 1992. Fitness consequences of potentially assortative mating inside and outside a hybrid zone in *Chorthippus parallelus* (Orthoptera: Acrididae): Implications for reinforcement and sexual selection theory. Biol. J. Linn. Soc. 45:219–234.

Sage, R.D., D. Heyneman, K.-C. Lim, and A.C. Wilson. 1986. Wormy mice in a hybrid zone. Nature 324:60–63.

Sanderson, N. 1989. Can gene flow prevent reinforcement? Evolution 43:1223–1235.

Searle, J. 1986. Factors responsible for a karyotypic polymorphism in the common shrew *Sorex araneus*. Proc. R. Soc. Lond. B. 229:277–298.

Spencer, H.G., B.H. McArdle, and D.M. Lambert. 1986. A theoretical investigation of speciation by reinforcement. Am. Nat. 128:241–262.

Stratton, G.E., and G.W. Uetz. 1986. The inheritance of courtship behavior and its role as a reproductive isolating mechanism in two species of *Schizocosa* wolf spiders (Araneae: Lycosidae). Evolution 40:129–141.

Vigneault, G., and E. Zouros. 1986. The genetics of asymmetrical male sterility in *Drosophila mojavensis* and *D. arizonensis* hybrids: Interactions between the Y-chromosome and autosomes. Evolution 40:1160–1170.

Virdee, S.R., and G.M. Hewitt. 1990. Ecological components of a hybrid zone in the grasshopper *Chorthippus parallelus* (Zetterstedt) (Orthoptera: Acrididae). Boll. San. Veg. Plagas 20:373–379.

———. 1992. Postzygotic isolation and Haldane's rule in a grasshopper. Heredity 69:527–538.

Wallace, A.R. 1989. Darwinism. Macmillan. London.

Williams, G.C. 1966. Adaptation and natural selection. Princeton University Press, Princeton, N.J.

Zouros, E. 1981. The chromosomal basis of sexual isolation in two sibling species of *Drosophila: D. mojavensis* and *D. arizonensis*. Genetics 97:703–718.

Zouros, E., and C.J. d'Entremont. 1980. Sexual isolation among populations of *Drosophila mojavensis:* Response to pressure from a related species. Evolution 34:421–430.

10

Interpretation of Mating between Two Bedbug Taxa in a Zone of Sympatry in KwaZulu, South Africa

MAUREEN COETZEE, RICHARD H. HUNT,
AND DEBRA E. WALPOLE

Medical Entomology, Department of Tropical Diseases
School of Pathology of the South African Institute for Medical Research
and the University of the Witwatersrand

THE COSMOPOLITAN BEDBUG, *Cimex lectularius* Linneaus, and the tropical bedbug, *Cimex hemipterus* (Fabricius), were originally assigned specific status on the grounds of external morphology alone. However, the value of a traditional taxonomic approach to the designation of species in the genus *Cimex* has been questioned by Usinger (1966) because of overlapping morphological differences. Cross-mating studies in the laboratory have shown that *C. lectularius* and *C. hemipterus* are largely intersterile (Omori 1939; Usinger 1966). It is extremely rare to find fertile eggs resulting from such crosses (Newberry 1988). The mating possibilities are shown in diagrammatic form in Figure 10.1. Usinger (1966) concluded that the populations represented two biologically distinct species, sterility being regarded as "a reproductive barrier," that is, a "postmating isolating mechanism" (*sensu* Mayr 1963). Omori (1939) reported that the crosses between *C. lectularius* females and *C. hemipterus* males were infertile and, furthermore, that the females' life spans were shortened. He suggested that these effects of the interspecific cross were the reason for the absence of *C. lectularius* in India, Formosa (Taiwan), and Southeast Asia. Omori's explanation for the distribution pattern of the two taxa implies that the *C. lectularius* females readily mate with *C. hemipterus* males under natural conditions.

The two bedbug forms occur in close sympatry in DDT-sprayed houses in northern KwaZulu, South Africa. It has been shown (Newberry et al. 1987)

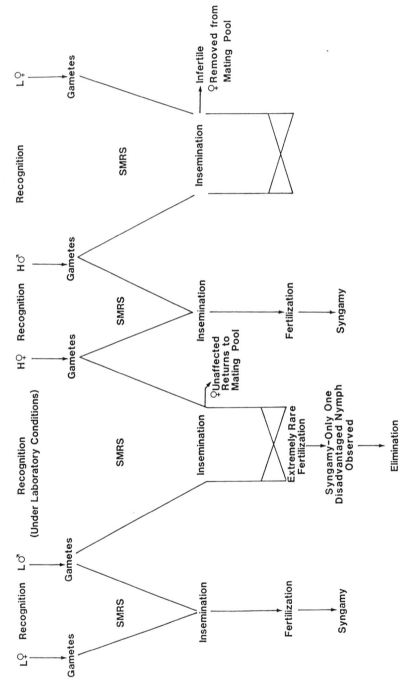

Fig. 10.1. Schematic representation of the mating behavior of two bedbug populations. L = *lectularius*, H = *hemipterus*; "syngamy" after King (1972).

that, in some cases, after a few months *C. lectularius* appears to be replaced by *C. hemipterus* in houses where they both occur. To explore these observations further, laboratory cross-mating experiments were set up. As a consequence of this work Walpole (1988) reported that the mating of *C. lectularius* females with *C. hemipterus* males resulted in the development of an easily distinguishable, permanent mark on the abdomen of the females. The mark was subsequently used to determine whether natural cross-mating was occurring in the area of sympatry in northern KwaZulu. Walpole and Newberry (1988) reported that a high percentage of a sample of *C. lectularius* females (69 percent, $N = 16$) collected from sympatric populations had, in fact, mated with *C. hemipterus* males. They suggested that this observation might be important in understanding the distribution of these two populations in South Africa. Newberry (1989*a*) provided further data supporting this suggestion, showing that in seven huts approximately 44 percent ($N = 219$) of female *C. lectularius* had cross-mated.

This chapter discusses the above results in terms of the two current concepts of species, these being the Isolation Concept (Dobzhansky 1947, 1970; Mayr 1963, 1985) and the Recognition Concept (Paterson 1978, 1985, 1988).

The Species Concepts

The phrase "species as a field for gene recombination" (Carson 1957) is not in question, but controversy has arisen as to how the limits to these fields are set and "how these limits arise ab initio" (Paterson 1985). A brief outline of the two current concepts of species and speciation is given below and then discussed further in the light of the observed cross-mating of the two bedbug taxa in nature.

The Isolation Concept of Species

Central to the Isolation Concept is the idea that species are reproductively isolated from one another. Mayr (1963) defined species as "groups of actually or potentially interbreeding natural populations which are reproductively isolated from other such groups." Gene flow from one species to another is prevented or limited by "ad hoc characters," the isolating mechanisms, which act to protect the genetic integrity of a species (Mayr 1963). These isolating mechanisms have been classified (Mecham 1961) into two groups (1) premating isolating mechanisms, those that prevent interspecific crosses from actually occurring, and (2) postmating isolating mechanisms, those that reduce success of interspecific crosses, should they actually occur, by gametic or zygotic mortality, hybrid inviability, or hybrid sterility. Dobzhansky (1970) described the premating isolating mechanisms as seasonal, habitat, ethological, or mechanical isolation. Mayr (1963) further suggested that, while premating barriers are very efficient at keeping one species isolated from another, should they fail the postmating mechanisms would act as a second set of barriers. Both Mayr (1963) and Paterson (1985) have pointed out that this concept is a relational one because, under the Isolation Concept, a species is defined purely in terms of its relationship to other species. Mayr (1963) stated that isolating mechanisms are ad hoc characters, that is, they are acquired through natural selection. This, however, was not in agreement with his view that speciation generally occurs in allopatry. Accordingly, he assumed that speciation occurred in allopatry as a consequence of pleiotropic change. Other isolating mechanisms could, he thought, then arise in sympatry as a result of natural selection (i.e., by reinforcement). Paterson (1985) has pointed out that referring to characters as mechanisms implies that they have evolved under natural selection, which, in turn, implies that they must have arisen due to natural selection in sympatry or at least parapatry. The problems of the reinforcement model have been dealt with extensively by Paterson (1978); Lambert and Paterson (1984); Lambert et al. (1984); Spencer et al. (1986); Masters et al. (1987); and Ferguson (1990).

Recognition Concept of Species

The Recognition Concept of species (Paterson 1978, 1985) defines a species as "that most inclusive population of individual biparental organisms which share a common fertilization system." This concept stresses the importance of how fertilization is brought about between individuals of the same species in their natural habitat. Paterson (1978) discusses a series of adaptations (*sensu* Williams 1966) that are involved in signaling between mating partners. The specific response of one partner to the specific signal from the other partner is regarded as an act of "recognition" (Paterson 1985). Thus, fertilization is preceded by recognition in this sense. The signals exchanged by mating partners have been collectively named the Specific-Mate Recognition System (SMRS) (Paterson 1978, 1985). Thus, in motile organisms, the SMRS is the major component of the fertilization system which brings about positive assortative mating between mating partners.

Paterson's concept defines the species in terms of individuals that share a common fertilization system and emphasizes that species are "incidental consequences of adaptive evolution" (Paterson 1985) during which the fertilization system and other adaptive characters adjust to the conditions of the environment to which they are restricted.

Speciation in Bedbugs

Under the Recognition Concept, to assess the genetical species status of *C. lectularius* and *C. hemipterus* populations that occur in sympatry in northern KwaZulu, one of the questions that needs to be answered is how frequently cross-mating is occurring in nature. The main difficulty in determining this is that female bedbugs mate several times during their lifetime (Usinger 1966). If one assumes that the two forms are present in equal numbers and that they mate only once, then one would expect 50 percent of the females to have been cross-mated. However, each female can mate several times, and, therefore, figures are difficult to interpret precisely and may not be a true reflection of actual mating frequencies. For example, in mixed populations with equal numbers of both taxa, if the probability of *C. lectularius* females mating with *C. lectularius* males is 80 percent, then calculating for four matings and assuming that the marked (Walpole 1988) females all survive, 59 percent of the females would be marked (Ferguson, pers. comm.). In fact, 69 percent (Walpole and Newberry 1988) and 44 percent (Newberry 1989*a*) of samples of female *C. lectularius* had cross-mated with *C. hemipterus* males in the field. Although no conclusions can be drawn as to random matings, it is clear that these crosses commonly occur in nature.

The frequency of homogamic matings involving *C. lectularius* females has not been given in the published data. Furthermore, it has not been determined whether male *C. lectularius* mate readily with female *C. hemipterus* under natural conditions. This is because crosses in this direction leave no detectable mark on the female (Walpole 1988; Newberry 1989*b*). However, this cross occurs readily in the laboratory, resulting in egg batches containing a high proportion of abnormal eggs (Omori 1939; Newberry 1988, 1989*b*).

Under the Isolation Concept, the frequency of mating is not important, only the barriers preventing introgression.

Another problem under the Recognition Concept is deciding if the environment in northern KwaZulu constitutes the bedbugs' "normal habitat," that is, can a habitat polluted with DDT be considered normal? The increase in bedbug numbers where malaria control programs are being carried out has been directly attributed to DDT in sprayed houses (Rafatjah 1971; Newberry et al. 1984). Bedbugs have been found in large numbers in DDT-sprayed houses while signs of infestation in unsprayed houses were significantly less (Newberry et al. 1984). Furthermore, DDT is thought to act as an irritant that stimulates bloodsucking insects to feed more often and to be more active (Rafatjah 1971). This may be important as there is evidence the mating in *C. lectularius* is positively associated with feeding (Usinger 1966). Under these circumstances it is debatable whether the bedbugs in northern KwaZulu are, in fact, under normal environmental conditions. One needs to know whether the presence of DDT in houses has a significant effect on mating and whether positive assortative mating occurs in the field. If it is shown that in unsprayed huts the two do not mate commonly, and that DDT not only produces greatly increased populations of bugs but also causes a breakdown of the SMRS which operates in unsprayed huts, then conclusions as to the specific status of the two taxa would change radically.

Under the Isolation Concept this, too, poses a problem as the definition of species applies to "natural" populations.

The results from KwaZulu still, however, need to be interpreted in terms of the genetic species status of the two populations. In order to do this we have assumed that the proportion of cross-matings of the bedbugs was not greatly influenced by the presence of DDT.

When applying the Isolation Concept, under which the genetic integrity of two species is protected by ad hoc isolating mechanisms (Mayr 1963), the data reported by Walpole and Newberry (1988) and Newberry (1989a) could be interpreted as follows. As no (or very weak) premating isolating mechanisms appear to be operating, the genetic integrity of the gene pools is being protected by postmating isolating mechanisms. The almost complete lack of viable egg production (i.e., sterility) is one of the postmating isolating mechanisms. No one has determined whether the isolating mechanism is gametic or zygotic, but the laboratory results suggest that it probably involves gamete mortality. Usinger (1966) reported that Davis showed the seminal fluid to be toxic to the *C. lectularius* females. Davis (1965), however, reported that artificial insemination of *C. lectularius* with sperm of this taxon and seminal fluid of *C. hemipterus* resulted in viable egg production. The "mark" reported by Walpole (1988) may possibly be a reaction to the sperm of *C. hemipterus*. Whatever the case, the use of the Isolation Concept would result in *C. lectularius* and *C. hemipterus* being recognized as two genetical species and that the mechanism responsible for their "reproductive isolation" is a postmating one, either gamete or zygote mortality. This interpretation would be supported by formal taxonomy in that they are morphologically distinct (see Usinger 1966; Mayr 1969:146) and possess different chromosome complements (Ueshima 1966), although *C. lectularius* is variable for the latter characteristic and overlaps with *C. hemipterus* in this respect (*C. lectularius*, $2n = 29-38$, *C. hemipterus*, $2n = 31$ [Usinger 1966]). Predictions made under the Isolation Concept would involve selection of characters leading to the establishment of premating isolating mechanisms and ultimately result in the sympatric coexistence of the two species with no interspecific mating. Data so far collected do not support this expectation.

The application of the Recognition Concept of species to the same set of data results in a very different interpretation. In motile animals the SMRS component of the fertilization system is of major significance in bringing about positive assortative mating. In the case of *C. hemipterus* females and *C. lectularius* males, mating has not been recognized in nature. However, in the case of *C. lectularius* females and *C. hemipterus* males, mating is certainly common. In other words, on evidence from field sites, *C. lectularius* females and *C. hemipterus* males, at least, share a common SMRS. This is so even though the cross-matings are ineffective since they do not result in successful offspring. Such matings also usually prevent the females being fertilized at subsequent matings, although this may depend on the age of the *C. hemipterus* males (Newberry 1989*b*). Under these circumstances one would predict from population genetic theory that the cause of the disadvantage would be eliminated under natural selection (Li 1955). This expectation seems to be borne out in tropical areas of the Oriental Region where *C. hemipterus* is the common form (Omori 1939). Although *C. lectularius* has been reported to have been introduced into this region by humans, it has not established itself there (Omori 1941; Usinger 1966). In KwaZulu *C. lectularius* seemingly disappears in the presence of sufficient numbers of *C. hemipterus* (Newberry et al. 1987; Newberry and Mchunu 1989), especially when *C. lectularius* constitutes less than 75 percent of a mixed infestation (Newberry 1989*a*).

Applying the theory of the Recognition Concept, these data, although incomplete, strongly support the view that *C. lectularius* and *C. hemipterus* are conspecific. For example, they appear to possess fertilization systems which have much in common as demonstrated by the mating that occurs in sympatry and their evident inability to coexist as species can.

Discussion

The postmating isolating mechanisms of the Isolation Concept involve "sterility" in one form or another. It has long been realized that sterility alone is an inadequate and unacceptable basis for delimiting a "field for gene recombination" (Darwin 1859:248; Mayr 1942:119; Mayr 1963:90; Paterson 1985, 1988). Nevertheless, some authors often write as if sterility alone is sufficient to delimit a species (e.g. Mayr 1963:551, 1985; Coyne et al. 1988; Newberry and Brothers 1990).

Mayr (1985) used the morabine grasshoppers of Australia as an example of insects in which "sterility is the principal or sometimes even exclusive isolating mechanism." However, Key (1974) pointed out that although some island forms of *"Moraba viatica,"* for example, did not produce fertile hybrids, there was gene exchange between each of these and other island and mainland populations. Key thus concluded that they should be regarded as "races" of the same species even though there were morphological and chromosomal differences between them. As he pointed out, "If the heterozygotes are at a significant selective disadvantage, as they tend to be with chromosomal rearrangements especially, the low vagility of the insects will ensure that individuals of the new form and their hybrid progeny will not be able to penetrate far into the solid phalanx of the old form before the invading gene or chromosomal rearrangement is eliminated from the lineage." Key based his conclusions on the population genetical theory of heterozygote disadvantage. White (1974), on the other hand, preferred to regard the *viatica* group as comprising five or six "good" species, although they, too, could not coexist for reasons similar to the ones prevailing in the *Cimex* case. These papers predate the publication of the Recognition Species Concept (Paterson 1978, 1985). Mayr, however, in 1985 still regarded the sterility of the morabine grasshoppers to be sufficient reason for calling the different populations good species. "In numerous cases, particularly in plants but also in certain groups of insects (morabine grasshoppers), sterility is the principal or sometimes even exclusive isolating mechanism." (Mayr 1985:307). This then became transformed to: "There is, for instance, no mate behavior recognition among morabine grasshoppers" (Mayr 1985:308), a wholly unjustifiable extension of anything White ever wrote about morabine grasshoppers. While it is possible for a species to lack "isolating mechanisms," mating in nature without a mate recognition system is impossible.

Mayr (1985) used the above arguments to reject Paterson's (1978, 1985) Recognition Species Concept, stating that the Isolation Concept and the Recognition Concept "are simply two sides of the same coin." He rejected recognition as the sole criterion of species delimitation on the grounds that it "postulates a degree of conscious cognitive activity that is not to be expected in 'lower' animals, and secondly because isolation among species is effected in numerous species or organisms by isolating mechanisms other than behavioral ones." The first of these arguments is invalid because the word *recognition*, as defined by Paterson (1982, 1985), is devoid of anthropomorphic connotations. The second argument in favor of sterility as the only "isolating mechanism" in some cases has been countered in Paterson (1988) and in this chapter. Mayr's use of phrases such as "to prevent hybridization" and "a protective device" when referring to characters that he calls "isolating mechanisms," is to assign unacceptable goals to them. Paterson (1985), on the other hand, has provided cogent arguments for the view that genetical species are the incidental consequences of a small population adapting to a new environment to which it has become restricted. Species are to be regarded as "effects" and not "adaptive devices" as Dobzhansky (1976) and many others thought they were. "Postmating isolating mechanisms" are also clearly effects (*sensu* Williams 1966) since they cannot be selected for, as Darwin (1859:245) noted long ago. This means that they are not ad hoc characters as Mayr (1963:109, 548) sometimes claimed isolating mechanisms to be.

The case of *Cimex lectularius* and *C. hemipterus* is an instructive one that merits more detailed research. On available evidence these two taxa mate frequently, at least in one direction, when in sympatry in northern KwaZulu. Yet they are intersterile, and consequently they appear to be unable to coexist for any length of time. The details from work in KwaZulu by Walpole and Newberry (1988) and Newberry (1989a) provide a ready explanation for why the two populations have maintained their allopatric distributions in Eurasia as described by Omori (1939, 1941).

A number of comments are appropriate in considering the evolutionary status of the two populations. At various times the two leaders of the evolutionary synthesis, which has prevailed since 1937 and which embraces the Biological Species Concept (or Isolation Concept), have viewed speciation as an adaptive process functioning toward the most efficient utilization of the environment (Mayr 1949:284; Dobzhansky 1976:104). Accordingly, it is not surprising that the "genetic integrity" of the resulting species should be seen to require protection, and that a class of "mechanisms" which achieves this end should have been identified ("isolating mechanisms"). Darwin in 1859 explained carefully why selection for any form of sterility could not occur, and, with only a few lapses, no geneticist has seriously challenged this view. With these points in mind it can be seen why these two bedbug populations provide an interesting test case. If it were not for the sterility, all would agree that two subspecies were being considered (Mayr 1963, 1969). Yet we can see that the sterility is a chance event. It can scarcely be claimed in this case that this is due to "isolating mechanisms" acting to protect the "genetic integrity" of the populations. We can be sure that the sterility has not this function (*sensu* Williams 1966) because (1) intersterility cannot arise under natural selection, and in any case (2) the barrier presumably arose in allopatry.

In contrast, Newberry and Brothers (1990) argued that "two intersterile populations cannot be part of the same field for gene recombination." Further, Brothers and Newberry (1990) stated, "Since the RC [Recognition Concept] holds that two totally intersterile populations between which no gene recombination can possibly occur, and which therefore comprise two distinct fields for gene recombination, are the same species if they have the same fertilization system, this aspect of the RC is incompatible with Carson's concept." It is perhaps important to point out what is meant by "recombination." There are three steps in the process of recombination: (1) independent assortment of chromosomes, (2) crossing over, and (3) chance recombination of gametes (Carson 1957:30). To transfer gametes, not as a single rare event but at the population level, is to provide the potential for gene recombination, that is, to share the same field. The frequent mating of the bedbug populations certainly allows for this potential, and they therefore share the same field for gene recombination. One might argue that while the potential for gene recombination is there, it is never actually realized (except for a single hybrid produced in the laboratory [Newberry 1988]) and, therefore, this is no different from species which only occasionally cross. We would like to point out that there is one major difference between the two situations: two sympatric species which only occasionally cross in nature continue to coexist with only the disadvantaged hybrid offspring going to extinction. In the case of the bedbugs, the one whole population, *C. lectularius,* goes to extinction very quickly in sympatry with *C. hemipterus*. The fact that in this case recombinants are very rarely produced is irrelevant as natural selection will act to eliminate any disadvantage whether it is manifested before or after recombination has taken place. The field data show that this is happening in KwaZulu.

Carson (1957) can be quoted on his views of "the field for gene recombination" as follows: "Reproducing individuals may be viewed as throwing random assortments of their genes, packaged in gametes, into the pool" (p. 26); "The fate of the genetic variability and of the continuing composition of the gene pool nonetheless is determined by the inexorable laws of population genetics and the directive factors of evolution working at the local level" (p. 29); most important for the case under consideration, "The potential [for recombination] may be there but never actually realized" (p. 31), a point he considered worthy of repetition on page 34: "The point should again be stressed that it is in almost every case necessary to speak in terms of potential rather than actual recombination." We go to such lengths to point out what is perhaps obvious because Newberry and Brothers (1990) used the title of Carson's paper as a "definition." Nowhere in Carson's paper does he "define" a species (in our opinion, the *as* in his title is not synonymous with *is*). Similarly, Brothers and Newberry's (1990) reference to Carson's "concept" is erroneous as Carson merely discusses Mayr's species concept in terms of population genetic theory. Furthermore, in neither paper by Newberry and Brothers do they refer to the limitations expressed by Carson in the body of the text. From the above quotations it appears that Brothers and Newberry (1990) are confusing "recombination" with "introgression." Their contention that "contact between the operational RC definition and its theoretical grounding appears here to have been lost" is wrong, both for the reasons pointed out by Ferguson (1990), that is, the field for gene recombination is an effect, and because their case is based on an apparent misunderstanding of the title of Carson's paper while ignoring the content.

Newberry and Brothers (1990), when accepting sterility as a means of delimiting the field for gene recombination, do not explain how their interpretation deals with intraspecific sterility, surely an important area to discuss if one defines species in this way. How do they view, for example, the sterility within *C. hemipterus* where completely sterile egg batches can be laid by females apparently "conspecifically" mated in the field (Newberry 1989*a*, Table 3)?

The case mentioned earlier of the morabine grasshoppers is a complex one with populations displaying varying degrees of differences in their karyotypes, fertility/sterility rates, and morphology. In contrast, the case of the bedbugs is the simplest form of natural selection acting on extreme genetic disadvantage: in the presence of sufficient numbers of *C. hemipterus, C. lectularius* goes to extinction because of the deleterious effects of mating on *C. lectularius* females (Omori 1939; Walpole and Newberry 1988; Newberry 1989a). It is quite comparable with the case of *Clarkia biloba* and *C. lingulata* mentioned by Paterson (1978, 1988). While accepting the common occurrence of intertaxon mating between the bedbugs, Newberry and Brothers (1990) ignored the population genetic prediction that natural selection would eliminate the cause of any disadvantage in a panmictic population. Their dismissal of the Recognition Concept on the basis that the dynamics of negative heterosis may be prevented by differences in, for example, ecological requirements, is ill founded. Spencer et al. (1986), in a computer simulation of the effects of changing parameters on the outcome of negative heterosis, clearly showed that ecological differences were irrelevant. The only way in which two such intersterile populations can persist for any length of time is in allopatry or parapatry. Masters and Spencer (1989) and Spencer et al. (1987) pointed out the differences between "persistence" and "permanence," the latter being the result of a speciation event. The persistence of the two bedbug taxa in allopatry or parapatry is, therefore, an ecological phenomenon, while the impermanence of *C. lectularius* when in sympatry with *C. hemipterus*, is a genetical one, directly relevant to their specific status.

Newberry and Brothers (1990) appear to think that, under the Recognition Concept, the fertilization system should include syngamy. They suggest that, if this were the case, the bedbugs do not share a common fertilization system as the sterility occurs before syngamy and should, therefore, be considered separate species. They stated: "Paterson states that the SMRS is the major part of the fertilization system in motile organisms but 'in sessile animals it may be restricted to the recognition of the sperm by the ovum' (p. 25), in which case the distinction between fertilization and syngamy seems to have disappeared." They see this as a "gray area" in the Recognition Concept. In fact, Paterson (1985) does not equate recognition of the sperm by the ovum with syngamy and quotes Lewis and John (1964) to clarify his use of the word *fertilization*. Also, Paterson's Figure 1 (1985:24) clearly shows that he considers syngamy to occur after fertilization. The answer to the question posed by Newberry and Brothers (1990): "How is one to judge whether fertilization [*sensu* Paterson 1985, not Brothers and Newberry 1990] systems are shared (held in common) or not?" may be found in simple observations of sympatric populations and the application of population genetic theory. The whole point about the bedbug populations in KwaZulu is that cross-mating occurs frequently in sympatry and therefore the effects on the one population, at least, are dramatic. In species where cross-mating occurs on only rare occasions to produce disadvantaged hybrids, neither population would be affected, and their evolutionary status would not be cause for debate.

Newberry and Brothers (1990) suggest that the Evolutionary Species Concept (Simpson 1951, 1961; Wiley 1978) may provide a "less restrictive and more heuristic theoretical definition" against which to test the bedbug situation. We agree with Mayr (1985) that there are major problems with this concept and with Vrba (1985) who prefers a more restrictive and genetical species concept.

It is obvious that more research is needed on the bedbug populations in order to answer, finally, some of the questions posed above. It should be possible to study the mating behavior of bedbugs in the laboratory because, as widely distributed ectoparasites, they are quite adaptable to differing conditions. Thus, their normal habitat may be easily simulated. Thereafter it would be relatively simple to study their courtship under natural conditions and compare it with their behavior in the presence of realistic quantities of DDT (as Omori's [1939] work in the Orient predates the use of DDT, we predict that DDT has no effect on the courtship behavior). It should be possible to study the randomness of mating in both directions with "choice experiments" using males of *C. lectularius* with their sperm marked with radiophosphorus, and using the mating "mark" together with untagged sperm to follow matings involving *C. hemipterus* males. Frequencies of sterile egg batches from wild-caught female *C. hemipterus* could be recorded from both mixed and single populations. Such studies should be done to clarify some aspects of the observations of Walpole and Newberry (1988) and Newberry (1989a,b). Biologists commonly use the ability of two populations to coexist independently over time as a criterion for specific status. Two forms with disadvantageous hybrids (or no offspring) will be capable of no more than parapatric distributions (e.g. Moran and Shaw 1977). It seems likely that restriction, by and large, to climatically distinct allopatric ranges (Usinger 1966) accounts for the long-term survival of the two bedbug forms. We suggest, therefore, that in the case of these two forms, divergence in allopatry was not great and did not involve the genes controlling the SMRS to any great extent. Because of this, the field for gene recombination, determined by the fertilization system, includes both populations. The answer to the question What keeps the two bedbug forms apart? is not sterility, but is more likely to be the climatological requirements of *C. hemipterus.*

The two bedbug forms already have formal taxonomic names at the species level. However, in the light of the foregoing arguments against their status as good biological species, they should not retain their present specific classification. Nevertheless, it is necessary to distinguish between the two populations for sensible discussion of the situation. We therefore purpose that the rank of subspecies be used, as defined by the International Code of Zoological Nomenclature (1985:265), that is, "the rank of the species group below species; the lowest rank at which names are regulated by the Code." Thus, we use *subspecies* "merely as a practical device of the taxonomist" (Mayr 1963:348) without inferring any evolutionary status of the two forms or invoking any species concept. In accordance with the rules of priority, the populations should be called *Cimex lectularius lectularius* and *C. lectularius hemipterus*.

Summary

Two taxa of man-biting bedbugs *Cimex lectularius* L. and *Cimex hemipterus* (F.) have been considered separate species on the basis of morphological differences between them and because of their cross-mating characteristics. The two bedbug taxa occur in sympatry in northern KwaZulu, South Africa, and previous studies have shown that *C. lectularius* females and *C. hemipterus* males mate readily in nature (69 percent, $N = 16$; 44 percent, $N = 219$). The reciprocal cross has been observed in the laboratory, but its frequency has not been assessed in nature. These results have been discussed in this chapter with reference to two current biological species concepts, the Recognition Concept of species and the Isolation Concept of species. Interpretation of the data in the light of the two concepts yields two very different views of the bedbugs' evolutionary status. When applying the Recognition Concept, the frequent mating that appears to occur in nature in northern KwaZulu between the two populations suggests that they share a common fertilization system and are, therefore, conspecific. The prediction, in accordance with population genetical theory, that natural selection will act to eliminate the cause of a disadvantaged gene or genome appears to be correct. In KwaZulu, where they occur in sympatry, female *C. lectularius* are adversely affected by matings with *C. hemipterus* males and as a result *C. hemipterus* seems to be displacing *C. lectularius*. In contrast, under the Isolation Concept, the two populations are clearly regarded as distinct biological species due to the existence of "postmating isolating mechanisms." However, the prediction under the Isolation Concept that in sympatry natural selection will act to reinforce the premating isolating mechanisms of both taxa does not appear to be taking place. We suggest that further studies be carried out to demonstrate the presence or absence of positive assortative mating to clarify the evolutionary status of these two taxa.

Acknowledgments

We dedicate this chapter to Hugh E. Paterson, who has been so influential in our education, and who has spent many hours in discussion with us on the whys and the wherefores of species. Drs. K. Newberry, W. Ferguson, and H. Spencer are thanked for constructive criticism of the manuscript. We thank

Keith Newberry for his time spent debating with us, thus helping to clarify some of our arguments.

References

Brothers, D.J., and K. Newberry. 1990. Interspecific mating in bedbugs does not support the Recognition Concept of species: Comments on Ferguson. S. Afr. J. Sci. 86:176–177.

Carson, H.L. 1957. The species as a field for gene recombination. Pp. 23–38 in E. Mayr, ed. The species problem. American Association for the Advancement of Science. Publication No. 50. Washington D.C.

Coyne, J.A., H.A. Orr, and D.J. Futuyma. 1988. Do we need a new species concept? Syst. Zool. 37:190–200.

Darwin, C. 1859. The origin of species by means of natural selection. John Murray, London.

Davis, N.T. 1965. Studies of the reproductive physiology of Cimicidae (Hemiptera). II. Artificial insemination and the function of the seminal fluid. J. Insect Physiol. 11:355–366.

Dobzhansky, T. 1947. Genetics and the origin of species. Columbia University Press, New York.

———. 1970. Genetics of the evolutionary process. Columbia University Press, New York.

———. 1976. Organismic and molecular aspects of species formation. Pp. 95–105 in F.J. Ayala, ed. Molecular evolution. Sinauer, Sunderland, Mass.

Ferguson, J.W.H. 1990. Interspecific hybridization in bedbugs supports the Recognition Concept of species. S. Afr. J. Sci. 86:121–124.

Key, K.H.L. 1974. Speciation in the Australian morabine grasshoppers: Taxonomy and ecology. Pp. 43–56 in M.J.D. White, ed. Genetic mechanisms of speciation in insects. Australia and New Zealand Book Co., Sydney.

King, R.C. 1972. A dictionary of genetics. 2d ed. Oxford University Press, London.

Lambert, D.M., M.R. Centner, and H.E.H. Paterson. 1984. Simulation of the conditions necessary for the evolution of species by reinforcement. S. Afr. J. Sci. 80:308–311.

Lambert, D.M., and H.E.H. Paterson. 1984. On "Bridging the gap between race and species": The Isolation Concept and an alternative. Proc. Linn. Soc. N.S.W. 107:501–514.

Lewis, K.R., and B. John. 1964. The matter of Mendelian heredity. J. and A. Churchill, London.

Li, C.C. 1955. Population genetics. University of Chicago Press, Chicago.

Masters, J.C., R.J. Rayner, I.J. McKay, A.D. Potts, D. Nails, J.W. Ferguson, B.K. Weissenbacher, M. Allsopp, and M.L. Anderson. 1987. The concept of species: Recognition versus Isolation. S. Afr. J. Sci. 83:534–537.

Masters, J.C., and H.G. Spencer. 1989. Why we need a new genetic species concept. Sys. Zool. 38:270–279.

Mayr, E. 1942. Systematics and the origin of species. Columbia University Press, New York.

———. 1949. Speciation and systematics. Pp. 281–298 in G.L. Jepsen, G.G. Simpson, and E. Mayr, eds. Genetics, paleontology and evolution. Princeton University Press, Princeton, N.J.

———. 1963. Animal species and evolution. Harvard University Press, Cambridge, Mass.

———. 1969. Principles of systematic zoology. McGraw-Hill, New York.
———. 1985. The species as category, taxon and population. Pp. 303–320 *in* S. Atran et al., eds. Histoire du Concept d'Espèce dans les sciences de la Vie. Foundation Singer-Polignac, Paris.
Mecham, J.S. 1961. Isolating mechanisms in anuran amphibians. Pp. 24–61 *in* W.F. Blair, ed. Vertebrate speciation. Texas University Press, Austin.
Moran, C., and D.D. Shaw. 1977. Population cytogenetics of the genus *Caledia* (Orthoptera: Acridinae). III. Chromosomal polymorphism, racial parapatry and introgression. Chromosoma 63:181–204.
Newberry, K. 1988. Production of a hybrid between the bedbugs *Cimex hemipterus* and *Cimex lectularius*. Med. Vet. Entomol. 2:297–300.
———. 1989a. The effects on domestic infestations of *Cimex lectularius* bedbugs of interspecific mating with *C. hemipterus*. Med. Vet. Entomol. 3:407–414.
———. 1989b. Aspects of the biology, specific status and control of the bedbugs *Cimex lectularius* and *Cimex hemipterus* in northern Natal and KwaZulu. Ph.D. dissertation, University of Natal, Pietermaritzburg.
Newberry, K., and D.J. Brothers. 1990. Problems in the Recognition Concept of species: An example from the field. S. Afr. J. Sci. 86:4–6.
Newberry, K., E.J. Jansen, and A.G. Quann. 1984. Bedbug infestation and intradomiciliary spraying of residual insecticide in KwaZulu, South Africa. S. Afr. J. Sci. 80:377.
Newberry, K., E.J. Jansen, and G.R. Thibaud. 1987. The occurrence of the bedbugs *Cimex hemipterus* (F.) and *Cimex lectularius* L. in northern Natal and Kwa-Zulu, South Africa. Trans. R. Soc. Trop. Med. Hyg. 81:431–433.
Newberry, K., and Z.M. Mchunu. 1989. Changes in the relative frequency of occurrence of infestations of two sympatric species of bedbug in northern Natal and KwaZulu, South Africa. Trans. R. Soc. Trop. Med. Hyg. 83:262–264.
Omori, N. 1939. Experimental studies on the cohabitation and crossing of two species of bed-bugs (*Cimex lectularius* L. and *C. hemipterus* F.) and on the effects of interchanging of males of one species for the other, every alternate day, upon the fecundity and longevity of females of each species. Acta Japan. Med. Trop. 1:127–154.
———. 1941. Comparative studies on the ecology and physiology of common and tropical bedbugs, with special references to the reactions to temperature and moisture. J. Med. Assoc. Formosa 60:555–729.
Paterson, H.E.H. 1964. Direct evidence for the specific distinctness of forms A, B and C of the *Anopheles gambiae* complex. Riv. Malar. 43:191–196.
———. 1978. More evidence against speciation by reinforcement. S. Afr. J. Sci. 74:369–371.
———. 1982. Perspective on speciation by reinforcement. S. Afr. J. Sci. 78:53–57.
———. 1985. The recognition concept of species. Pp. 21–29 *in* E.S. Vrba, ed. Species and speciation. Transvaal Museum Monograph No. 4. Pretoria.
———. 1988. On defining a species in terms of sterility: Problems and alternatives. Pacific Sci. 42:65–71.
Rafatjah, H. 1971. The problem of resurgent bed bug infestation in malaria eradication programmes. J. Trop. Med. Hyg. 74:53–56.
Simpson, G.G. 1951. The species concept. Evolution 5:285–298.
———. 1961. Principles of animal taxonomy. Columbia University Press, New York.
Spencer, H.G., B.H. McArdle, and D.M. Lambert. 1986. A theoretical investigation of speciation by reinforcement. Am. Nat. 128:241–262.
Spencer, H.G., D.M. Lambert, and B.H. McArdle. 1987. Reinforcement, species and speciation: A reply to Butlin. Am. Nat. 130:958–962.

Ueshima, N. 1966. Cytology and cytogenetics. Pp. 183–237 *in* R.L. Usinger, ed. Monograph of Cimicidae. Thomas Say Foundation. Vol. 7. Publication of the Entomological Society of America, Baltimore.

Usinger, R.L. 1966. Monograph of Cimicidae. Thomas Say Foundation. Vol. 7. Publication of the Entomological Society of America, Baltimore.

Vrba, E.S. 1985. Introductory comments on species and speciation. Pp. ix–xviii *in* E.S. Vrba, ed. Species and speciation. Transvaal Museum Monograph No. 4. Pretoria.

Walpole, D.E. 1988. Cross-mating studies between two species of bedbugs (Hemiptera: Cimicidae) with a description of a marker of interspecific mating. S. Afr. J. Sci. 84:215–216.

Walpole, D.E., and K. Newberry. 1988. A field study of mating between two species of bedbug in northern Kwa-Zulu, South Africa. Med. Vet. Entomol. 2:293–296.

White, M.J.D. 1974. Speciation in the Australian morabine grasshoppers—the cytogenetic evidence. Pp. 57–68 *in* M.J.D. White, ed. Genetic mechanisms of speciation in insects. Australia and New Zealand Book Co., Sydney.

Wiley, E.O. 1978. The evolutionary species concept reconsidered. Syst. Zool. 27:17–26.

Williams, G.C. 1966. Adaptation and natural selection: A critique of some current evolutionary thought. Princeton University Press, Princeton, N.J.

11

Species Concepts and the Nature of Ecological Generalizations about Diversity

GIMME H. WALTER

Laboratory of Ecological Zoology, Department of Biology, University of Turku
and
Department of Entomology, the University of Queensland

ECOLOGY IS A complex discipline, and even within a narrowly defined area of investigation opinion may be diverse. Often though, even different opinions may be related through their common dependence on a particular underlying principle or assumption (although this dependence may be tacit or perhaps even unrecognized). In general terms I believe this is the current situation in interpretation of community structure.

Should alternative direction for ecological understanding and research be sought, it may therefore come more readily from a consideration of alternative underlying principles. Employment of an alternative basis for asking questions would mean that the questions given prominence will inevitably be different, and so will the nature of the interpretations.

In this chapter I concentrate on community structure. It seems an appropriate time to examine the foundation principles of this aspect of community ecology because direction is being sought to overcome fundamental problems (Ricklefs 1987) and rules of community assembly are being reconsidered, with recommendations that communities as a whole should be the object of study (Drake 1990). Here I summarize difficulties in current approaches and provide justification for an alternative approach based on different premises, which derive from the Recognition Concept of species.

Primarily I have sought clarity and brevity. As a consequence my statements may be direct. I hope that any bluntness will be construed in no way other than as an effort to communicate precisely and effectively in a complex discipline. Also, I have used citations to illustrate that criticized interpretations are indeed current.

Introduction

Understanding the diversity of species is a central goal of biology, and explanatory theory derives from two perspectives. Interpretation of the generation of diversity (through speciation) is the realm of evolutionary study and is the older endeavor. Community ecologists have a different goal, and they concentrate on the factors that influence local diversity. These are also seen as problems of community structure and of the limitation (or maintenance) of diversity of species on a local scale (May 1986; Roughgarden and Diamond 1986). These two broad approaches to understand diversity are considered to be complementary and worth of fuller synthesis (Futuyma 1986: 483; Ricklefs 1987, 1989a: 616-617).

A synthesis of species theory and community ecology seems justified by the existence of ideas common to both areas of investigation of diversity. But what is not clear is the form the synthesis should take, and here interpretation is made difficult by significant recent developments and altered perceptions in both species theory and community ecology. These changes should have profound implications for the theory and method of community ecology and for interpretation of its precise relationship with species theory. The aim of this chapter is to explore this relationship to establish what nature it should have.

A problem in discussing community theory lies in the diversity of uses of the terms *community* and *community structure* and in the diversity of aims in the science of community ecology. I therefore restrict my comments to one particular area of community ecology, that which aims at developing a general explanatory theory of community structure (in the spirit of Ricklefs' [1987] synthesis, for example).

Evolutionary thought is at present widely influenced by the Isolation Concept of species, a situation reflected in ecological theory by the tacit acceptance of that concept. Indeed, the Isolation Concept plays an important role in ecological research, but this has to be inferred because the relevance of species theory in ecological research is almost never made explicit.

In the first section of this chapter I demonstrate that the influence of the Isolation Concept has been pervasive in the development of ecological theory, and I draw attention to the internal logical consistency between the Isolation Concept of species and interpretations of community structure.

Ecologists have long been aware of the multiplicity of factors that influence community structure. Attention has been given for some time to disturbance; regional or global influences; and recruitment, predation, interspecific competition, and other interactions (e.g. Ricklefs 1987, 1989a; Underwood and Fairweather 1989; Menge and Olson 1990), thus conferring plurality on explanations of community structure (Schoener 1986a). Clearly it is no longer believed that interspecific competition is the sole structuring force in communities.

But how is interpretation and understanding of such ecological complexity arranged, and, consequently, how do we ask questions about such complexity? It seems that here interspecific competition (and the competitive exclusion principle) still plays an important role in giving direction to particular aspects of ecological research and interpretation. Usually community studies set out to explain what it is that prevents certain species from outcompeting other species that are present. This is implicit when it is claimed, for instance, that predation structures a particular community (e.g. Ricklefs 1989b). The role of interspecific competition and the competitive exclusion principle in ecological investigation and explanation is frequently tacit (Walter 1988), probably because such assumptions and their influences are easily overlooked.

I examine the aims and assumptions of pluralistic community ecology in the second section of the chapter. The persistent importance of ideas about interspecific competition and competitive exclusion in directing research and explanation seems to be a consequence (at least in part) of the resource being the focal point in explanations of community structure. Basing general theories of community structure on resource use tends to circumscribe the type of generalization that can be made, because interpretation is centered around an implied role for interspecific competition.

The Isolation Concept and modern interpretations of community structure have long been associated through several common ideas. Interspecific competition and competitive exclusion have been important in interpreting the generation of diversity and the structure of communities. The resultant combination of ideas about reproductive isolation, ongoing adaptation within large populations, and interspecific competition promotes a particular approach in ecological theorizing about diversity.

The attention of evolutionists has been drawn to the deficiencies of the Isolation Concept of species (by Paterson 1981, 1982a, 1985, 1988), and evolutionists are aware of the competing Recognition Concept (see Otte and Endler 1989). The Recognition Concept of species is not simply a variant of the Isolation Concept; it is substantially different (Paterson 1984). Yet it not only eliminates the logical inconsistencies in the Isolation Concept identified by Paterson (1981), it also has little-appreciated but far-reaching consequences for theory and practice in ecology (Paterson 1973, 1985; Walter et al. 1984; Hulley et al. 1988a,b). Acceptance of the Recognition Concept implies that attempting to generalize about community structure has certain limitations, especially if one wishes to reduce reliance on interspecific competition and the competitive exclusion principle in asking questions and developing generalizations.

Section three outlines the consequences of acceptance of the Recognition Concept for theories of ecology that deal with local diversity and community structure. Attention is drawn to the way in which the form of ecological generalizations derived from the Recognition Concept differ from generalizations currently sought in ecology.

The Isolation Concept and Ecological Theory

For a long time the Isolation Concept was the only genetical species concept. Its wide acceptance meant that interpretation of adaptation in ecological theory was consistent with interpretation of adaptation under the Isolation Concept. Nevertheless, the important transcending links between species concept, the interpretation of adaptation, and the development of theory in ecology have been largely missed by ecologists. These relationships have, for instance, not been mentioned in the ecology chapter of a recent compilation (Otte and Endler 1989) that deals with speciation and its consequences.

The protracted prominence of the Isolation Concept may even have created the impression that interpretation of the origin of species is a closed book. It has been stated that "understanding the essentials of how species originate was the most important intellectual achievement of the 19th century," and that for this century and perhaps the next one "the largely unanswered question . . . is to understand how *many* species there are" (italics in original) (May 1986).

Attention has been drawn, in the ecological literature, to the implications of the Recognition Concept for interpreting ecological phenomona linked to interspecific competition (Walter et al. 1984). But in response the evolutionary significance of interspecific competition has still been considered "without commenting at all on the appropriateness of the speciation mechanism" (Schoener 1984).

The important link between species concept and ecological interpretation is therefore considered further.

Adherents of the Isolation Concept, besides stipulating an allopatric model of speciation (Mayr 1963; Futuyma and Mayer 1980), accept that populations of species continue to adapt to prevailing (and inevitably changing) circumstances in their postallopatric history (Mayr 1963:66, 547). Although periods of stasis may also be expected (Charlesworth et al. 1982; Futuyma 1987), the central tenet is continual adaptation, if not continuous adaptation, in populations of species. Adaptation to local circumstances of environment and community, by character displacement, for example, is expected in populations.

Considering the wide acceptance of the Isolation Concept, it is not surprising that recent statements by ecologists about adaptation indicate they accept the interpretation of adaptation outlined above. Colwell (1984:391), for example, states that "short-term evolution of a meaningful sort is not only possible but commonplace." And it is presumably implicit in statements about "population processes, including habitat specialization and within habitat specialization" (Ricklefs 1989a:600) because the circumstances under which such adaptations are believed to be acquired are not clarified. In common with the Isolation Concept, ongoing adaptation is seen in ecology to be a central attribute of populations in different areas. In any case, there has apparently been no attempt by ecologists to develop an alternative basis for interpreting adaptation.

Adaptations in two categories are given special prominence by the Isolation Concept, those associated by biologists with hybridization and those involved in resource use (Mayr 1963:66). The postallopatric phase of speciation is interpreted as one of "fitting into" both the community of breeding species and the community of species that makes demands on the same resources. The community of breeding species would be composed of closely related congenerics, but the ecological community is seen to be composed of closely related organisms and organisms of different phyla or kingdoms (e.g. Brown et al. 1979; Hochberg and Lawton 1990).

Whereas Mayr (1963:82, 547) was unequivocal about the importance of adaptation to the ecological community, Futuyma (1986:502-3) has been more cautious. Futuyma reasoned that changes in population size occur more rapidly then genetic changes, so it is likely that ecological processes such as competitive exclusion contribute more to the formation of "stable associations" than do evolutionary processes. How much more is not specified. It appears most ecologists would expect some combination of ecological influences and local adaptation. Roughgarden (1989:210), for example, states, "The differences actually exhibited by coexisting species must somehow represent both the differences they had upon entry into the community and differences evolved while in the community."

Whatever their persuasion in this regard, a pervasive and still highly influential idea to both evolutionists and ecologists is that of competitive exclusion (e.g. Futuyma 1986:30ff., 35, 483-484, 488, 491, 502; Ricklefs 1989a:612, 615-617; Roughgarden 1989:204, 209-211; Aarssen 1989). The competitive exclusion principle (or law; see Murray 1986; Walter 1988) stipulates what is possible in terms of additional species joining a community or guild. Conversely, it has also long been accepted that the intensity of competition may be ameliorated by various ecological factors, thus allowing coexistence of species with similar requirements (e.g. Hutchinson 1948). Natural selection is also considered to have molded resource-partitioning mechanisms to promote coexistence of species, in some communities at least. Resources believed to be partitioned in this way include even "enemy-free space" (Ricklefs 1989b).

Understanding community composition through explanation of coexistence thus derives from expectations generated by the exclusion principle itself. There remains, in other words, a conceptual reliance on interspecific competition when "communities" are investigated and explained in this way (Walter 1988). In any case, communities are invariably defined or specified with reference to common resource use and, at least, the potential for interspecific competition among member species.

The Isolation Concept of species, and its consequent interpretation of speciation, was not developed independently of the notion that competitive exclusion is a pervasive influence in nature. Indeed, Mayr (1963:88) postulated that speciation itself was driven by interspecific competition and could not proceed unless a vacant niche was available for the incipient species to colonize. Later in his book he stated explicitly: "Each species is an independent genetic system, which has the properties of being reproductively isolated from the ecologically compatible with other sympatric species. Speciation means the acquisition of these properties" (Mayr 1963:546). The relationship between the Isolation Concept and community theory is further illustrated by interpretation of adaptive radiations relying on competition being a strong ecological force (e.g. Mayr 1942:271, 1963:574, 584; Futuyma 1986:32, 503).

More recently, Mayr (1982) incorporated into his species concept the idea of a specific niche (Hengeveld 1988), and his thinking in this area reveals how close is the interplay between ideas in evolution and ecology. "[T]he major biological meaning of reproductive isolation is that it provides protection for a genotype adapted for the utilization of a specific niche. Reproductive isolation and niche specialization (competitive exclusion) are, thus, simply two sides of the same coin" (Mayr 1982:275). Hengeveld (1988) has drawn attention to the typology inherent in interpretations of this nature.

The advantages proposed for reproductive isolation are similar in concept to those proposed for resource partitioning (*sensu stricto:* see Walter 1991). Both involve selection pressures from impinging species, and both have postulated advantages for the species as a group. Reproductive isolation is seen to maintain the genetical integrity of species (Dobzhansky 1935, 1951, 1976; Mayr 1970; all cited by Paterson 1980, 1981; Masters et al. 1984), and resource partitioning reputedly protects species from the exploitations of other species (e.g. Schoener 1965:199). Although the authors cited in this paragraph may accept that group selection is not an important process in evolutionary change, it remains that they do refer to group benefits (even if it is only implicit in the terminology). Postulated group benefits of this nature cannot be explained as functional traits by reference to individual selection.

In summary, this section draws attention to the internal consistency between suites of evolutionary ideas derived from the Isolation Concept and ecological ideas about diversity. That such consistency should exist is not unexpected because developments in each area have influenced developments in the other. Indeed, anything else would have been surprising. The point does serve as a reminder that a change in fundamental ideas in one area must be followed through into related or derived areas of endeavor to assess their impact seriously and objectively.

Resources and Generalizations about Communities

Ecology has been described as the scientific discipline that attempts to determine the causes of patterns in the distribution, abundance, and dynamics of the Earth's biota (Tilman 1989), and patterns are expected to be determined predominantly by interactions between organisms. Species influence each other both through direct pairwise interactions and through indirect interactions (Tilman 1989). "Community ecology" is sometimes more tightly circumscribed than specified here, but not necessarily so, because some also include such topics as species abundance relations, food web patterns, and geographical trends in life history traits (Roughgarden and Diamond 1986). Much of what follows, although specifically about those aspects of community ecology that seek generalizations about local diversity and community structure, is pertinent to the broader perceptions of ecology outlined above.

Aims and Generalization in Community Ecology

Generalizing about the practice of community ecology is not easy. Terms frequently have several meanings. Even the root terms *community* and *community structure* are not used with consistency (see MacFadyen 1963; Poore 1964; Underwood 1986; Roughgarden and Diamond 1986:334), and their operative definition is only infrequently specified at the outset of any study. Despite this, and despite the boundaries to guilds and communities having to be arbitrarily set, the phrases have widespread acceptance and are used even by critical analysts in their own work.

Notwithstanding these inconsistencies of definition, community ecologists believe it should be possible to arrive at general principles that underlie "community structure," and thence to make generalizations (evolutionary and ecological) about causality in different categories of communities. This means that generalized explanations are sought for the "variety and abundance of organisms at any place and time" (Roughgarden and Diamond 1986:333), because in any circumscribed area or on any specified resource there is "limited membership of species" (Elton 1950:29; Roughgarden and Diamond 1986:333; Roughgarden 1989:203). What is anticipated, therefore, with regard to community structure, are the assembly rules that are believed to underlie the tangle of species' interactions in nature (Lawton 1987, 1989; Drake 1990). Despite communities being known to vary in their species composition from site to site, such unifying principles are still expected, for "this natural variety does not imply that each community must be considered as one of a kind or that a comparison of communities will not reveal common themes in their composition and in the processes that occur within them" (Roughgarden and Diamond 1986:333). The aim is to identify a system of rules that can be extrapolated from one situation to another in nature. In other words, generalized community-level properties are sought, and it has been emphasized that investigation should be at the level of the community (Drake 1990).

It should not be forgotten that the generalizations expected concern a postulated entity (the community or guild) despite its being impossible to specify the limits to such an entity in nature in a nonarbitrary way.

Implicit in the aim and method is that ecological interpretation from research in one area would translate directly to explain ecology in other areas that are deemed to be sufficiently similar. Explanation and prediction could thus be transferred. For example, MacArthur (1972) said: "for organisms of type A, in environment of structure B, such and such relations will hold" (cited by both Colwell [1984:394] and May [1986:1120]). In other words, it is felt that species could be categorized into uniform ecological groupings or that communities can be classified into types. Such thoughts are in line with suggestions, made sporadically in the ecological literature, of the possibility of a periodic table equivalent for ecology (e.g. Southwood 1980).

Comparative studies are recommended for detecting potentially important patterns or trends in communities, because patterns may provide clues about underlying structuring mechanisms (e.g. Ricklefs 1989*a*). Within-community analyses also rely on comparison, because resource use among the species is compared or relative abundances of the species are compared. Pattern analysis among species that use the same resource is a central technique in studying local diversity. The regularity perceived in many patterns provides confidence in the comparative approach to this problem (Lawton 1989), and it is believed that this is an important step toward revealing the "mechanisms" that structure communities.

Although the study of species' biology has been encouraged by some community ecologists (e.g. Colwell 1984), it is apparent that species are more usually seen in a functional role (the "fixed ecological role" of May and Seger [1986]). Each species is seen as a unit and its function is played out in a relational way to other species. Species (and sometimes even aggregations of species called *trophic species* [Paine 1988]) are thus the vehicles for energy flow, resource partitioning, and the interactions that are seen to be the essential attributes of communities and ecosystems (Walter 1993). This thinking is employed, too, when the term *assembly rule* is "used to describe the mechanics of how the species of a community fit together" (Drake 1990). Common roles or common interactions will, it is believed, point to the "mechanisms" that may be significant in generating community structure. Interactions are considered to have primary significance in this way because species in nature are believed to have a place in a complex web of interactions (Walter 1993). Communities are thus seen to represent higher-order phenomena that merit generalized explanation at the community level (e.g. Cody 1989:227).

The acquisition of a specific role in a community is achieved in a relational way. Species adapt to the template set by the type and availability of resources in the community, and resource availability is profoundly influenced by the suite of other species that use the resource. Such local adaptation in populations is expected under the Isolation Concept of species (see the above section "The Isolation Concept and Ecological Theory").

In summary, community ecologists generalize about aspects that "communities" are perceived to have in common, despite each "community" being arbitrarily specified. Much attention is directed at interactions between species, and it is believed that generality, rules, or principles will emerge from comparison of the influence of interactions in different communities. Once the common ground between different communities has been identified, it is thought, then explanations of causality will be possible, and prediction at the community level will follow. Species are seen to play a significant role here, but in a particular way. They are frequently seen as unitary entities that fulfill certain functional obligations within a matrix whose operations are expected to be more revealing about the complexity of nature than the study of species and their adaptations and requirements per se. In any case, which particular species has the role is not of great consequence, because competitive exclusion should ensure sole occupancy of a niche (or role).

Resources and Pluralistic Community Ecology

Pluralistic ecology is portrayed as new, and it is pluralistic, because the emphasis is claimed to have been taken from interspecific competition as the major motive force in ecology and evolution. It is accepted now that one or more of many factors may influence local diversity and thus "structure" communities or guilds in nature (e.g. Price 1984; Schoener 1986*a,b*; McIntosh 1987).

Both Colwell (1984) and May and Seger (1986) have refuted the novelty of these developments, a view that is readily appreciated. Hutchinson (1948) long ago pointed out that predation, physical disturbance, and other factors could subjugate the dominating role of competition. But whatever the outcome of the debate about the novelty of these developments in ecological theory (see McIntosh 1987), interspecific competition has a central role in interpreting certain aspects of community ecology, even in pluralistic community ecology, for reasons that follow.

To investigate the special subset of organisms that constitute the community, the properties believed to allow local coexistence are scrutinized (see Roughgarden and Diamond [1986] and Roughgarden [1989] for details). The problem of coexistence has also been given a more global context: "In addition to contemplating coexistence in local areas of habitat, ecologists must examine patterns of coexistence within regions. The partitioning of habitats and of geographical area within regions has as much relevance for community structure as the partitioning of resources within habitats" (Ricklefs 1987).

The properties investigated by ecologists to explain coexistence include at least two classes.

1. Ecological factors, such as predation and disturbance, are those external features that reduce the impact of competition and are therefore believed to explain the coexistence of species in certain circumstances.

2. Characters of organisms form the second class, and here there are two qualitatively different subgroups.

a. Characters of organisms may be mechanisms that evolved to permit coexistence and so serve a specific function in this regard, that of resource partitioning.

b. Characters may have evolved for purposes other than coexistence, but they may or may not be important in allowing coexistence (Walter 1991). If they evolved for purposes unrelated to coexistence, they are evolutionary byproducts (or effects: Williams 1966) in relation to coexistence. By definition they are effects in this context even should they be demonstrably important in allowing coexistence. However, unequivocal demonstration of this point is not simple, and whether such a demonstration has general significance in the species' distribution or just local significance needs to be considered (Sale 1988; Walter 1988).

The above breakdown suggests that if species occur together regardless of differences in resource use (which is likely if interspecific competition is not a major ecological force), drawing attention to differences in resource use as "coexistence mechanisms" will inevitably mislead attempts to understand diversity and the adaptations of species (Walter 1991). The distinction of Williams (1966) between function (evolved mechanism) and effect (evolutionary by-product) is almost never made in ecological theorizing, although it is significant in understanding species and adaptation, and thus diversity and community structure. The point is raised again later.

Questions about coexistence generate explanations of why competitive exclusion is not occurring among species that make demands on the same resource. The circumvention of competitive exclusion reputedly allows more species to occur (coexist) in an area (or on a particular resource) than would otherwise be expected. Questions about community structure are usually of a similar nature to coexistence questions.

Although coexistence may be explained in terms of predation, physical disturbance, and so on, its relevance derives from the radical competitive exclusion, and so does the relevance of "community structure," by which "we usually mean patterns of resource use and species interactions, as well as community composition" (Colwell 1984:390). There is no sense in examining resource use relationally among species unless one has expectations about interspecific competition having the general importance that was once ascribed to it (Walter 1988). Ideas about coexistence are otherwise in danger of being trivial.

These statements do not imply that interspecific competition does not occur in nature, simply that organizing research and interpretation around interspecific competition may not be the best way to approach an understanding of diversity and "community structure."

Attention has been drawn to flaws in competition theory over a long period (e.g. Andrewartha and Birch 1954; Simberloff 1976, 1980, 1982, 1983; Heck 1976; Wiens 1977; Connell 1980; Lawton and Strong 1981; Walter et al. 1984; Hulley et al. 1988a,b), and disillusion with the restricted explanatory power and limited achievements of the theory has been expressed (e.g. Heck 1980; Brown 1981). But community ecologists who deal with coexistence (and community structure) have, in general, not diverged significantly from the original preoccupation with interspecific competition, as outlined above (Walter 1988).

The terminology of current ecological theory is derived, together with its analogical status, from economics. The derivation of Darwin's economics metaphor is well known (Montagu 1952; Schweber 1977, 1980; Brady 1982; Riley 1986), and words like *competition, community, guild,* and *resource* are inevitably related. They constitute part of the word field (Lambert and Hughes 1988) of community ecology.

The economics metaphor is therefore deeply rooted, and recent attempts to give other ecological factors a more realistic status in relation to interspecific competition have not been successful and, it seems, cannot be achieved within that conceptual framework, because the concept of shared resources or common demands on resources is still central in community ecology (e.g. Brown 1981; Pianka 1983:253; Price 1984; Price et al. 1984; Wiens 1984; Abrams 1988). The resource, abstracted in this way, represents a common currency. Species are seen relationally to other species that use the same resource, and the perceived "cohesion" of communities is thus bolstered, despite their epiphenomenal status (Simberloff 1980), and their consequently arbitrary delimitation (Roughgarden and Diamond 1986; Underwood 1986). That is why pluralistic ecology is not so significantly different from early developments in community ecology.

In practice, community ecologists who seek generalized models of community structure begin by delimiting a resource that is of interest, usually because it is expected, for some reason, to be in short supply, or to be the basis of interspecific (competitive) interactions. In many such community studies attention is focused on an area or a particular environment. Often in these cases space is a resource for territorial or sessile organisms. Thus the set of questions, as well as the range of possible answers, is delimited by the attention inevitably drawn in this way to competition (even if it is through emphasis on the lack of competition). So interspecific competition is preeminent even to pluralistic community ecology because the theory is inevitably resource based, and the emphasis is on interactions, mainly those between species that are potential competitors (e.g. Price et al. 1984:9; Roughgarden 1989:220).

It is evident that the Isolation Concept of species has been one of the principal vehicles for maintaining this ecological approach to the study of diversity. Both Mayr and Dobzhansky (cited by Paterson 1982a, 1986) saw species as "adaptive devices" that are molded so that resources in the environment will be most efficiently used. Similar group selection is implicit in ecological thought when species are seen as adaptive units. Concentration on the resource implies that limitation, at some point, will occur as more and more *species* are added. The issue is *species* packing and coexistence of *species*. Mayr (1982:296) makes the point explicit when he says, "A species, regardless of the individuals composing it, interacts as a unit with other species with which it shares the environment . . . This interaction of species is the principal subject of ecology." Concentration on species as units speciously lends weight to the idea that only a restricted number of species can coexist on a particular resource. These ideas accord well with the "limited membership" of Elton (1950) and Roughgarden and Diamond (1986), and they reveal an important link between the resource-based ecology of today and the theory of evolution. There is thus reinforcement from the Isolation Concept of species for the continued stress on resources and interspecific competition as a basis for interpretation in ecology.

Despite the problems faced in community ecology, some believe that a change in paradigm is not needed, although a shift in emphasis has been recommended (e.g. Colwell 1984:392; Ricklefs 1987, 1989a). But what must be confronted is:

1. The evidence demonstrating that competition is not a pervasive ecological or evolutionary influence in nature;

2. The fact that communities are designated arbitrarily according to common resource use, with the aim of making generalized statements about community structure or assembly;

3. The unrealistic conclusions that derive when species are considered as unitary entities;

4. The logical problems in the Isolation Concept of species, which underpins current generalizations about adaptations related to community structure;

5. An alternative species concept with different implications for the interpretation of adaptation and diversity (see below).

A decision is then possible on whether a limited shift such as that described by Colwell (1984) is sufficient. It seems to me that the difficulties inherent in resource-based ecological theory could not be adequately dealt with by modification of the current framework. The search within these confines (e.g. Wiens 1983, 1984; Price 1984; Ricklefs 1987, 1989*a*; Drake 1990) has not produced an alternative at a time when one needs to be articulated.

To illustrate the above point, it is perhaps a preoccupation with resources that causes theorists to depend for their interpretations of diversity on a postulated homogeneous habitat as a unit (e.g. Murray 1986; Ricklefs 1989*a*:599), despite the criticisms of extrapolating the predictions of the Lotka-Volterra competition equations to nature. Knowledge of agronomy, vegetation, climate, and the local ecology of organisms all suggest that any identifiable area of homogeneous habitat would be so small as to be meaningless in the explanation of diversity.

The beginnings of any alternative theory in ecology must lie outside a resource-based foundation. This does not imply a denial of competition ever occurring in nature, or that the activities of individuals of one species do not influence the ecology of organisms of other species in the field, or even that resources are not important to organisms. It implies that we need a new basis from which to ask questions about the diversity of organisms in nature and about their use of the environment and resources. Before suggesting some guidelines for development of alternative theory in this area, I examine the method by which generalizations about communities are sought.

Shortcomings in the Study of Community Structure

Community structure (and local diversity) would inevitably be influenced by more than one factor or cause, most likely by many. Hilborn and Stearns (1982) have analyzed the problem of multiple causes of observed phenomena in ecological and evolutionary investigation. The task of assigning causality in communities would seem to be an obvious place for application of their suggested methods.

Analysis of community structure is akin to an example Hilborn and Stearns (1982) discuss in some detail, that of the postulated influence of intraspecific competition in returning populations to an equilibrium. They use the example to demonstrate that simple hypotheses (or single-factor hypotheses) may necessarily rest on a long series of complex assumptions, whereas complex hypotheses may rest on only a few simple assumptions. Their recommendation is for preference of hypotheses with simpler assumptions (which I take to mean more realistic assumptions about nature), but these may be multiple-factor hypotheses. An hypothesis based on the joint influence of more than one factor may, indeed, be less complex overall than the related single-factor hypothesis, because the former may be based on more realistic (simpler) assumptions (Hilborn and Stearns 1982).

Outwardly, multiple-factor hypotheses would seem to be ideal for pluralistic community ecology. But examination of the assumption is necessary before a decision should be made. I detail below some of the important assumptions of pluralistic community ecology.

First, it is assumed that species, as units, can be used to analyze community properties (see above). Any evolutionary statements made on this basis must invoke group selection (for which there is no good evidence from nature [Wilson 1983]). This implies that a species cannot have a unitary role in nature (see below), and analyses that involve species as units are not realistic biologically (Walter 1993).

Second, communities and guilds are assumed to be reasonable abstractions of nature (e.g. Tilman 1989:98). The acknowledged arbitrary specification of any community or guild (see above) makes this doubtful. The arbitrariness stems from resource use being the usual basis for delineating communities. Further, a limited range (usually only one) of the resources used by organisms is presumed to delimit communities realistically.

Third, competitive exclusion is assumed to be an important organizing force in nature. This is doubtful. Since the earliest interpretations of competitive exclusion were made, no unequivocal evidence for its postulated significance has been gathered (Heck 1980; den Boer 1986; Walter 1988). Hutchinson (1961) and others realized that the exclusion principle may be unrealistic, but they felt that it generated questions that yielded good information. I have argued that the data generated are too superficial to justify maintenance of this approach (Walter 1988).

Fourth, in studies of community structure, information about ecological processes is often translated directly into information about evolutionary processes (see Hulley et al. 1988a,b). Interspecific competitive interactions are frequently assumed to generate directional selection pressures, and adaptation to the community is thus presumed to be ongoing. Recent developments in species theory suggest that the most significant natural selection after speciation may be stabilizing selection (Paterson 1985, 1986; see below).

Fifth, although adaptation is a central concept in community analysis, the distinction is rarely made between evolutionary function and the by-products of adaptive evolution (effects). Williams (1966:8–14) and Paterson (1982b, 1984, 1985, 1986) have clarified and elaborated on the significance of the distinction, and its relevance for certain aspects of community ecology have been specified (Walter 1991, 1993).

Sixth, interactions between species on the same trophic level are assumed to have repeatable consequences that can be detected by analyzing patterns of species' characteristics and patterns of species' abundance. Considering the variable nature of the environment, this is not likely to be, and it is known that different species respond independently and differently to environmental change (e.g. Chapin and Shaver 1985), so season-to-season changes in weather and long-term climatic change must sooner or later disrupt any patterns ("community regularity") that may be detected.

Seventh, it is assumed that "community regularity" (e.g. Lawton and Gaston 1989) can be defined in a nonarbitrary way and can be measured.

It is apparent that investigation of community structure is based upon several unrealistic premises. A search for simpler and more realistic assumptions is warranted.

Another important requirement in the investigation of multiple causation is the certain knowledge that the subject of inquiry is a single phenomenon (Hilborn and Stearns 1982). If phenomena are subdivided we lose generality, and if they are combined we sacrifice accuracy, if not all explanatory power. From the above assumptions it seems clear that community studies combine phenomena (function and effect, species of diverse origin, ecological and evolutionary processes, for example) in search of a generalized explanation of causality. Community studies overgeneralize, so the possibility exists that more is obscured than is revealed about species' adaptations and about the presence or absence of species in local areas, for example.

The interpretation that community studies overgeneralize uncovers a further flaw in method. The challenge to ecologists to reveal alternatives causes for particular patterns of resource use in communities is a diversion, because such patterns are usually generated by comparisons suggested by competition theory (Hulley et al. 1988*a,b*; Walter 1988).

To summarize this section, theories that deal with community structure have not been able to deal effectively with the major criticisms raised over the past three or four decades. Direction is still largely determined by earlier beliefs about interspecific competition. There is a consequent neglect of the unique origin and adaptations of each species and the impact these adaptations have on the local abundance of organisms. Comparison of species on the basis of their common resource requirements, and analysis of other patterns (some on a grander scale), are the major tools of community ecology despite the conceptual and methodological shortcomings.

A Change of Focus: The Recognition Concept and Theory in Ecology

The earlier sections draw attention to two points. Logical consistency must be maintained between evolutionary principles and ecological theory. And since the concept of adaptation is central to ecological theory, the lead should come from evolutionary principles. What individual organisms do in nature is dependent on their genetic makeup, and that is derived largely through the combined effects of recombination from the species gene pool in sexual organisms and of natural selection. In short, theory in ecology must have its origins in species theory and must be consistent with the principles of that theory.

In the subsection that follows I give a brief outline of the Recognition Concept and certain consequences that relate to current theory and practice in ecology. Some questions about local adaptation are also raised, and tests are discussed. Then I consider points that should be seriously considered in theoretical development in a derivative discipline such as ecology. The emphasis placed by the Recognition Concept on individual organisms is considered in relation to ecological theory, and the final subsection makes some points about the form of general theory in ecology.

The Recognition Concept: Consequences and Questions

The Recognition Concept of species suggests avenues for the investigation and interpretation of ecological phenomena and, particularly, sets different constraints as to the level at which realistic generalizations can be made in ecology. Brief outlines in this direction have already been published (Paterson 1973, 1985; Walter et al. 1984; Hulley et al. 1988a,b).

The Recognition Concept predicts that complex adaptations in sexual species are acquired only when a small, isolated population is brought under directional selection, through confinement to an environment that is different from the one to which the parental population is adapted (Paterson 1984, 1985). That is how specieswide characters come to be fixed throughout the species gene pool. Usually, in populations within their normal habitat, stabilizing selection is conferred by the coadapted sexual communication system, by characters of the fertilization mechanism being adapted to the usual habitat and normal way-of-life of individuals of the species, and by the difficulties of fixing complex genetically based changes in a sizable and widespread population (Paterson 1986).

Once adaptation to the new environment has been achieved in geographical isolation, speciation may or may not have occurred, depending on whether the fertilization mechanism (or the Specific-Mate Recognition System [SMRS]) was significantly altered or not (Paterson 1986). Species, therefore, are effects of local adaptive evolution and are not adaptive devices deployed for efficient use of the environment (Paterson 1985). Individuals from the newly adapted populations may migrate into areas of similar suitable habitat should it be available and accessible, and the distribution of the species may thus expand.

With regard to current views of community structure, several significant consequences emerge from this interpretation of the origin of adaptation.

Because species have unique and nonrelational origins (with reference, at least, to most of the potential competitors alongside which they currently exist), and because that origin may have been influenced by chance events, it is apparent that species do not serve an evolved functional role, and a species is not evolved to be a component of a community or ecosystem (Paterson 1973, 1985; Walter et al. 1984; Walter 1993). Particular care is warranted when communities are considered on a local scale (Walter et al. 1984). In interpreting the possible adaptive significance of characters (for resource partitioning, for instance) it cannot be assumed that they evolved in response to local "community structure" or, indeed, or any other local factor that may be perceived to confer selective advantages. The character under consideration must be demonstrated not to be specieswide if there is any possibility it evolved in response to the local conditions of interest. Questions about local adaptations are raised again later.

The only other way that particular roles could be conferred on species would be through the sort of screening envisaged by the competitive exclusion principle. But evidence for the effective operation of interspecific competition in this way is questionable, and examples from nature are open to alternative explanation (Walter 1988). In any case, field studies demonstrate that interspecific competition is seldom a strong or persistent ecological influence (Andrewartha and Birch 1954; Wiens 1977; Connell 1983; Hulley et al. 1988a,b), and it is not surprising that there are no unequivocal examples of competition-induced character displacement (Grant 1972, 1975; Connell 1980; Grine 1981). Several examples are persistently cited in evidence (e.g. Schoener 1984), yet no study has satisfied the minimum criteria of Connell (1980) for a strong case to be made.

Phenomena investigated by ecologists are variable in many ways, and it is undoubtedly that that has significantly impeded successful generalization. For example, after adaptive change takes place in a small allopatric population, the individuals of an expanding population are likely to encounter different suites of species in different areas, and they may live alongside them if the environment (which includes other organisms and their influences) is suitable. All suitable areas may not be colonized. Therefore, the variability of associations mentioned by Roughgarden and Diamond (1986), and others, is not unexpected.

There are yet other sources of variability. Organisms of any species may be influenced by a diversity of external factors, such as air or water currents, predation, parasitism, competition from other individuals, local change to the environment, and so on. Within any area, several such ecological factors may simultaneously be operative on individuals within the population. Disentangling these influences and interactions to establish the cause of the organisms' abundance is notoriously complicated. In addition, the magnitude of the above factors, and even the various combinations in which they occur, may be different in different areas or under different environmental conditions. If one considers simultaneously the factors that may influence individuals of a second species that use similar resources, and that the abundance of this potential competitor may be influenced by different factors from those that influence the first, the level of complexity and variability that one confronts in taking a community approach is greatly increased.

These points illustrate, in part, why attempts to produce generalized explanations on the basis of community pattern analysis have not been so successful and why attempts to generalize explanation from one community to another have invariably run into the problem of exceptional cases. Even generalization about species that have particular common ecological characteristics (e.g. early colonizing species) yields exceptions (e.g. Grubb 1987). On this basis one should also question the validity of attempts at classifying communities for purposes of generalized explanation of their structure.

Genetical differences among individuals of a species present ecologists with another source of variability. The Isolation Concept emphasizes this intraspecific variation and, therefore, the potential for adaptation in different populations within the extended distribution of a species. This line of thought stems from "population thinking," which has been discussed by Mayr (1982). It may be this emphasis that has prevented ecologists from establishing an effective theoretical basis around the Isolation Concept of species, by drawing attention from speciation as the source of adaptive novelty. In other words, the "eclipse of history" (see Ricklefs 1987) was not the sole oversight of community ecologists; it was built into the Isolation Concept. In any case variability among individuals has attracted concern among ecologists, and efforts to deal with it have been made (Orians 1962; Lomnicki 1978, 1988; Hassell and May 1985; May 1986).

Although genetical variation among individuals of a population occurs inevitably, undue emphasis of this feature in interpreting adaptation under the Isolation Concept seems to have directed attention away from certain particularly important adaptations. The Recognition Concept, in contrast, focuses on the significance of specific adaptations (see below), and these are ones that are predicted to be under stabilizing selection. For ecologists, this introduces a significant advantage because the stability expected in such characteristics provides at least a measure of constancy among all the variability that confounds efforts at generalization.

Of primary importance are the complex adaptive characters that individuals carry. Although they derive ultimately from the time of speciation, they dictate what is currently possible in terms of the ecology of individuals, and they thus influence presence and abundance in a fundamental way. Before considering the implications of this interpretation for generalization in ecology (in the subsection that later deals with individuals), some points about local adaptation are discussed.

The complex adaptations of individuals of a species are predicted to be similar over the geographical range of the species (Paterson 1985, 1986). The same may be true of adaptations that are simpler, such as those based on substitution of single alleles, but these simpler adaptations will not inevitably be species wide (e.g. sickle-cell anemia). A point so crucial for ecologists warrants wide and stringent testing, and this requires careful consideration. The complicating factors are subtle, and their circumvention depends on a clear understanding of the Recognition Concept of species and consistent application of the concept.

To illustrate, many situations exist in which populations of a species ostensibly have different adaptations in different areas. Typically, the assumption is that the species, defined taxonomically, represents one genetically defined species, but this is not inevitably true. The possibility must be tested on the basis of the appropriate questions (Paterson 1991). There are many cases in which much-studied problem species, defined by taxonomic criteria, turned out to be several genetical entities. It is likely, on theoretical grounds, that this will be the case in certain other instances, and the assumption that only one species is under consideration needs to be critically assessed in appropriate tests (Paterson 1991). This is one reason why Paterson (1981) dissociates *species* in taxonomy (which names and categorizes) from *species* in evolutionary genetics (which deals with gene flow in nature).

If intraspecific differences in complex adaptations are discovered, the origin of the adaptation requires further consideration. Did it arise and spread through a large population, or did it arise in a small, isolated population in which all individuals came under the same directional selection for a protracted period (Paterson 1986)? Every possibility must be scrutinized before different complex adaptations can justifiably be seen to evolve and then spread through extended populations of a species' distribution.

Considerations for Alternative Theories in Ecology

There are several fundamental points that should be allowed to guide development of theory in ecology, because their implications ramify to dependent disciplines.

First, the distinction between function and effect is fundamental in the investigation of adaptation (Williams 1966; Paterson 1984). Strictly applied, it generates guidelines for ecological theory and practice (Walter 1991).

Second, an important "level" of functional organization (*sensu* Williams 1966) in nature is represented by the fertilization mechanism (or SMRS) of sexual organisms. Such mechanisms have the consequence that mating occurs in a positively assortative way and that species' gene pools result (as an effect). There are no such mechanisms to promote structure or organization in communities (Walter 1993).

Third, interactions between individuals of different species and between individuals and the environment may take place because complex adaptations ensure their occurrence, or even prevent their occurrence. Such functional interactions profoundly influence the ecology of organisms, and although they confer a measure of organization in nature, it is not community-level organization or structure (Walter 1993). Other interactions occur despite there being no mechanisms to ensure (or prevent) their occurrence (e.g. competitive interactions). They are incidental interactions and do not confer functional organization.

Fourth, stasis of complex adaptations is expected between speciation events (Paterson 1985, 1986). Examples of complex ecological adaptations are given elsewhere (Hengeveld 1987; Walter 1993).

Fifth, each genetical species has a unique origin, and each has at least some characteristics or adaptations that are different from those of close relatives. Because adaptation is acquired in response to specific environmental contingencies and may be subject to chance influences, it is likely that ecological generalizations among species will be of only limited utility.

Sixth, the evolution of organisms is inextricably linked with their ecology (and vice versa), but ecological processes do not necessarily translate directly into the continuously strong directional selection implicit in Hutchinson's (1965) title *The Ecological Theater and the Evolutionary Play* and in Haldane's (1956) discussion of ecology and natural selection. In the extended populations that ecologists usually study, if only in part, stabilizing selection may be of overriding importance.

These guiding points, in combination with criticisms of much current practice in community ecology, suggest that ecological theory that deals with community structure needs a different direction. That it should stem from species theory has been justified earlier.

The emphasis derived from the Recognition Concept suggests that, in developing general theories to aid in understanding organisms and diversity in nature, the questions believed to be relevant need to be seriously reconsidered. Questions about community structure and assembly have relevance to postulated entities above the species level. They do not have relevance to an understanding of the origins of species and adaptation, or to the ecology of individuals (and, therefore, to questions of abundance and distribution). In other words, when we change the underlying premises with which we deal in ecology, we must reconsider our questions (Walter 1988, 1993) so that they are realistically related to our premises.

The Recognition Concept draws attention to the relationship of organisms to their environment, but it does so in a particular way (Paterson 1986). The complex adaptations of species (including the mechanisms for habitat location and habitat recognition) were acquired in relation to specific localized conditions of environment and habitat. Moreover, that relationship persists. Individuals of the species are consequently found in habitats that closely resemble their speciation habitat in aspects that are of functional significance to them. For example, the fertilization mechanism is closely coadapted to the habitat occupied by individuals of the species, and it is unlikely to operate effectively in other, different, situations.

As a consequence of the behavior of individuals that ties them to specific habitats, species distributions change in concert with the distribution of suitable habitat (Coope 1978; Paterson 1985, 1986). Thus, the independently acquired adaptations of species on the same trophic level account for species "falling out" of communities independently and at different times as climatic conditions change (e.g. Davis 1986; Hengeveld 1987; Graham and Grimm 1990). Because individuals of many species require, as part of their environment, the presence of individuals of a particular species (as a host, for example), the dependent species would go together with the host species.

Attention is thus drawn to the behavioral and physiological adaptations of individuals and to understanding their origin. Because such adaptations are predicted to have evolved at speciation, our questions about their origins are about historical events. In many ways this requires considerable reorientation of ecological research on adaptations, because historical events cannot be investigated directly by means of experiment or by comparison of competitors or potential competitors. Sometimes evidence of the origin of characters or species may not exist, but that must be accepted.

Individuals and Theory in Ecology

The Recognition Concept draws attention to the behavior of individuals and to the adaptations possessed by individuals. Increasingly it is recognized in ecology that the morphological, behavioral, and physiological properties of individuals are central to understanding collectives like populations and communities (Andrewartha and Birch 1954; Watt 1964; Lomnicki 1978, 1988; Hassell and May 1985; May 1986; Walter and Zalucki, in press). As hinted above, this undoubtedly warrants development of novel theory; information from the study of individuals is not likely to be readily compatible with older research programs oriented at a different level (Walter and Zalucki, in press). It is in this area that the Recognition Concept provides further direction, by raising the question of the population consequences of the species-specific primary adaptations that individuals have. This insight provides new perspective to the old question of abundance. Predation, competition, and so on, thus become secondary considerations in the sense that they modify a pattern derived from the expression of specific adaptations in a particular situation. This does not imply, however, that these ecological influences are inconsequential.

The approach to ecology that emerges is therefore not organized around interspecific competition or the detection of ecological patterns among species (not even if they are only potential competitors). There is also less emphasis placed on local directional selection, because any local deviations are likely to be subject later to stabilizing selection. Consequently, the study of ecology loses a certain appealing cohesion that the study of communities possessed. But at the same time it also discards the teleology and idealism inherent in resource partitioning and coexistence studies (Simberloff 1980; Walter 1993) that parallels that found in the Isolation Concept of species (Paterson 1982a).

For practical purposes, Andrewartha and Birch (1984) speak of an average individual. Under a given set of environmental conditions, we can expect a consistent response from individuals of a given species. But environmental conditions change over a short distance in nature, and they may affect the ecology of conspecifics in different ways (e.g. Dobkin et al. 1987). In any specified area, environmental conditions also change through time, whether seasonally, unpredictably over a short period, or inevitably (but still unpredictably) over a longer period (Andrewartha and Birch 1954; Coope 1978; Graham and Grimm 1990). Genetically equivalent individuals of the same species under different environmental conditions would have different survival chances and different reproductive potential. Abundance is thus a consequence (or *effect*) of the adaptations of the particular individuals that are living, reproducing, and migrating under inevitably changing environmental conditions (Walter and Zalucki, in press). The pattern produced is, of course, open to further change by other ecological factors such as predation.

Organisms may alter the characteristics of the environment around them. Such changes often affect other organisms negatively, and they have frequently been described as competitive interactions. But (without evidence to the contrary) it is only incidentally (or as an *effect*), that the environment has been rendered unsuitable for individuals of another species (Birch 1957; Connell 1987; Connell et al. 1987; Walker and Chapin 1987). The new conditions may, at the same time, be suitable for establishment of individuals of yet other species (Connell 1987). Such situations, commonly called inhibition and facilitation, respectively, are clearly frequent in nature, of significance to the presence and existence of organisms and the ecology of any specified area (Connell 1987), and need to be accommodated in ecological theory.

Treatment of organism-induced changes to the environment has, so far, been in generalized models based on such features as inhibition, facilitation, or disturbance, or combinations of these features (Connell and Slatyer 1977; Noble and Slatyer 1980; Walker and Chapin 1987). But almost every area has large numbers of species present, with others having the potential to be present, so vast number of colonizations and interactions by members of different species become possible. It is questionable, therefore, whether effective generalizations can be developed around the changed environment to explain community structure. Any change can potentially generate innumerable indirect effects, which seems to set very definite limits to generalization. However, the models and the research initiated by them have been useful in drawing attention to the suite of interactions that may possibly occur as diversity of vegetation and other organisms changes. But there does, nevertheless, seem to be cause to rethink organism-induced change to the environment, for theoretical purposes, in relation to individual organisms and their adaptations. This perspective suggests that true interspecific competition may be most realistically included as a special subset of environment modification.

Much of the behavior and its ecological consequences discussed so far may be influenced unpredictably by external influences, so chance events need to be considered as well. This clearly complicates the task of producing effective generalizations, and the level of complexity must increase exponentially as more and more species are considered simultaneously. In any case, this is why Simberloff (1980) speaks of probabilism in dealing with certain aspects of ecology.

Bearing in mind all the points raised in this section about sources of variability in natural situations implies that establishing whether a species can or cannot exist in a specified area is not a simple matter. And neither is establishing what factors are most significant in influencing its abundance. In investigating and interpreting either of these situations, the primary consideration must be the adaptations of those organisms to the environment (including food and other requirements, as well as climate). Among other aspects, understanding the sensory capabilities of the organisms and then relating them to the environment and to the observable behavior of the organisms is a complex undertaking that is not well developed, especially not in an ecological sense. The sensory adaptations and the requirements of immature stages may be quite different from other life stages and yet may have important consequences for the life of the resultant adults (Sale 1990*a,b*). These aspects need to be integrated in a generalized and realistic way into theory that attempts to understand the ecology of individuals, and consequently abundance and local diversity. Indeed, the integration of physiology and behavior in this way should undoubtedly occur at a fundamental level because these aspects mediate functional interactions which are, in turn, fundamental to the existence and reproduction of organisms (Walter 1993).

In short, the Recognition Concept suggests an approach to ecology in which there is a strict limit on generalizing at "community" or "ecosystem" levels. Extrapolation of interpretations from one "system" or "community" to another that is deemed sufficiently similar (see the above section, "Resources and Generalizations about Communities") is more likely to mislead than to enlighten. These limits to generalization are far exceeded by current practice. Emphasizing the potential for competitive interactions between species that use the same resource as a primary focus, as in theories of community structure, must pay the above costs.

General Theories in Ecology

The points made in this section do not imply that general theories are taboo; the Recognition Concept is itself a general theory, one that interprets the adaptations of species and the acquisition of adaptation. In the same vein, a general theory of similar format needs to be articulated to interpret occurrence and abundance. Note that the nature of these general theories does not allow extrapolation of explanation to different situations in the way that community theories are intended. They are generalized explanatory statements developed from those aspects of the acquisition of adaptation that species have in common. The theories therefore suggest ways in which to approach interpretation of each species' evolution and ecology (each of which would have its own solution). In other words, the Recognition Concept, besides drawing attention to defining generalizations about species, emphasizes the species-specific qualities of individuals. Adaptations of congenerics, for instance, are usually different, and as a consequence their ecology is likely to be different. For each species studied ecologically, we should erect a corresponding theory of its ecology, but allowance must be made for differences in ecology in different situations and in different areas.

Application of the above guiding strictures in ecological investigation and understanding of what has already occurred in nature presents the usual problems encountered in historical research. For predictive ecology, on the other hand, there are the foreseeable difficulties of dealing with unpredictably changing conditions. The situation is even further compounded by the presence of numerous species (each with unique adaptations). Any one of these species may be so influenced by the new conditions that their perceived "function" or "status" in the "ecosystem" would have to be revised. Good examples are provided by the unanticipated outbreaks and pest status of phytophagous mites on orchards and other crops when DDT was introduced (DeBach 1974), by parasitism unexpectedly reducing damaging densities of echinoids in American kelp beds (Johnson and Mann 1988), and by the sudden, unpredicted appearance in human populations of a virus with lethal consequences that has already dramatically influenced human ecology in certain areas. Such changes are usually unanticipated, but they do contribute considerably to ecological understanding (Lehman 1986) or, at least, to an appreciation of one more ecological influence or outcome out of a range that is likely to remain unknowable.

Discussion

Species and Interactions

The suggestions made for ecological theory in the previous section are different from recommendations for understanding local diversity made recently in the literature by others. The approach and questions are different, and the method is not a comparative one. If the aim is to understand ecology from the perspective of general principles, so that extrapolation from one area or situation to another is possible, then the strictures specified by recent advances in understanding adaptation must be considered seriously. From area to area, the complex adaptations of species should remain constant, but environmental conditions and the suite (and relative abundance) of "competitors" and so on will inevitably be different. Of primary significance to interpreting local presence and abundance would be physiological and behavioral adaptations for locating habitats, hosts, or other sites of special significance to individuals of the species. Tolerance levels are likely to be important too. Different growth stages frequently have different tolerances and may be adapted to a different environment from that inhabited by older or younger conspecifics. In certain organisms learning may also be an important consideration in understanding local ecology.

Theory with such a basis does not deny a role for competition, predation, and so on, but the ecological influences of these interactions are seen to have a different context, one that relates to their variable influence in nature. Revision of their context should also be consistent with the likelihood that many of the interactions whose influence is measured are, themselves, incidental effects of other processes (Walter 1993). The independent evolution of most potential competitors, and the vast number of different competitors, and the vast number of different competitive interactions that are possible for individuals of any species, must preclude the possibility that competitive interactions have functional significance. Although predators and parasites are certainly adapted to locate and exploit their prey, the population consequences of these factors are open to modification from many angles. Although here we deal with functional interactions (Walter 1993), the variability of their ecological impact must also preclude them from having functional significance in a community or ecosystem context.

The impact of evolutionary by-products cannot be generalized from one area to another, even on a local scale, as is frequently the case in current ecological explanation. In time, the impact of any such influence may change in one site as well. For these reasons the guidelines suggested in the previous section are not aimed at community- or ecosystem-level research.

Do such alternatives to community-level study necessarily miss factors of significance to the explanation of diversity and abundance? Communities have no objective reality because their delineation remains arbitrary, so it is not inevitable that a species-based approach would overlook factors of significance. Alternative approaches that have been developed to explain diversity or abundance (e.g. Andrewartha and Birch 1954, 1984; Paterson 1985; Hulley et al. 1988*a,b*) do incorporate comprehensively the influence of other species, whether on the same or on different trophic levels, and they do stipulate that investigation of their influence may be of fundamental importance. Influences from all spheres are thus considered, but within a different conceptual framework. The important point is that a different approach, with different basic questions, is employed to abstract information from the complexities of nature.

Alternatives to the community or ecosystem approach are frequently portrayed unrealistically. It seems unlikely that anyone would recommend that the ecological investigation of diversity and abundance "could be reduced to the simultaneous, but separate and independent, study of single-species populations" (Roughgarden and Diamond 1986:334). Another extract *reads:* "The populations at a site consist of all those that happened to arrive there" (Roughgarden 1989). But this also seems to be an unreasonable representation of alternative approaches because it is a statement with no explanatory power, and one which suggests no line of research that would be meaningful.

The Entangled Bank and Local Ecology

A common justification for current approaches taken in community studies seems to stem from Darwin's (1859:489–490) closing comments about the complexity of the entangled bank and from the significant (sometimes indirect) influence of individuals of one species on the abundance of one or more other species (e.g. Paine 1974). It is surely this sort of demonstration that leads ecologists, who may even be skeptical of the community concept, to deem it acceptable to study "assemblage" structure or organization even if integrated communities do not exist (e.g. Underwood 1986:355–356).

But studies that have yielded insights in community ecology have seldom been comparative ones, and they have concentrated on explaining the presence and abundance of a limited number of species (often only one) as influenced by a limited number of processes (e.g. Dayton 1973; Paine 1974; Underwood 1984). Such studies may be more appropriately seen as investigations of the abundance of species in a localized area. The insights they have yielded are notable and are illustrated by one of the better known of such studies. Predator removal led to interpretations of competitive exclusion of barnacles by mussels in the intertidal zone (Paine 1974). A subsequent study recorded a "postexclusion" elevation of population densities of the "inferior competitors," owing to the barnacles settling in large numbers on the mussels (Lee and Ambrose 1989). Also, replication of the manipulation by Dayton (1971) returned different results from those obtained originally by Paine (1974) because recruitment at his site followed a different pattern (Underwood and Denley 1984).

Here we are dealing with a limited number of species, yet a given change to the local ecology may have several different outcomes. These may be modeled successfully in a probabilistic way, but the various outcomes themselves do not yield underlying principles that have relevance to communities in general or even communities of a particular type. To understand the biological basis of what is occurring, we need to understand more fully the adaptations, requirements, and tolerances of the organisms involved (for both survival and reproduction); the source of recruits and the local influences on their ecology; and how the characteristics of the organisms match the local conditions of interest. That species composition of any area is not necessarily inevitable is widely known (e.g. Glenn-Lewin [1980] for plants; Underwood and Denley [1984] and Keough [1984] for the intertidal zone), and this is another area in which probabilism will need to be incorporated.

Circumscription on the basis of interaction (of potential competitors on a resource) or local ecology does therefore have a cost, and in particular one of the principal aims of community ecology may have to be sacrificed. It may not be possible to develop successful generalizations on this basis, at least not of the type generally sought in community ecology (see the above section "Resources and Generalizations about Communities"). Whether studies of local communities (even those that incorporate regional or global influences) can ever expose general principles of evolution and ecology must therefore be scrutinized. Information gained about adaptation is likely to be limited in studies of local communities because the method is inappropriate to the study of adaptation (see the above section, "The Recognition Concept and Theory in Ecology"). And whether sufficient detail will be obtained from community studies to understand fully the abundance of organisms is also questionable (Walter 1988). Conversely, a knowledge of the specific adaptations of the species present would aid interpretation of their presence and their abundance in the local area.

The numbers of individuals of each species in a local area will depend on their adaptations, the local environment, the individuals of other species that are present, and the way in which the environment is changed by those other organisms. So, the changing abundance of each species will be affected by several factors (and their abundance should be seen as an *effect* of these influences). Thus, the changing impacts of incidental interactions (such as competitive ones) are also incidental. Changing environmental conditions and other extraneous influences add to difficulties of coping with local communities. It is, therefore, not inevitable that local circumstances will be repeated elsewhere, even were the species composition of that area to be identical. Results from studies of local communities are, therefore, not necessarily applicable elsewhere and can justifiably be seen as anecdotal. Other recent reanalyses of "local" ecology (Ricklefs 1989*a*) lend support to that conclusion. This does not imply that questions about local areas do not have relevance, because they certainly do (e.g. Lehman 1986). But whether a general theory of local community structure is possible needs to be questioned.

In summary, there is a need to assess critically all intellectual advances that have been made in the fundamental areas of biological interpretation and to apply acceptable developments logically in dependent disciplines that are as complicated as ecology. At the same time, rejection of older, deficient approaches and hypotheses must be pursued actively. Removal of their influence must extend to the questions asked, as well as to the method and the explanations offered. Community ecology has not developed in this way, and this may explain why it has, over the years, become an accretion of ad hoc additions to the original competition-based interpretations. Severe criticisms of competition theory and community concepts, and the recent development of the Recognition Concept of species, suggest that alternative theories of ecology warrant development and scrutiny.

Acknowledgments

My education owes an inestimable debt to the dual influences of Pat Hulley and Hugh Paterson, whose high standards of scholarship, wide-ranging interests, and strong sense of justice have been an inspiration. Their interest and their friendship, shared unselfishly by Mary Hulley and Shirley Paterson, are continuing highlights. I appreciate, too, the discussions with Martin Benfield, Tony Clarke, Paul Flower, Errki Haukioja, Kalevi Kull, Kari Lehtilä, John Milne, Seppo Neuvonen, Anna Nilson, Chris Pavey, Kai Ruohomäki, Stefan Schmidt, Juha Tuomi, Helen Wallace, Myron Zalucki, and Martin Zobel, and the many comments they made on the manuscript. Jocelyne Campbell and Gail Walter kindly directed preparation of the manuscript and typed it. Financial assistance to visit Turku was made available by the University of Queensland.

References

Aarssen, L.W. 1989. Competitive ability and species coexistence: A "plant's-eye" view. Oikos 56:386–401.
Abrams, P.A. 1988. How should resources be counted? Theor. Pop. Biol. 3:226–242.
Andrewartha, H.G., and L.C. Birch. 1954. The distribution and abundance of animals. University of Chicago Press, Chicago.
———. 1984. The ecological web: More on the distribution and abundance of animals. University of Chicago Press, Chicago.
Birch, L.C. 1957. The meanings of competition. Am. Nat. 91:5–18.
Brady, R.H. 1982. Dogma and doubt. Biol. J. Linn. Soc. 17:79–96.
Brown, J.H. 1981. Two decades of homage to Santa Rosalia: Toward a general theory of diversity. Am. Zool. 21:877–888.
Brown, J.H., D.W. Davidson, and O.J. Reichman. 1979. An experimental study of competition between seed eating desert rodents and ants. Am. Zool. 19:1129–1143.
Chapin, F.S., and G.R. Shaver. 1985. Individualistic growth response of tundra plant species to environmental manipulations in the field. Ecology 66:564–576.
Charlesworth, B., R. Lande, and M. Slatkin. 1982. A neo-Darwinian commentary on macroevolution. Evolution 36:474–498.
Cody, M.L. 1989. Discussion: Structure and assembly of communities. Pp. 227–241 in J. Roughgarden, R.M. May, and S.A. Levins, eds. Perspectives in ecological theory. Princeton University Press, Princeton, N.J.
Colwell, R.K. 1984. What's new? Community ecology discovers biology. Pp. 387–396 in P.W. Price, C.N. Slobodchikoff, and W.S. Gaud, eds. A new ecology: Novel approaches to interactive systems. John Wiley, New York.
Connell, J.H. 1980. Diversity and the coevolution of competitors; or, the ghost of competition past. Oikos 35:131–138.
———. 1983. On the prevalence and relative importance of interspecific competition: Evidence from field experiments. Am. Nat. 122:661–696.
———. 1987. Change and persistence in some marine communities. Pp. 339–352 in A.J. Gray, M.J. Crawley, and P.J. Edwards, eds. Colonization, succession and stability. Twenty-sixth symposium of the British Ecological Society. Blackwell, London.
Connell, J.H., I.R. Noble, and R.O. Slatyer. 1987. On the mechanisms producing successional change. Oikos 50:136–137.
Connell, J.H., and R.O. Slatyer. 1977. Mechanisms of succession in natural communities and their role in community stability and organization. Am. Nat. 111:1119–1144.
Coope, G.R. 1978. Constancy of insect species versus inconstancy of quaternary environments. Pp. 176–187 in L.A. Mound and N. Waloff, eds. Diversity of insect faunas. Ninth symposium of the Royal Entomological Society of London. Blackwell, London.
Darwin, C. 1859. On the origin of species by means of natural selection. A facsimile of the first edition, 1964. Harvard University Press, Cambridge, Mass.
Davis, M.B. 1986. Climatic instability, time lags, and community disequilibrium. Pp. 269–284 in J.M. Diamond and T.J. Case, eds. Community ecology. Harper and Row, New York.
Dayton, P.K. 1973. Two cases of resource partitioning in an intertidal community: Making the right prediction for the wrong reason. Am. Nat. 107:662–670.
DeBach, P. 1974. Biological control by natural enemies. Cambridge University Press, London.

den Boer, P.J. 1986. The present status of the competitive exclusion principle. Trends Ecol. Evol. 1:25–28.
Dobkin, D.S., I. Olivieri, and P.R. Ehrlich. 1987. Rainfall and the interaction of microclimate with larval resources in the population dynamics of checkerspot butterflies (*Euphydryas editha*) inhabiting serpentine grassland. Oecologia 71:161–166.
Dobzhansky, T. 1935. A critique of the species concept in biology. Philos. Sci. 2:344–355.
———. 1951. Genetics and the origin of species. 3d ed. Columbia University Press, New York.
———. 1976. Organismic and molecular aspects of species formation. Pp. 95–105 *in* F.J. Ayala, ed. Molecular evolution. Sinauer, Sunderland, Mass.
Drake, J.A. 1990. Communities as assembled structures: Do rules govern pattern? Trends Ecol. Evol. 5:159–164.
Elton, C. 1950. The ecology of animals. 3d ed. Methuen, London.
Endler, J.A. 1989. Conceptual and other problems in speciation. Pp. 625–648 *in* D. Otte and J.A. Endler, eds. Speciation and its consequences. Sinauer, Sunderland, Mass.
Futuyma, D.J. 1986. Evolutionary biology. Sinauer, Sunderland, Mass.
———. 1987. On the role of species in anagenesis. Am. Nat. 130:465–473.
Futuyma, D.J., and G.C. Mayer. 1980. Non-allopatric speciation in animals. Syst. Zool. 29:254–271.
Glenn-Lewin, D.C. 1980. The individualistic nature of plant community development. Vegetatio 43:141–146.
Graham, R.W., and E.C. Grimm. 1990. Effects of global climate change on the patterns of terrestrial biological communities. Trends Ecol. Evol. 5:289–292.
Grant, P.R. 1972. Convergent and divergent character displacement. Biol. J. Linn. Soc. 4:39–68.
———. 1975. The classical case of character displacement. Pp. 237–337 *in* T. Dobzhansky, M.K. Hecht, and W.C. Steere, eds. Evolutionary biology. Plenum, New York.
Grine, F.E. 1981. Trophic differences between "gracile" and "robust" Australopithecines: A scanning electron microscope analysis of occlusal events. S. Afr. J. Sci. 77:203–230.
Grubb, P.J. 1987. Some generalizing ideas about colonization and succession in green plants and fungi. Pp. 81–102 *in* A.J. Gray, M.J. Crawley and P.J. Edwards, eds. Twenty-sixth symposium of the British Ecological Society. Blackwell, London.
Haldane, J.B.S. 1956. The relation between density regulation and natural selection. Proc. R. Soc. Lond. B 145:306–308.
Hassell, M.P., and R.M. May. 1985. From individual behavior to population dynamics. Pp. 3–32 *in* R.M. Sibly and R.H. Smith, eds. Behavioral ecology: Ecological consequences of adaptive behavior. Twenty-fifth symposium of the British Ecological Society. Blackwell, Oxford.
Heck, K.L. 1976. Some critical considerations of the theory of species packing. Evol. Theor. 1:247–258.
———. 1980. Competitive exclusion or competitive delusion? Paleobiology 6:241–242.
Hengeveld, R. 1987. Scales of variation: Their distinction and ecological importance. Anna. Zool. Fennici 24:195–202.
———. 1988. Mayr's ecological species criterion. Syst. Zool. 37:47–55.
Hilborn, R., and S.C. Stearns. 1982. On inference in ecology and evolutionary biology: The problem of multiple causes. Acta Biotheor. 31:145–164.
Hochberg, M.E., and J.H. Lawton. 1990. Competition between kingdoms. Trends Ecol. Evol. 5:367–371.
Hulley, P.E., G.H. Walter, and A.J.F.K. Craig. 1988*a*. Interspecific competition and com-

munity structure. I. Shortcomings of the competition paradigm. Riv. Biol.–Biol. Forum 81:57–71.

———. 1988*b*. Interspecific competition and community structure. II. The recognition concept of species. Riv. Biol.–Biol. Forum 81:261–283.

Hutchinson, G.E. 1948. Circular causal systems in ecology. Ann. N.Y. Acad. Sci. 50:221–246.

———. 1961. The paradox of the plankton. Am. Nat. 95:137–145.

———. 1965. The ecological theater and the evolutionary play. Yale University Press, New Haven, Conn.

Johnson, C.R., and K.H. Mann. 1988. Diversity, patterns of adaptation, and stability of Nova Scotian kelp beds. Ecol. Mono. 58:129–154.

Keough, M.J. 1984. Dynamics of the epifauna of the bivalve *Pinna bicolor:* Interactions among recruitment, predation, and competition. Ecology 65:677–688.

Lambert, D.M., and A.J. Hughes. 1988. Keywords and concepts in structuralist and functionalist biology. J. Theor. Biol. 133:133–145.

Lawton, J.H. 1987. Are there assembly rules for successional communities? Pp. 225–244 *in* A.J. Gray, M.J. Crawley and P.J. Edwards, eds. Colonization, succession and stability. Twenty-sixth symposium of the British Ecological Society. Blackwell, London.

———. 1989. Book review: Disentangling the bank. Nature 339:517.

Lawton, J.H., and K.J. Gaston. 1989. Temporal patterns in the herbivorous insects of bracken: A test of community predictability. J. Anim. Ecol. 58:1021–1034.

Lawton, J.H., and D.R. Strong. 1981. Community patterns and competition in folivorous insects. Am. Nat. 118:317–338.

Lee, H., and W.G. Ambrose. 1989. Life after competitive exclusion: An alternative strategy for a competitive inferior. Oikos 56:424–427.

Lehman, J.T. 1986. The goal of understanding in limnology. Limnol. Oceano. 31:1160–1166.

Lomnicki, A. 1978. Individual differences between animals and the natural regulation of their numbers. J. Anim. Ecol. 47:461–475.

———. 1988. Population ecology of individuals. Princeton University Press, Princeton, N.J.

MacArthur, R.H. 1972. Coexistence of species. Pp. 253–259 *in* J. Behnke, ed. Challenging biological problems. Oxford University Press, Oxford.

MacFadyen, A. 1963. Animal ecology: Aims and methods. Sir Isaac Pitman, London.

Masters, J., D.M. Lambert, and H.E.H. Paterson. 1984. Scientific prejudice, reproductive isolation, and apartheid. Persp. Biol. Med. 28:107–116.

May, R.M. 1986. The search for patterns in the balance of nature: Advances and retreats. Ecology 67:1115–1126.

May, R.M., and J. Seger. 1986. Ideas in ecology. Am. Sci. 74:256–267.

Mayr, E. 1942. Systematics and the origin of species. Columbia University Press, New York.

———. 1963. Animal species and evolution. Harvard University Press, Cambridge, Mass.

———. 1970. Populations, species and evolution. Harvard University Press, Cambridge, Mass.

———. 1982. The growth of biological thought. Harvard University Press, Cambridge, Mass.

McIntosh, R.P. 1987. Pluralism in ecology. Annu. Rev. Ecol. Syst. 18:321–341.

Menge, B.A., and A.M. Olsen. 1990. Role of scale and environmental factors in regulation of community structure. Trends Ecol. Evol. 5:52–57.

Montagu, A. 1952. Darwin competition and cooperation. Schuman, New York.
Murray, B.G. 1986. The structure of theory, and the role of competition in community dynamics. Oikos 46:145–158.
Noble, I.R., and R.O. Slatyer. 1980. The use of vital attributes to predict successional changes in plant communities subject to recurrent disturbances. Vegetatio 43:5–21.
Orians, G.H. 1962. Natural selection and ecological theory. Am. Nat. 96:257–263.
Otte, D., and J.A. Endler, eds. 1989. Speciation and its consequences. Sinauer, Sunderland, Mass.
Paine, R.T. 1974. Intertidal community structure: Experimental studies on the relationship between a dominant competitor and its principal predator. Oecologia 15:93–120.
———. 1988. Food webs: Road maps of interactions or grist for theoretical development? Ecology 69:1648–1654.
Paterson, H.E.H. 1973. Animal species studies. J.R. Soc. West. Aust. 56:31–36.
———. 1980. A comment on "mate recognition systems." Evolution 34:330–331.
———. 1981. The continuing search for the unknown and unknowable: A critique of contemporary ideas on speciation. S. Afr. J. Sci. 77:113–119.
———. 1982a. Darwin and the origin of species. S. Afr. J. Sci. 78:272–275.
———. 1982b. Perspective on speciation by reinforcement. S. Afr. J. Sci. 78:53–57.
———. 1985. The recognition concept of species. Pp. 21–29 in E.S. Vrba, ed. Species and speciation. Transvaal Museum Monograph No. 4. Pretoria.
———. 1986. Environment and species. S. Afr. J. Sci. 82:62–65.
———. 1988. On defining species in terms of sterility: Problems and alternatives. Pacific Sci. 42:65–71.
———. 1991. The recognition of cryptic species among economically important insects. Pp. 1–10 in M.P. Zalucki, ed. *Heliothis:* Research methods and prospects. Springer-Verlag, New York.
Paterson, H.E.H., and M. Macnamara. 1984. The recognition concept of species. S. Afr. J. Sci. 80:312–318.
Pianka, E.R. 1983. Evolutionary ecology. 3d ed. Harper and Row, New York.
Poore, M.E.D. 1964. Integration in the plant community. Pp. 213–226 in A. MacFadyen and P.J. Newbould, eds. British Ecological Society Jubilee Symposium, London, 28–30 March 1963. Blackwell, Oxford.
Price, P.W. 1984. Alternative paradigms in community ecology. Pp. 353–383 in P.W. Price, C.N. Slobodchikoff, and W.S. Gaud, eds. A new ecology: Novel approaches to interactive systems. John Wiley, New York.
Price, P.W., W.S. Gaud, and C.N. Slobodchikoff. 1984. Introduction: Is there a new ecology? Pp. 1–11 in P.W. Price, C.N. Slobodchikoff, and W.S. Gaud, eds. A new ecology: Novel approaches to interactive systems. John Wiley, New York.
Ricklefs, R.E. 1987. Community diversity: Relative roles of local and regional processes. Science 235:167–171.
———. 1989a. Speciation and diversity: The integration of local and regional processes. Pp. 599–622 in D. Otte and J.A. Endler, eds. Speciation and its consequences. Sinauer, Sunderland, Mass.
———. 1989b. Nest predation and the species diversity of birds. Trends Ecol. Evol. 4:184–186.
Riley, P.A. 1986. The origin of the principle of competitive exclusion: Was Darwin influenced by Sismondi? The Linnean 2:20–22.
Roughgarden, J. 1989. The structure and assembly of communities. Pp. 203–226 in J. Roughgarden, R.M. May, and S.A. Levins, eds. Perspectives in ecological theory. Princeton University Press, Princeton, N.J.

Roughgarden, J., and J. Diamond. 1986. Overview: The role of species interactions in community ecology. Pp. 333–343 *in* J.M. Diamond and T.J. Case, eds. Community ecology. Harper and Row, New York.

Sale, P.F. 1988. Perception, pattern, chance and the structure of reef fish communities. Environ. Biol. Fish. 21:3–15.

———. 1990*a*. Recruitment of marine species: Is the bandwagon rolling in the right direction? Trends Ecol. Evol. 5:25–27.

———. 1990*b*. Reply from Peter Sale. Trends Ecol. Evol. 5:231.

Schoener, T.W. 1965. The evolution of bill size differences among sympatric congeneric species of birds. Evolution 19:189–213.

———. 1984. Counters to the claims of Walter et al. on the evolutionary significance of competition. Oikos 43:248–250.

———. 1986*a*. Overview: Kinds of ecological communities—ecology becomes pluralistic. Pp. 467–479 *in* J.M. Diamond and T.J. Case, eds. Community ecology. Harper and Row, New York.

———. 1986*b*. Resource partitioning. Pp. 91–126 *in* J. Kikkawa and D.J. Anderson, eds. Community ecology: Pattern and process. Blackwell, London.

Schweber, S.S. 1977. The origin of the Origin revisited. J. Hist. Biol. 10:229–316.

———. 1980. Darwin and the political economists: Divergence of character. J. Hist. Biol. 13:195–289.

Simberloff, D. 1976. Trophic structure determination and equilibrium in an arthropod community. Ecology 57:395–398.

———. 1980. A succession of paradigms in ecology: Essentialism to materialism and probabilism. Pp. 63–99 *in* E. Saarinen, ed. Conceptual issues in ecology. Reidel, Dordrecht.

———. 1982. The status of competition theory in ecology. Ann. Zool. Fennici 19:241–253.

———. 1983. Competition theory, hypothesis-testing, and other community ecological buzzwords. Am. Nat. 122:626–635.

Southwood, T.R.E. 1980. Ecology—a mixture of pattern and probabilism. Pp. 203–214 *in* E. Saarinen, ed. Conceptual issues in ecology. Reidel, Dordrecht.

Tilman, D. 1989. Discussion: Population dynamics and species interactions. Pp. 89–100 *in* J. Roughgarden, R.M. May, and S.A. Levin, eds. Perspectives in ecological theory. Princeton University Press, Princeton, N.J.

Underwood, A.J. 1984. Vertical and seasonal patterns in competition for microalgae between intertidal gastropods. Oecologia 64:211–222.

———. 1986. What is a community? Pp. 351–367 *in* D.M. Raup and D. Jablonski, eds. Patterns and processes in the history of life. Springer-Verlag, Berlin.

Underwood, A.J., and E.J. Denley. 1984. Paradigms, explanations, and generalizations in models for the structure of intertidal communities on rocky shores. Pp. 151–180 in D.R. Strong, D. Simberloff, L.G. Abele, and A.B. Thistle, eds. Ecological communities: Conceptual issues and the evidence. Princeton University Press, Princeton, N.J.

Underwood, A.J., and P.G. Fairweather. 1989. Supply-side ecology and benthic marine assemblages. Trends Ecol. Evol. 4:16–20.

Walker, L.R., and F.S. Chapin. 1987. Interactions among processes controlling successional change. Oikos 50:131–135.

Walter, G.H. 1988. Competitive exclusion, coexistence and community structure. Acta Biotheor. 37:281–313.

———. 1991. What is resource partitioning? J. Theor. Biol. 150:137–143.

———. 1993. The concept of interaction in ecological theory. Pp. 133–148 *in* K. Kull and

T. Tiivel, eds. Lectures in theoretical biology: The Second Stage 1, Vol. 2. Estonian Academy of Sciences, Tallinn.

Walter, G.H., P.E. Hulley, and A.J.F.K. Craig. 1984. Speciation, adaptation and interspecific competition. Oikos 43:246–248.

Walter, G.H., and M.P. Zalucki. In press. Rare butterflies and theories of evolution and ecology. *In* R.L. Kitching, R. Jones and N. Pierce, eds. Biology of Australian butterflies. CSIRO, Melbourne.

Watt, A.S. 1964. The community and the individual. Pp. 203–211 *in* A. MacFadyen and P.J. Newbould, eds. British Ecological Society Jubilee Symposium, London, 28–30 March 1963. Blackwell, Oxford.

Wiens, J.A. 1977. On competition and variable environments. Am. Sci. 65:590–597.

———. 1983. Avian community ecology: An iconoclastic view. Pp. 355–403 *in* A.H. Brush and G.A. Clark, eds. Perspectives in ornithology: Essays presented for the centennial of the American Ornithologists' Union. Cambridge University Press, Cambridge.

———. 1984. Resource systems, populations, and communities. Pp. 397–436 *in* P.W. Price, C.N. Slobodchikoff, and W.S. Gaud. eds. A new ecology: Novel approaches to interactive systems. John Wiley, New York.

Williams, G.C. 1966. Adaptation and natural selection: A critique of some current evolutionary thought. Princeton University Press, Princeton, N.J.

Wilson, D.S. 1983. The group selection controversy: History and current status. Anna. Rev. Ecol. Syst. 14:159–187.

12

Heterogeneity in Distribution of Fixed Chromosomal Inversions in Ancestor and Descendant Species in Two Groups of Mosquitoes: Series Myzomyia and Neocellia of *Anopheles* (*Cellia*)

CHRISTOPHER A. GREEN

Department of Medical Parasitology
London School of Hygiene and Tropical Medicine

> What is needed first is to steep oneself in the *events*, to approach the phenomena with as few preconceptions as possible, to take a naturalist's observational, descriptive approach to these events, and draw forth those low-level inferences which seem most native to the material itself.
>
> Carl R. Rogers, 1961

AN INTRIGUING evolutionary puzzle has emerged from anopheline cytogenetic data—to explain how two groups of species show a significant difference in the relative proportions of inversions unique to single species and those shared by two or more species; the ratios are 75:12 and 26:44 (homogeneity chi-square = 38.58 [df = 1]; $p \ll .01$). One group of twenty species has a total of 87 fixed inversions needed to account for the rearrangement of their chromosomes from a single, common ancestor. The second group of eighteen species has 70 inversions. The two sets of figures, number of species versus number of inversions of each group, are similar (homogeneity chi-square = 0.02 [df = 1]; $p = .90$). What kind of events could be responsible for the difference in unique: shared inversions in the two groups?

The chromosomal data are a by-product of a bread-and-butter application of population genetics to disentangle gene pools of anopheline vectors of malaria parasites from those of harmless relatives. This was necessary to better understand the dynamics of transmission of malaria parasites because anopheline vectors often share an indistinguishable morphology with harmless, sympatric relatives (Paterson 1963). Coetzee and Hunt (this volume) deal extensively with this problem. Anopheline chromosome variation was used to look for evidence of the population genetic consequences of sympatric mixtures of different, specific, mate recognition systems, that is, positive assortative mating within two or more chromosomally marked groups. The chromosome variants are alternatives for paracentric inversions where both homozygotes as well as heterozygotes are clearly visible in easily obtained polytene chromosomes. In the course of these studies, data accumulated about species-specific chromosomal rearrangements, and these were fitted together into a phylogeny to which cladistic principles were applied (Green 1982*a,b*; Green et al. 1985*b*).

The Mosquitoes

The majority of the two hundred or so species in *Anopheles* (*Cellia*) are confined to the old world tropics. Their taxonomy is well understood because they include major vectors of malarial parasites. Taxonomic reviews include Gillies and de Meillon (1968) and Gillies and Coetzee (1987) for the Afrotropical region; Rao (1984) for south Asia; Reid (1968) and Harrison (1980) for southeast and east Asia. All these workers recognize six informal species groups or series within *Cellia:* Cellia, Myzomyia, Neocellia, Neomyzomyia, Paramyzomyia, and Pyretophorus. All six series occur in the Afrotropical region and four in the Oriental region. The majority of species are endemic to single, zoogeographic regions. Neomyzomyia occurs throughout the geographical distribution of the subgenus and is dominant on Madagascar and in the Australasian region, suggesting that it is closest to the common ancestor of the subgenus *Cellia*. Gillies and de Meillon (1968) came to the same conclusion judged from the morphology of *An. wilsoni,* an Afrotropical member of Neomyzomyia. Apart from this suggestion of Neomyzomyia being older than the other series, there is no clear idea of phylogenetic relationships between the series. The chromosome data come from series Neocellia and Myzomyia. A species of Myzomyia, *An. majidi,* resembles the general adult morphology of Neocellia, suggesting a close relationship between the two series. Unfortunately, the chromosomes of *An. majidi* have not been described.

There are five species of Neocellia and forty-five of Myzomyia in the Afrotropical region and twenty-four of Neocellia and thirteen of Myzomyia in the Oriental region. Twenty species of Neocellia and eighteen of Myzomyia have been chromosomally sampled and, together with their gross, geographic distributions, are shown in Figure 12.1a and 12.1b.

The Chromosomes

All species in *Cellia* have three pairs of chromosomes, two pairs of autosomes and a sex-associated pair, males being heterogametic. "Readable" polytene chromosomes occur in larval salivary glands and in the ovarian nurse cells of some species during egg development. The latter are used here. Whole-arm, autosomal translocations have produced all three combinations of the four autosomal arms (Green and Hunt 1980). All the species sampled have an autosomal constitution of Arm 2 + Arm 5 and Arm 3 + Arm 4 except the *An. funestus/parensis/confusus/vaneedeni* group, which has 2 + 4, 3 + 5. All other rearrangements of the DNA molecules so far recorded from microscopy are due to paracentric inversions. Data from Neocellia (Fig. 12.1a) come from Coluzzi et al. (1970), Green (1982a), Green and Baimai (1984), and Green et al. (1985a,b; 1992). Data from Myzomyia (Fig. 12.1b) come from Green (1982b), Green and Miles (1980), and Subbarao et al. (1983). Data for *An. minimus, An. flavirostris, An. mangyanus,* and *An. aconitus* are unpublished.

Rearrangements of X chromosomes are not included in these data because homologies are difficult to determine between species even within single groups (e.g. the Maculatus group). Furthermore, they have extensive blocks of heterochromatin judged from mitotic chromosomes (e.g. the Maculatus group, Green et al. 1985a) which would incline this arm to parallelism, that is, different inversions with apparently the same break points judged from polytene chromosomes. Conventions used in designating the inversions are given in Green (1982a,b), Green and Hunt (1980), and Green et al. (1985b).

Application of cladistics to chromosome data is presented in general by Farris (1978); specifically for anophelines in Green (1982b) and Green et al. (1985b); and for *Drosophila* in Lemeunier and Ashburner (1984). Using cladistic principles, one may guess which alternative of an inversion is the ancestral, uninverted sequence (plesiomorphic) and which the derived, inverted sequence (apomorphic). Apomorphs are indicated in Figure 12.1 by either single black squares per row (autapomorphies) or more than one per row (synapomorphies). White squares indicate fixation of plesiomorphs. Half-white and half-black squares indicate that both alternatives float within a single species, (i.e., $5+c/5c$ in *An. stephensi* and $3+o/3o$ in *An. wellcomei*). There are no data for Arm 5, Myzomyia, and Arm 3 in *An. demeilloni,* shown as white blocks of squares in Figure 12.1b.

Discussion

Synapomorphies in Figure 12.1 indicate the phylogenetic relationships shown in Figure 12.2. These hypothetical statements, in Figures 12.1 and 12.2, are supported by morphological and zoogeographic data. The species appear

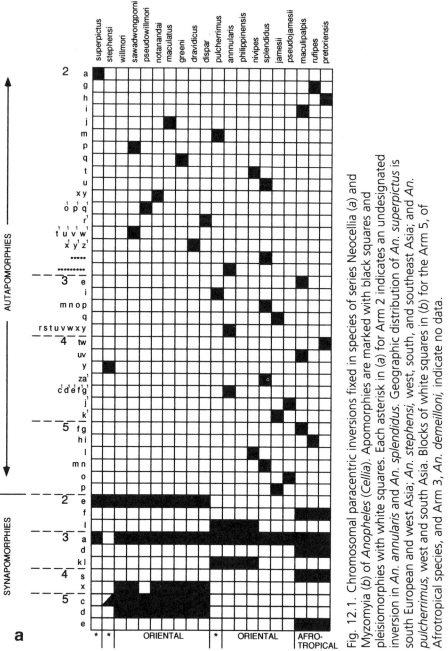

Fig. 12.1. Chromosomal paracentric inversions fixed in species of series Neocellia (a) and Myzomyia (b) of Anopheles (Cellia). Apomorphies are marked with black squares and plesiomorphies with white squares. Each asterisk in (a) for Arm 2 indicates an undesignated inversion in An. annularis and An. splendidus. Geographic distribution of An. superpictus is south European and west Asia; An. stephensi, west, south, and southeast Asia; and An. pulcherrimus, west and south Asia. Blocks of white squares in (b) for the Arm 5, of Afrotropical species, and Arm 3, An. demeilloni, indicate no data.

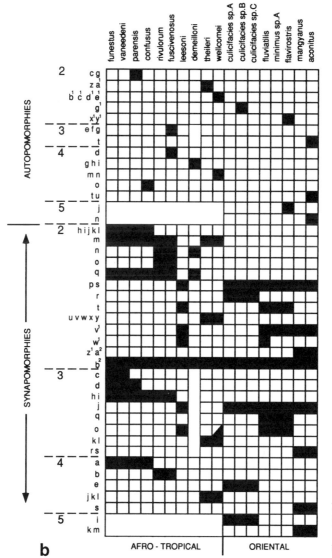

Fig. 12.1b

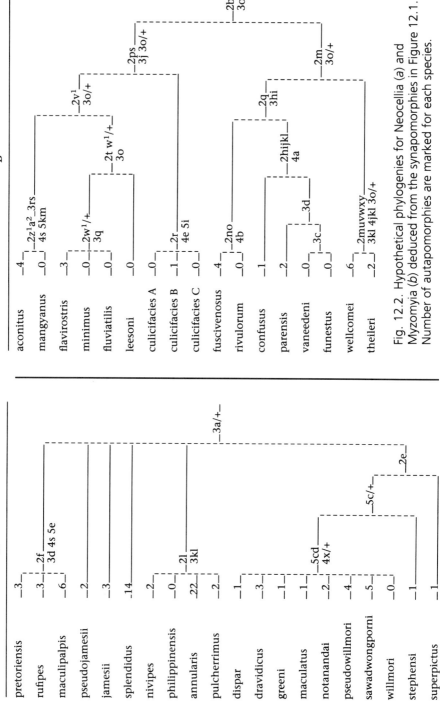

Fig. 12.2. Hypothetical phylogenies for Neocellia (a) and Myzomyia (b) deduced from the synapomorphies in Figure 12.1. Number of autapomorphies are marked for each species.

sorted with respect to the chromosome data, particularly in Figure 12.1b, that is, a great frequency of contiguous black squares along rows and blocked together with similar rows down the columns. In fact, species are arranged solely on the basis of relationships suggested by taxonomists judging from morphology and the species' geographic provenance. Therefore, there is a good deal of conformity between morphological, zoogeographical, and chromosomal data—each mutually supportive of a single scheme of relationship. Notice that *An. leesoni* and *An. demeilloni* are exceptions. The chromosome data indicate the former belongs to the Oriental *An. minimus* group and was misplaced within the Afrotropical *An. funestus* group, but *An. demeilloni* is, cladistically, inexplicable (Green 1982b).

Let us return to the puzzle, which can now be rephrased as a difficulty in explaining the significant difference between the ratio of autapomorphies to synapomorphies in the two clades: Neocellia, 75:12, and Myzomyia, 26:44. There are two kinds of events involved in these data—the mutational events that produce inversions and the population genetic events that fix inversion alternatives as species characteristics. A test of between clade differences for the latter events requires pooling of the inversions. The character matrices in Figure 12.1 have been pooled row-wise where two or more inversions, within a single arm, are fixed in a single species or the same group of species. The new pair of ratios becomes Neocellia, 35:11; Myzomyia, 14:28 (homogeneity chi-square = 14.58 [df = 1]; $p \ll .01$). A further collapse of similar inversions from all arms so that any species will contribute either 1 or 0 toward the autapomorph score, as in Figure 12.2, gives ratios: Neocellia, 18:5; Myzomyia, 8:13 (homogeneity chi-square = 5.76 [df = 1]; $p = .02$). The three ratios of autapomorphic to synapomorphic inversions are surprisingly homogeneous within each series (Neocellia, $p = .31$, and Myzomyia, $p = .90$)—surprising, because I expected that removal of individual contributions by *An. annularis* and *An. splendidus* would significantly change the ratio in Neocellia by reduction of the autapomorphs and that removal of the number of synapomorphs of the *An. wellcomei/theileri* pair would significantly alter the ratio in Myzomyia in the opposite direction.

The puzzle becomes more tangible in evolutionary terms in Figure 12.2. Why are there relatively so few ancestors in the Neocellia phylogeny? Relative to Myzomyia, there is no shortage of inversion events. To better understand the puzzle, consider only the Myzomyia data. These are consistent with a generally accepted view about the association of inversion fixation with speciation events which involves the sequence:

1. An inversion occurs as a unique mutational event. Multiple inversions may occur in one event when three or more breaks occur together.

2. The frequency of the new chromosome increases by one or more effects; heterosis, "hitchhiking" on linked (enclosed) genes, random stochastic

events associated with small populations, and simple change within a large population. The means of increase are not important here except that they are independent of the speciation event.

3. By chance, during subdivision/migration/reduction of parts of this polymorphic population, a speciation event fixes the inverted chromosome in a descendant species. The ancestor may have several polymorphic inversions, which arose from independent mutations, from which more than one inverted sequence may be fixed in a single speciation event.

Assume that the fixation of inversions is an accidental accompaniment of speciation events so that during evolutionary time they accumulate in a phylogeny more or less at random. The ratio of fixed inversions in extant species to those in ancestors is 26:44. The assumption of randomness gives an expectation that dividing the data between the Afrotropical and Oriental clades would give similar ratios. This expectation is realized: Afrotropical = 16:26 and Oriental = 8:14 ($p = .89$). According to the random hypothesis, the larger number of synapomorphies indicates a larger number of ancestors than extant species, but Figure 12.2 does not support this—extant: ancestors = 18:14. However, this is accounted for if ancestors included the same relative proportion of species, lacking autapomorphies, as occur in extant species, that is, 0.5. Such extinct species would be "hidden" in a phylogeny that depended on inversion differences between species. The adjusted ratio, 18:28, is homogeneous with the autapomorph: synapomorph ratio, 26:44 ($p = .98$).

I supposed, from this experience, that inversion fixation in anophelines would be a good, general indicator of their evolutionary history *until* the data accumulated from Oriental Neocellia.

How does one account for the Neocellia data from a phylogenetic point of view? One contributing possibility is that the group is polyphyletic, in which case some ancestors are "missing." Whether this is true will be revealed by the sampling of more species of *Cellia*. However, there are data to suggest that this is not the case. Arm 4 of *An. stephensi* differs from that of the hypothetical ancestor of Myzomyia by four inversions: the autapomorph of *An. stephensi*, 4*y*, and 4*pqr* (Green et al. 1985b). Furthermore, parts of the clade are, each, monophyletic—the *An. annularis*, the Afrotropical, and *An. stephensi* groups—and, apart from the Afrotropical species, each shows relatively greater proportions of autapomorphs to synapomorphs.

Given there are few ancestors, why did these have such a relatively smaller number of autapomorphs than extant species? It looks as though each of the most recent ancestors went through a burst of speciation events, most of which were accompanied by relatively greater chromosomal rearrangement than had occurred during speciation events leading to their own origin. An intriguing fact is that the species with the maximum number of autapomorphies, *An. annularis*, has *An. philippinensis*, with no autapomorphies, for one of its nearest relatives.

The nine Arm 2 inversions of *An. annularis* are undesignated in Figure 12.1 (Green et al. 1985*b*). There are many break points that are shared by postulated inversions such that several different inversions can be postulated which make up more than one sequence of a minimum of nine inversions. It seemed silly to pick one of these several sequences by arbitrarily designating inversions in one, rather than another, of these sequences. The same is true for the inversions on Arms 3 and 4 in *An. annularis*—these were designated before the logic of leaving them undesignated occurred to me. A phylogenetic explanation of this large number of autapomorphs is that *An. annularis* represents a group of unknown species, each taking a share of the autapomorphs. These species may be extant or extinct or a mixture of both. The presence of extant species seems unlikely since the inversions are fixed in samples taken throughout the geographic distribution of *An. annularis:* Bangladesh, Thailand, Java, Taiwan, and the Philippines.

An alternative explanation is that the genome in *An. annularis* suffered some form of instability associated with its origin (incidentally, why no rearrangement of Arm 5?). Could the puzzle be explained by differences in the origin of inversions between Neocellia and Myzomyia or, more likely, differences in the relative frequency of different modes of the origin of inversions that are fixed by speciation events? Shaw et al. (1983) give a new idea about the mode of origin of inversions, alternative to that given before.

Two races of the grasshopper *Caledia captiva* are characterized by genomic rearrangement due to many translocations. They form a hybrid swarm of structural heterozygotes where they meet in Queensland. Laboratory-produced hybrids give rise to a high incidence of novel chromosomal rearrangements, seen in their backcross progeny, in which they are confined to the parental hybrid chromosomes. The same rearrangements may occur repeatedly, and some of these have been found in the natural hybrid swarm. As Shaw et al. (1983) say, "The nonrandom and recurrent nature of these chromosomal mutations at high frequencies provides a plausible explanation for the establishment and fixation of chromosomal rearrangements in natural populations." This genomic instability is comparable to that seen in many other organisms.

Over the past fifteen years, the cytogeneticists' "black-box" term, *heterochromatin,* has become a matter for great excitement since some phenomena associated with it involve mobile DNA sequences. Mobile sequences are highly heterogeneous in their structure and mode of transposition. Some of them stimulate chromosomal rearrangements including inversions, for example, Spm elements of *Zea mays* (Nevers and Saedler 1977); and P elements (Engels and Preston 1984), fold-back transposons (Bingham and Zachar 1989) and I factors (Finnegan 1989) of *D. melanogaster*. Some elements are known to occur widely in natural populations, for example, P and I elements in *D. melanogaster* (Ronsseray and Anxolabehere 1986).

Phylogenetic distribution of different kinds of transposons vary. P elements are restricted to some species of the *D. melanogaster* group (Engels 1989); I factors occur in twenty of the twenty-one species so far tested in the *D. melanogaster* group (Finnegan 1989); and the retrotransposons extend from the *D. melanogaster* group into *D. pseudoobscura* (Bingham and Zachar 1989). An oddity about P element distribution is the presence of nearly homologous sequences in *D. melanogaster* and *D. willistoni*, a relatively distant relative, but a total absence in *D. simulans*, a much closer relative of *D. melanogaster*, suggesting that P elements represent recent "infection" or genetically horizontal transfer (Ronsseray and Anxolabehere 1986; Engels 1989).

Retrotransposon sequences have been found in *An. gambiae* (Besansky 1990). Like similar elements in *Drosophila*, homologous sequences are dispersed throughout the genome, which is a criterion that was used to suppose that they are transposons. *An. annularis* is characterized by a high degree of ectopic pairing in polytene chromosome spreads relative to *An. philippinensis*. A common sense inference is that this is evidence for pairing between homologous DNA sequences dispersed throughout the genome. Tests for this in cases of ectopic pairing in *Drosophila melanogaster* failed to demonstrate homologous sequences at the sites of pairing (Cohen 1976; Pardue et al. 1977) though researchers did not exclude the possibility (Cohen 1976). Strobel et al. (1979) showed three sites of insertion of the retrotransposon *copia* about the same sites and in the same strain of *D. melanogaster*, which also showed ectopic pairing (Barr and Ellison 1972), though this correlation may seem spurious because of the large number of insertion sites shown for transposons in the *D. melanogaster* genome. Ajioka and Hartl (1989) give figures of 115–120 independent sites per chromosome arm at which insertion can occur in *D. melanogaster*, but occupancy of such sites is generally low.

Transposons have no known beneficial effects, generally causing harmful mutations during transposition except for those involved in regular production of antigenic variation in *Trypanosoma* and antibody variation in mammalian B-cells. These beneficial effects are an end result of interclone selection among randomly generated variants. In gametogenic cells, they are an enormously rich source of variation for evolutionary change but, as Ajioka and Hartl (1989:947) point out, "the entire population might benefit in some way from the mutagenic effects of transposable elements. However, hypotheses that invoke group selection are subject to many other problems and reservations."

The puzzle may be explained by relatively common bouts of transposition of mobile DNA during speciation events within the Neocellia clade in situations of parapatry and variation for transpositional-stimulating factors comparable to that in the grasshoppers studies by Shaw et al. (1983). This suggestion is available to observational corroboration because it predicts that transposonlike DNA sequences exist at some inversion break points, in particular, at those in *An. annularis*. Research into this possibility has practical importance because malaria control agencies are showing interest in the idea of manipulation of populations of vectors through genetic transformation. The ideal scenario would involve, first, discovery of a gene that reduces or eliminates the vector capacity of a population; second, mounting this in a transposon; third, transfecting the target population where the vehicle behaves like an infective pathogen, transcending Mendelian and population genetic constraints, and the rate of "infection" outpaces the rate of extinction of factors promoting transposition. The search is on to find an anopheline transposon suitable as a vehicle—a mosquito equivalent of the retrovirus responsible for human immuno-deficiency?

Acknowledgments

Thanks to Elisabeth Vrba, who, mystified by the difference between the Neocellia and Myzomyia data, first made me think about intrinsic, genomic differences and got me through the disappointment at the lack of phylogenetic information in the Neocellia data.

References

Ajioka, J.W., and D.L. Hartl. 1989. Population dynamics of transposable elements. Pp. 939–958 in D.E. Berg and M.M. Howe, eds. Mobile DNA. American Society for Microbiology, Washington D.C.

Barr, H.J., and J.R. Ellison. 1972. Ectopic pairing of chromosome regions containing chemically similar DNA. Chromosoma 39:53–61.

Bingham, P.M., and Z. Zachar. 1989. Retrotransposons and the FB transposon from *Drosophila melanogaster*. Pp. 485–502 in D.E. Berg and M.M. Howe, eds. Mobile DNA. American Society for Microbiology, Washington D.C.

Besansky, N.J. 1990. Evolution of the T1 retroposon family in the *Anopheles gambiae* complex. Mol. Biol. Evol. 7:229–246.

Cohen, M. 1976. Ectopic pairing and evolution of 5S ribosomal RNA genes in the chromosomes of *Drosophila funebris*. Chromosoma 55:349–357.

Coluzzi, M., G. Cancrini, and M. Di Deco. 1970. The polytene chromosomes of *Anopheles superpictus* and relationships with *Anopheles stephensi*. Parasitologia 12:101–112.

Engels, W.R. 1989. P elements in *Drosophila melanogaster*. Pp. 437–484 in D.E. Berg and M.M. Howe, eds. Mobile DNA. American Society for Microbiology, Washington, D.C.

Engels, W.R., and C.R. Preston. 1984. Formation of chromosome rearrangements by P factors in *Drosophila*. Genetics 107:657–678.

Farris, J.S. 1978. Inferring phylogenetic trees from chromosome inversion data. Syst. Zool. 27:275–284.
Finnegan, D.J. 1989. The I factor and I-R hybrid dysgenesis in *Drosophila melanogaster*. Pp. 503–518 *in* D.E. Berg and M.M. Howe, eds. Mobile DNA. American Society for Microbiology, Washington, D.C.
Gillies, M.T., and B. de Meillon. 1968. The Anophelinae of Africa south of the Sahara. Publications of the South African Institute for Medical Research, No. 54, Johannesburg.
Gillies, M.T., and M. Coetzee. 1987. A supplement to the Anophelinae of Africa south of the Sahara (Afrotropical region). Publications of the South African Institute for Medical Research, No. 55, Johannesburg.
Green, C.A. 1982*a*. Polytene-chromosome relationships of the *Anopheles stephensi* species group from the Afrotropical and Oriental regions (Culicidae, *Anopheles* (*Cellia*) series Neocellia). Pp. 49–61 *in* W.W.M. Steiner, W.J. Tabachnick, K.S. Rai, and S. Narang. Recent developments in the genetics of insect disease vectors. Stipes Publ., Champaign, Ill.
———. 1982*b*. Cladistic analysis of mosquito chromosome data (*Anopheles* (*Cellia*) Myzomyia). J. Heredity 73:2–11.
Green, C.A., and V. Baimai. 1984. Polytene chromosomes and their use in species studies of malaria vectors as exemplified by the *Anopheles maculatus* complex. Pp. 89–97 *in* V.L. Chopra, B.C. Oshi, R.P. Sharma, and H.C. Bansal. Genetics: new frontiers. Vol. 3. Proc. Fifteenth Int. Cong. Genet. Oxford and IBH Publ. Comp., New Delhi, Bombay, Calcutta.
Green, C.A., V. Baimai, B.A. Harrison, and R.G. Andre. 1985*a*. Cytological evidence for a complex of species within the taxon *Anopheles maculatus* (Diptera: Culicidae). Biol. J. Linn. Soc. 24:321–328.
Green, C.A., B.A. Harrison, T. Klein, and V. Baimai. 1985*b*. Cladistic analysis of polytene chromosome arrangements in anopheline mosquitoes, subgenus *Cellia*, series Neocellia. Can. J. Genet. Cytol. 27:123–134.
Green, C.A., and R.H. Hunt. 1980. Interpretation of variation in ovarian polytene chromosomes of *Anopheles funestus* Giles, *A. parensis* Gillies and *A. aruni*? Genetica 51:187–195.
Green, C.A., and S.J. Miles. 1980. Chromosomal evidence for sibling species of malaria vector *Anopheles culicifacies* Giles. J. Trop. Med. 83:75–78.
Green, C.A., R. Rattanarithikul, and A. Charoensub. 1992. Population genetic confirmation of species status of the malaria vectors *Anopheles willmori* and *An. pseudowillmori* in Thailand and chromosome phylogeny of the Maculatus group of mosquitoes. Med. Vet. Entomol. 6:335–341.
Harrison, B.A. 1980. Medical entomology studies. XIII. The Myzomyia series of *Anopheles* (*Cellia*) in Thailand, with emphasis on intra-interspecific variations (Diptera: Culicidae). Contributions of the American Entomological Institute. Vol. 17.
Lemeunier, F., and M. Ashburner. 1984. Relationships within the *melanogaster* species subgroup of the genus *Drosophila* (*Sophophora*). IV. The chromosomes of two new species. Chromosoma 89:343–351.
Nevers, P., and H. Saedler. 1977. Transposable genetic elements as agents of gene instability and chromosomal rearrangements. Nature 268:109–115.
Pardue, M.L., L.H. Kedes, E.S. Weinberg, and M.L. Birnstiel. 1977. Localisation of sequences coding for histone messenger RNA in the chromosomes of *Drosophila melanogaster*. Chromosoma 63:135–151.

Paterson, H.E.H. 1963. The species, species control and anti-malarial spraying campaign: Implications of recent work on the *Anopheles gambiae* complex. S. Afr. J. Med. Sci. 28:33–44.

Rao, T.R. 1984. The Anophelines of India. Malaria Research Centre, New Delhi.

Reid, J.A. 1968. Anopheline mosquitoes of Malaya and Borneo. Studies from the Institute for Medical Research, Malaysia, No. 31. Government of Malaysia.

Rogers, C.R. 1961. On becoming a person. Constable, London.

Ronsseray, S., and D. Anxolabehere. 1986. Chromosomal distribution of P and I elements in a natural population of *Drosophila melanogaster*. Chromosoma 94:433–440.

Shaw, D.D., P. Wilkinson, and D.J. Coates. 1983. Increased chromosomal mutation rate after hybridization between two subspecies of grasshopper. Science 220:1165–1167.

Strobel, E., P. Dunsmuir, and G.M. Rubin. 1979. Polymorphism in the chromosomal locations of elements of the *412, copia* and *297* dispersed repeated gene families in *Drosophila*. Cell 17:429–439.

Subbarao, S.K., K. Vasantha, T. Adak, and V.P. Sharma. 1983. *Anopheles culicifacies* complex. Evidence for a new sibling species, species C. Ann. Entomol. Soc. Am. 76:985–988.

13

Biological Function: Two Forms of Explanation

DAVID M. LAMBERT

Evolutionary Genetics Laboratory, Ecology and Evolution
School of Biological Sciences, University of Auckland

THERE HAVE BEEN two great intellectual traditions that have run their own courses, largely in parallel and largely unrecognized, I believe, throughout the history of evolutionary biology. These traditions have had, at their centers, quite distinct conceptions of explanation. Within one tradition the contemporary view of biological function still sits comfortably and, indeed, represents a form of explanation. In terms of the alternative tradition, that concept is foreign; not so much actively rejected, it is rather ignored as an explanatory mode. It is not merely of "historical interest" that I trace this schism back to the greatest of all biological debates, that between the two great French biologists Geoffroy Saint-Hilaire and Georges Cuvier in 1830. I do this because, to my mind, the views of these two antagonists represent the finest of the two forms of explanation.

Geoffroy and Cuvier

Cuvier (Fig. 13.1) was a precocious child. He excelled in the career-oriented training he received at a militarized German academy, and the tenth edition of Linnaeus's *Systema Naturae* was the main guide in his early researches. Indeed, by the age of nineteen he had read most of the then classics in natural history. Geoffroy became an enthusiastic supporter of Cuvier after learning of the latter's researches in zoology, and while some colleagues warned Geoffroy not to create a potential competitor, he ignored their advice and played a leading role in bringing Cuvier to the Muséum d'Histoire Naturelle in 1795.

In contrast to Cuvier, Etienne Geoffroy Saint-Hilaire (Fig. 13.2) received a

Publication 50 from the Evolutionary Genetics Laboratory, University of Auckland.

Fig. 13.1. Georges Cuvier, the founder of functional explanations.

broad philosophical education typical of clerically run schools of that era. He obtained degrees in philosophy and law, and under the pretext of studying medicine, took courses in the natural sciences at the Jardin des Plantes in Paris. Geoffroy was greatly influenced by abbé Réne-Just Haüy, a distinguished mineralogist who had transformed that science by demonstrating that all crystals could be constructed from a combination of a small number of basic types.* After Geoffroy was instrumental in Haüy's release from prison, just before the September massacre of 1792, Haüy became one of Geoffroy's principal supporters. Through Haüy, Geoffroy was appointed to the post of "subkeeper and subdemonstrator of the Cabinet" at the Jardin des Plantes. As luck would have it, shortly after this the Jardin was transformed into the Muséum d'Histoire Naturelle and Geoffroy, along with Lamarck, was appointed as a professor of zoology (for a full discussion, see the marvellous book by Toby Appel [1987]).

*This is of particular interest since there are considerable parallels between this approach and that of Goethe, with whom Geoffroy was later to have much in common. It also has parallels with the contemporary views of dynamic structures in biology.

Fig. 13.2. Geoffroy Saint-Hilaire, the founder of generative explanations.

For thirty-seven years Geoffroy and Cuvier both lived, taught, and worked at the Muséum d'Histoire Naturelle. Although being initial collaborators in an exciting new biology of revolutionary France, their paths diverged, and tension resulted between them. This strain derived, in a broad sense, from Geoffroy's predisposition to regard biology as being essentially a philosophical enterprise, while Cuvier had a particular aversion to ideas. This is not surprising. Cuvier was disgusted at the senseless violence of initial phases of the French Revolution and believed that ideas were at the root of such events. This almost certainly led to his fear of what he considered to be unbridled generalizations in science. No wonder then that he should so vehemently reject the approach of Geoffroy Saint-Hilaire (Appel 1987).

Geoffroy was the originator of a novel, and highly ingenious, framework for biology. At its heart Geoffroy argued that "there is, philosophically speaking only a single animal" (Russell 1982:54–55). So like Goethe, who argued that "everything is a manifestation of something more universal" (Arber 1946), Geoffroy sought to understand the unity in biological diversity. In this regard he developed the concept of what was to later become "homology," *sensu* position in a set of relations (as distinct from the *explanation* of such positional relations offered by Darwin, namely history). Geoffroy, who maintained that an emphasis on function acted to obscure homologies, argued for a "pure morphology uncontaminated by functional considerations" (quoted in Webster and Goodwin 1982). The explanation, for a diverse array of characters, each with a different function, for Geoffroy was not the function itself, but the underlying processes that generate it. For him there was one underlying process, and hence all the diverse forms represented in a sense "the one animal." This view infuriated Cuvier because, for him, function represented an explanation, the "reason for existence," and the animal's need was sufficient to determine its structure. It is my own view, however, that the debate between these two great biologists should not be regarded as that between the primacy of structure or function. The debate was itself a manifestation of those using distinct explanatory modes, and these continue to be used in contemporary biology.

I am aware that the "great debate" is variously interpreted by biologists and historians in a diverse array of ways. For example, E. S. Russell (1982) regarded it as the "common sense" teleological view and the abstract transcendal. Others thought that it represented a confrontation between those who asserted that animal form could best be explained by reference to function or by morphological laws. Goethe himself regarded it as an even more fundamental conflict. He viewed it as a debate between those whose concentration and emphasis was on facts and analysis versus those who regarded ideas as fundamental and who placed an emphasis on synthesis. This most fundamental debate in the history of biology was, in fact, as Appel (1987) has argued, all of these. However, the issue of explanation lies deep at the roots of this conflict.

Cuvier regarded every part of an animal as designed to contribute to the functional integrity of the whole. For him the functional unity of the organism is primary. That is, for Cuvier function was the manifestation of design, and design is itself a response to the "needs" of the organism. Cuvier marveled at the internally adapted characteristics, each to the other. This gave rise to his famous "principle of correlations," which is the cornerstone of his life's work. The carnivore, for example, is an integrated whole, in which the sense organs, alimentary system, muscles, claws, and teeth are all dependent on each other. He said: "the form of the tooth implies the form of the condyle; that of the shoulder blade that of the claws, just as the equation of a curve implies all its properties." So Cuvier was a functionalist of the whole. Indeed, the "harmony of the whole" was paramount (see Russell 1982:38–39).

Hence it is clear that it would be a mistake to suggest that the Geoffroy/Cuvier debate was one about the primacy of structure over function, or vice versa. Cuvier regarded function as an explanation of the existence of forms, while Geoffroy had an implicit idea of explanation based on an underlying unity of process.

Contemporary Functionalism

This great tradition, in which Cuvier played such a vital role, continues today, but with an important difference. Contemporary functionalist views carry the legacy of their Darwinian ancestry. They are not, like Cuvier's form of the conception, about the internally adapted integrity of the whole organism. Today functionalism is concerned with the trait, and not the organism. Although the concept of "need" is still paramount, in today's functionalism need is imposed by the environment, not resulting from the integrity of the whole, as in Cuvier's conception.*

The publication in 1966 of *Adaptation and Natural Selection* by G. C. Williams marked the beginning of an new stage in the development of the study of evolutionary biology, and yet it represented, in some sense, a return to earlier views of Cuvier. Williams's (1966) book was initially influential because it represented a well-formulated and cogent attack on the concept of group selection. However, years later, it came to be widely regarded as a seminal work because of its thorough analysis of a range of the concepts of evolutionary biology, particularly function.

Williams (1966) tells us that the writing of his book was stimulated by a then-current opposition to natural selection that was both overt and cryptic. In this regard Waddington's (1957) claims about the inadequacy of natural selection, unless supplemented by genetic assimilation, were singled out for special mention, as were Darlington's (1958) comments regarding the properties of genes and chromosomes. Williams (1966:4) lamented our inability to *imagine* that "the blind play of genes can produce man." He continued: "It is difficult for many people to imagine that an individual's role in evolution is entirely contained in its contribution to vital statistics. It is difficult to imagine that an acceptable moral order could arise from vital statistics, and difficult to dispense with belief in a moral order in living nature. It is difficult to imagine that the blind play of the genes could produce man."

Yet Williams also made an honest and sober confession, remarking: "Major difficulties also arise from the current absence of *rigorous criteria* for deciding whether a given character is adaptive, and, if so, to precisely what it is an adaptation" (italics added). He went on to accept a "ground rule" or "doctrine," namely, that "adaptation is a special and onerous concept that should be used only where it is really necessary." Indeed, this represents one of the

---

*I would like to note here the power of functionalism. E. S. Russell, who was a great supporter of functionalism under the particular manifestion of Lamarckism, so admired Cuvier. Likewise, a contemporary functionalist like Dawkins (1986) appreciates the writings of the Reverend William Paley. In each case an evolutionist finds a particular sympathy with a nonevolutionist. The sympathies are a result of the sharing of a "way of seeing" which actually predominates over whether one believes in evolution or not.

most memorable points to come from this landmark volume and represented a prescient viewpoint, later to be mirrored by a host of contemporary authors.

Williams rejected the idea that current utility was either necessary or sufficient to recognize an adaptation. He invoked an old dichotomy, in a way very similar to the Reverend William Paley, remarking: "A benefit to an organism might result from chance instead of design." He continued: "The decision as to the purpose of a mechanism must be based on an examination of the machinery and an argument as to the appropriateness of the means to the end. It cannot be based on value judgments of actual or probable consequences." Williams then *imagined* a scene in which a fox is on its way to a henhouse for the first time after a heavy snowfall. He suggests that it might encounter considerable difficulty in forcing its way through the snow. On subsequent trips, however, it may follow the same path and have a much easier time of it, because of the already existing track. Williams argued that this may result in a considerable saving of time and food energy for the fox and that such savings might be crucial for survival. He then asked the rhetorical question: "Should we therefore regard the paws of a fox as a mechanism for constructing paths through snow?" Answering himself, he declared: "Clearly we should not. It is better, because it avoids the onerous biological principles of adaptation and natural selection, to regard the trail-blazing as an incidental effect of the locomotor machinery, no matter how beneficial it may be." But what are the fox's feet for? Williams explains: "An examination of the legs and feet of the fox forces the conclusion that they are designed for running and walking, not for the packing or removal of snow . . . Although the construction of a path through the snow should not be considered a function of the activities that have this effect, the fox does adaptively exploit the effect by seeking the same path on successive trips to the henhouse. The sensory mechanisms by which it perceives the most familiar and least obstructed routes and the motivation to follow the path of least effort are clearly adaptations."

But where are the "rigorous criteria" by which we are able to make such a judgment? I would contend that there are none. In Williams's own words "function" is simply "obvious." Yet Williams's volume is explicitly dedicated to the search for the "rigorous criteria for deciding whether a given character is adaptive, and, if so, to precisely what it is an adaptation." He (Williams 1966:9) admitted that "it is often easy, in practice, to perceive functional design intuitively, but unfortunately disputes sometimes arise as to whether certain effects are produced by design or merely as by-products of some other function." Williams regarded design as being indicated "whenever the function is served with sufficient precision, economy, efficiency, etc., to rule out pure chance as an adequate explanation." He also advocated a "not infallible rule," that of recognizing an adaptation when it shows a clear analogy with human implements. Hence the "purpose" of the biological character is obvious. But adaptation can also be assumed on the basis of the "indirect evidence of complexity and constancy." A benefit to the organism is not, for Williams, a sufficient reason for recognizing an adaptation: "A benefit can be the result of chance instead of design."

In this essay I want to explore some of the implications of these ideas, particularly insofar as they relate to the origin and development of the Recognition Concept, but also because it seems to me that they have a range of implications for evolutionary biology as a whole.

Function and the Recognition Concept

Paterson applied Williams's notion that a character could be regarded as an adaptation only if there was clear evidence that it had been selected *for* that *particular* function. Based on his studies of a broad range of insect species, particularly in Africa (see the review by Hunt and Coetzee, this volume), Paterson (1981) was convinced that speciation generally occurred in allopatry. Therefore, following Williams's logic, he concluded that species should not be defined in terms of "isolating mechanisms" (1985), since neither "premating" nor "postmating isolating mechanisms" could have been directly selected for the *function* of isolating one species from another. In allopatry, whatever reason there was for changes in the male-female communication systems, it was not for the "purpose" or "function" (*sensu* Williams 1966) of isolating one species from another.

This, in turn, led Paterson to embark upon an investigation to formulate what he considered to be a more appropriate way to define and discuss species. I would suggest that initially, the Recognition Concept was such a powerful impetus to the investigation of biological phenomena because it focused attention on a particular instance of this general notion of function.

Of course, there is a logical connection between the reinforcement model of speciation and the idea of "isolating mechanisms." It is certainly no surprise that Dobzhansky, who first used the isolation terminology, should have been such an ardent supporter of reinforcement, as Paterson and I have previously pointed out (Lambert and Paterson 1984). To many, reinforcement is an appealing model in that there should logically be a *direct* relationship between speciation and natural selection. And reinforcement is an appealing argument also because it was easily able to be translated into what we could call the *generalized selection argument*. This argument takes the form: imagine that there is heritable variation in a population with respect to some character (principles of inheritance and variation) and that some forms are more successful than others in terms of their ability to survive and reproduce (principle of struggle for existence) as a result of the possession of that character. In order, however, to distinguish between evolution as a phenomenon and evolution as a process,

Darwin required a further principle, the principle of natural selection, in which the cause of differential reproduction was the character under consideration. This being the case, evolution (*sensu* changes in the composition of a population) must *inevitably* take place. Hence the character under consideration can be regarded as an adaptation, because it has been selected for a particular function (the cause of the differential reproduction).

Let us consider an example. We could say that the phenomena of "isolating mechanisms" are adaptations to keep species isolated and consequently that they represent functional design for that purpose. Any organism which has the genetic predisposition to mate with an individual from another species is likely to produce offspring that have a poorly coadapted genetic constitution and hence will be selectively disadvantageous (e.g. they may be sterile). Importantly, we have designated the "need" to be solved as that of gametic wastage. Why was reinforcement such an appealing argument? Simply because it can easily be translated into this generalized selection argument, one in which there is an immediately obvious and credible "need" which organisms face, namely excessive gametic wastage. Surely if some individuals are less wasteful, then selection will favor them? That is, any individual which we can *imagine* that has a predisposition to mate with those of the same group will have a greater genetic contribution in the next generation. Having so translated the particular argument into this general format, it now becomes immediately convincing.

How Can We Determine Any Particular Function?—The Search for Williams's "Rigorous Criteria"

Williams's 1966 volume represents such a significant publication because it acted to develop the concept of function as a thoroughly historical notion. Adaptations, according to this view, are "molded" as a result of a long series of independent selective acts that increase in frequency a newly arisen mutation in a population. Løvtrup (1988) asserted that, if the macromutation theory is correct, we cannot ascribe purpose to the various attributes of living forms. This is because the capacity to perform the functions of the organism preexisted its preservation by selection (for that reason?). (This must always be the case because in order for selection to act there must be forms that are functionally advantageous over other concurrently existing forms.) However, from a neo-Darwinian point of view the operative word in such scenarios is *mold*, or perhaps *fashioned*, for example, as used by Williams (1966). As Williams (1966) said: "The designation of something as the means or mechanism for a certain goal or function or purpose will imply that the machinery involved was fashioned by selection for the goal attributed to it." Apart from illustrating an almost overt teleology, it is clear that the idea being entertained here is that there is a long succession of mutations which have a small effect on the character under consideration, and that each of these increases in frequency in the population and eventually becomes fixed. Hence the character is "molded" through many generations. Each act of fixation by selection must be because the mutations are all affecting the character in the same way, in that they are all increasing efficiency with respect to some environmental "need."

If I were to assert that a character has a particular function (*sensu* Williams 1966), how might it be possible for me to reject this claim? As in the case of reinforcement it might be possible to show that selection is highly unlikely to produce the phenomenon under consideration (Paterson 1978, 1982; Lambert et al. 1984; Spencer et al. 1986). This now seems to have been widely recognized in evolutionary biology, even though it is an immediately appealing verbal argument. I might also be able to reject the suggestion if I know that the character, in fact, predated the function. For example, since we know that feathers arose before birds flew, then they could not have evolved for this function. Second, I might simply argue that it is extremely unlikely. For example, the idea that flamingos are pink because they are cryptically colored at sunset may be dismissed as simply an unreasonable argument and therefore highly unlikely. We would also reject a proposed function if we could show that this is not the current utility of the character under investigation. Current utility is not necessary evidence for function according to Williams, but its absence must exclude the possibility. That is, current utility is a necessary, but not a sufficient, explanation of current function. Alternatively, we might also escape some of the difficulties with particular functional claims by changing the "character" that is thought to be adapted. For example, in the case of the fox, discussed above, Williams (1966) simply changes the character to which snow removal is an adaptation. Instead of it being the paws themselves, it becomes the "sensory mechanisms by which it perceives the most familiar and least obstructed routes and the motivation to follow the path of least effort" (p. 13).

All these represent some of the difficulties with the designation of functions. Even if the general perspective and the assumptions on which the argument is based were correct (both of which I dispute), it could properly be said that functions simply get "lost in history." It becomes an exercise in *imagination* (and consequent *belief*) to be able to justify the existence of characters as representing the remnants of past historical forces or contingencies. Hence it is necessary to *imagine* a series of historical events in order to validate the biology we seek to explain. Some consider that resistance to Darwinian ideas represents evidence for a lack of ability to imagine. Williams (1966:4), for example, said: "I believe that modern opposition, both overt and cryptic, to natural selection, still derives from the same sources that led to the now discredited theories of the nineteenth century. The opposition arises, as Darwin himself observed, not from what reason dictates but from the limits of what the *imagination* can accept."

Imagination and Belief

Indeed, imagination and belief are powerful things, and the ability to imagine is as important for a scientist as it is for a philosopher, artist, or writer. Evolutionary biology, perhaps more than any other discipline, is dependent upon these important human activities. In respect to imagination Charles Darwin himself was clearly very gifted. He readily acknowledged this, remarking that he had a "fair share of inventiveness." His confession was, in fact, a marked understatement. Even in his childhood he confessed an inclination to invent stories about almost any topic. This tendency was to become a lifelong characteristic. Darwin's "one long argument" presented in the *Origin* represented, in fact, a new style of imaginative, inventive argument. It was one which emerged from a shared discourse between himself and other writers of the time such as John Milton, with whom he was familiar (Bratchell 1981; Levine 1988). Hence not only ideas, but metaphors, myths, and, most important, narrative patterns were exchanged freely. Darwin's argument in the *Origin* was preeminently a narrative in both style and content. It represents a form of imagined history, but its emphasis on time and change make it a narrative. Through manifold particular instances the general point of Darwin's theory slowly becomes apparent in the mind of the reader, in the same way that the general message of one of Darwin's most frequently read authors—Charles Dickens—also becomes clear. In her stimulating analysis, Gillian Beer (1985) pointed out that Darwin's views reached their full authority as a result of an acceptance of the congruity between theory and nature not only because of a revolution in the minds of scientists, but also, significantly, because of a revolution in the *beliefs* of "other inhabitants of the same culture."

However, Darwin has, over the years, been criticized by a number of authors for this novel style of argument so dependent upon imagination. As early as 1864 William Whewell, for example, objected to the notion in that if we are able to *imagine* a series of transitional forms from one condition to another, then this should be accepted as evidence for *believing* that such a transition had, in fact, taken place. He said: "For it is assumed that the mere possibility of imagining a series of steps of transition from one condition of organs to another, is to be accepted as a reason for believing that such transition has taken place. And next, such a possibility being thus imagined, we assume an unlimited number of generations for the transition to take place in, and that this indefinite time may extinguish all doubt that the transitions really have taken place" (Whewell 1864:xvii–xviii, quoted in Himmelfarb 1968).

Further, the imagining of this possibility, together with the assumption that there may have been a very large number of generations during which this transition could have occurred, should extinguish all doubts that the transitions really did take place. In so doing Darwin was asking us to accept a kind of "promissory note" for the future (Foss and Rothenberg 1988). The promise was that all phenomena will eventually yield to his explanation.* In this way Darwin took multiple possibilities, and in contrast to conventional logic where these result in reduced probability, he manages to convert them into increased probabilities (Himmelfarb 1968). So liabilities become assets. Darwin achieved this by encouraging us to accept a "constantly mounting obligation." As Gertrude Himmelfarb (1968) pointed out: "Having first agreed to the theory in cases where only some of the transitional stages were missing, the reader was expected to acquiesce in those cases where most of the stages were missing, and finally in those where there was no evidence of stages at all" (Himmelfarb 1968:335). Our imagination of particular forms makes them possible forms in our minds; more than that, it makes them possible forms that might conceivably have actually existed. At the same time our ability to imagine makes extant forms appropriate to their conditions of life because we can imagine impossible and inappropriate forms too.

There has always been a degree of opposition to this conception of the world. Among others, Samuel Butler and Bernard Shaw (see Loewenberg [1957] for discussion) both declared it contrary to all observation, sense, and experience. However, given its modern formulation, which owes more to a version of sociobiology than to strict neo-Darwinism, it is not difficult to see why young people today would probably disagree with Butler and Shaw. In a Britain still intellectually dominated by Thatcherism and, in many other parts of the world where even "socialist" governments assert confidence in the hidden hand of Adam Smith, young people must now "know" that the world is a hostile place of self-interest where, if you are competitive, efficient, and productive, then you individually, and society as a whole, will prosper (Millar et al. 1986). It should not be surprising to these young people to learn that the biological world is a mirror image of the social, political, and economic world in which they live.

Yet there is another major reason for the success of Darwin's ideas, one which is very important to our discussion. In a very real sense Darwin made every origin historical. According to his theory, biological form, and its contemporary geographical distribution, can be explained only in historical terms. It represents a series of contingent possibilities (now thought of in terms of mutations and dispersal events, for example) which have been frozen in

---

*Wenner (1989) pointed out that a theory which offers such promissory notes and which is eventually able to provide supporting evidence in return is greatly strengthened in the minds of supporters.

time. Just as he did for coral reefs, Darwin claimed that history orders the forms that we see in nature (see Lambert, Millar, and Michaux [1989] for a detailed discussion; cf. Gould 1986). This historification of biology is at the center, not only of Darwinism, but also of contemporary biology. It is so because the framework of Darwinian theory is based not on rationality but on imagined (and perhaps, at least in some cases, imaginary) historically contingent events. Hence biology has now become an historicist science, in the sense that "the history of life sufficiently accounts for its nature" (*Funk and Wagnalls International Edition* 1959).

Within this historification of biology, the concept of function has held a central place.

The Historification of Function and Biology

For Cuvier, function was "reason for existence," but when function became an historically generated "reason for existence" this was accompanied by the notion that it was also the result of some externally generated (environmental) "need." Hence function, in this sense, acts to traitize the organism. It is an indispensable procedure, within a neo-Darwinian framework, to traitize organisms, and such traits are then, initially at least, assumed to be independent of each other. This is a necessary assumption, and it lies at the heart of the whole Darwinian approach, despite the fact that it is sometimes acknowledged that characters are often not independent. The argument is that this dependence simply acts to reduce the effectiveness of selection.

Within this framework, correlations between traits which subsequently become apparent require an explanation based upon historical principles. Consider the case of polyandry in the dunnock (*Prunella modularis*). The dunnock is a small passerine bird with a variable mating system. In this species some female territories are defended by one male bird (monogamy), while others are defended by two "unrelated" males (polyandry). In an important study Burke et al. (1989) recently used DNA fingerprinting to link observations of mating behavior and parental care with precise knowledge of reproductive success. They found that, although males are not able to discriminate between their own offspring and those of other males in a multiply sired brood, they nevertheless "increase their own reproductive success by feeding offspring in relation to their access to the female during the mating period, which is a good predictor of paternity" (Burke et al. 1989:249). That is, males which had the most "access" to the female, and therefore those which are presumed to have had the greater number of matings with her (and therefore probably sired more offspring), spend more time feeding the offspring of that female. This correlation between two traits, mating behavior and parental care, is then resolved by reference to a form of historical logic. Why do individuals feed the offspring of the female more when they have had "access" to her during mating? Burke et al. (1989) argued that "males apparently use mating access to determine whether they feed young." They further suggest that females benefit by copulating with two males because chicks are fed more, and that more chicks survive with the help of two males rather than one, and indeed females are known to attempt to escape the alpha males' close attentions and to encourage beta males to mate (Davies 1985; see also the recent and interesting paper by Davies et al. 1992).

While this was a significant and innovative study, there is an alternative interpretation. Mating and feeding are not separate traits that are unconnected in terms of the temporal changes within an organism's lifetime. In many bird species feeding is a fundamental part of courtship. For example, in species such as the crested grebe, Huxley, as long ago as 1914, showed the importance of reciprocal feeding (Huxley 1914). In addition, the nine species of lovebirds belonging to the genus *Agapornis* all engage in courtship feeding, transferring regurgitated food from one member of the pair to the other (Dilger 1962). Hence the correlation that Burke and Davies established is perfectly intelligible in terms of the biology of the mating-feeding continuum.

In fact, the same hormones that are involved in courtship and copulation are involved in the development of the behavior of feeding young (Marler 1968). These are not separate behaviors, but part of an ontogenetic sequence. Hence, it is therefore not surprising that a male which courts a female and mates with her is then likely to continue that series of behaviors and to feed the young.

The interactive relationship between feeding and mating is no more graphically illustrated than by the innovative studies by Lehrman and his colleagues on the reproductive behavior of the ring dove (Fig. 13.3). A male begins to court the female as a result of the action of androgens. Male courtship stimulates the pituitary release of the hormone FSH in the female dove. In turn, FSH results in follicle development. The follicles secrete estrogen, which affects uterine growth and development. In the next one to two days both males and females begin to construct a nest, and during this phase copulation occurs. Lehrman's experiments show that the presence of the nest stimulates the production and secretion of progesterone in females. This hormone promotes incubation behavior in both males and females after the eggs are laid. Egg laying is activated, in part, by the action of the LH hormone, which is secreted from the pituitary gland of the female. As a result of the presence of eggs, combined with the stimulation from incubation behavior, the pituitary gland in both the male and the female secretes prolactin. This hormone, in turn, acts to inhibit FSH and LH secretion, and reproductive behavior stops. Prolactin also stimulates crop development and the production of crop milk and may also help to maintain incubation behavior. As the adults continue to feed the

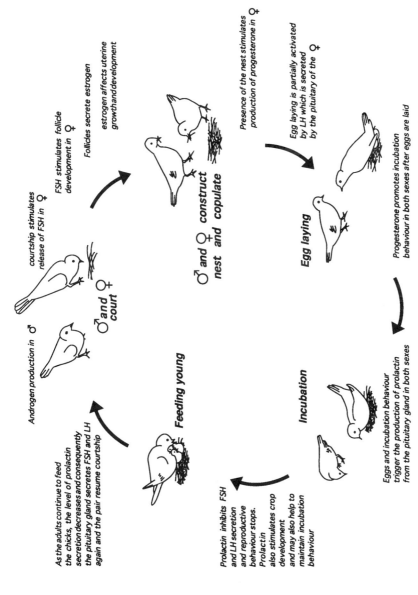

Fig. 13.3. The integration of mating and feeding behavior in the ring dove. Reproduced with permission from *Animal Behavior*, 2d ed. 1986. By L.C. Drickamer and S.H. Vessey. Wadsworth Publishing Company, Belmont, Calif.

chicks, the level of prolactin secretion decreases, and consequently the pituitary secretes FSH and LH. The pair then resumes courtship, and the sequence begins again (Drickamer and Vessey 1986). Hence this set of highly integrated behaviors depends on a double set of reciprocal interrelations. Changes in the activity of the endocrine system are induced by stimuli from the environment (these include those that arise from the behavior of an individual and its mate). There is a reciprocal relation between the behavior of one mate on the endocrine system of the other and the effects of the behavior of the second bird on the endocrine system of the first. The behaviors that result from these interactions include both mating *and* feeding (Fig. 13.4). Indeed, they are part of a continuum, and they emerge out of this set of reciprocal interactions (Lehrman 1964).

In a sense, the prior analysis represents an instance of historical explanation based on the separation of "mating" and "feeding" behaviors. However, we need to ask if this approach to explaining biology simply rationalizes the particular phenomenon in accordance with an historical narrative. Have we then sufficiently accounted for its nature? I suggest that we have not. I contend

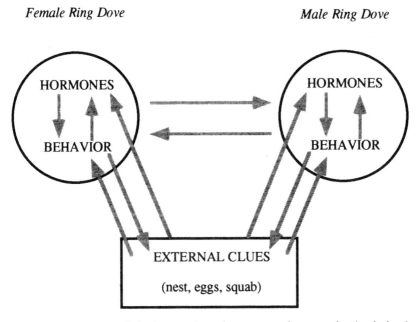

Fig. 13.4. Summary of the interactions that govern the reproductive behavior of the ring dove. Hormones regulate behavior and are themselves affected by behavioral and other stimuli. In turn, the behavior of each bird affects the hormones and the behavior of its mate.

that we have created a kind of "taxonomy of knowledge," that is, a system for the organization of what we know. Using this procedure we can potentially explain *any* example. Moreover, it cannot be argued that, using the approach I have outlined, I have merely elucidated the "proximal cause" of the behaviors, for which there is, in turn, an "ultimate cause," for example, behaving so as to maximize reproductive success. If one suggests that the set of behavioral and hormonal interactions themselves are the product of natural selection, it is important to realize that there has been a fundamental shift in the target of explanation. The phenomenon which is now being explained has been moved backward from the behavior itself to the developmental processes which give rise to it. So the biological knowledge has forced the target of explanation to change. In turn, as more data become available the target may well shift again, yielding to this other form of explanation.

An Alternative View of Function

Prior to and subsequent to Williams's (1966) analysis, function has been regarded in a quite different way. According to the alternative tradition, function is simply the "activity of structure." "The actions of living matter," wrote Thomas Henry Huxley in 1875, "are termed its functions" (quoted in Coleman 1977:143). And for Cannon (1958), function was simply the "mode of action of structure." In Woodger's (1929) most important attempt at resolution of antitheses in biology he discussed the separation of structure and function. "Biological thought has been harassed by this antithesis throughout its history," he remarked (Woodger 1929:326). There is, according to this view, a fundamental interdependence of structure and function. It represented, not a concept of spatial structure, but what we can call a dynamic structure (see essays in Goodwin et al. 1989; Lambert and Hughes 1988; Jantsch 1980). Of course, defined spatial structures result from the interaction of processes underlying any system. Indeed, the idea that activity (function) and structure are complementary is an expression of process thinking. Such dynamic or "space-time" structure includes the function of the system and thus also its organization and its relation with the environment (Jantsch 1980). One natural consequence of the notion of function as an integral part of structure is that there will be an interdependence of function within an organism. This is because organisms operate on system-level principles of organization.

Such a holistic, integrated notion was central to Cuvier's approach. For him the functions of the parts are all intimately bound up with one another, and one function could not vary without corresponding modifications in the others. So it is clear that for Cuvier structure and function were bound up together. This emphasis of the relation of parts to the whole is a hallmark of the contemporary interest in dynamic structures.

Self-organization

The particular concept of function we use has an important influence on our thought processes about biology. The sense discussed by Woodger, for example, is one in which we consider biological structures in a dynamic sense and where we attempt to unify "activity" (i.e., function in the sense of the active phase of structure) with dynamic structures. Recognition of this leads us to ask how interactions that occur within a developing organism can "trigger" change. Unities (self-regulating entities *sensu* Maturana and Varela [1987]) undergo continuous change either as a result of their interaction with the "environment" or as a result of their own internal dynamic. Indeed, such continuous interactions cease only when the dynamic itself ceases. The unity and its interaction can be represented as follows:

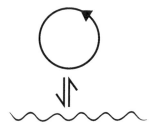

Specific-Mate Recognition Systems (SMRSs) represent structurally coupled unities that also interact in a particular context, for example, via habitat choice. When two or more unities are continually interacting they will therefore undergo coupled ontogenies. This is because their interactions are *recurrent* interactions. Every organism reciprocally perturbs an interacting conspecific during the activation of the process of mate recognition.

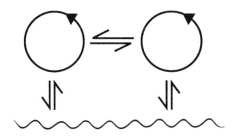

In these interactions the "environment" does not direct or specify the changes in the unity (e.g., an organism); it simply triggers them. So the organisms, via their interactions, effect mutually congruent structural changes as long as the interactions (e.g., mate recognition) continue. This is what Maturana and Varela (1987) called *structural coupling*. They would speak of structural coupling whenever there are recurrent interactions leading to the structural congruence between two (or more) systems.

The concept of structural coupling provides an alternative to the idea of coadaptation. The former concept is advantageous in that it gives us a language for discussion in which the underlying dynamic is emphasized. It is, of course, explicitly ahistorical. When there are recurrent interactions between unities (organisms in this case) within a particular context (environment), these will consist of reciprocal perturbations. The nature (structure) of the context simply triggers structural changes in autopoietic (self-renewing) unities, but does not specify or direct them.

In the case of mate recognition, "males" and "females" reciprocally perturb their own internal states, as we see so clearly in the case of the ring doves. This results in a series of mutually congruent structural changes in the unities. Fertilization is in some cases the final state of the dynamic (e.g. in species without parental care), and, of course, the initial state of a new and novel unity. Of course, when we are dealing with organisms that show parental care, the structural coupling continues. The reciprocal perturbation also continues, and again there is congruence in that organisms coordinate to feed the offspring.

The SMRS, then, represents a second-order unity. Species are consequently a third-order unity, which is constituted as a result of the sharing of communication among individuals. Species generate a particular phenomology, "one in which the individual ontogenies of all the participating organisms occur fundamentally as part of the network of co-ontogenies that they bring about in constituting third order unities" (Maturana and Varela 1987). Any individual organism, then, is part of a particular species only as long as it is part of that structural coupling.

There is, however, in many species a necessary coupling of fertilization and parental care. Of course, there is a great diversity of forms of such coupling. Nevertheless, the important relationship between these is evidenced by many species in which fertilization is intimately associated with care of offspring. For example, in the classic stickleback the nest is constructed by the male, and egg-laying and fertilization takes place only within the context of the nest. Similarly, many birds use feeding as an important part of mate recognition. Clearly these aspects of fertilization and parental care are part of a continuum.

Conclusions

I would suggest that Williams's volume should be regarded as a major influence in the historification of the concept of function in evolutionary biology.

This view is similar to that of Cuvier but with an underlying historical theme. In parallel has been the idea that function represents the "active phase of structure." Of course, structure and function must be separate if one concept is historical and the other purely logical.* I argue here for a reunification of these concepts, but not in their currently accepted forms.

Despite my objections to Williams's influence, I think that Paterson's investigation of the nature of species in biology, using Williams's ideas as a critical tool, was an important advance. (I suspect that this is true for a good many evolutionary biologists, certainly those represented in this volume.) Yet it served to demonstrate even more general problems, not just contemporary views of species, but with the whole of Darwinian biology (White et al. 1990). Based on the arguments outlined in this contribution, I would suggest that it is not appropriate to claim that the function of the SMRS is efficient fertilization. Of course it regularly results in such fertilization, but the efficiency of the process (or lack of it, for that matter) does not explain its existence. Functions (*sensu* Williams 1966) simply get lost in history. Function as we know it today represents a concept gone wrong. But more than that, it is a particular manifestation of a broad trend in a biology gone wrong, one in which explanation resides, not in the logic of underlying processes, but in historically imagined events which lead to an "adapted" world.

For an historical point of view in biology we need to believe in the relevance of our imagination to tell us something about the nature of the world we live in. Because we can imagine an almost infinite set of possible worlds, we regard this world as being contingently unique. The world does not have to be the way it is, it just is. It is then made intelligible only through enunciating and reconstructing its contingently driven historical path.

In contrast, we need to develop an alternative biology in which structures are again active and one in which their generative processes represent the driving force of nature. The generative view outlined here has a long intellectual tradition, the foundations of which were vividly enunciated by Goethe and the great structuralist biologist Geoffroy Saint-Hilaire. It is one in which biological processes, because they arise through the operation of generative principles, act to "explore" the world through the manifestation of logical forms, not *driven* but *triggered* by contingent circumstance. By then "asking questions of nature," and by analyzing the "riddles" in which it answers, we can elucidate the intelligibility. The great Swedish evolutionary biologist Antonio Lima-de-Faria (1983) puts it well: "An organism is just one of the mirrors that the universe uses to look at itself."

---

*By this comment I do not mean to suggest that history is synonymous with time. History is a narrative of past events. Hence atemporal views and historical views are not opposites (Lambert et al. 1989).

Acknowledgments

Of course, in a sense, all the chapters in this volume are in appreciation of Hugh Paterson. Nevertheless, I would like to specifically express my gratitude for his friendship and the richness of his always provocative attitudes, which have been so stimulating and influential for me personally. I wish Hugh and Shirley a long, productive, and happy retirement.

Once again I thank my colleagues in the Evolutionary Genetics Laboratory, particularly Craig Millar and Peter Ritchie, for many helpful discussions. The manuscript benefited from their comments, and those from Hamish Spencer, Peter Taylor, and Søren Loevtrup. It is also a pleasure to acknowledge the influence of Emeritus Professor John Morton. I hope that he will recognize that influence and tolerate that with which he disagrees. I am indebted to Iain McDonald and Vivian Ward for the production of plates. My research is supported by the University of Auckland Research Committee, the Vice Chancellors Committee, and the New Zealand Lottery Board (Scientific).

References

Appel, T.A. 1987. The Cuvier-Geoffroy debate: French biology in the decades before Darwin. Oxford University Press, Oxford.

Arber, A. 1946. Goethe's botany. Chronica Botanica 10:67–126.

Beer, G. 1983. Darwin's plots: Evolutionary narrative in Darwin, George Eliot, and nineteenth-century fiction. Routledge and Kegan Paul, London.

Bratchell, D.F. 1981. The impact of Darwinism: Texts and commentary illustrating nineteenth century religious, scientific and literary attitudes. Avebury Publishing Co., England.

Burke, T., N.B. Davies, M.W. Bruford, and B.J. Hatchwell. 1989. Parental care and mating behavior of polyandrous dunnocks *Prunella modularis* related to paternity by DNA fingerprinting. Nature 338:249–251.

Cannon, H.G. 1958. The evolution of living things. Manchester University Press, Manchester.

Coleman, W. 1977. Biology in the nineteenth century: Problems of form, function and transformation. Cambridge University Press, Cambridge.

Darlington, C.D. 1958. Evolution of genetic systems. Oliver and Boyd, London.

———. 1960. Darwin's place in history. Oxford Basel Blackwell, Oxford.

Davies, N.B. 1985. Cooperation and conflict among dunnocks *Prunella modularis* in a variable mating system. Anim. Behav. 33:628–648.

Davies. N.B., B.J. Hatchwell, T. Robson, and T. Burke. 1992. Paternity and parental effort in dunnocks *Prunella modularis:* How good are male chick-feeding rules? Anim. Behav. 43:729–745.

Dawkins, R. 1986. The blind watchmaker. Norton, New York.

Dilger, W.C. 1962. The behavior of lovebirds. Sci. Am. 206:3–11.

Dobzhansky, T. 1976. Organismic and molecular aspects of species formation. Pp. 95–105 *in* F.J. Ayala, ed. Molecular evolution. Sinauer, Sunderland, Mass.

Drickamer, L.C., and S.H. Vessey. 1986. Animal behavior: Concepts, processes and methods. Wadsworth Publishing Co., Belmont, Calif.

Foss, L., and K. Rothenberg. 1988. The second medical revolution: From biomedicine to infomedicine. New Science Library, Shambhala Publication, Boston.

Goodwin, B.C., A. Sibatani, and G. Webster, eds. 1989. Dynamic structures in biology. Edinburgh University Press, Edinburgh.

Gould, S.J. 1986. Evolution and the triumph of homology; or, why history matters. Am. Sci. 74:60–69.

Himmelfarb, G. 1968. Darwin and the Darwinian revolution. Norton and Co., New York.

Hughes, A.J., and D.M. Lambert. 1984. Functionalism, structuralism and "Ways of Seeing." J. Theor. Biol. 111:787–800.

Huxley, J. 1914. The courtship habits of the Great Crested Grebe (*Podiceps cristatus*); with an addition to the theory of sexual selection. Proc. Zool. Soc. No. xxxv.

Jantsch, E. 1980. The self-organising universe. Pergamon Press, Oxford.

Lambert, D.M., M. Centner, and H.E.H. Paterson. 1984. Simulation of the conditions necessary for the evolution of species by reinforcement. S. Afr. J. Sci. 80:308–311.

Lambert, D.M., and A.J. Hughes. 1984. Misery of functionalism: biological function, a misleading concept. Riv. Biol.–Biol. Forum 77:477–501.

———. 1988. Keywords and concepts in structuralist and functionalist biology. J. Theor. Biol. 133:133–145.

Lambert, D.M., C.D. Millar, and B. Michaux. 1989. On the contingent and the inevitable. Riv. Biol.–Biol. Forum 82:335–337.

Lambert, D.M., and H.E.H. Paterson. 1984. On "Bridging the gap between race and species": The Isolation Concept and an alternative. Proc. Linn. Soc. N.S.W. 107:501–514.

Lima-de-Faria, A. 1983. Molecular evolution and organisation of the chromosome. Elsevier, Amsterdam.

Levine, G. 1988. Darwin and the novelists: Patterns of science in Victorian fiction. Harvard University Press, Cambridge, Mass.

Loewenberg, B.J. 1957. Darwinism reaction or reform? Reinhart and Co., New York.

Løvtrup, S. 1988. Design, purpose and function in evolution: meditations on a classical problem. Environ. Biol. Fish. 22:241–247.

Millar, C.D., N.R. Phillips, and D.M. Lambert. 1986. Evolution—the struggle continues. Nature 321:475.

Marler, P. 1968. Mechanisms of animal behavior. John Wiley, New York.

Maturana, H.R., and F.J. Varela. 1987. The tree of knowledge: The biological roots of human understanding. New Science Library, Shambhala Publication, Boston.

Paterson, H.E.H. 1981. The continuing search for the unknown and the unknowable: A critique of contemporary ideas on speciation. S. Afr. J. Sci. 77:113–119.

———. 1982. Perspectives on speciation by reinforcement. S. Afr. J. Sci. 78:53–57.

———. 1985. The recognition concept of species. Pp. 21–29 *in* E.S. Vrba, ed. Species and speciation. Transvaal Museum Monograph No. 4. Pretoria.

Russell, E.S. 1982. Form and function: A contribution to the history of animal morphology. University of Chicago Press, Chicago.

Spencer, H.G., B.H. McArdle, and D.M. Lambert. 1986. A theoretical investigation of speciation by reinforcement. Am. Nat. 128:241–262.

Waddington, C.H. 1957. The strategy of the genes. Allen and Unwin, London.

Webster, G., and B.C. Goodwin. 1982. The origin of species: A structuralist approach. J. Social. Biol. Struct. 5:15–47.

Wenner, A.M. 1989. Concept-centered versus organism-centered biology. Am. Zool. 29:1177–1197.
White, C.S., Michaux, B., and D.M. Lambert. 1990. Species and neo-Darwinism. Syst. Zool. 39:399–413.
Williams, G.C. 1966. Adaptation and natural selection. Princeton University Press, Princeton, N.J.
Woodger, J.H. 1929. Biological principles: A critical study. Routledge and Kegan Paul, London.

Properties of Specific-Mate Recognition Systems

14

Pheromone Communication in Moths and Its Role in the Speciation Process

CHARLES E. LINN, JR., AND WENDELL L. ROELOFS
Department of Entomology
New York State Agricultural Experiment Station, Cornell University

One of the most important properties of any species, regardless of one's definition or preference for a particular speciation process, is the set of activities by which individuals in a population reproduce (Mayr 1970; White 1978; Templeton 1981; Futuyma 1986; Vrba 1985; Paterson 1985). In a population of mobile sexually reproducing organisms, mating may be the only time in their life history that individuals encounter one another (Kemp 1985), and considering the salience of successful mating in the life of an individual organism, one is not surprised that most mating systems constitute a coordinated, and often complex, set of adaptive traits that allow individuals to locate and recognize prospective mates (for the case in insects see Matthews and Matthews 1978; Blum and Blum 1979; Thornhill and Alcock 1983; Eberhard 1985, 1990). The uniqueness and complexity of each communication and mating system depend in part on the phylogenetic history of the group, the complexity and stability of the habitat (both biotic and abiotic) in which the organisms live, the mobility of the individuals and the sensory systems and modalities utilized in mate location, courtship, and copulation (Wilson 1975; Smith 1977; Sebeok 1977; Paterson 1985).

Paterson's Recognition Concept represents a set of ideas on speciation and the evolution of mating systems that has challenged the more prominent Biological Species Concept (Paterson 1978, 1980, 1981, 1982a,b, 1985). The differences between these two concepts are not trivial, each being rooted in very different ways of explaining the diversity found in the organic world. The details of this debate are lengthy and can be found elsewhere (see other chapters in this volume as well as Vrba 1985; Lambert and Paterson 1984; Lambert et al. 1987; Coyne et al. 1988); here we will present only a brief overview of Paterson's major points.

Paterson's major argument is that a species should be defined genetically by the presence of a "shared fertilization system" between bisexual members of a population, rather than by the presence of prezygotic or postzygotic elements preventing mating mistakes that may or may not result in hybridization. Although Paterson is in agreement with Mayr (and others) that speciation occurs primarily in allopatry, he argues that characters involved in reproductive isolation could not have been selected to function, in an adaptive sense, as reproductive isolating mechanisms, but rather that reproductive isolation is an incidental by-product or "effect" (following the argument by Williams 1966) of the divergence of the fertilization system (Paterson 1985; Lambert and Paterson 1984).

An important feature of the shared fertilization system is a set of signals that are recognized by conspecifics and constitute the Specific-Mate Recognition System (SMRS). The "raison d'être of the SMRS is to ensure effective syngamy within a population of organisms occupying their preferred habitat. The characters of the SMRS are adapted to function efficiently in this preferred habitat" (Paterson 1982a). It is critical to recognize that the SMRS represents a coadapted complex of traits and that stabilizing selection will act to maintain the integrity of the system. The Recognition Concept thus favors the punctuated equilibrium model of evolution (see Vrba 1985), and accordingly, change in a species' SMRS will occur only in response to an alteration in the habitat, either when a portion of the population emigrates to a new habitat or if there is a fragmentation of the existing habitat, for example, as a result of climatic change. For Paterson, "speciation and the acquisition of a new niche are part of the same process of adaptation to a new habitat. There is no special process of cladogenesis; speciation is an incidental *effect* resulting from adaptation of the characters of the fertilization system, among others, to a new habitat" (Paterson 1985).

These ideas represent a major break from the more "Darwinian"-based Biological Species Concept, in which variability in character traits is seen as an important source upon which selection (natural and sexual) acts continuously to modify the mating phenotype. In the Recognition Concept variability within a particular trait is recognized, but is superseded by the integrative properties of the system as a whole. Natural selection is seen to play only a minor role, principally during the period when a new SMRS may arise (Paterson 1985).

The purpose of this chapter is to characterize the mating system of the group of insects known as the moths, a group in which mate location, courtship, and copulation are achieved primarily by the use of chemical signals (Baker 1985, 1989a; Cardé and Baker 1984). We will explore a number of the features of this mating system as they relate to the Recognition Concept and other arguments, as presented above, concerning the origin and adaptive function of reproductive character traits. These will include the concept of signal specificity and variability observed in the communication channel, the general problem of what constitutes the *habitat* in discussions on speciation, the importance of "noise" (or interspecific interactions in general) in the development and functioning of the communication channel, and the genetics of the production and reception of pheromone signals.

The Moth Mating System

In a majority of moth species the mating system involves long-distance attraction and short-range courtship, both involving the use of chemical signals (Baker 1989a,b; Cardé and Baker 1984). Typically, the long-distance attractant is produced and released by females, with males perceiving the signal and flying upwind from some distance to where the female is located (see Baker [1989c] for an account of the orientation behaviors involved in the upwind flight sequence). With very few exceptions, this phase of the pheromone-mediated sequence is found in all of the moths, from ancestral to more derived forms.

In contrast, the close-range courtship sequence varies considerably in its complexity, both interspecifically and intraspecifically (Baker 1989b; Birch et al. 1990). In some species the close-range sequence involves a chemical dialogue between the sexes, with males executing complex stereotypic displays in which they release their own chemical signals from often elaborate and specialized scent structures (Birch et al. 1990), and with females exerting a selective action on these displays (Baker 1989b). In many others courtship involves very little display and no apparent input from the female. In addition, whereas the behaviors involved in close-range courtship are dominated by the use of chemical signals, they also can involve several other sensory modalities as well. Vision is often involved in the courtship display, and copulatory movements may include important tactile cues (Baker 1985, 1989b; Birch et al. 1990). In a number of species, sound is involved in the close-range orientation of males to females (Spangler 1985; Krasnoff and Roelofs 1990).

The majority of moth pheromone components involved in long-range attraction are composed of even-numbered 10-18 carbon-chain acetates, alcohols, and aldehydes, typically with one or two positions of unsaturation along the chain. Longer-chain epoxides and ketones also are found in selected groups (such as the Lymantriidae and Arctiidae) (Tamaki 1985). The analysis of several hundred pheromones indicates further that, in almost all cases, they are multicomponent blends of chemicals. These blends are typically composed of one abundant (or major) component and several additional chemicals (or minor components) present in much lower quantities (Tamaki 1985). Closely related groups (family or subfamily) often use the same major component, as a result of possessing a common biosynthetic pathway that produces the pheromone (Roelofs and Bjostad 1984; Bjostad 1989). Minor components can arise from the various intermediates available in the gland that are part of the pathway that produces the major components and can vary in number and functional group depending on the starting materials and the number of steps in the pathway (see Bjostad 1989). The pheromones of closely related species may thus differ as a result of the complement of minor components in the blend, as well as in the ratio of components produced. One common finding is the presence of a specific ratio of geometric isomers, usually involving the major components and its complementary isomer (Roelofs and Brown 1982; Cardé and Baker 1984; Bjostad 1989).

In addition to the chemical signal itself, it is well recognized that the communication channel also includes important environmental and physiological factors that influence the mating system (Cardé 1979; Cardé and Webster 1981; Cardé and Baker 1984; McNeil 1990). For example, in each species female release of pheromone and peak male responsiveness occur at a characteristic time of the day, normally in a habitat where females would oviposit. Temporal specificity is governed by photoperiodic cues, but also can be modulated by temperature, relative humidity, and other exogenous factors (Cardé and Baker 1984; McNeil 1990). Habitat specificity may be influenced by host odors or visual cues (see Bell and Cardé 1984). In addition, hormonal and neuromodulatory actions influence both female pheromone production and male responsiveness, and there are both age- and density-dependent effects on the sensitivity of males to the signal and the production of the signal in females (McNeil 1990; Linn and Roelofs 1987; Tang et al. 1989; Raina and Menn 1987; Raina et al. 1989; Teal et al. 1989).

From the above it is clear that in its broad outline the mating system of moths displays two important features relevant to a discussion of the evolution of mating systems. First, mating behavior involves a complex sequence of coordinated steps between male and female that functions to bring the two sexes together. Because it is *required* for the long-range attraction of males and, thus, for successful mate location, it is reasonable to propose that according to Paterson's Recognition Concept the pheromone signal is an adaptive character and should be considered as part of the coadapted set of traits that constitute the SMRS. Second, the communication channel is adapted to the appropriate habitat in which the organisms live and develop. In the discussion to follow we will focus on the long-range communication system, as the evolution of close-range courtship displays has recently been reviewed in considerable detail by Baker (1989*b*), Birch et al. (1990), Krasnoff and Roelofs (1990), and Eisner and Meinwald (1987).

Signal Specificity and Intrapopulation Variation in the Long-Distance Pheromone Signal

One of the prominent points of debate between the Biological Species and Recognition concepts concerns the importance of variation, both within and between populations of organisms, on the stability of species and in the speciation process. Paterson (1978, 1985) has argued that variability in the mating phenotype, and especially in that *set* of characters critical for mate location, will be low because of stabilizing selection acting to preserve the integrity of the coadapted set of characters that constitute the SMRS. Variation will certainly occur in the expression of any particular character, but this will not necessarily result in selection altering the character as an element of the mating system. The emphasis is on a signaling *system* displaying a high degree of specificity, providing for efficient transfer of information (see White et al., this volume; Shannon and Weaver 1949; Baker 1985; Smith 1977; Lambert and Levey 1979).

In contrast, whereas the Biological Species Concept also recognizes that the sender and receiver components of the communication channel must be tuned to a common signal, mating is viewed more as a competitive activity among individuals for a limited resource, and thus there is a greater emphasis on the role of natural and sexual selection acting on individuals within a population who exhibit variation in traits composing the mating phenotype. Less emphasis is placed on the mating phenotype as an integrated system, with more interest in the power of selection to act on variability in specific character traits to produce change. Signal specificity is viewed not only as a mechanism for preventing mating mistakes, but also as a mechanism for enhancing the ability of individuals to maximize their own reproductive effort by locating a prospective mate before conspecifics (Thornhill and Alcock 1983; Cardé and Baker 1984; Cardé 1986).

Signal specificity is, in fact, one of the fundamental concepts to have emerged from numerous early field studies on sex pheromones, in which the importance of multicomponent blends as species-specific signals became evident. This, in turn, has generated great interest in determining the intrapopulation variation in blend composition produced by female moths and the corresponding variation in male response (see Schlyter and Birgersson 1989; Löfstedt 1990). We will first examine three species in which a fairly low degree of variability has been found. In the first, with the redbanded leafroller, *Argyrotaenia velutinana,* Miller and Roelofs (1980) analyzed the ratio of (Z/E11-14:OAc) in glands from individual female moths collected from the field, as well as from an established laboratory colony. The mean ratio for the laboratory colony was 7.0 percent E to Z, whereas for the field population it was 9.1 percent. The coefficient of variation (C.V.) for both populations, however, was only 9.7 percent. In field tests with this species (Roelofs et al. 1975) males exhibited peak response to a narrow range of ratios (7–9 percent E) centered on the female-produced ratio.

A second species that has been examined in some detail is the pink bollworm moth, *Pectinophora gossypiella*. Collins and Cardé (1985) analyzed the ratio of Z,E/Z,Z 7,11-16:OAc in individual females and found a mean of 44.2 ± 2.3 percent Z,E in Z,Z with a C.V. of 5.3 percent. In a separate study Haynes et al. (1984) reported a mean of 38.3 ± 4.22 percent (SD) for this species. Field studies have shown that males can respond in high numbers to a relatively wide range of the two isomers that compose this species' pheromone (Flint et al. 1977). Flight-tunnel studies confirmed this but also showed that the wide response range of males may be a result of higher temperatures (> 26°C), as this species does exhibit greater response specificity if tested at lower (20–21°C) temperatures (Linn and Roelofs 1985; Linn et al. 1988).

The third example exhibiting low variability is in the Oriental fruit moth, *Grapholita molesta*. Unfortunately, as a result of the low quantities found in females, individual variation in female production has not been examined. However, based on the high degree of specificity exhibited by males of this species we would expect females to show a low degree of variability in ratios of components produced. In an extensive series of flight-tunnel, field-trapping, and field-observation studies it was found that male specificity was very high for the natural Z/E8-12:OAc isomer ratio in the blend (Z8-12:OAc + 6 percent E8-12:OAc + 3–30 percent Z8-12:OH, Cardé et al. 1979), with significantly lower response levels to 2 percent or 10 percent E ratios compared with the natural 6 percent E (Baker 1989b; Baker et al. 1981; Linn and Roelofs 1983, 1989; Linn et al. 1987, 1990).

Other examples illustrate that some species exhibit considerably more variability in the production of, and response to, ratios of components in their pheromone blends. As we will show, these examples often involve pheromone blends composed of monounsaturated compounds that are not geometric isomers or are compounds that are derived from plant precursors via separate metabolic pathways, suggesting that there is a strong link between the variability in a species' pheromone signal and the ability to regulate the production of a precise ratio of components. For example, in a Swedish population of the turnip moth, *Agrotis segetum*, Löfstedt et al. (1985b) showed that the coefficients of variation for the three main components of the pheromone were 14.8 ± 127 percent for Z5-10:OAc, 55.6 ± 32 percent for Z7-12:OAc, and 29.6 ± 59 percent for Z9-14:OAc. Field studies have shown that males respond in high proportions to a range of mixtures of the three principal pheromone acetates. Similarly, in the almond moth, *Ephestia cautella*, individual variation ranged from 63:27 to 97:3 (Z,E)-9,12-14:OAc to Z9-14:OAc, with a mean of 88:12 (Barrer et al. 1987), and in the summerfruit tortrix, *Adoxophyes orana* F. v R., individual female production of Z9-14:OAc and Z11-14:OAc ranged from 3.5:1 to 11:1 (Guerin 1986). In the latter species male capture occurred over a range of ratios, from 5 to 25 percent Z11- in Z9-14:OAc, matching the range produced by females (Guerin 1986). Finally, in the potato tuberworm, *Phthorimaea operculella*, populations in California and Japan were analyzed for ratios of (E,Z)-4,7-tridecadienyl acetate and (E,Z,Z)-4,7,10-tridecatrienyl acetate. In the California population the amount of triene ranged from 27 to 88 percent, with a mean of 56 percent and a C.V. of 23 percent, whereas in Japan the range was 16–71 percent, with a mean of 42 percent and a C.V. of 31 percent (Ono et al. 1990). In agreement with data from female gland analyses in *P. operculella*, males were found to respond to a very wide range of component ratios (from 9:1 to 1:9) in several populations sampled throughout the world (Ono et al. 1990).

Variation in any phenotypic character among a group of individuals has both genetic and nongenetic components to it. Individuals will vary with respect to their own genotypic makeup, and the particular phenotype that an organism displays will be a product of the interaction of its genotype with the environment in which it develops (Mayr 1970; Futuyma 1986). With respect to the variation in ratios of components produced by females, we believe that, as was suggested above, this largely reflects constraints imposed by the biosynthetic pathways for producing and regulating a complex blend of components. For example, in *G. molesta*, the three compounds that compose the pheromone are products of a single biosynthetic pathway, and the Z/E isomeric mixture is controlled by one enzymatic step at the end of the pathway (see Roelofs and Bjostad 1984). Thus, in this system there is the potential for tight control in the production of this ratio, and, as indicated above, this appears to be the case as there is a high degree of signal and response specificity in this species (Baker et al. 1981; Linn and Roelofs 1983; Linn et al. 1985, 1987, 1990). In contrast, in a species such as the potato tuberworm, *P. operculella*, the two pheromone components ([E,Z]-4,7-tridecadienyl acetate and [E,Z,Z]-4,7, 10-tridecatrienyl acetate) are biosynthesized from two different precursors, linoleic and linolenic acid, respectively, and thus two pathways are involved in pheromone production (Ono et al. 1990). In addition, the precursors for the *P. operculella* pheromone are dietary constituents, and there may be differences in availability of these compounds leading to interindividual variability in production of the two compounds.

The relationship between genetic and environmental influences on female production can also be elucidated by making repeated measurements from individuals. As Löfstedt (1990) noted, "Repeatability in a ratio between pheromone components can be looked upon as the fraction of the total variance that is due to differences between individuals and it sets an upper limit for the heritability that the character can have." Du et al. (1987), with an yponemeutid moth, *Y. padellus,* showed that the repeatability of emitted ratios was between 0.82 and 0.90. Individual females emitted approximately the same blend at subsequent samplings, but interindividual variation in component ratios was significantly lower for E to Z11-14:OAc (C.V. = 15 percent) than for other components in the blend (C.V. = 46 percent). Similarly, Witzgall and Frérot (1989) showed that blend ratios of *Cacoecimorpha pronubana* were very consistent over three calling periods, with the ratio of Δ11 components (Z11-14:OAc, E11-14:OAc, and Z11-14:OH) displaying small interindividual variation. These studies suggest that repeatability in female production, and thus the heritability for this character, is fairly high, especially with respect to Z/E isomeric ratios. Unfortunately, studies have not as yet been conducted with a species, such as *A. segetum,* which uses a mixture of compounds that are not geometric isomers and for which the ratio is less easily controlled.

With respect to variability in male response, although very few studies have examined the behavior of individual males to a series of different blends, it is generally held that the observed variability is the result of some males possessing broader response ranges, rather than different blend preferences (see Linn et al. 1991). One field study, involving mark-recapture with *G. molesta* (Cardé et al. 1976), concluded that males do not have different preferences for various blend ratios. In a second study, with *O. nubilalis,* the repeatability of male response to a range of blends was tested (Glover et al. 1991). In this species, distinct strains utilizing a 97:3 or 1:99 mix of Z/E11-14:OAc are found in sympatry in the United States. Males of the parent populations exhibit relatively narrow response windows with peak levels occurring to the mean ratio produced by females. However, in crosses between these two strains, hybrid males responded to a much wider range of blends. In repeated flight-tunnel exposures males could be characterized in one of three ways: (1) those responding only to the predominant female-produced blend, (2) those responding to a wide range of ratios, and (3) those that did not respond to any blend.

Several studies also have shown, however, that environmental factors can significantly affect male response thresholds, resulting in dramatic changes in response specificity exhibited by a population of males. One of the most important factors is temperature, affecting both the diel timing of mating activity as well as male response thresholds and specificity. For most species there is a preferred temperature range that is optimal for mating, and a number of studies have shown that the timing of mating behavior in the diel cycle can be advanced or delayed, so as to occur during the most favorable conditions. In *A. velutinana,* for example, lowering the temperature from 24 to 16°C shifts the mating period into the photophase, and males become responsive to pheromone after as little as fifteen minutes of exposure to the lower temperature (Cardé et al. 1977). The adaptiveness of this response relates to the seasonal activity of these insects. In the summer months *A. velutinana* exhibits a typical scotophase activity period when temperatures are optimal for mating. In the spring, however, scotophase temperature may be prohibitively low, and thus with the temperature-influenced shift, mating activity can occur during the afternoon hours, when temperatures are more favorable.

Equally dramatic is the effect of temperature on male perception of odor quality. If male *G. molesta* and *P. gossypiella* are acclimated and tested in the flight tunnel at higher temperatures than normal (26°C compared with 20–21°C), the response specificity of males broadens to include off-ratios at higher, lower, and optimal release rates (Linn et al. 1988; Baker et al. 1988; Baker 1989a,c; Linn et al. 1990). We should note, however, that in the case with *G. molesta*, whereas the temperature effect was also observed in field observations of individual males (Linn et al. 1990), field-trapping studies with this species over a summer period indicate a high degree of specificity for the natural Z/E8-12:OAc isomeric ratio (Linn and Roelofs 1983), suggesting that mating behavior in this species may shift on a daily basis so as to occur at a time when temperature conditions are optimal for male specificity to be expressed.

Given the observed variation and phenotypic plasticity in the pheromone system we can ask: Does the observed variability in long-distance attraction in some moth pheromone systems render them less efficient, or mean that they are not part of the recognition system? We would argue against this, for two reasons. The first relates to the fact that signal specificity can vary in at least two different ways: (1) in the composition of the blend, that is, the number and type of components produced; and (2) in the ratios of components produced by females. In the discussion above we emphasized the latter aspect, but with respect to the former there appears to be very low variability, as in almost all species that have been examined a high proportion of females produce the full complement of components, and in all cases reported, peak levels of male response are dependent on male perception of the female-produced mixture. Thus, even in those cases where a large variation exists in the ratio of components produced by females, signal specificity is still very high because of the requirement for male perception of the complete set of components released (for *A. segetum* see Löfstedt et al. [1985a]).

The second reason relates to the role of the habitat in the signaling process. The available data suggest that whereas male-response specificity can be altered dramatically by, for example, temperature, mating behavior generally occurs in the preferred habitat at a time when conditions are optimal for signaling. These observations stress the importance of the idea that the recognition system functions as a *unit* in effecting mate location and syngamy and that this unit includes all of the components of the signaling system, as well as the habitat in which the signal operates, an idea that is in agreement with the Recognition Concept, and one that we believe is fundamental to an understanding of the role of multicomponent blends in moth pheromone communication systems (see the section below on signal specificity in a noisy environment).

Given this argument, however, we also recognize the importance of variation in the population and the importance of addressing the question of whether this variation *is* a potential source upon which selection can act, allowing local breeding groups to adapt to slightly different environments. In the following section we will explore a number of cases suggesting that this can in fact happen, resulting in differentiation of populations over a species' geographic range.

Geographic Variation in the Long Distance Signal

In addition to the variation that exists within a local breeding population, it is important to consider the question of how a species' signaling system varies over its geographic range. Paterson has maintained that the geographic range of a species will be a function of the distribution of the specific habitat in which the SMRS evolved, and that the cohesiveness of the recognition system, which is a central feature of the mating system, will be strongly influenced by stabilizing selection, rather than gene flow. The strong relationship between the structure of the recognition system and the habitat leads to the conclusion that species will generally track environmental oscillations with their geographic distributions, rather than evolve (see Vrba 1985).

In contrast, many discussions associated with the Biological Species Concept argue that the situation is considerably more dynamic. As Mayr (1970) argued, "We can conclude that virtually all species are composed of numerous local populations, each adapted to their local environment, yet sharing much of their genetic system with other conspecific populations and retaining contact with them through gene flow." Because the environment, with the exception of the rarest cases, is also variable and unpredictable, and because of differential survival and reproductive success among individuals that vary in their mating phenotypes, it is proposed that natural selection will act continuously to modify the fitness of individuals to the environment and the composition of each population's gene pool. The integrity of a species over geographic space, as well as time, will thus be the result of a balance between forces acting to constrain divergence among populations, such as gene flow, and those that promote divergence, such as localized natural and sexual selection, drift, and mutations (Mayr 1970; Slatkin 1987; Barton and Hewitt 1989; Hewitt 1988). Over its geographic range a species can thus be thought of as a mosaic of populations that may very genotypically and phenotypically. Depending on local environmental pressures, gradual change in these populations may occur over time, resulting in some divergence and, if appropriate barriers arise, speciation (Barton and Hewitt 1989). It should also be clear from this discussion that variation among individuals in a population with respect to parameters of the mating system plays a central role.

In the moth long-distance pheromone communication systems that have been studied, results favoring stasis as well as local divergence among populations have been found. With respect to stasis, Haynes and Baker (1988) showed that in a series of populations of *P. gossypiella* in the United States, the variation in blend ratio for > 70 percent of females analyzed was within 10 percent of the mean value. They also showed that the blend of this species was very consistent on a worldwide basis. In only one population, from China, was there a difference, although even this difference was considered very small (see also Flint et al. 1979). In another study, with the cabbage looper, *Trichoplusia ni*, Haynes and Hunt (1990a) analyzed gland contents from individual females from three localities distributed across the United States, as well as a long-maintained laboratory colony. The pheromone of this species is a six-component mixture, and, in all populations analyzed, the qualitative as well as quantitative parameters of the signal were very similar, although, interestingly, as with *P. gossypiella*, the greatest difference was observed between the field populations and a laboratory colony.

There also are several excellent examples that demonstrate a greater degree of geographic variation in a pheromone system. The first is that involving several populations of the turnip moth, *A. segetum*, in Europe (Löfstedt et al. 1986b). The pheromone of the turnip moth consists of a mixture of three homologous acetates: Z5-10:OAc, Z7-12:OAc, and Z9-14:OAc. Studies by Arn et al. (1983), Löfstedt et al. (1986b), and Hansson et al. (1990) have demonstrated that there are three distinct groups within Europe and Western Asia: A French population (in France, Denmark, and Switzerland) in which the ratio of components is 47:40:13; a Swedish population (in Sweden and Britain) producing a 4:52:44 mixture; and Armenian/Bulgarian populations that produce ratios of 1:52:47 and 1:42:57, respectively. Whereas there is overlap among these groups as a result of high levels of intrapopulation variation (see Löfstedt et al. 1986b), there also is evidence for their being considered as distinct populations. First, in field trials males of the French population could be attracted to Z5-10:OAc alone, a result never obtained with the Swedish populations. Second, in Bulgaria males were trapped with Z7-12:OAc and Z9-14:OAc together, supporting the idea that Z5-10:OAc is of less importance for these populations. Third, and perhaps most important, in Hungary both the Swedish and the Armenian/Bulgarian populations coexist, apparently as parapatric populations.

The second example occurs in what are now considered to be sibling species complexes of New Zealand leafroller moths, previously classified as *Planotortrix excessana* (Walker, *sensu* Dugdale 1966) and *Ctenopseustis obliquana* (Walker, *sensu* Green and Dugdale 1982; Foster et al. 1986; Foster and Roelofs 1987,

1988; Foster et al. 1987, 1989, 1990, 1991). Results of a number of studies show that these two complexes consist of two and three sibling species, respectively (Foster et al. 1991; Dugdale 1990). With respect to *P. excessana*, two forms have been characterized: one utilizing Z8-14:OAc and 14:OAc (now *P. octo*, Dugdale 1990) and a second utilizing Z5-14:OAc and Z7-14:OAc (now *P. excessana*, Dugdale 1990). The *P. excessana* population exhibits considerable variation in the ratio of the two components over its range. Foster et al. (1989), in a detailed analysis of this form, collected individual females from around the country and found that they produced pheromone blends ranging from 3:97 to 71:29 of Z5- to Z7-14:OAc. Laboratory populations established from the extremes of this geographic range produced mean ratios and total quantities that were significantly different, but in which the ranges of ratios and quantities did overlap. Male response occurred to the entire range of ratios produced (Clearwater and Foster, unpublished), although there was a slight bias for males of both localities (Waikato, North Island, and Dunedin, South Island) to mate preferentially with females from the Dunedin population (Foster et al. 1989). It was concluded on the basis of male cross-attractions and successful mating that the observed variation among populations was intraspecific. However, Foster et al. (1990) also showed that, whereas *P. excessana* exhibited intraspecific variation, other *Planotortrix* species also use these same two components and in much narrower ratios.

In the case of *C. obliquana*, a study by Clearwater et al. (1991) has demonstrated that two population types can be characterized: one (Type I) utilizing Z8- and Z5-14:OAc (now *C. obliquana*) and a second (Type II) utilizing Z5-14:OAc alone (now *C. herana*). The first population type was previously split into two types (I and III, see Foster et al. 1986) on the basis of differences in mean proportions of components and quantities of pheromone produced. More extensive sampling (Clearwater et al. 1991) demonstrated that the two populations can still be characterized by different mean ratios of Z8- and Z5-14:OAc produced, but that there is overlap between the populations. The results of field cage studies indicate that males of both populations are attracted to females of one of the populations, and thus it appears that there is considerable intrapopulation variation in this form in New Zealand. This contrasts with the situation between individuals of Type I or III and Type II, in which both male and females can be characterized along a number of parameters (see Clearwater et al. 1991), substantiating their status as sibling species.

One of the conclusions from the previous section on intrapopulation variation was that the observed variability in ratios of components produced by females was the result of inherent problems with controlling the biosynthetic pathways for these components. It is interesting to note that, in the two cases discussed above, wide geographic variation occurs in species that also exhibit wide intrapopulation variability as well and use combinations of components whose biosynthetic pathways are not linked and are therefore more difficult to regulate (see Roelofs and Bjostad 1984; Roelofs and Wolf 1988; Foster et al. 1990). This supports the idea that, even though signal specificity is relatively high because of the requirement for perception of a particular mixture of components (whatever the specificity for the ratio), variation within the channel can provide a source upon which selection can act. Given the almost certain instability of the habitat over long periods of time, it appears that selection on the communication channel can result in divergence of populations over a species' geographic range. Future studies with *A. segetum* and the New Zealand leafrollers should provide a greater understanding of these forces, as well as those involving other features of the habitat that may contribute to divergence.

Another interesting case that, at first, appears to involve geographic variation within a species is in the European corn borer, *Ostrinia nubilalis*. This insect was introduced into North America on at least two and probably three separate occasions during the twentieth century (Caffrey and Worthley 1927) and consists of three distinct races in New York State based on voltinism and sex pheromone blend differences. In New York, the bivoltine populations exhibit moth flight periods in June and August, whereas univoltine moth flight periods occur in July. Specific pheromone blends distinguish a Z race (97:3 mix of Z/E11-14:OAc's) and an E race (1:99 mix). The two pheromone races both occur in Europe and North America (Klun et al. 1975; Anglade et al. 1984; Roelofs et al. 1985). The pheromonally distinct populations exhibit strong ethological isolation in the field, but have enough genetic compatibility to produce viable and fertile hybrids. Pheromone production for a specific blend in each race, however, is highly canalized for that blend with very little variation in blend ratio (Roelofs et al. 1987). The blend of each race was found to be controlled by a major autosomal gene (Klun and Maini 1979; Roelofs et al. 1987). Recent discovery (Glover et al. 1990) of a sex-linked marker enzyme, triose phosphate isomerase, that is fixed for one allele in the E race and is polymorphic in the Z race, has allowed for studies of gene flow among the races in areas of sympatry. In New York preliminary evidence (Glover et al. 1991*b*) shows that gene flow is unidirectional from the E to the Z race since field populations of the E race throughout the state remain fixed for the single allele. A contributing factor for this observation is that hybrid males from both reciprocal crosses respond to a range of pheromone blends from the 35:65 Z/E ratio produced by hybrid females to the 97:3 ratio produced by the Z parents (Roelofs et al. 1987; Glover et al. 1991*a*). Since they do not respond to the 1:99 Z/E ratio of the E parents, the incidental hybrids produced in the field populations would more likely mate with a Z female than with an E female.

The accumulated evidence on the European corn borer races in New York State suggests that they should perhaps be viewed as two sibling species, rather than as one species with a significant degree of variability in the pheromonal mating system. We noted above that in the case of *A. segetum* or the New Zealand leafrollers the occurrence of sibling species could be the result of selection on a large variable population, and that this, in turn, could be the result of an inability to tightly control the qualitative aspects of pheromone production. In contrast, *O. nubilalis* uses a mix of geometric isomers, with males displaying a high specificity for their conspecific female ratios. This system thus represents one in which the pheromone signal is tightly regulated and there is little potential for local divergence as a result of selection acting on variability in the population. In a later section we will explore in greater detail how this system might have arisen (see also Löfstedt 1990).

A final example involves the oblique-banded leafroller moth, *Choristoneura rosaceana* Harris. Hill and Roelofs (1979) reported that a mixture of Z- and E11-14:OAc plus Z11-14:OH in a 90:5:5 ratio was the pheromone of *C. rosaceana* in New York and that addition of a fourth component, Z11-14:ALD, had no effect on the level of trap capture. In contrast Vakenti et al. (1988) showed that in British Columbia an 85:15 ratio of Z11-14:OAc and Z11-14:OH was the best lure for this species. This blend, however, is also the pheromone of the European leafroller, *Archips rosanus*. Traps baited with this two-component mix attracts males of both species. Vakenti et al. (1988) showed further that the three-component mixture of Hill and Roelofs (1979) did not increase trap capture of *C. rosaceana,* nor did it increase trap specificity between this species and *A. rosanus*. However, addition of 1 to 4 percent of the Z11-14:ALD component in British Columbia to the three-component blend significantly enhanced trap capture of *C. rosaceana* and eliminated cross-attraction of *A. rosanus*. There, thus, appeared to be a geographic difference in this species' pheromone specificity because of the presence of a competing species in the western region. More recently, Thomson et al. (1990) conducted additional field tests with *C. rosaceana* in British Columbia and Quebec, Canada. A series of ratios of the three- and four-component blends was tested in each area. In British Columbia again the four-component blend (Z/E11-14:OAc + Z11-14:OH + Z11-14:ALD) was superior in terms of trap capture and trap specificity, with a 100:2:1:1.5 ratio proving to be the best. In contrast, in Quebec, there was no difference in trap capture among a series of ratios of the three- and four-component blends or among any of the ratios of the four-component system tested.

This last example of geographic variation in a species' pheromone is of special interest because it raises the question as to what factors are involved in the differentiation of populations within a species' range, and specifically whether interspecific competition has played a role in the divergence of, for example, the *C. rosaceana* pheromone system. This case, in fact, has been discussed as a potential case of character displacement, a concept that has received much attention, but for which there is very little hard evidence (see Butlin 1987a,b). In the following section we will examine this problem in greater depth, focusing on a particular aspect of communication systems that we feel has not been adequately addressed by proponents of the Recognition Concept.

Signal Specificity in a "Noisy" Environment

The studies presented above indicate that there may be considerable geographic variation in moth mating systems, to the extent that sibling species status may be warranted solely on the basis of differences in the pheromone communication system. One possible, and widely accepted, explanation for these differences is that populations have adapted to local requirements of their habitat (see Mayr 1970; Futuyma 1986). This conclusion by itself is not entirely incompatible with the Recognition Concept. According to Paterson (1985; see also Kemp 1985), slight regional differences in elements of an SMRS could arise due to differences in the local habitat, and this would appear to be one of the adaptive features of interindividual variation in the population.

The critical question, however, is: What does "adapted to local requirements of the habitat" really mean? or, more generally, What are the forces that are involved in the initial structuring, and subsequent modification, of a mating system?

As we have noted several times, Paterson has argued that the SMRS comprises a signaling system that has evolved in and is adapted to a specific habitat (Paterson 1985). In addition, statements in Paterson (1985), Kemp (1985), and Vrba (1985) indicate that whereas biotic elements are not ruled out as agents for selection, there is clearly an emphasis on the *physical* environment as the most important selection force in shaping the specific elements of the SMRS. Similarly, Paterson and colleagues have forcefully argued against any type of reinforcement or reproductive character displacement (see Butlin 1987a) as mechanisms whereby populations can continue to diverge in sympatry as a result of secondary contact (Paterson 1978, 1981, 1985; Harper and Lambert 1983; Lambert and Paterson 1984; Lambert et al. 1987; Spencer et al. 1986; Carson 1982; see, however, Mayr 1970; Dobzhansky 1970; Ayala et al. 1974; Butlin 1987a,b; Ritchie et al. 1989; Coyne et al. 1988; Coyne and Orr 1989; Hewitt 1988; Barton and Hewitt 1989). According to the Recognition Concept, the SMRS that characterizes any species is habitat specific and will change only in response to a change in that habitat.

The question of exactly what the term *habitat* means, however, is of great importance in understanding the selection forces that influence the evolution of a new chemical communication system. *Allopatry* does not mean a biological, or chemical, vacuum, and it is difficult to imagine a situation in which a species, either as an established population (with species status) in a stable habitat, or a recently isolated one that is in a suboptimal environment (and thus in which the mating system is undergoing change), is not confronted by other species of varying relatedness that use common pheromone components or other odors (e.g. from host plants) that might interfere with or adversely affect the communication channel. We would thus argue in favor of the hypothesis that, in addition to physical environmental pressures, there is also the potential for biotic factors, and specifically forms of interspecific competition, to be adaptive selective forces on pheromone systems, because of the problems associated with a "chemically noisy" environment (see Cardé and Baker 1984; Baker 1985, 1989a; Linn et al. 1984; Cardé 1986).

The importance of noisy environments can, in fact, be found in discussions of all major communication systems (see Smith 1977; Wiley and Richards 1982; Morris and Fullard 1983), and has been a prominent theme in pheromone studies as well (see Cardé 1986). From the earliest field studies recognizing the importance of specific mixtures it was hypothesized that closely related species have "partitioned" a common communication channel so that each has a unique mating system (Cardé and Baker 1984; Baker 1985; Greenfield and Karandinos 1979). Whereas partitioning can occur along a variety of dimensions in the communication channel, including temporal differences in responsiveness and spatial differences in mating behavior within the habitat structure, it has usually been inferred that this is the result of pressure on the chemical signal itself (Cardé and Baker 1984; Cardé 1986).

Two examples illustrate why this concept has played such a prominent role in discussions of the evolution of moth pheromones. The first is found in the tortricid moths that form a species complex on apple trees in the northeast United States (Roelofs and Brown 1982; Chapman and Lienk 1971). Two subfamilies compose this group: the Olethreutinae, a group of relatively host-specific, internal feeders, using 12-carbon acetates as the backbone of their pheromones; and the Tortricinae, a group of foliage feeders, using 14-carbon acetates in their pheromones. Many of the tortricine species use specific ratios of Z- and E11-14:OAc in their sex pheromone, with additional minor components varying among the group. For example, *Archips semiferanus* uses a 30:70 blend of Z- and E11-14:OAc, and *A. cerasivoranus* employs a 20:80 blend of the same components, whereas *A. argyrospilus* possesses a 60:40 mix of the two components, and the sibling species, *A. mortuanus,* is optimally attracted to a 90:10 blend. These species often cohabit the same trees and are active as adults at the same time of the night and the season (Chapman and Lienk 1971), and yet several field-trapping studies have demonstrated that each species responds selectively to its own pheromone blend (see Cardé et al. 1977).

The second example occurs in the European small ermine moths, Yponomeutidae (Löfstedt and Van der Pers 1985). This genus of moths has a broad distribution, with nine forms occurring in sympatry in the west palearctic region. The forms can be separated into two species groups on morphological criteria, and further into four distinct species and five semispecies. The members of these groups show varying degrees of similarity, and all have considerable diel and seasonal overlap in their mating activity. Thus, the group has been the focus of a study investigating the steps in a speciation process (Menken 1982). Pheromone studies with four of the species (*Y. evonymellus, Y. cagnagellus, Y. vigintipunctatus,* and *Y. padellus*), which occur in southern Sweden, showed that (1) Z11-14:OAc is the major component for all four species, (2) this compound alone does not attract males, (3) all four species require varying proportions of E11-14:OAc, and (4) with the exception of *Y. cagnagellus,* all require a third component, Z11-14:OH or Z11-16:OAc (Löfstedt and Van der Pers 1985). Specificity in *Y. cagnagellus* is the result of a specific Z/E isomer ratio, whereas in *Y. padellus* Z11-16:OAc is required, and, interestingly, this compound inhibits the response of male *Y. evonymellus* and *Y. vigintipunctatus.* These latter two species also possess the most similar pheromones and in the field, as well as in flight-tunnel tests (Hendriske 1986), show cross-attraction. They also show, however, the greatest differences in diel and seasonal periodicities in response. It was concluded that partitioning has occurred in this group along both chemical and temporal "niche" dimensions (as well as possible host plant differences, spatial location, or additional minor components; see Löfstedt and Van der Pers 1985), resulting in the observed specificity in these species' communication systems.

Throughout this chapter we have emphasized the importance of long-distance communication as an essential component of the mating system, and therefore an essential component of the SMRS. Given this, the question must be asked: What possible linkage could there be between a property of the physical habitat and selection for a specific isomeric blend of pheromone components? Rather, it would seem that the Recognition Concept must accommodate the potential for biological effects to influence the divergence of a new SMRS. We believe the above-cited examples support the hypothesis that interspecific competition *can* be a selective force on a species' communication system and that, in fact, this may have been a driving force in the early evolution of multicomponent blends in these insects (see Baker 1985; Cardé 1986; Greenfield 1981). As we noted earlier there are serious biosynthetic constraints on the diversity of individual compounds that can be used by closely related species in their pheromones. This means that with closely related species occupying the same environmental space there is a significant potential for interference in the communication channel. The result would be a significant loss of efficiency in the communication pathway (meaning not only that individuals in a population are unable to locate one another and mate in a frequency sufficient to maintain a viable population [Paterson 1985; Kemp 1985], but also from the aspect of an individual's mating success the ability to recognize the appropriate signal and reach a female before competing, conspecific males. Prevention of "mating mistakes" is thus a potentially important selective force, and one solution to this problem is for male-response thresholds to be dependent on perception of a mixture of compounds and, if possible, a precise ratio. Perception of a specific mixture dramatically enhances the signal-to-noise ratio and thus the specificity of the channel (see Baker 1985; Cardé and Baker 1984).

Additional support for this hypothesis concerning the evolution of multicomponent blends comes from experimental results demonstrating three important ways in which pheromone components influence male behavior: (1) through male response thresholds, (2) through redundancy in the signal, and (3) as interspecific behavioral antagonists.

Pheromone Components and Male Response Thresholds. Defining the response specificity of males to a blend of pheromone components has been of great interest in pheromone research (Linn and Roelofs 1989). In the extensive flight-tunnel, field-trapping, and field observations with *G. molesta* cited earlier, two important principles emerged: First, peak levels of response, as well as optimal flight behavior over the entire response sequence from activation through upwind flight to copulatory attempts, are dependent on perception of the natural blend of components. Second, in the presence of the blend, males display their lowest response thresholds. The importance of this observation lies in the fact that in some species, although males may respond to high concentrations of the major component alone (in contrast to single minor components), at the female-release rate of this component they do so in very low numbers and in a manner qualitatively different from the response to the blend. Thus, addition of minor components lowers the threshold for response to the blend, and thus the highest probability for successful mate location is dependent on male perception of the natural blend of components.

The major conclusion from the *G. molesta* studies is that multicomponent pheromone blends act as collective units to produce optimal response thresholds and flight behavior in males. The advantage of this system is that the effect of noise from interfering signals is reduced by the requirement that males detect and integrate the information from a unique blend of compounds in order for neural pathways involved in initiation of the flight motor program to be activated. This hypothesis also has been supported in studies with several other moth species including, *A. velutinana* and *T. ni* (Linn et al. 1985; Linn and Roelofs 1989), *A. segetum* (Löfstedt et al. 1985a; Van der Pers and Löfstedt 1986), and *O. nubilalis* (Glover et al. 1987).

Pheromone Components and Redundancy in the Signal. The second important relationship among pheromone components, male-response thresholds, and interspecies interactions is evidenced by the redundancy observed in the pheromone systems of several insects (see Linn and Roelofs 1989; Dunkelblum et al. 1987). The pheromone of the cabbage looper, *T. ni*, is a six-component blend identified from females on the basis of predictions made from an analysis of the biosynthetic pathway for this species (Bjostad et al. 1984; Roelofs and Bjostad 1984). Flight-tunnel studies showed that this six-component blend was equivalent in activity to females releasing pheromone as well as to gland extracts. However, when single components were removed from the blend, a decrease in activity was observed only when the major component was missing. A complete set of subtraction tests showed that minor components in the blend could substitute for one another, but only in specific ways, to produce optimal male response. Thus, whereas none of the minor components alone was essential for male activity, it was clear that males could detect the presence of each of the minor components by the way in which they could be substituted for one another.

Linn et al. (1984) suggested that the results observed with the *T. ni* pheromone represented a form of redundancy in the signal, a phenomenon not uncommon to many communication systems (see Wiley and Richards 1982; Smith 1977; Miller 1982; Doherty 1985), and that this redundancy could be adaptive in terms of interspecific interference between pheromone blends containing common components. Baker (1989*b*) later suggested that if a minor component were present in the environment in excess, for instance, if it were emitted by another species as its major component, it could result in habituation of sensory pathways in individuals that were sensitive to its presence. By substituting for one another, minor components could provide an alternative source for stimulation, thus maintaining optimal male-response thresholds that are dependent on perception of an appropriate mixture of compounds.

One criticism of the redundancy hypothesis is that it is "adaptationist" in nature, invoking a causal explanation for the action of a component simply because of its presence in the female-released signal. Thus, an alternative explanation is that some of the minor components are neutral in their effect. While we agree that caution must be exercised in invoking a function for a compound based solely on its presence, we would argue in this case that, because of the specific substitutions allowed to achieve maximal response with incomplete blends, it is clear that males can detect all of the minor components. This has, in fact, recently been verified in a study by Todd et al. (1991) in which single antennal receptor responses for all of the *T. ni* components were achieved, with specific interactions among minor components occurring as predicted by the results of the behavioral studies.

Pheromone Components as Interspecific Behavioral Antagonists. This phenomenon has been found in a number of species (Cardé et al. 1977; Linn et al. 1984; Glover et al. 1989), with one of the most complete studies involving two noctuid moths, *T. ni* and the soybean looper, *Pseudoplusia includens* (Linn et al. 1988; Grant et al. 1988). These two moths are often sympatric, with overlapping temporal activity patterns. They possess a common major pheromone component, Z7-12:OAc, and two minor components, 11-12:OAc and 12:OAc, but differ in their remaining complements of minor components. Flight-tunnel and field trapping studies have shown that although there is a small degree of cross-attraction, it is largely prevented by the actions of two mechanisms. First, male *T. ni* appear to be optimally responsive to their blend because of a lower threshold for response to the blend than to Z7-12:OAc alone. The few *T. ni* males that do respond to the *P. includens* blend do so in a qualitatively similar manner to those few that respond to Z7-12:OAc alone. Thus, it was concluded that they represent a small portion of the male population that have broad-response profiles with respect to odor quality and, therefore, do not discriminate between their pheromone blend and the major component of the blend alone.

Male *P. includens*, in contrast, are prevented from cross-attraction to the *T. ni* blend by the action of an interspecific behavioral antagonist. This compound is Z5-12:OAc, a minor component of the *T. ni* blend that is *not* produced by *P. includens* females. When it is added to the *P. includens* blend, however, it causes a dramatic reduction in trap catch and a significant level of arrested upwind flight in the flight tunnel (see Baker 1989*a*). It further has been demonstrated that male *P. includens* possess a specific, and very sensitive, antennal receptor cell for the Z5-12:OAc compound (Grant et al. 1988).

We believe that the presence of behavioral antagonists suggests that adaptive changes in the communication signal (or fine-tuning) can occur as a result of secondary contact between two species that share common elements in their pheromones. The presence of a specific receptor cell that is "hard-wired" to detect a heterospecific pheromone component as an interspecific behavioral antagonist can be viewed as an adaptive response, because it provides an additional mechanism by which efficient signal recognition occurs and mating mistakes are prevented. In a number of documented cases the antagonistic action is between two species that share to some extent common elements in their pheromones, as well as a significant probability of encountering each other in the environment during the mating period. These interactions suggest that the antagonist action is not a random one, but rather that it evolved as an adaptive device to prevent interspecific mating mistakes and to fine-tune an already functional pheromone signal. We emphasize that this fine-tuning need not involve closely related species, for which a mating probability might exist, but can also involve more distantly related forms within or between families that use common pheromone components (i.e., the presence of chemical noise in the environment).

In summary, we would argue that multicomponent blends have evolved and can be fine-tuned over time (as especially in the case with behavioral antagonists) to counteract the problem of chemical noise in the environment and that the specificity provided by these blends serves two purposes: (1) selective response to conspecific signals resulting in successful mate location and (2) detection of an antagonistic response to heterospecific signals that might cause interference. As Becker (1982) noted, in terms of an *individual's* reproductive success, "interspecific discrimination of this sort minimized unnecessary expenditure of energy, the potential for hybridization, and in general makes intraspecific communication more efficient" (see Butlin 1989*b;* Doherty 1985; Ryan 1990; Wiley and Richards 1982; Miller 1982).

Genetics of Pheromone Production and Response

Any meaningful theory of species formation must incorporate an understanding of the way in which genetic elements change. Studies on the genetics of pheromone production and response in the moths have increased significantly in the last decade, and in at least one species, *O. nubilalis,* there is now considerable detail concerning the modes of inheritance of the factors controlling the various elements of the system (Roelofs et al. 1987; Glover and Roelofs 1988; Glover et al. 1990; Roelofs and Glover 1990). We will first, however, present information from two other species that provides important insights into the genetic control of these communication systems and its relevance to speciation processes, as well as the discussions presented above.

In studies with two monomorphic species, attempts were made to alter the communication system by selection of females, within laboratory populations, that produced higher or lower than normal ratios of components. The first (Roelofs et al. 1986; Sreng et al. 1989) involved *A. velutinana,* an insect that uses a seven-component blend with Z- and E11-14:OAc's as the principal components. In a series of exploratory studies, several selection protocols were applied to selected family lines, with the results initially indicating considerable additive genetic variability in the factors controlling blend ratio. Further selection, however, failed to move the ratio of the two major components beyond the range of the base population, indicating that the Z/E11-14:OAc ratio is strongly canalized. Additional studies showed, however, that the ratio of 12-carbon minor components to the 14-carbon acetates responded quickly to direct selection and that it was possible to produce a "high" line with 42 percent 12-carbon acetates and a "low" line of 14 percent. These lines, however, also stabilized after several generations, and it was concluded that changes in the sex pheromone blend of female *A. velutinana* resulting in a biologically differentiated population would be difficult to achieve via gradual selection of small changes in laboratory populations.

The second study involved the pink bollworm moth, *P. gossypiella.* The two pheromone components of this species are (Z,Z)- and (Z,E)-7,11-16:OAc. In a laboratory colony Collins and Cardé (1989a,b) reported that these two components had a genetic correlation of 0.99 and that the heritability of the total amount of pheromone and the ratio of the blend in this colony were 0.42 and 0.34, respectively. In addition, the coefficient of variation for the component blend was 5.3 percent, indicating a relatively low overall level of variability, and thus a low potential for shift in the chemical signal of this species. Artificial selection was applied to this colony for twelve generations, producing a slight shift in mean blend ratio from 42.94 ± 0.97 percent (Z,E) isomer to 48.23 ± 1.25 percent. Although this change was statistically significant it is doubtful that it would cause a change in the mating system. Selection for a lowered blend was not successful. In addition, total amount of pheromone was reduced in the selected lines and was attributed to inbreeding depression.

The above studies demonstrate the difficulty in gradually selecting for altered isomeric ratios of components in pheromone blends, and as suggested several times above, we believe this is because of the way in which these components are synthesized. The Z/E ratio is a product of a single enzymatic step at the end of a pathway that produces both isomers (Roelofs and Bjostad 1984). This means that there is potentially a single gene controlling the ratio of isomers, which under stabilizing selection would display very little potential variability. Because of this low level of genetic variability there would presumably be very little possibility for selection to produce a gradual shift in the ratio to a point where divergence would occur.

Evidence for this comes from an alternative set of selection experiments on populations of species displaying polymorphisms. Whereas a number of studies have been undertaken (see Glover and Roelofs 1988; Löfstedt 1990), we will focus here on work done with *O. nubilalis*. As noted earlier, in the northeastern United States three races of this species can be differentiated on the basis of pheromone blend composition and voltinism: a bivoltine population utilizing predominantly Z11-14:OAc in its blend, a bivoltine population utilizing predominantly E11-14:OAc, and a univoltine population utilizing predominantly Z11-14:OAc (Roelofs et al. 1985). Pure cultures of these races were established in the laboratory, and the inheritance patterns for sex pheromone production by females, detection of the pheromone components at the antennal receptor level in males, and the flight response of males in the sustained-flight tunnel were studied. Analyses were carried out on progeny from reciprocal crosses, as well as progeny from subsequent F2 and maternal and paternal backcrosses. The results (Roelofs et al. 1987) showed that (1) female production of the pheromone blend is controlled primarily by a single autosomal factor, (2) the spike-amplitude of antennal olfactory cells is controlled by another autosomal factor, and (3) selective response to blends resulting in sustained upwind flight and source location over a 1.5-m distance in the flight tunnel is controlled by a sex-linked gene. Further analyses suggested that, although it is not known whether the major factor on the Z chromosome is a single gene or a set of closely linked genes, there is a minimum interaction of the major factor on the Z chromosome and modifying factors elsewhere in the genome. Finally, a recombination study was carried out to determine whether there was linkage between the autosomal genes coding for pheromone production and male olfactory receptor response, with the results showing conclusively that linkage was not a factor (Löfstedt et al. 1988). Z-linked inheritance of male antennal neuronal characters has also been demonstrated in the New Zealand *Ctenopseustis* species discussed earlier (Hansson et al. 1989). In this case antennal receptor properties were involved, with evidence for a single gene with two alleles on the Z chromosome. Analysis of hybrids showed, however, that additional genes also were involved. Similarly, in two species of sulfur butterfly regulatory genes on the Z chromosome that are specific for each species have been proposed to explain the large influence of this chromosome on most of the phenotypic characters that distinguish the species, including visual and olfactory signals of the courtship communication system (Grula and Taylor 1980).

The studies with *O. nubilalis* support the hypothesis that a single genetic factor controls the production of a specific Z/E isomer mix by influencing the activity of a single enzyme at the end of the biosynthetic pathway. This means that, as was suggested above, it is unlikely that the communication system can be changed by gradual shifts in production and preference for a new ratio of components, but rather that evolution of a new signal must begin with a major (mutational) change in the pathway. The results also demonstrate the important point that the male-female signaling system is not genetically coupled (Butlin and Ritchie 1989; Boake 1991), but rather appears to be under the influence of at least three independent genetic elements. This means that certain characters in the sender or receiver (such as production of or response to minor components) may be changed, or even lost, without affecting the opposite partner. These elements potentially can remain in the communication system as vestigial, or neutral, elements, with the possibility of acquiring new roles at a future time.

The Evolution of a New Sex Pheromone

In this final section we will explore several aspects of the evolution of a new pheromone system, given the information provided above concerning the importance of multicomponent blends, noisy environments, and the genetic control of pheromone production and reception. Our "exploratory" discussion will focus on the problem of altering a highly canalized system (such as that of *A. velutinana* or *O. nubilalis*), the potential for gradual, and perhaps frequent modifications in the effects of minor components, and the importance of "chemical noise" in influencing both the evolution of a new signal and its stability over time.

Paterson (1985) proposed in his development of the Recognition Concept that a species passes through two stages in its life history. The first, as has been noted several times in this chapter, can be characterized as a period of stasis, when there is a stable breeding population and stabilizing selection is operative on the mating phenotype. The second period is characterized by genetic and phenotypic change, when either the environment becomes fragmented as a result of climatic or geological factors, or a portion of the population migrates to a new environment (see Kemp 1985). This will, in either case, result in an unstable population of smaller size being located in a new (suboptimal) environment. In this situation it is proposed that the genetic cohesiveness of the population will become disorganized (see Carson 1982; Barton and Charlesworth 1984; Fabian 1985), and the variability in the mating phenotypes will increase with a subsequent reduction in response specificity, allowing individuals greater variability in the type of mate they can locate. Under these conditions, directional (or sexual; see Kemp 1985) selection can act on the mating system and in the event that a new stable SMRS evolves and is adapted to the new environment, and if this new SMRS is sufficiently different from the parent SMRS to ensure efficient syngamy in the new populations, then speciation will have occurred (Paterson 1985). Paterson (1978) emphasized, however, that because of the coadapted nature of the system, it is "unlikely that the characters of the SMRS can adapt quickly to new conditions. This is because the coadaptation that exists between the mating partners with respect to signals and receivers constitutes an effective buffer to change. The signal and receiver can only change in small steps, with coadaptation being reestablished after each step."

As noted previously, the above argument indicates that the Recognition Concept favors the punctuated equilibrium model for speciation (see Eldredge and Gould 1972; Paterson 1985; Vrba 1985; Williamson 1987; however, see also Carson 1987), and, accordingly, for change to occur in a pheromone system there would appear to be two necessary criteria: (1) stabilizing selection would have to be significantly disrupted, and (2) because of the coadapted nature of the recognition process, directional selection would have to act in gradual steps on the available variability in the signaling system to form a new set of characters that would constitute a new SMRS. However, as noted above in the case of pheromones that are composed of geometric isomers, the observed female-produced ratio is the product of a single enzymatic step at the end of the biosynthetic pathway, and selection experiments with at least two insects have demonstrated that the ratio is a highly fixed trait. This means that even if stabilizing selection was significantly relaxed, there is so little variability within the population in the single gene that codes for the enzyme that directional selection would not result in a significant change. Rather, for change to occur in a pheromone system, such as that found in *O. nubilalis* or *A. velutinana*, the initial step in the process would have to be a major (mutational) change (Roelofs and Glover 1990).

There are at least two examples in the pheromone literature lending support to the idea that a single major change in the pheromone system can result in a radically new signal. The first involves the three races of the European corn borer moth, *O. nubilalis*, discussed above. Roelofs et al. (1987) proposed that a single gene effect could alter the biosynthetic pathway if it acted at the final reduction step in the sequence. As noted, in many species of moths this last reduction step appears to be the critical step for the determination of the final Z/E isomer ratio. In *O. nubilalis*, for example, both E- and Z- parent strains, as well as F1 females, have intermediate unsaturated 14-carbon acyl compounds with similar E/Z ratios of approximately 70:30. Thus the blend differences in the races of *O. nubilalis* are due to a single change in the activity of the reductase controlling the final step in the production of pheromone.

The second example is with the cabbage looper, *T. ni*. The pheromone of this species is composed of six compounds: Z7-12:OAc (100 percent); 12:OAc (6); 11-12:OAc (2); Z5-12:OAc (8); Z7-14:OAc (0.8); and Z9-14:OAc (0.6) (Bjostad et al. 1984). In the course of a study involving collections of volatiles released by individual females, Haynes and Hunt (1990*b*), discovered three individuals that released (1) much reduced levels of the major component, Z7-12:OAc; (2) excessive amounts of the Z9-14:OAc component; and (3) a near absence of Z5-12:OAc. These individuals were mated, and from the analysis of the crosses it was determined that one recessive autosomal gene was involved in producing the mutant blend.

A consideration of the biosynthetic pathway for the *T. ni* pheromone indicates that a single mutation can generate the three major differences between the mutant and normal individuals. The biosynthesis of the *T. ni* pheromone involves two different pathways (Roelofs and Bjostad 1984). The major component Z7-12:OAc, as well as Z9-14:OAc, are derived from 16:Acyl via the action of Δ11-desaturase plus two chain-shortening steps to produce Z9-14:Acyl and Z7-12:Acyl. The Z5-12:OAc component is derived from 18:Acyl by the same enzymatic steps to produce the cascade of intermediates Z11-18, Z9-16, Z7-14, and Z5-12:Acyl. The acids then are reduced and acetylated to produce the pheromone components. It has been found that the mutant pheromone blend is the result of a mutational change in enzymes controlling chain shortening of the two Z9 acyl intermediates to produce Z7-12:Acyl and Z7-14:Acyl, respectively (Jurenka, Haynes, and Roelofs, unpublished). Without this chain-shortening step, the Z9-14 and Z9-16 intermediates build up, and the chain-shortened intermediates are not produced. This reduces or eliminates the already low levels of Z7-14:OAc and Z5-12:OAc (derived from the Z9-16:Acyl intermediate) in the pheromone, greatly increases the amount of Z9-14:OAc component, and also lowers the relative amount of the major components Z7-12:OAc (derived from the Z9-14:Acyl intermediate), satisfying the three conditions found in the mutant.

The above examples illustrate the potential for a single mutational effect to alter the pheromone system, with three important outcomes: (1) the formation of a new ratio of geometric isomers, or (2) addition of a major component, derived from the incorporation of a new pathway, and (3) minor components being lost or added to the female-produced blend. At this point, however, it is important to remember that, even if a major change were to occur in an allopatric population, for the new signal to operate it must not only be distinguishable from the parent signal, but also be different enough to prevent interference with signals from other species in the habitat. If this is not the case, the population will probably not diverge to the point of becoming a distinct species, because there is little possibility for additional, slight shifts to occur in the isomeric ratio, or the complement of major components, of the blend.

Suppose, however, that the signal *is* compatible with the new environment, giving rise to a stable communication system. We can then ask: Is it possible for additional changes to occur in the production of minor components, and perhaps in the way they are perceived, as a result of secondary contacts and the associated changes in the impact of a noisy habitat? We believe this is the case and that the example involving the loss of a minor component is particularly interesting, because it provides a potential explanation for the evolution of a behavioral antagonist. The reason for this, as was noted above, is that the male and female elements of the signaling system are not genetically linked in some species, allowing males to retain the receptor-mediated pathway to the brain for a component that was lost. An example of this may be found in *O. nubilalis*. Glover et al. (1989) showed in flight-tunnel tests with a series of related monounsaturated 12- and 14-carbon acetates that the upwind flight response of males of each race was most dramatically affected in a negative way by the presence of 1 percent Z9-14:OAc added to the normal Z/E11-14:OAc ratios for each race. In addition, Hansson et al. (1987) showed that males possess specific antennal receptors for the Z9-14:OAc component, even though it is not a pheromone component. The most interesting point, however, is that whereas females do not produce the Z9-14:OAc components as part of the natural pheromone, they do possess a large quantity of the intermediate precursor, Z11-16:Acyl, which by a chain-shortening step could be converted to Z9-14:Acyl. Female *O. nubilalis* may have lost the ability to carry out this chain-shortening step, but males still have the ability to perceive this component. We do not know whether this compound was once a part of the *O. nubilalis* pheromone, but the information above suggests this as a possibility.

How then, does the "vestigial," or "neutral," pheromone receptor pathway become a route for antagonistic input? Possibilities can be proposed, based on a consideration of the way in which processing of olfactory information occurs in the insect brain (Christiansen and Hildebrand 1987a). Pheromone components interact with specialized receptor molecules on the male antennae, activating neural pathways that go directly into the deutocerebrum. First-level convergence and processing occurs in a specialized structure, the macroglomerular complex, with local interneurons making connections with other individual glomeruli in the antennal lobe, and projection interneurons then carrying information out of the deutocerebrum to higher brain centers. It is well established that GABA-mediated inhibitory synaptic connections are prevalent in the local interneuronal pathways in the deutocerebrum (Waldrop et al. 1987). It also is known in the tomato hornworm, *Manduca sexta,* that of the several types of projection interneurons one, in particular, responds to the pheromone blend in an off-on-off manner when the odor signal is pulsed onto the antennae, and it has been proposed that this reflects an interaction between the activity of inhibitory and excitatory pathways in the deutocerebrum (Christiansen and Hildebrand 1987b). The importance of this finding lies in the fact that sustained upwind-oriented flight in male moths is dependent on detection of pulses of pheromone as they exist in the filamentous structure of an odor plume (see Baker 1985, 1989c).

Based on this information concerning neural processing, Baker (1989a), using data from Akers and O'Connell (1988) and Baker et al. (1976) for *A. velutinana,* has proposed that if inputs from specific pheromone receptor pathways differentially affect the proportion of excitatory and inhibitory pathways in the deutocerebrum, this could result in changes in the observed output of the projection interneurons. Off-blends would result in lower or higher proportions of inhibitory inputs, resulting in an inability to track the filamentous structure of the odor plume and loss of output in the projection interneurons. Given this, we would need only to propose that a potential behavioral antagonist be a minor component that now contributes sensory input to these inhibitory pathways, possibly as a result of a (gradual) change in threshold for initiation of action potentials.

Alternatively, the effect of the antagonist component could occur directly in a brain "recognition center," presumably in the protocerebrum (Roelofs and Glover 1990). In this case, sensory profiles that included the antagonist would be "interpreted" by the brain recognition center as being different from the profile for the conspecific blend, resulting in males not flying upwind to heterospecific females. One point of interest in support of this idea concerns the fact that perception of the pheromone blend by male *O. nubilalis,* leading to completed flights to a source in the flight tunnel, is genetically controlled by a sex-linked factor (Roelofs et al. 1987). Since there are numerous projection neurons, of several types, sending information to higher brain centers, and all capable of being affected by deutocerebral pathways, this finding supports the hypothesis that a specific neuronal site in the central nervous system (CNS) to which all of these neurons send projections may be involved in ultimately controlling male perception of and response to multicomponent blends.

The hypothesis that minor components may exert differential effects over time as a result of secondary contact also is supported by the presence of redundant components in the signal. Production of these "extra" compounds represents a very low energetic commitment on the part of females because the compounds are present in low amounts and are already part of the biosynthetic pathway. A change in the sensitivity of males to a particular component, as the result of interference in the form of noise, could alter blend specificity while maintaining optimal male thresholds. One example from the pheromone literature that has been proposed as evidence for this concerns the tortricid moth *Archips argyrospilus* (Cardé et al. 1977; Cardé 1986). This species uses a four-component blend composed of a 90:10 ratio of Z/E11-14:OAc, with Z9-14:OAc and 12:OAc. Field tests with this species in New York show that the Z9-14:OAc component is necessary for peak attraction, whereas in British Columbia this was not the case, even though females in each region produce the component. The enhanced sensitivity to Z9-14:OAc in the New York populations may be the result of the higher density of tortricines in this region, and thus the requirement for a more specific signal in the more "noisy" habitat. The Z9-14:OAc component may, in conjunction with 12:OAc, be thought of as a redundant component in the pathway that can vary in its impact on male behavior as a result of fluctuations in selection pressures for a specific signal. Note again that this requires no change in the females, as they are already producing the small amount of this minor component as a consequence of the pathway for production of the Z/E11-14:OAc ratio (Roelofs and Bjostad 1984). Rather, the presence of redundant components allows for a "fine-tuning" of the recognition system in males to prevent interference between these species.

In the above discussion we have emphasized change in the female production of a new pheromone blend, assuming subsequent change in male perception to achieve a stable recognition system. The question of how male perception of odor quality is dramatically altered so that females who have changed from a 90:10 to a 10:90 ratio can attract mates is more problematic, because much less is known about the higher CNS elements involved in blend perception compared to our understanding of female biosynthesis. The studies with *O. nubilalis* indicate that a single sex-linked factor influences this aspect of the system, but the exact nature of this factor is unknown and should be the subject of intense study in the future (see Löfstedt 1990).

Summary

As we noted at the beginning of the last section, the evolutionary scenario provided above is intended to focus on what we believe, at the present time, are important points that must be addressed by the Recognition Concept. The first is the problem of a noisy environment, and in general the question of what factors in the "habitat" are involved in influencing the formation of a new recognition system. A reading of the literature on the Recognition Concept does not provide a great deal of detail concerning this question, and in our opinion it is only one of a number of points that needs more development. The second point concerns the role of intraspecific competition in influencing the development of the signaling system. There is at present a great body of literature concerning the role of sexual selection on mating systems, and again this is dealt with in only a cursory manner by the Recognition Concept. The third point concerns the difficulties with the necessity that changes in the recognition system during species formation occur in a gradual manner. The fourth point concerns the potential for modification, or fine-tuning, in the pheromone blend by secondary contact (by, for example, the evolution of interspecific behavioral antagonists). We recognize and acknowledge the difficulties with reinforcement and character displacement models, as well as problems with models involving interspecific competition. However, we believe that the evidence for the action of multicomponent blends, redundant components, and behavioral antagonists constitutes an excellent test case for reinforcement and reproductive character displacement. A final point, relating to all of the above, concerns the importance of variation in the mating phenotype.

It is our belief that at the present time there are a number of systems, including those involving moth pheromone communication, that offer excellent opportunities to examine the issues raised above, as well as others involved in the debate between the Biological Species and Recognition concepts. What is needed, however, is a more detailed account of the ideas encompassed by the Recognition Concept, so that the appropriate experiments can be designed.

Acknowledgments

This project began as a consequence of several stimulating discussions with Dr. David Lambert, in which it became obvious to at least one of us (CEL) that a number of "paradigms" in evolutionary biology that we had taken for granted needed greater consideration. Further investigation revealed not only that this would be beneficial in terms of our own personal understanding, but also that it could impact on our approach to research on sex pheromones. If nothing else, we feel that a sharper focus has been achieved. We would also thank Drs. Thomas Baker, Stephen Foster, Thomas Glover, and Russell Jurenka and Ralph Charlton for their thorough and critical reviews.

References

Akers, R.P., and R.J. O'Connell. 1988. The contribution of olfactory receptor neurons to the perception of pheromone component ratios in male redbanded leafroller moths. J. Comp. Physiol. A 163:641–650.

Anglade, P., J. Stockel, and I.W.G.O. Cooperators. 1984. Intraspecific sex pheromone variability in the European corn borer, *Ostrinia nubilalis* (Lepidoptera, Pyralidae). Agronomie 4:183–187.

Arn, H., P. Esbjerg, R. Bues, M. Tòth, G. Szöcs, P. Guerin, and S. Rauscher. 1983. Field attraction of *Agrotis segetum* males in four European countries to mixtures containing three homologous acetates. J. Chem. Ecol. 9:267–276.

Ayala, F.J., M.L. Tracey, D. Hedgecock, and R.C. Richmond. 1974. Genetic differentiation during the speciation process in *Drosophila*. Evolution 28:576–592.

Baker, T.C. 1985. Insect mating and courtship behavior. Vol. 9. Pp. 621–672 in G.A. Kerkut and L.I. Gilbert, eds. Comprehensive insect physiology, biochemistry, and pharmacology. Pergamon Press, New York.

———. 1989a. Sex pheromone communication in the Lepidoptera: New research progress. Experientia, Basel 45:248–262.

———. 1989b. Origin of courtship and sex pheromones of the oriental fruit moth and a discussion of the role of phytochemicals in the evolution of lepidopteran male scents. Pp. 401–418 in C.H. Chou and G.R. Waller, eds. Phytochemical ecology: Allelochemicals, mycotoxins and insect pheromones and allomones. Institute of Botany, Academia Sinica Monograph Series No. 9, Taipei, China.

———. 1989c. Pheromones and flight behavior. Pp. 231–255 in G.G. Goldsworthy and C. Wheeler, eds. Insect flight. CRC Press, Boca Raton, Fla.

Baker, T.C., R.T. Cardé, and W.L. Roelofs. 1976. Behavioral responses of male *Argyrotaenia velutinana* (Lepidoptera: Tortricidae) to components of its pheromone. J. Chem. Ecol. 2:333–352.

Baker, T.C., B.S. Hansson, C. Löfstedt, and J. Löfqvist. 1988. Adaptation of antennal neurons in moths is associated with cessation of pheromone-mediated upwind flight. Proc. Natl. Acad. Sci. 85:9826–9830.

Baker, T.C., W.L. Meyer, and W.L. Roelofs. 1981. Sex pheromone dosage and blend specificity of response by Oriental fruit moths. Entomol. Expt. Appl. 30:269–279.

Barrer, P.M., M.J. Lacey, and A. Shani. 1987. Variation in relative quantities of airborne sex pheromone components from individual female *Ephestia cautella* (Lepidoptera: Pyralidae). J. Chem. Ecol. 13:639–653.

Barton, N.H., and B. Charlesworth. 1984. Genetic revolutions, founder effects, and speciation. Annu. Rev. Ecol. Syst. 15:133–164.

Barton, N.H., and G.M. Hewitt. 1989. Adaptation, speciation and hybrid zones. Nature 341:497–503.

Becker, P.H. 1982. The coding of species-specific characteristics in bird sounds. Pp. 213–252 in D.E. Kroodsma and E.H. Miller, eds. Acoustic communication in birds. Vol. 1. Academic Press, New York.

Bell, W.J., and R.T. Cardé, eds. 1984. Chemical ecology of insects. Sinauer, Sunderland, Mass.

Birch, M.C., G.M. Poppy, and T.C. Baker. 1990. Scents and eversable scent structures of male moths. Annu. Rev. Entomol. 35:25–58.
Bjostad, L.B. 1989. Chemical characterization of sex pheromones and their biosynthetic intermediates. Chem. Sen. 14:411–420.
Bjostad, L.B., C.E. Linn, J.-W. Du, and W.L. Roelofs. 1984. Identification of new sex pheromone components in *Trichoplusia ni*, predicted from biosynthetic precursors. J. Chem. Ecol. 10:1309–1323.
Blum, M.S., and N.A. Blum. 1979. Sexual selection and reproductive competition in insects. Academic Press, New York.
Boake, C.R.B. 1991. Coevolution of senders and receivers of sexual signals: Genetic coupling and genetic correlation. Trends Ecol. Evol. 6:225–227.
Butlin, R.K. 1987a. Speciation by reinforcement. Trends Ecol. Evol. 2:8–13.
———. 1987b. Species, speciation, and reinforcement. Am. Nat. 130:461–464.
Butlin, R.K., and M.G. Ritchie. 1989. Genetic coupling in mate recognition systems: What is the evidence? Biol. J. Linn. Soc. 37:237–246.
Caffrey, D.J., and L.H. Worthly. 1927. A progress report on the investigations of the European corn borer. U.S. Dept. Agric. Tech. Bull. 1476.
Cardé, A.M., T.C. Baker, and R.T. Cardé. 1979. Identification of a four-component sex pheromone of the female Oriental fruit moth, *Grapholitha molesta* (Lepidoptera: Tortricidae). J. Chem. Ecol. 5:423–427.
Cardé, R.T. 1979. Behavioral responses of moths to female-produced pheromones and the utilization of attractant-baited traps for population monitoring. Pp. 286–315 *in* R.L. Rabb and G.G. Kennedy, eds. Movement of highly mobile insects: Concepts and methodology in research. University Graphics, North Carolina State University, Raleigh, N.C.
———. 1986. The role of pheromones in reproductive isolation and speciation in insects. Pp. 303–317 *in* M.D. Huettel, ed. Evolutionary genetics of invertebrate behavior. Plenum Press, New York.
Cardé, R.T., and T.C. Baker. 1984. Sexual communication in insects. Pp. 355–386 *in* W. Bell and R.T. Cardé, eds. Chemical ecology of insects. Sinauer, Sunderland, Mass.
Cardé, R.T., T.C. Baker, and W.L. Roelofs. 1976. Sex attractant responses of male oriental fruit moths to a range of component ratios: Pheromone polymorphism? Experientia, Basel 32:1406–1407.
Cardé, R.T., A.M. Cardé, A.S. Hill, and W.L. Roelofs. 1977. Sex pheromone specificity as a reproductive isolating mechanism among the sibling species *Archips argyrospilus* and *A. mortuanus* and other sympatric tortricine moths (Lepidoptera: Tortricidae). J. Chem. Ecol. 3:71–84.
Cardé, R.T., and R. Webster. 1981. Endogenous and exogenous factors controlling insect sex pheromone production and responsiveness, particularly among the Lepidoptera. Pp. 978–991 *in* Scientific papers of the Institute of Organic and Physical Chemistry of Wroclow Technical University. No. 22, Conference 7.
Carson, H.L. 1982. Speciation as a major reorganization of polygenic balances. Pp. 411–433 *in* C. Barigozzi, ed. Mechanisms of speciation. Liss, New York.
———. 1987. Population genetics, evolutionary rates and neo-Darwinism. Pp. 209–217 *in* K.S.W. Campbell and M.F. Day, eds. Rates of evolution. Allen and Unwin, London.
Chapman, P.J., and S.E. Lienk. 1971. Tortricid fauna of apple in New York (Lepidoptera: Tortricidae); including an account of apples' occurrence in the state, especially as a naturalizing plant. Spec. Publ. N.Y.S. Agric. Exp. Sta. Geneva, New York.

Christensen, T.A., and J.G. Hildebrand. 1987a. Functions, organization, and physiology of the olfactory pathways in the lepidopteran brain. Pp. 457–483 in A.P. Gupta, ed. Arthropod brain: Its evolution, development, structure, and functions. John Wiley, New York.

———. 1987b. Male-specific sex pheromone-selective projection neurons in the antennal lobes of the moth *Manduca sexta*. J. Comp. Physiol. A 160:553–569.

Clearwater, J.R., S.P. Foster, S.J. Muggleston, J.S. Dugdale, and E. Priesner. 1991. Intraspecific variation and interspecific differences in the sex pheromones of sibling species in the *Ctenopseustis obliquana* Complex. J. Chem. Ecol. 17:413–429.

Collins, R.D., and R.T. Cardé. 1985. Variation in and heritability of aspects of pheromone production in the pink bollworm moth, *Pectinophora gossypiella* (Lepidoptera: Gelechiidae). Ann. Entomol. Soc. Am. 78:229–234.

———. 1989a. Selection for altered pheromone-component ratios in the pink bollworm moth, *Pectinophora gossypiella* (Lepidoptera: Gelechiidae). J. Insect Behav. 2:609–621.

———. 1989b. Heritable variation in pheromone response of the pink bollworm, *Pectinophora gossypiella* (Lepidoptera: Gelechiidae). J. Chem. Ecol. 15:2647–2659.

Coyne, J.A., and H.A. Orr. 1989. Patterns of speciation in *Drosophila*. Evolution 43:362–381.

Coyne, J.A., H.A. Orr, and D.J. Futuyma. 1988. Do we need a new species concept? Syst. Zool. 37:190–200.

Dobzhansky, T. 1970. Genetics of the evolutionary process. Columbia University Press, New York.

Doherty, J.A. 1985. Trade-off phenomena in calling song recognition and phonotaxis in the cricket, *Gryllus bimaculatus* (Orthoptera, Gryllidae). J. Comp. Physiol. A 156:787–801.

Dugdale, J.S. 1966. A new genus for the New Zealand "elusive *Tortrix*" (Lepidoptera: Tortricidae: Tortricinae). N. Z. J. Sci. 9:391–398.

———. 1990. Reassessment of *Ctenopseustis* Meyrick and *Planotortrix* Dugdale with descriptions of two new genera (Lepidoptera: Tortricidae). N. Z. J. Zool. 17:437–465.

Dunkelblum, E., R. Snir, S. Gothilf, and I. Harpaz. 1987. Identification of sex pheromone components from pheromone gland volatiles of the tomato looper, *Plusia chalcites* (Esp). J. Chem. Ecol. 13:991–1003.

Du, J.-W., C. Löfstedt, and J., Löfqvist. 1987. Repeatability of pheromone emissions from individual female ermine moths *Yponomeuta padellus* and *Y. rorellus*. J. Chem. Ecol. 13:1431–1441.

Eberhard, W.G. 1985. Sexual selection and animal genetalia. Harvard University Press, Cambridge, Mass.

———. 1990. Animal genitalia and female choice. Am. Sci. 78:134–141.

Eisner, T., and J. Meinwald. 1987. Alkaloid-derived pheromones and sexual selection in Lepidoptera. Pp. 251–269 in G.D. Prestwich and G.L. Blomquist, eds. Pheromone biochemistry. Academic Press, New York.

Eldredge, N., and S.J. Gould. 1972. Punctuated equilibria: An alternative to phyletic gradualism. Pp. 82–115 in T.J.M. Schopf, ed. Models in paleobiology. W.H. Freeman, San Francisco.

Fabian, B. 1985. Ontogenetic explorations into the nature of evolutionary change. Pp. 77–85 in E.S. Vrba, ed. Species and speciation. Transvaal Museum Monograph No 4. Pretoria.

Fisher, R.A. 1958. The genetical theory of natural selection. 2d ed. Dover, New York. First published in 1930.

Flint, H.M., M. Balsubramanian, J. Campero, A.F. Strickland, A. Ahmad, J. Barral, S. Barosa, and A.F. Khail. 1979. Pink bollworm moth: Response of native moths to ratios of (Z,Z)- and (Z,E)-isomers of gossyplure in several cotton growing regions of the world. J. Econ. Entomol. 72:758–762.

Flint, H.M., R.L. Smith, D.E. Forey, and B.R. Horn. 1977. Pink bollworm: Response of males to (Z,Z)- and (Z,E)- isomers of gossyplure. J. Econ. Entomol. 71:664–666.

Foster, S.P., J.R. Clearwater, and S.J. Muggleston. 1989. Intraspecific variation of two components in sex pheromone gland of *Planotortrix excessana* sibling species. J. Chem. Ecol. 15:457–465.

Foster, S.P., J.R. Clearwater, S.J. Muggleston, and J.S. Dugdale. 1986. Probable sibling species complexes within two described New Zealand leafroller moths. Naturwissenschaften 73:156–158.

Foster, S.P., J.R. Clearwater, S.J. Muggleston, and P.W. Shaw. 1990. Sex pheromone of a *Planotortrix excessana* sibling species and reinvestigation of related species. J. Chem. Ecol. 16:2461–2474.

Foster, S.P., J.R. Clearwater, and W.L. Roelofs. 1987. Sex pheromone of *Planotortrix* species found on mangrove. J. Chem. Ecol. 13:631–637.

Foster, S.P., J.S. Dugdale, and C.S. White. 1991. Sex pheromones and the status of greenheaded and brownheaded leafroller moths in New Zealand. N. Z. J. Zool. 18:63–74.

Foster, S.P., and W.L. Roelofs. 1987. Sex pheromone differences in populations of the brownheaded leafroller, *Ctenopseustis obliquana*. J. Chem. Ecol. 13:623–629.

———. 1988. Sex pheromone biosynthesis in the leafroller moth *Planotortrix excessana* by Δ10 desaturation. Arch. Ins. Biochem. Physiol. 8:1–9.

Futuyma, D.J. 1986. Evolutionary biology. 2d ed. Sinauer, Sunderland, Mass.

Glover, T.J., M.G. Campbell, C.E. Linn, and W.L. Roelofs. 1991a. Unique sex-chromosome mediated behavioral-response specificity of hybrid male European corn borer moths. Experientia, Basel 47:980–984.

Glover, T.J., M.G. Campbell, and W.L. Roelofs. 1990. Sex-linked control of sex pheromone behavioral responses in European corn borer moths confirmed with TPI marker gene. Arch. Insect Biochem. Physiol. 15:67–77.

Glover, T.J., J.J. Knodel, P.S. Robbins, C.J. Eckenrode, and W.L. Roelofs. 1991b. Gene flow among three races of European corn borers (Lepidoptera, Pyralidae) in New York State. Environ. Entomol. 20:1356–1362.

Glover, T.J., N. Perez, and W.L. Roelofs. 1989. Comparative analysis of sex pheromone response antagonists in three races of European corn borer. J. Chem. Ecol. 15:863–873.

Glover, T.J., and W.L. Roelofs. 1988. Genetics of lepidopteran sex pheromone systems. ISI Atlas of Science: Anim. Plant Sci. 1:279–282.

Glover, T.J., X.-H. Tang, and W.L. Roelofs. 1987. Sex pheromone blend discrimination by male moths from E and Z strains of European corn borer. J. Chem. Ecol. 13:143–151.

Grant, A.J., R.J. O'Connell, and A.M. Hammond. 1988. A comparative study of pheromone perception in two species of noctuid moths. J. Insect Behav. 1:75–96.

Green, C.J., and J.S. Dugdale. 1982. Review of the genus *Ctenopseustis* Meyrick, with reinstatement of two species. N. Z. J. Zool. 9:427–436.

Greenfield, M.D. 1981. Moth sex pheromones: An evolutionary perspective. Florida Entomol. 64:4–17.

Greenfield, M.D., and M.G. Karandinos. 1979. Resource partitioning of the sex communication channel in clearwing moths (Lepidoptera: Sesiidae) in Wisconsin. Ecol. Monogr. 49:403–426.

Grula, J.W., and O.R. Taylor. 1980. The effect of X-chromosome inheritance on mate-selection behavior in the sulfur butterflies, *Colias eurytheme* and *C. philodice*. Evolution 34:688–699.

Guerin, P.M., H. Arn, H.R. Buser, and P.J. Charmillot. 1986. Sex pheromone of *Adoxophyses orana*: Additional components and variability in ratio of (Z)-9- and (Z)-11-tetradecenyl acetate. J. Chem. Ecol. 12:763–772.

Hansson, B.S., C. Löfstedt, and S.P. Foster. 1989. Z-linked inheritance of male olfactory response to sex pheromone components in two species of tortricid moths, *Ctenopseustis obliquana* and *Ctenopseustis* sp. Entomol. Exp. Appl. 53:137–145.

Hansson, B.S., C. Löfstedt, and W.L. Roelofs. 1987. Mendelian inheritance of sex pheromone reception in male European corn borers. Naturwissenschaften 74:497–499.

Hansson, B.S., M. Tòth. C. Löfstedt, G. Szöcs, M. Subchev, and J. Löfqvist. 1990. Pheromone variation among eastern European and a western Asian population of the turnip moth *Agrotis segetum*. J. Chem. Ecol. 16:1611–1622.

Harper, A.A., and D.M. Lambert. 1983. The population genetics of reinforcing selection. Genetica 62:15–23.

Haynes, K.F., and T.C. Baker. 1988. Potential for evolution of resistance to pheromones: Worldwide and local variation in chemical communication system of pink bollworm moth, *Pectinophora gossypiella*. J. Chem. Ecol. 14:1547–1560.

Haynes, K.F., and R.E. Hunt. 1990a. Interpopulational variation in emitted pheromone blend of the cabbage looper moth, *Trichoplusia ni*. J. Chem. Ecol. 16:509–519.

———. 1990b. A mutation in the pheromonal communication system of the cabbage looper moth, *Trichoplusia ni*. J. Chem. Ecol. 16:1249–1257.

Haynes, K.F., L.K. Gaston, M.M. Pope, and T.C. Baker. 1984. Potential for evolution of resistance to pheromones: Interindividual and interpopulational variation in chemical communication system of the pink bollworm moth. J. Chem. Ecol. 10:1551–1565.

Hendriske, A. 1986. Intra- and interspecific sex pheromone communication in the genus *Yponomeuta*. Physiol. Entomol. 11:159–169.

Hewitt, G.M. 1988. Hybrid zones—natural laboratories for evolutionary studies. Trends Ecol. Evol. 3:158–167.

Hill, A.S., and W.L. Roelofs. 1979. Sex pheromone components of the oblique-banded leafroller moth *Choristoneura rosaceana*. J. Chem. Ecol. 5:3–11.

Hughes, A.J., and D.M. Lambert. 1984. Functionalism, structuralism, and "Ways of Seeing." J. Theor. Biol. 111:787–800.

Kemp, A.C. 1985. Individual and population expectations related to species and speciation: The case of raptorial birds. Pp. 59–69 *in* E.S. Vrba, ed. Species and speciation. Transvaal Museum Monograph No. 4. Pretoria.

Klun, J.A., et al. 1975. Insect sex pheromones: Intraspecific pheromonal variability of *Ostrinia nubilalis* in North America and Europe. Environ. Entomol. 4:891–894.

Klun, J.A., and S. Maini. 1979. Genetic basis of an insect chemical communication system: The European corn borer. Environ. Entomol. 8:423–426.

Krasnoff, S.B., and W.L. Roelofs. 1990. Evolutionary trends in the male pheromone systems of arctiid moths: Evidence from studies of courtship in *Phragmatobia fuliginosa* and *Pyrrharctia isabella* (Lepidoptera: Arctiidae). Zool. J. Linn. Soc. 99:319–338.

Lambert, D.M., and B. Levey. 1979. The use of discriminant function analysis to investigate the design features of specific mate recognition systems. Presented to the Zoological Society of Southern Africa Symposium on Animal Communication. Cape Town, South Africa.

Lambert, D.M., B. Michaux, and C.S. White. 1987. Are species self-defining? Syst. Zool. 36:196–205.

Lambert, D.M., and H.E.H. Paterson. 1984. On "Bridging the gap between race and species": The Isolation Concept and an alternative. Proc. Linn. Soc. N.S.W. 107:501–514.

Linn, C.E., L.B. Bjostad, J.-W. Du, and W.L. Roelofs. 1984. Redundancy in a chemical signal: Behavioral responses of male *Trichoplusia ni* to a 6-component sex pheromone blend. J. Chem. Ecol. 10:1635–1658.

Linn, C.E., M.G. Campbell, and W.L. Roelofs. 1985. Male moth sensitivity to multicomponent pheromones: the critical role of the female-released blend in determining the functional role of components and the active space of the pheromone. J. Chem. Ecol. 12:659–668.

———. 1987. Pheromone components and active spaces: What do male moths smell and where do they smell it? Science 237:650–652.

———. 1988. Temperature modulation of behavioral thresholds controlling male moth sex pheromone response specificity. Physiol. Entomol. 13:59–67.

———. 1991. The effects of different blend ratios and temperature on the active space of the oriental fruit moth sex pheromone. Physiol. Entomol. 16:211–222.

Linn, C.E., A. Hammond, J.-W. Du, and W.L. Roelofs. 1988. Specificity of male response to multicomponent pheromones in noctuid moths *Trichoplusia ni* and *Pseudoplusia includens*. J. Chem. Ecol. 14:47–57.

Linn, C.E., and W.L. Roelofs. 1983. Effect of varying proportions of the alcohol component on sex pheromone blend discrimination in male oriental fruit moths. Physiol. Entomol. 8:291–306.

———. 1987. Effects of octopaminergic agents on the sex pheromone mediated behavior of adult lepidoptera. Pp. 162–173 in R.M. Hollingworth and M.B. Green, eds. Sites of action for neurotoxic pesticides. ACS symposium series No. 356. American Chemical Society.

———. 1989. Response specificity of male moths to multicomponent pheromones. Chem. Sen. 14:421–437.

Linn, C.E. Jr., and W.L. Roelofs. 1985. Response specificity of male pink bollworm moths to different blends and dosages of sex pheromone. J. Chem. Ecol. 11:1583–1590.

Löfstedt, C. 1990. Population variation in moth pheromone communication systems and its genetic control. Entomol. Exp. Appl. 54:199–218.

Löfstedt, C., A. Elmfors, M. Sjögren, and E. Wijk. 1986a. Confirmation of sex pheromone biosynthesis from (16-D_3) palmitic acid in the turnip moth using capillary gas chromatography. Experientia, Basel 42:1059–1061.

Löfstedt, C., B.S. Hansson, W.L. Roelofs, and B.O. Bengtsson. 1988. The linkage relationship between factors controlling female pheromone production and male olfactory response in the European corn borer *Ostrinia nubilalis*. In Reproductive isolation by sex pheromones in some moth species. Thesis. B.S. Hansson, Lund University, Lund, Sweden.

Löfstedt, C., B.S. Lanne, J. Löfqvist, M. Appelgren, and G. Bergstrom. 1985b. Individual variation in the pheromone of the turnip moth, *Agrotis segetum*. J. Chem. Ecol. 11:1181–1196.

Löfstedt, C., C.E. Linn, and J. Löfqvist. 1985a. Behavioral responses of male turnip moths, *Agrotis segetum*, to sex pheromone in a flight tunnel and in the field. J. Chem. Ecol. 11:1209–1221.

Löfstedt, C., J. Löfqvist, B.S. Lanne, J.N.C. Van der Pers, and B.S. Hansson. 1986b. Pheromone dialects in European turnip moths *Agrotis segetum*. Oikos 46:250–257.

Löfstedt, C., and J.N.C. Van der Pers. 1985. Sex pheromones and reproductive isolation in four European small ermine moths. J. Chem. Ecol. 11:649–666.

Matthews, R.W., and J.R. Matthews. 1978. Insect behavior. John Wiley, New York.

Mayr, E. 1970. Populations, species, and evolution. Harvard University Press, Cambridge, Mass.

McNeil, J. N. 1990. Behavioral ecology of pheromone-mediated communication in moths and its importance in the use of pheromone traps. Annu. Rev. Entomol. 36:407–430.

Menken, S.B.J. 1982. Biochemical genetics and systematics of small ermine moths. Z. Zool. Syst. Evol. 20:131–143.

Miller, E.H. 1982. Character and variance shift in acoustic signals in birds. Pp. 253–295 in D.E. Kroodsma and E.H. Miller, eds. Acoustic communication in birds. Vol. 1. Academic Press, New York.

Miller, J.R., and W.L. Roelofs. 1980. Individual variation in sex pheromone component ratios in two populations of the redbanded leafroller moth, *Argyrotaenia velutinana*. Environ. Entomol. 9:359–363.

Morris, G.K., and J.H. Fullard. 1983. Random noise and congeneric discrimination in *Conocephalus* (Orthoptera: Tettigoniidae). Pp. 73–96 in D.T. Gwynne and G.K. Morris, eds. Orthopteran mating systems: Sexual competition in a diverse group of insects. Westview Press, Boulder, Colo.

Ono, T., R.E. Charlton, and R.T. Cardé. 1990. Variability in pheromone composition and periodicity of pheromone titer in the potato tuberworm moth, *Phthorimaea operculella*. J. Chem. Ecol. 16:531–542.

Paterson, H.E.H. 1978. More evidence against speciation by reinforcement. S. Afr. J. Sci. 74:369–371.

———. 1980. A comment on "mate recognition systems." Evolution 34:330–331.

———. 1981. The continuing search for the unknown and unknowable: A critique of contemporary ideas on speciation. S. Afr. J. Sci. 77:113–119.

———. 1982a. Perspective on speciation by reinforcement. S. Afr. J. Sci. 78:53–57.

———. 1982b. Darwin and the origin of species. S. Afr. J. Sci. 78:272–275.

———. 1985. The recognition concept of species. Pp. 21–29 in E.S. Vrba, ed. Species and speciation. Transvaal Museum Monograph No. 4. Pretoria.

Raina, A.K., H. Jaffe, T.G. Kempoe, P. Keim, R.W. Blacher, H.M. Fales, C.T. Riley, J.A. Klun, R.L. Ridgway, and D.K. Hayes. 1989. Identification of a neuropeptide hormone that regulates sex pheromone production in female moths. Science 244:796–798.

Raina, A.K., and J.J. Menn. 1987. Endocrine regulation of pheromone production in Lepidoptera. Pp. 159–174 in G.D. Prestwich and G.J. Blomquist, eds. Pheromone biochemistry. Academic Press, Orlando, Fla.

Ritchie, M.G., R.K. Butlin, and G.M. Hewitt. 1989. Assortative mating across a hybrid zone in *Chorthippus parallelus* (Orthoptera: Acrididae). J. Evol. Biol. 2:339–352.

Roelofs, W.L., and L.B. Bjostad. 1984. Biosynthesis of lepidopteran pheromones. Bioorg. Chem. 12:279–298.

Roelofs, W.L., and R.L. Brown. 1982. Pheromones and evolutionary relationships of Tortricidae. Annu. Rev. Ecol. Syst. 13:395–422.

Roelofs, W.L., J.-W. Du, C.E. Linn, T.J. Glover, and L.B. Bjostad. 1986. The potential for genetic manipulation of the redbanded leafroller moth sex pheromone. Pp. 263–272 in M.D. Huettel, ed. Evolutionary genetics of invertebrate behavior. Plenum Press, New York.

Roelofs, W.L., J.-W. Du, X.-H. Tang, P.S. Robbins, and C.J. Eckenrode. 1985. Three European corn borer populations in New York based on sex pheromones and voltinism. J. Chem. Ecol. 7:829–836.

Roelofs, W.L., and T.J. Glover. 1990. Genetics of a moth sex pheromone system. Pp. 279–282 in C.J. Wysocki and M.R. Kare, eds. Genetic variation in chemical sensing systems. Marcel Dekker, New York.

Roelofs, W., T. Glover, X.-H. Tang, I. Sreng, P. Robbins, C. Eckenrode, C. Löfstedt, B.S. Hansson, and B.O. Bengtsson. 1987. Sex pheromone production and perception in European corn borer moth is determined by both autosomal and sex-linked genes. Proc. Natl. Acad. Sci. 84:7585–7589.

Roelofs, W.L., A. Hill, and R. Cardé. 1975. Sex pheromone components of the redbanded leafroller, *Argyrotaenia velutinana*. J. Chem. Ecol. 1:83–89.

Roelofs, W.L., and W.A. Wolf. 1988. Pheromone biosynthesis in Lepidoptera. J. Chem. Ecol. 14:2019–2031.

Ryan, M.J. 1990. Signals, species, and sexual selection. Am. Sci. 78:46–72.

Schlyter, F., and G. Birgersson. 1989. Individual variation in bark beetle and moth pheromones—a comparison and an evolutionary background. Hol. Ecol. 12:457–465.

Sebeok, T.A. 1977. How animals communicate. Indiana University Press, Bloomington.

Shannon, C.E., and W. Weaver. 1949. The mathematical theory of communication. University of Illinois Press, Urbana.

Slatkin, M. 1987. Gene flow and the geographic structure of natural populations. Science 236:787–792.

Smith, W.J. 1977. The behavior of communicating. Harvard University Press, Cambridge, Mass.

Spangler, H.G. 1985. Sound production and communication by the greater wax moth (Lepidoptera: Pyralidae). Ann. Entomol. Soc. Am. 78:54–61.

Spencer, H.G., B.H. McArdle, and D.M. Lambert. 1986. A theoretical investigation of speciation by reinforcement. Am. Nat. 128:241–262.

Sreng, I., T.J. Glover, and W.L. Roelofs. 1989. Canalization of the redbanded leafroller moth sex pheromone blend. Arch. Insect Biochem. Physiol. 10:73–82.

Tamaki, Y. 1985. Sex pheromones. Pp. 145–191 in G.A. Kerkut and L.S. Gilbert, eds. Comprehensive insect physiology, biochemistry, and pharmacology. Vol. 9. Pergamon Press, New York.

Tang, J.D., R.E. Charlton, R.A. Jurenka, W.A. Wolf, P.L. Phelan, L. Sreng, and W.L. Roelofs. 1989. Regulation of pheromone biosynthesis by a brain hormone in two moth species. Proc. Natl. Acad. Sci. 86:1806–1810.

Teal, P.E.A., J.H. Tumlinson, and H. Oberlander. 1989. Neural regulation of sex pheromone biosynthesis in *Heliothis* moths. Proc. Natl. Acad. Sci. 86:2488–2492.

Templeton, A.R. 1981. Mechanisms of speciation—a population genetic approach. Annu. Rev. Ecol. Syst. 12:23–48.

Thomson, D.R., N.P.D. Angerilli, C. Vincent, and A.P. Gaunce. 1991. Evidence for regional differences in the response of obliquebanded leafroller (*Choristoneura rosaceana* [Lepidoptera: Tortricidae]) to sex pheromone blends. Environ. Entomol. 20:935–938.

Thornhill, R., and J. Alcock. 1983. The evolution of insect mating systems. Harvard University Press, Cambridge, Mass.

Todd, J.L., K.F. Haynes, and T.C. Baker. 1992. Antennal neurones specific for redundant pheromone components in normal and mutant *Trichoplusia ni* males. Physiol. Entomol. 17:183–192.

Vakenti, J.M., A.P. Gaunce, K.N. Slessor, G.G.S. King, S.A. Allan, H.F. Madsen, and J.H. Borden. 1988. Sex pheromone components of the oblique-banded leafroller moth, *Choristoneura rosaceana*, in the Okanagan Valley of British Columbia. J. Chem. Ecol. 14:605–621.

Van der Pers, J.N.C., and C. Löfstedt. 1986. Signal-response relationships in sex pheromone communication. Pp. 235–242 *in* T.L. Payne, M.C. Birch and C.E.J. Kennedy, eds. Mechanisms of insect olfaction. Clarendon Press, Oxford.

Vrba, E.S. 1985. Introductory comments on species and speciation. Pp. ix–xviii *in* E.S. Vrba, ed. Species and speciation. Transvaal Museum Monograph No. 4. Pretoria.

Waldrop, B., T.A. Christensen, and J.G. Hildebrand. 1987. GABA-mediated synaptic inhibition of projection neurons in the antennal lobes of the sphinx moth, *Manduca sexta*. J. Comp. Physiol. A 161:23–32.

White, M.J.D. 1978. Modes of speciation. W.H. Freeman, San Francisco.

Wiley, R.H., and D.G. Richards. 1982. Adaptations for acoustic communication in birds: Sound transmission and signal detection. Pp. 131–181 *in* D.E. Kroodsma and E.H. Miller, eds. Acoustic communication in birds. Vol. 1. Academic Press, New York.

Williams, G.C. 1966. Adaptation and natural selection. Princeton University Press, Princeton, N.J.

Williamson, P.G. 1987. Selection or constraint?: A proposal on the mechanism of stasis. Pp. 129–142 *in* K.S.W. Campbell and M.F. Day, eds. Rates of evolution. Allen and Unwin, London.

Wilson, E.O. 1975. Sociobiology. Harvard University Press, Cambridge, Mass.

Witzgall, P., and B. Frérot. 1989. Pheromone emission by individual females of the carnation tortrix, *Cacoecimorpha pronubana*. J. Chem. Ecol. 15:707–717.

15

Chemical Signals and the Recognition Concept

CHRIS S. WHITE, DAVID M. LAMBERT

Evolutionary Genetics Laboratory, School of Biological Sciences
University of Auckland

AND STEPHEN P. FOSTER

DSIR Plant Protection, Palmerston North, New Zealand

RECOGNITION IS A fundamental property of the world we live in. In the inanimate world, recognition occurs in a diversity of ways. The chemical composition of a crystal acts as a template for the addition of further crystalline layers. Through a process of recognition, the most familiar crystal, the snowflake, assembles itself and grows (Fig. 15.1). In the animate world, recognition is also a central property. DNA replicates through complementary chemical structures, a process similar to that for crystals; enzymes recognize a substrate; our immune systems recognize invading molecules; and cells in multicellular organisms recognize each other through ontogeny. For example, the ordered self-assembly of the bacteriophage from each of its constituent elements is possible only given the complementary nature of those elements (Fig. 15.1).

Sexual reproduction is dependent upon the most basic of complementary relationships, that between the process which produces the sexes (or their gametes) and the process of fertilization. Sexual life cycles are possible only given this complementary sequence of events. Hence, through the uniting operation of these complementary processes, biological forms belonging to the same inclusive system are always the same, and yet they are always different, in the sense that these same processes generate variation. This then gives

Presented in part at the International Society of Chemical Ecology, Sixth Annual Meeting, 7–11 August, 1989, at the University of Göteborg, Göteborg, Sweden. Publication 49 from the Evolutionary Genetics Laboratory, University of Auckland.

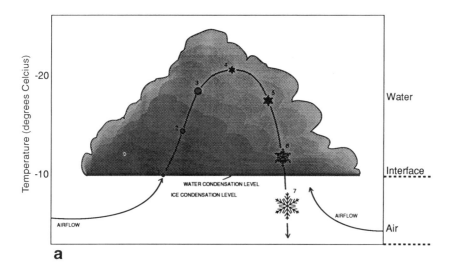

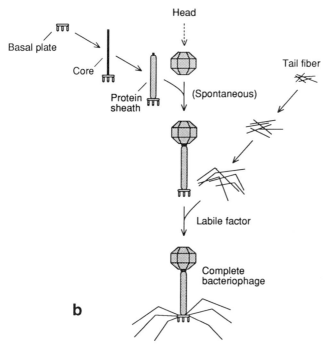

Fig. 15.1. The formation of a snowflake within a cloud (a) and the self-assembly of a bacteriophage (b)—examples of recognition in nature.

rise to the phenomenon of evolution, or descent with modification as Darwin described it.

The Recognition Concept

The Recognition Concept of species emphasizes the fundamental complementary property of biological forms, placing a central emphasis on the importance of sexual reproduction. Hugh Paterson (1978, 1981, 1982, 1985) presented the idea that, in sexual organisms, species are groups of individuals that share a common fertilization system. Central to this idea is the operation of a conspecific recognition system which is expressed between males and females, the Specific-Mate Recognition System (SMRS). This system consists of a set of interactions between males and females, often described as signals and responses, that take place within a particular context of habitat, season, internal state, etc. Under the appropriate conditions, fertilization is a regular consequence of the operation of the system.

Mate recognition is the sequence of events resulting from the operation of the SMRS. It is an expression of the complementary properties of males and females that allow the interactions between them to take place. An obvious physical example is the genitalia of many organisms that exhibit a "lock and key" complementary nature. Thus mate recognition is to be understood in the sense of recognition between enzyme and substrate mentioned earlier. This differentiates it from recognition in the sense of "to know again," an act of memory which many organisms are capable of. For example, monogamous birds are able to recognize (know again) a single individual mate but could potentially accomplish "mate recognition" with any conspecific of the opposite sex.

The Nature of Systems

A system is a network of interdependent and interacting parts that form a collective whole. Therefore a system must be understood in terms of the whole, not in terms of its constituent parts. Vrba (this volume) argues that species should properly be regarded as complex systems in which the properties of the systems as a whole are a nontrivial result of the properties of its constituent parts and the laws of their interaction (Simon 1962). In particular, we consider that SMRSs need to be understood as wholes, not as a list of the individual behavioral, chemical, or visual factors that compose them. We suggest that many of the disagreements about the nature of the mate recognition systems, voiced in the recent literature (e.g. Verrell 1988) derive from the fact that some authors use the concept explicitly in the holistic sense (Lambert et al. 1987) and others do not. By treating the SMRS as a collection of characters, the latter have often mistakenly equated a few elements or a single element of the system with the system as a whole. As a consequence of this, they often assert that, by demonstrating particular properties of individual elements they have shown that the system itself has these properties. This misunderstanding of the nature of SMRSs (and, indeed, biological systems in general) results in part from the fact that the elements of systems are more conveniently studied than the whole. This results in an attempt to understand such elements independently from the context within which they operate. We will consider this later, using examples from the literature dealing with pheromones of Insecta.

The recognition system is often used synonymously with only the sequence of events (signals and responses) leading to fertilization. We argue here for a broader approach. We suggest that the SMRS should be regarded as the complete set of complementary properties of the male and female that make the signals and responses possible, together with the context (environmental, internal state) in which they are expressed (*sensu* Lambert et al. 1989). We need to recognize that it is this *system* that has the properties that enable recognition to occur, whereas the particular events that constitute a mate recognition process are manifested as the result of these properties.

Chemicals and Mate Recognition

Chemical communication among the Insecta has evolved to a considerable degree. Chemicals released by plants and animals are perceived by and modify the behavior of all species of insects. Of greatest concern to this discussion are the chemical messages that pass between conspecific mates and in particular those used in mate location and recognition (i.e., sex pheromones).

Pheromone systems have been most extensively studied in the Lepidoptera and, more specifically, in moths. The topic of sex pheromones of moths has been reviewed in detail elsewhere (Linn and Roelofs, this volume), and we do not propose to repeat this. The objective of this discussion is to demonstrate how the concepts of recognition and system can be applied to the investigation and understanding of mate recognition in insect species that use sex pheromones. In particular, we emphasize the interactions that occur between the components or "elements" of the system and the environment in which mate recognition occurs.

Pheromones and the Specific-Mate Recognition System

To anyone who has observed the powerful effect of a sex pheromone on an insect's behavior it is easy to consider the SMRS of these insects as simply the release of the sex pheromone by one sex and the response to this pheromone by the other. However, the release and perception of the pheromone does not constitute the mate recognition system but, rather, particular elements of the system. The mate recognition system of a species is made up of the interactions that contribute to and organize the process of recognition by conspecific mates within a natural context.

Somewhat arbitrarily, we define the term *element* as a component of the SMRS that has been, or can be, studied as a unit. This could be a particular type of pheromone signal, its production and release, or perception of the signal by the receiver. This definition is flexible because of the many different levels at which the SMRS operates. An element may therefore be physiological, behavioral, or even ecological in nature, or include aspects of all of these. A more precise definition is not needed, provided the boundaries of any element under study are defined and its linkages with the system as a whole are recognized. We do not wish to emphasize an exhaustive list of the individual elements of SMRSs, but rather to show how the interactions of constituent elements are a manifestation of the complementary properties that conspecific males and females possess.

Principles of the System

Although the elements of insect SMRSs show great diversity, those systems that utilize pheromones to elicit orientation over a distance generally exhibit a common *modus operandi*. While variations on this theme can be found, we will restrict our discussion to those that follow the general sequence of steps summarized as follows:

1. One sex (more commonly the female) produces *de novo*, or sequesters from other sources, the sex pheromone and releases it, often from a specialized gland.

2. The other sex detects the pheromone, and by various maneuvers, orients and moves toward the pheromone source.

3. At very close range, during the so-called courtship phase of mating, other sensory modalities, including visual, tactile, auditory, and chemical (an additional sex pheromone or environmental chemical) interact with the longer-range sex pheromone.

4. Copulation and fertilization of gametes are an outcome (i.e., mate recognition) of the preceding steps.

There is clear evidence that the elements of a mate recognition system are not independent but interact and that these interactions organize the operations of the SMRS both temporally (the order of events) and spatially (where they occur). For example, in species that use chemicals that have been sequestered from host plants as their sex pheromones, mate recognition is dependent on the interaction of elements including the environment, host location, and feeding behaviors (perhaps even in larval stages as in the courtship pheromones produced by males of several arctiid moths; Krasnoff and Roelofs [1990]), as well as physiological processes that allow the sequestration or modification of the appropriate chemicals. The interactions of these elements are organized such that when the insect releases the pheromone, not only has it sequestered the necessary chemicals, but it is also physiologically and behaviorally capable of mating.

Similarly, *de novo* pheromone production, such as the fatty-acid-derived compounds used by most species of moths (Bjostad et al. 1987), is mediated in many species by neuroendocrine factors such as the pheromone-biosynthesis-activating neuropeptide (PBAN) (Raina and Menn 1987). The interactions of neuroendocrine factors with other elements can lead to organization of the system. For example, in the true armyworm, *Pseudaletia unipuncta,* juvenile hormone not only controls ovarian development, but also apparently acts to induce both calling behavior and the release of PBAN, which stimulates pheromone production (Cusson and McNeill 1989). Other (nominally) internal (e.g. temporal periodicity) and environmental (e.g. temperature, light intensity) factors interact throughout the organism's life such that the behavior of release of the pheromone is in an appropriate context (in both a physiological and spatio-temporal sense).

With regard to the response to the chemical signal (i.e., the second step), unquestionably the most studied systems are those of moths (see Linn and Roelofs, this volume). However, the types of responses that occur in these systems do not appear to be restricted to moths, but are used by species in a range of insect orders (e.g. Birch and White [1988] and Harris and Foster [1991]). Upon detecting a conspecific female's pheromone blend, an insect such as a male moth will typically fly upwind toward the odor source exhibiting behavioral patterns including optomotor anemotaxis and an in-built program of counterturns. Both of these responses are elicited by the interaction of sensory input through different modalities, including olfactory (the pheromone), visual, and mechanical (windspeed) (Baker 1989). Additionally, the behavioral responses of male moths to pheromone are known to be influenced by other factors such as light intensity, temperature (Linn et al. 1988*b*), pheromone blend composition (see Linn and Roelofs, this volume), and the level of sensory habituation to the pheromone (Linn and Roelofs 1981). The interactions of these factors with the physiology and behavior of the male organize his complementary response to the female signal. However, the interactions between elements, and their organization, are not restricted to those elements of the mate recognition system in the same step; they also occur between steps. Thus, insects utilizing sex pheromones do not randomly exhibit location behaviors, but rather these behaviors are exhibited after release (i.e., step 1) and detection (i.e., step 2) of the pheromone. In this way, the interaction of production and release with detection and behavioral response synchronizes (and hence organizes) these steps.

The interactions between elements of the mate recognition system are more obvious at closer distances as the organisms themselves directly interact (i.e., during courtship). Throughout pheromone-mediated flight many species of insects use visual and other cues primarily for steering with respect to orientation and groundspeed (Baker 1989). At closer range, however, visual cues, in combination with olfactory cues, influence landing behavior (Foster and Harris 1992), exhibition of courtship behaviors (Baker and Cardé 1979), and, of course, mate recognition. The few studies that have investigated the importance of visual recognition of mates during pheromone-mediated courtship have shown that visual cues need to be linked to olfactory (pheromone) cues for mate recognition to occur (Baker and Cardé 1979).

During the courtship stage of pheromone-mediated behavior, mates can interact behaviorally through tactile (Baker and Cardé 1979) and auditory (Krasnoff and Yager 1988) as well as visual and olfactory sensory modalities. Mate recognition does not result from input through just one of these modalities, but through the interaction of the different sensory modalities involved. As a consequence, the behavioral sequences (interactions) need not be exhibited rigidly or in a strict hierarchical sequence. Just as observed for the three-spined sticklebacks (Tinbergen 1951), male and female Oriental fruit moths exhibit a complex sequence of interactive courtship behaviors, including protrusion of the male's hair pencils and release of volatile chemicals, as well as tactile contact between the exhibiting male and the female. However, again as for the stickleback, the sequence can be flexible; another male approaching a male and female exhibiting these courtship behaviors receives their visual and perhaps olfactory stimuli and does not exhibit full courtship behaviors but proceeds directly to attempt copulation with the female (Baker 1982).

In the final step, the influence of elements in the preceding steps is apparent. Thus, in many species of insects using sex pheromones, following copulation females stop releasing and producing pheromone as well as become unresponsive to further copulatory attempts by males (see, for example, *Heliothis zea*; Raina 1989). In a number of species of insects (e.g. *Drosophila melanogaster*; Chen et al. 1988) male accessory gland secretions have been directly implicated in these changes, while in others, such as *Lymantria dispar*, the presence of sperm in the spermatheca apparently provides a signal, transmitted through the nervous system, to stop pheromone production (Giebultowicz et al. 1991).

Recognizing the complexity of the SMRS allows the description of it to be simplified. This is because the common principles of interaction and organization are emphasized rather than the description of functionality for many individual elements. These principles can and should be applied to the study of any SMRS. For example, bark beetles, such as *Dendroctonus brevicomis*, not only feed upon their host trees but also use them in mate recognition. Females of this species, after landing on a host tree, release two bicyclic ketals along with the terpene myrcene (released from the tree) that cause other males and females to aggregate on the tree or surrounding trees prior to feeding and mating (which takes place in this environment) (Wood 1982). The description of these systems is complicated for two reasons. First, this type of pheromone is strictly considered an aggregation rather than a sex pheromone, and consequently its role in mating is obscured. Second, the chemical myrcene is not strictly considered a pheromonal but rather an allomonal compound (as it is not directly released by the insect). When mate recognition is viewed as the result of the operation of a mating system, the environment in which the outcome of fertilization occurs is a necessary part of that mating system. Thus, the so-called aggregation pheromone (including the chemical myrcene) is considered part of the SMRS, for without it mates would not be brought within proximity. By adopting this view, the system is emphasized rather than the complex and often ambiguous roles of individual elements.

We have identified the conceptual links between elements of the SMRS and their organization in systems using pheromones. It would be incorrect to assume from this that either the interactions or the organization were fully understood, or that elements or systems had been studied intentionally to reveal these aspects. In fact, much of our argument is directed at ensuring this work is done in the future. This involves a new methodology of approach which sets different questions, framed in different terms, even though the techniques of investigation may be similar. The questions should be directed toward the interpretation of biological interactions during reproduction within a larger whole (the system). They should also address interaction and organization within that whole, rather than placing priority on the variation and dynamics of single, isolated elements.

Mate Recognition and "Variation" in Systems

A key attribute of biological systems is the nature of their response to variation, and this affects how variation is interpreted in relation to their operation. For mating systems this has been the source of a major controversy centered the around wider theories and beliefs about the role of variation in biology.

The importance of variation in biology and in understanding mate recognition depends on the approach one takes to explaining diversity. In a selectionist approach, variation is considered *the* basis of most evolutionary change; it is necessary for selection to occur (Coyne et al. 1988). Thus, variation in the characteristics of organisms of the same species is emphasized and is used to account for most, if not all, of the diversity in these characteristics among species. This approach is the basis of many accounts of the origin of the characteristics of mating systems by sexual selection (see Carson, this volume). By contrast, viewing these characteristics as interdependent elements of systems places little emphasis on this type of variation as a means to explain the diversity of mate recognition systems. The diversity of communication systems among species is undeniable, as is the fact that there is variation in the elements of individual systems. However, it is fundamental to a proper understanding of systems and their properties that the concept of variation cannot be applied to systems as a whole.

This difference in perspective has led to confusion in many debates over the Recognition Concept, with opponents claiming that variation per se is evidence of variation in mating systems. The root of the confusion lies in the fact that variation must relate to a particular variable or element; it cannot relate to a whole system. Thus the concept of variation, which is expressed in terms of individual characteristics such as size, shape, pheromone quantity or ratio, and the concept of system, which comprises all these elements and their organization, are mutually exclusive. One can discuss variation within the elements of the system and differences between systems, but not the variation of a system.

Both Carson (this volume) and Butlin (this volume) confuse these two fundamentally different concepts. Carson, for example, says, "Paterson admits the possibility of the existence of only a small genetic variance on each of the *components* of the system. Lambert (1982) and Millar and Lambert (1984) have also stressed what they see as the invariability and therefore species diagnostic nature of a mate recognition *system*" (italics added). Apart from the absence of any reference to "invariability" in Millar and Lambert (1984), the equating of "components" with the "system" illustrates the kind of confusion to which we refer. It is also this type of thinking that has led to the mistaken claims that the Recognition Concept includes the notion that mating systems are "fixed" or "rigid" within species, and that the concept is typological in relation to the role of these characteristics for species identification (Carson, this volume).

If mating systems are, as we claim, either the same (within a species) or different (between different species), but cannot be considered in the selectionist sense of variation, then there are three important implications:

1. Variation in the elements of a recognition system should not affect its outcome in its natural environment. Thus, for example, conspecific females that produce different ratios of pheromone components should all be able to attract conspecific males.

2. What is presumed to be variation within the recognition system of a supposed species, if it has any effect on mate recognition, may in fact represent differences between the mating systems of different species.

3. Mate recognition systems should change abruptly, rather than by the gradual change in elements through the action of selection.

In sexual species, mate recognition and its outcome (i.e., fertilization of gametes) occur only if mates (or in some instances only their gametes) occupy a common space at a common time. Where mates are located remotely from each other, then they must come together; if they do not, then clearly they will not mate. Therefore, the element or elements used in bringing mates to the same place is an integrated and necessary part of the SMRS. In this discussion we are concerned with mate location utilizing sex pheromones; however, visual and auditory sensory modalities are also used by many species of insects for mate location, and the points we make are pertinent to these systems as well.

Not surprisingly, considering its importance to the selectionist view, there are many reports of "variation" in SMRSs (for reviews on variation in pheromone elements and its possible evolutionary role, see Löfstedt [1990, 1991]). Because of the self-regulating properties of systems, especially those such as mating systems that depend on the complementary properties of male and female, the stability of the system in response to variation in elements is emphasized. Variation is important only where it has an immediate effect on the operation of the system, such as assortative mating. Thus, the issue raised by Butlin (this volume), of whether the degree of variability in the elements of mating systems is larger or smaller than that of other characteristics of organisms, is irrelevant. It is only the organism's perception of the variance that is important. In any case, describing variance as "small" or "large" in the absence of any comparisons with variance in other characteristics is purely arbitrary.

The selectionist view of mating systems has also resulted in confusion over the effects of variation in elements of the SMRS. It has led Ritchie and Hewitt (this volume) to the mistaken conclusion that Butlin's voluminous (and somewhat uncritical) records of variation in elements is evidence against the stability of mating systems. In fact, there is no relationship between stability and variation. The basis of Ritchie and Hewitt's and Butlin's argument arises from a fundamentally incorrect view of the Recognition Concept derived from their assumptions about biological change. As we discuss later, both we and Paterson consider that the SMRS is generally very stable (not invariant). Under special conditions it may change rapidly and then once again remain stable over a very long period. Thus, Butlin's expected correlation between local environmental parameters and variation in SMRS components is clearly not a prediction of the Recognition Concept. Further, in this scenario variation is unrelated to change, which is what Butlin and his colleagues find difficult to reconcile with their selectionist approach to biology.

If SMRSs do not vary, there are two possible interpretations of observed "variation" in the elements of SMRS.

1. Where the variation in an element or elements of the system does not result in any partitioning of mate recognition (i.e., males or females of the species are able to recognize conspecific mates producing the observed range of the variable[s] in question) then the variability is within the system and is therefore intraspecific.

2. If what appears to be "variability" in the element or elements does relate to partitioning of mate recognition (i.e., males or females of one taxon do not recognize females or males of another taxon as mates), then the variability is not within the system, and is indeed not variability, but rather is indicative of a difference between mate recognition systems of different species. In practice, confusion from this situation is likely only to arise if the two or more taxa have been previously conflated as a single species.

Variation and Diversity in Pheromone Systems

In terms of pheromone production and response, probably all species exhibit some intraspecific variability. Thus, even in a relatively tightly regulated ratio of geometric isomers, such as that of (Z)- and (E)-11-tetradecenyl acetate produced by female red-banded leafroller moths, *Argyrotaenia velutinana*, there is a coefficient of variation between individuals of 9.7 percent (Miller and Roelofs 1980). Species using sex pheromone components that are not as closely biosynthetically related as geometric isomers generally exhibit even greater variability in ratios (e.g. see Foster et al. 1989). Likewise, behavioral responses to different pheromone ratios and quantities can also show considerable variability, usually over a range of ratios and quantities of components at least as great as that produced by conspecific females (e.g. see the study of Roelofs et al. [1986] on *A. velutinana* males).

While most studies have investigated the variability of elements in the population as a whole, few have examined whether the variability observed is between individuals or is contained within the response of the individual. The few studies that have addressed variability of elements in individuals have found that in females the variability appears to be between individuals rather than within an individual over time (Du et al. 1987; Witzgall and Frérot 1989). Studies on variability of responses of individual males are complicated by the individual's responses varying with respect to both temperature (Linn et al. 1988a) and previous exposure to pheromone (Linn and Roelofs 1981). However, Cardé et al. (1976), by trapping male Oriental fruit moths to different ratios of its two main pheromone components, (E)- and (Z)-8-dodecenyl acetates, marking these males, and releasing and recapturing them, were able to show that the variability in response to different ratios was principally within an individual. Therefore, from these limited data it appears that interindividual variability in production is complemented by individual variability in response, the result being that the system is stable.

Another example illustrates this point further. Females of two populations of the New Zealand tortricid moth *Planotortrix excessana* produce significantly different quantities and ratios of the two most abundant compounds, (Z)-5- and (Z)-7-tetradecenyl acetates (Foster et al. 1989). Likewise, males of the two populations also exhibit significantly different responses to different ratios of these two components (Clearwater and Foster, unpublished). In a field cage trial, however, males of either population were found to mate in equal numbers with tethered females of their own and the other population (Foster et al. 1989). Thus, while the elements of a SMRS vary, the interactions of the elements stabilize the system.

The second situation, that is, where the outcome (fertilization of gametes) is affected by variability in elements of the SMRS, highlights the difference between the selectionist view of mate recognition and an analysis based on systems. The properties of systems suggest that such assortative mating is the result of the operation of the distinct mate recognition systems of (by most accepted definitions) different species. In contrast, the selectionist view focuses on the significance of the variability itself and often overlooks the likely biological distinctness of the groups concerned. Concomitant with the selectionist view is the almost tacit acceptance of the current taxonomic description of the group. In effect, it is assumed that the methods by which the taxonomic description has been arrived at (usually abstract criteria based on morphology) have greater significance than the implications of variation for the biological process of mate recognition.

In most cases (at least to the tribal and usually to most generic level differences), species are readily differentiated, and hence the usual taxonomic description concurs with patterns observed in the pheromonal communication systems on the status of the taxon. It is in the cases of morphologically very similar or indistinguishable (i.e., sibling or cryptic) species that the investigation into differences between mate location mechanisms and consequently mate recognition systems can be of great insight. Despite this, there are a number of cases in the pheromone literature where possible differences between taxa have not been explored more extensively, and the phenomenon has been ascribed to variation within a single species. By doing so, not only is the consequence of the "variation" on mate recognition overlooked, but, by a twisted argument, it is used as evidence that systems are variable.

One of the first examples where two morphologically similar species were shown to be able to be distinguished by differences in their sex pheromone systems was in the tortricids, *Archips mortuanus* and *A. argyrospilus* (Roelofs and Comeau 1969). However, this was a case where the species were already recognized through morphological differences as being distinct; the pheromone study merely confirmed this. Studies on the sex pheromones of some New Zealand tortricid pests led to the discovery that the described species of *Planotortrix excessana* (*sensu* Dugdale 1966) and *Ctenopseustis obliquana* (*sensu* Green and Dugdale 1982) actually consisted of two and two or three taxa, respectively. These taxa were elucidated by the differences in their sex pheromone systems, in both production by the females and the complementary responses by the respective males. Subsequent morphological and allozyme analyses of these taxa supported their distinction and led to the taxa being classified as discrete, cryptic species (Dugdale 1990; Foster et al. 1991) (see also the discussion by Lambert and Paterson 1982).

In other cases, despite the emphasis on variation in an element or elements, the consequences of this variation for the SMRS have been overlooked. We believe an emphasis on understanding the system rather than on the detailed description of variation would allow the biological consequences of such variation to be better understood. For example, quite major pheromone differences between two colormorphs of larvae of the larch budmoth, *Zeiraphera diniana,* have been found (Guerin et al. 1984; Baltensweiler and Priesner 1988), with adult moths resulting from one morph producing and responding mainly to (E)-9-dodecenyl acetate and adults from the other morph producing and responding mainly to (E)-11-tetradecenyl acetate. Additionally, female moths that resulted from intermediate (in color) larvae were found to produce both compounds in approximately equal quantities. Implicit in this study was that the larch budmoth comprised a single species and hence the term "pheromone polymorphism" was applied to the phenomenon. From a systems perspective, however, it appears that this apparent "polymorphism" or "variation" could be the result of two, perhaps three, distinct species, each with a distinct SMRS. Whether this is the case, or whether it is truly a polymorphism within a single species, is unknown. What is clear, however, is that it cannot be resolved by examining the elements in isolation (e.g. by examining, say, production of different components by the respective females). Rather, the SMRS of the larch budmoth(s) must be examined as a whole. Thus the responses (or lack of them) of males of one morph to females of the other morph and vice versa should be determined within a natural context. Additionally, as a polymorphism refers specifically to "the occurrence of two or more genetically different individuals in the same breeding population" (Stenesh 1989), it would seem appropriate to test for divergence in specific-mate recognition between adults of the two morphs.

A similar, but perhaps more complicated, situation exists in the turnip moth, *Agrotis segetum*. A number of studies have shown that there is considerable "variation" in the SMRS of *A. segetum,* both within and between populations. Some populations, such as those from Sweden, England, and Hungary, are similar in both production of ratios of (Z)-5-decenyl acetate, (Z)-7-

dodecenyl acetate, and (Z)-9-tetradecenyl acetate by females, and response to these ratios by males. Females of a French population though, produce much greater relative amounts of (Z)-5-decenyl acetate. Males of this population also have a much greater proportion of receptor cells on their antennae sensitive to this compound and, in contrast to the other populations, can be caught in traps baited only with this compound (Arn et al. 1983; Löfstedt et al. 1986). Populations of *A. segetum* in Armenia and Bulgaria are even more distinct from these populations; females produce very small amounts of (Z)-5-decenyl acetate (approximately 1 percent of total blend) and large amounts of the other two components, (Z)-7-dodecenyl acetate and (Z)-9-tetradecenyl acetate. Males of these populations also have very few receptor cells for (Z)-5-decenyl acetate and can be caught in traps baited only with (Z)-7-dodecenyl acetate and (Z)-9-tetradecenyl acetate (Subchev et al. 1986; Hansson et al. 1990).

The significance of the distinctness of the various populations is as yet unclear (Hansson et al. 1990) as only the variability or differences in elements have been studied. It is clear that the interactions of these and other elements of SMRSs of the populations concerned need to be examined in order to better understand the significance of this variation for the operation of these systems. At present it would seem possible that the concept of *A. segetum* could contain a complex of cryptic species which, on the one hand, have distinct (from each other) SMRSs and, on the other, have elements within each of their SMRSs that are variable. As a final point, to emphasize the lack of understanding of the effect of differences in SMRSs, the previous concept of *A. fucosa* from Japan has recently been synonymized with *A. segetum* despite its pheromone consisting of a mixture of (Z)-5-decenyl acetate and (Z)-7-decenyl acetate (Löfstedt et al. 1986). We emphasize again the necessity of investigating systems as a whole so that the variability in elements of a system (such as production of ratios of components) can be distinguished from the diversity of different systems.

Interactions between Pheromone Systems

There are two types of interactions between pheromone systems of different species that have been noted. The first is cross-attraction between species that share similar pheromone signals or responses (e.g. Booij and Voerman 1984; Walgenbach et al. 1983). The second is what has been termed "antagonistic" interactions, where the pheromone of one species inhibits the sexual response of the other (e.g. Cardé et al. 1977; Glover et al. 1989). Such cases have been given great significance by various critics of and commentators on the Recognition Concept (Coyne et al. 1988; Linn and Roelofs, this volume) as they appear both to contradict the species specificity of the SMRS and to confirm the existence of isolating mechanisms, which are incompatible with the idea of mate recognition and the Recognition Concept.

The answer to these contradictions lies once more in analyzing the events of mate recognition in terms of the properties of the system that produce them. Production and perception of pheromones are elements within the SMRS of the species that possesses them. Thus, it is not problematic that interspecific interactions between such pheromonal elements of distinct systems occur. The crucial point is that the operation and outcome of each respective system is not influenced by such interactions.

Cross-attraction has been supposed, by those who consider mating systems as a collection of independent elements, to demonstrate a lack of specificity in the SMRS and thus a lack of species cohesion. However, the validity of many supposed cases of cross-attraction between species, especially those with long-distance chemical signaling, is questionable. In many examples the behavior of cross-attraction between the different species has not been observed in a natural context, but rather in an artificial, constrained context, or even to synthetic versions of the pheromone. In such examples, it is not the mate-recognition system that is being studied but the response of isolated elements to an abstract set of conditions.

No examples of cross-attraction have demonstrated a breakdown in the operation of an SMRS. However, it should be noted that there are no particular properties of SMRSs that would specifically preclude cross-attraction from occurring in nature. Within a group of animals such as moths, for example, the different specific pheromone signals are derived from the modification of a limited number of biosynthetic pathways (see Linn and Roelofs, this volume). It is therefore likely that certain species will share common pheromone signaling and receiving elements in their respective mate-recognition systems. In these cases, the systems will have the potential to interact via the common elements. However, inability to express the full set of complementary properties will preclude the complete operation of the respective systems. Thus, such interactions of one or a few elements will usually terminate quickly and do not result in fertilization. An excellent example of this is the cross-attraction between *Polistes* wasps which breaks down in proximity due to conspecific recognition of cuticular chemical signals (Post and Jeanne 1984).

If pheromones are regarded as part of a mating system, cross-attraction is not an unexpected natural phenomenon. It is, however, peripheral to the understanding of species and their mating systems as it does not relate to the central defining property of species, namely, conspecific recognition (Lambert et al. 1987). Ironically, although wrongly put forward as evidence against the Recognition Concept, cross-attraction is difficult to accommodate under a selectionist view that emphasizes interspecific reproductive isolation. Instead, it should be selected against and eliminated over time, to promote efficient reproductive isolation. Its occurrence, therefore, has to be accounted for by postulating that divergence of the respective, cross-attractive signals is a relatively recent and dynamic event.

In the same category as cross-attraction is interspecific inhibition of the sexual response, or behavioral antagonism, recently quoted as a mechanism for ensuring reproductive isolation between species (e.g. Christensen et al. 1990). A few examples have been noted where species appear to possess receptors specifically for components of the pheromone blend of another species as in the case of *Trichoplusia ni* and *Pseudoplusia includens* (Linn et al. 1988a; Linn and Roelofs 1992). As for cross-attraction, this demonstrates the interactions between elements of two SMRSs, in particular, the behaviorally antagonistic effect of a pheromone component produced by *T. ni* females on the mate location behavior of *P. includens* males. It is suggested that the presence of the receptor cell for the *T. ni* pheromone component in the *P. includens* males is an example of an isolating mechanism since *P. includens* females do not produce the component in question. This assumes that a receptor in an unstimulated state does not in itself form part of a signal. The work of Priesner (1986) suggests that maximal male response in *Polia pisi* was elicited using a pheromone blend that stimulated some receptor cells but not others. Stimulation of all receptors terminated the male response. The critical point in all such cases is that, as for cross-attraction, the significance of this type of interaction must be considered in terms of its effect on the outcome of the operation of the whole SMRS. It seems very unlikely, in this or any other case, that, in the absence of the antagonistic effect, mate recognition would proceed to interspecific fertilization. Thus, the antagonistic response in moths may have no greater significance than the reaction of humans to the chemical signals of a male goat.

Changes in Specific-Mate Recognition Systems

The nature of SMRSs is not consistent with a gradual modification of one mating system into another. This is because changes in any element of the system, such as the production of a new blend by the female, without concurrent change in the male antennal receptors or response would result in a breakdown of the integrity of the system and, hence, its successful operation. This integrity is dependent on the complementary properties of male/female sexuality, that is, on such properties of separate and different organisms during adult life. These properties, in turn, are the product of the process of sexual differentiation during development. It is because of this biologically stabilized nature of the SMRS that we consider the typical picture of a continuum of minutely different mating systems lying along some historical lineage connecting one species with another to be most unlikely. Indeed, as Turner (this volume) reminds us, Eldredge and Gould (1972) originally considered that an understanding of the stability of species lay in a view of species and individuals as homeostatic systems.

Paterson and other proponents of the Recognition Concept have generally favored a model of speciation involving rapid change in small populations adapting to a new environment (Paterson 1982). Given the improbability of independent, simultaneous change in complementary male and female elements, it has been postulated that the respective male and female components of the system may be inherited as coadapted complexes (Coyne et al. 1988). However, genetic studies on the European corn borer, *Ostrinia nubilalis*, have shown that, for this species at least, genes affecting male perception, behavioral response, and female production are located on different chromosomes (Roelofs et al. 1987; Löfstedt et al. 1989).

To understand change in mating systems, we consider it most useful to focus on the processes that produce the complementary male and female properties during development. Any alteration of the adult properties must arise as a result of a change in this developmental pathway. It is well known that such changes in development can be induced by varying the physical conditions under which development takes place, resulting in significantly altered adult forms known as phenocopies (Lambert et al. 1989). These changes are often referred to as transformations because they result in a new form that can be understood as a rational modification of the original according to certain rules.

The concept of transformation, although it occurs rapidly, is not a concept allied to sympatric speciation. If induced transformations, such as phenocopies, are an example of transformations in nature they will generally be associated with environmental change. Logically this will most often arise in a spatially separate population of a species. The association of transformation and environmental change should also in no way be conflated with the concept of selection. Transformation is rational, following the rules dictated by the nature of the system being transformed, and is generationally instantaneous.

A classic example of this is the work of Goldschmidt (1935a,b), who showed that exposing developing *Drosophila* eggs or larvae to abnormal heat regimes produced a range of altered adult forms often resembling genetic mutants. Less varied, but perhaps more pertinent, are the examples such as the work of Standfuss (1896), who found that low temperatures applied to developing individuals of the central European butterfly *Vanessa urticae* produced forms indistinguishable from *V. polaris* found in Lapland. In more recent studies by Ho et al. (1983) the bizarre *Drosophila* phenocopy, bithorax, has been found to persist, in the absence of any inducing stimulus, when induced in a number of consecutive generations.

A coordinated change producing a new, stable SMRS requires a transformation that preserves the complementary nature of the male-female sexual communication system. Such a transformation is most likely where it occurs in the process of sexual differentiation. As mentioned earlier, sexual differentiation is the complement of the integration that occurs between male and female during mate recognition. That is, the different elements of males and females that are involved in the operation of the SMRS arise during this common developmental process.

A number of studies have suggested the nature of the developmental links between the sexes with regard to the determination of male and female characteristics. A classic example of this is the phenomenon of intersexuality in *Lymantria dispar* originally by Goldschmidt (1931) and later by Clarke and Ford (1980) and others. These experiments produced end points of a perturbed developmental process that were stable adult forms possessing a mosaic of male and female sexual characteristics. The actual parts of the adult individual affected were shown to depend on the time during development at which the environmental perturbation occurred. The resulting forms were seen to have sexual morphology recognizable as combined components of the original adult sexes.

The nature of the process of sexual differentiation in many species also offers support for the close association of male and female development. Investigation of this process in *Drosophila* has shown that all individuals possess male and female primordia prior to differentiation. Female development is largely dependent on the presence of a specific protein produced by the tra^{2+} gene (Belote and Baker 1982). In its absence, a male develops regardless of chromosomal sex. In humans, the unusual case of the "penis at twelve" syndrome shows similar patterns. In members of a particular family, some morphologically female children are indistinguishable from others until age twelve, at which time they develop all typical male secondary sexual characteristics and become fertile, heterosexual males members of their society (Imperato-McGinley et al. 1974).

These examples illustrate some key features of the sexual differentiation process:

1. Sexual differentiation can be disrupted, resulting in a number of rational end points that differ from the typical adult sexes, including a range of states intermediate to them.

2. It is the developmental process that determines sex, not the genetic sex of the zygote.

3. The development of the sexual attributes associated with one sex can be linked with and affect the expression of the sexual attributes associated with the other.

These properties of sexual differentiation may assist the investigation of possible pathways for change in SMRSs involving sex pheromones. Insect pheromone systems involve the action of two physically distinct, sex-specific, and apparently unliked sets of structures and processes that must change to produce a new system. These are, in the female, the pheromone-producing structures and pheromone synthesis and release and, in the male, the peripheral perception structures, central nervous system (CNS) processing, and resultant behavioral responses to the pheromone. In order to reveal possible pathways from one stable pheromone system to another (i.e., how systems might change), the links between the development of female signaling and male perception need to be identified and described. Such links could occur at two levels as shown diagrammatically in Figure 15.2.

1. A link between the control of signaling in the female and the perception or response in the male (Link 1, Fig. 15.2). Both of these processes are controlled by the CNS and other common neural structures; a transformation of these control structures in both sexes could affect both signaling and perception.

2. A link between the differentiation of the female pheromone gland and the male antennae (Link 2, Fig. 15.2). This is a more speculative suggestion as such a connection between these quite separate adult structures has not yet been looked for.

Certain links between male and female pheromonal elements are already known, and these suggest that this area warrants further investigation. A common element in the male and female systems is the sub-oesophageal ganglion (SOG). In female moths, the neuropeptide PBAN, which controls pheromone biosynthesis, is released from the SOG (Raina and Menn 1987), while a FMRFamide peptide thought to be involved in flight behavior for mate-seeking by males has been identified in brain-SOG extracts of the moth *Manduca sexta* (Kingan et al. 1990).

Developmental links in sexual differentiation of *Manduca* have also been identified. When male larval antennal imaginal discs of *M. sexta* are transplanted onto a female, the antennal lobes of the female brain differentiate according to the male pattern and vice versa. In addition, when stimulated by sex pheromone through the male antenna, transformed adult females exhibit orientation and flight behaviors characteristic of males (Schneiderman et al. 1982; Hildebrand and Montague 1986). Thus, the sexual state of one element can be altered by a change in the sexual state of another. Hence, the interaction of these male and female elements during development may provide some clue as to the nature of the transformations that are necessary to change from one SMRS to another.

It is obvious that pheromone systems can change to give rise to new SMRSs. It seems equally clear that this is unlikely to occur by gradual change in individual elements, such as the female signal, when these are viewed in relation to the rest of the system. We have outlined a means by which the synchronous change in male and female attributes might occur and have attempted to identify the most likely place to look for the evidence to support this.

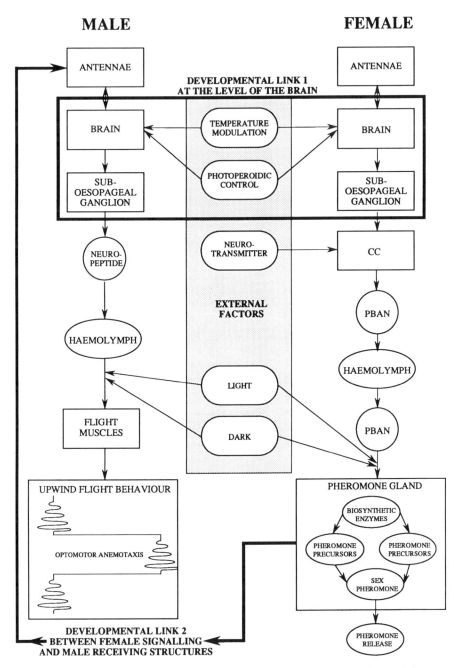

Fig. 15.2. A conceptual pheromone communication system as a subset of the SMRS and possible links between male and female properties (female system adapted from Raina and Menn [1987]).

Discussion

Pheromones demonstrate both the obstacles and the opportunities for the understanding of SMRSs inherent in the study of particular elements of mate recognition. Because pheromones are such a conspicuous and apparently dominant element of some SMRSs, there is a tendency to consider the sending and reception of this chemical signal as if it were the *whole* SMRS. This is an obstacle to understanding both the significance of the element and the nature of the system of which it is a part. However, as summarized in this chapter, a closer study of the pheromones in relation to the organisms themselves reveals that they are integrated elements of a complex, complementary system operating between conspecific males and females in an appropriate environment. This ability to precisely describe not only the signal and receiver, but also the integration of these elements into the mate recognition process provides an excellent opportunity for developing a broader understanding of SMRSs.

Using the example of pheromone systems, we have shown that variation occurs in elements of mate-recognition systems, but that the concept of variation does not relate to the systems themselves. If this is accepted, then it can be seen that variation and selection do not necessarily provide a basis for change in mating systems. Gradual, independent change of elements in a system that requires integration of adult male and female properties seems unlikely. As an alternative to gradual selection, we suggest that the ontogenetic processes of sexual differentiation and development of complementary sexual properties may provide the basis for an understanding of the processes by which SMRSs might change. The developmental links and interdependence of apparently separate parts of adult sexuality suggest a potential for synchronous change in male and female aspects of the pheromone communication system.

Understanding pheromone communication within the context of the mating system offers insight into the understanding of species in relation to our interpretation of the Recognition Concept. Examples of cross-attraction and behaviorally antagonistic responses between species can consequently be understood as the expected epiphenomena among a large number of species using a limited range of signals and receivers in their SMRSs. Such examples have not shown the presence or absence of such interspecific interactions to affect the integrity of the SMRSs of the species involved.

There is currently a range of distinct approaches to the study of sex and sexual systems in biology. The traditional view is that sex, exemplified here by pheromone signals and receivers, represents an ancient solution to the problem of survival, either for the short or long term. The argument is that this is obviously the case because the evolution of sex is so common it must represent a successful solution to a common problem. In contrast, a range of authors, such as Margulis and Sagan (1986), have argued that the evolution of this phenomenon can be understood only by a thorough analysis of the biology of existing sexual systems found in nature. Specifically, with regard to chemical signals and SMRSs, we have argued in this chapter that these systems must also be understood through analysis of their biology rather than through speculation concerning their past or future. Specific-mate recognition, species, sexual differentiation, and sexuality are all inextricably locked together. To understand any one it will be necessary to understand and encompass the whole.

Acknowledgment

The authors' work is supported by the grants from the Auckland University Research Committee to D.M. Lambert and C.S. White.

References

Arn, H., P. Esbjerg, R. Bues, M. Tòth, G. Szöcs, P. Guerin, and S. Rauscher. 1983. Field attraction of *Agrotis segetum* males in four European countries to mixtures containing three homologous acetates. J. Chem. Ecol. 9:267–276.

Baker, T.C. 1982. Variations in male Oriental fruit moth courtship patterns due to male competition. Experientia, Basel 39:112–114.

———. 1989. Sex pheromone communication in the Lepidoptera: New research progress. Experientia, Basel 45:248–262.

Baker, T.C., and R.T. Cardé. 1979. Courtship behavior of the oriental fruit moth (*Grapholitha molesta*): Experimental analysis and consideration of the role of sexual selection in the evolution of courtship pheromones in the Lepidoptera. Ann. Entomol. Soc. Am. 72:173–188.

Baltensweiler, W., and E. Priesner. 1988. Studien zum Pheromon-Polymorphismus von *Zeiraphera diniana* Gn. (Lep., Tortricidae). J. Appl. Entomol. 106:217–231.

Belote, J.M., and B.S. Baker. 1982. Sex determination in *Drosophila melanogaster:* Analysis of transformer-2, a sex transforming locus. Proc. Natl. Acad. Sci. 79:1568–1572.

Birch, M.C., and P.R. White. 1988. Responses of flying male *Anobium punctatum* (Coleoptera: Anobiidae) to female sex pheromone in a wind tunnel. J. Insect Behav. 1:111–115.

Bjostad, L.B., W.A. Wolf, and W.L. Roelofs. 1987. Pheromone biosynthesis in lepidopterans: Desaturation and chain shortening. Pp. 77–120 *in* D. Prestwich and G.J. Blomquist, eds. Pheromone biochemistry. Academic Press, New York.

Booij, C.H.J., and S. Voerman. 1984. New sex attractants for 35 tortricid and 4 other lepidopterous species found by systematic field screening in the Netherlands. J. Chem. Ecol. 10:135–144.

Cardé, R.T., and T.C. Baker. 1984. Sexual communication with pheromones. Pp. 355–383 *in* W.J. Bell and R.T. Cardé, eds. Chemical ecology of insects. Chapman and Hall, London.

Cardé, R.T., T.C. Baker, and W.L. Roelofs. 1976. Sex attractant responses of male Oriental fruit moths to a range of component ratios: Pheromone polymorphism? Experientia, Basel 32:1406–1407.

Cardé, R.T., A.M. Cardé, A.S. Hill, and W.L. Roelofs. 1977. Sex pheromone specificity as a reproductive isolating mechanism among the sibling species *Archips argyrospilus* and *A. mortuanus* and other sympatric tortricine moths (Lepidoptera: Tortricidae). J. Chem. Ecol. 3:71–84.

Carlson, D.A., P.A. Langley, and P. Huyton. 1978. Sex pheromone of the tsetse fly: Isolation, identification and synthesis of contact aphrodisiacs. Science 201:750–753.

Chen, P.S., E. Stumm-Zollinger, T. Aigaki, J. Balmer, M. Bienz, and P. Böhlen. 1988. A male accessory gland peptide that regulates reproductive behavior of female *D. melanogaster*. Cell 54:291–298.

Christensen, T.A., S.C. Geoffrion, and J.G. Hildebrand. 1990. Physiology of interspecific chemical communication in *Heliothis* moths. Physiol. Entomol. 15:275–284.

Clarke, C., and E.B. Ford. 1980. Intersexuality in *Lymantria dispar* (L.). A reassessment. Proc. R. Soc. Lond. B 206:381–394.

Coyne, J.A., H.A. Orr, and D.J. Futuyma. 1988. Do we need a new species concept? Syst. Zool. 37:190–200.

Cusson, M., and J.N. McNeil. 1989. Involvement of juvenile hormone in the regulation of pheromone release activities in a moth. Science 243:210–212.

Du, J.-W., C. Löfstedt, and J. Löfqvist. 1987. Repeatability of pheromone emissions from individual female ermine moths *Yponomeuta padellus* and *Y. rorellus*. J. Chem. Ecol. 13:1431–1441.

Dugdale, J.S. 1966. A new genus for the "elusive *Tortrix*" (Lepidoptera: Tortricidae: Tortricinae). N. Z. J. Sci. 9:391–398.

———. 1990. Reassessment of *Ctenopseustis* Meyrick and *Planotortrix* Dugdale with descriptions of two new genera (Lepidoptera: Tortricidae). N. Z. J. Zool. 17:437–465.

Eldredge, N., and S.J. Gould. 1972. Punctuated equilibria: An alternative to phyletic gradualism. Pp. 82–115 *in* T.J.M. Schopf, ed. Models in paleobiology. W.H. Freeman, San Francisco.

Foster, S.P., J.R. Clearwater, and S.J. Muggleston. 1989. Intra-specific variation of two components in the sex pheromone gland of a *Planotortrix excessana* sibling species. J. Chem. Ecol. 15:457–465.

Foster, S.P., and M.O. Harris. 1992. Factors influencing the landing of male *Epiphyas postvittana* (Walker) exhibiting pheromone-mediated flight. J. Insect Behav. 5:699–720.

Foster, S.P., C.S. White, and J.S. Dugdale. 1991. Sex pheromones and the status of greenheaded and brownheaded leafroller moths in New Zealand. N. Z. J. Zool. 18:63–74.

Giebultowicz, J.M., A.K. Raina, and R.L. Ridgway. 1991. Role of the spermatheca in the termination of sex pheromone production in mated gypsy moth females. Pp. 101–104 *in* I. Hrdy, ed. Insect chemical ecology. Academia, Prague.

Goldschmidt, R.B. 1931. Analysis of intersexuality in the gypsy moth. Q. Rev. Biol. 6:125–142.

———. 1935*a*. Gen Und Asseneigenschaft. I. Zeitschrift fver Induktive Abstammungs—und Vererbungslehre 69:38–69.

———. 1935*b*. Gen Und Asseneigenschaft. II. Zeitschrift fver Induktive Abstammungs—und Vererbungslehre 69:70–131.

Green, C.J., and J.S. Dugdale. 1982. Review of the genus *Ctenopseustis* Meyrick (Lepidoptera: Tortricidae), with reinstatement of two species. N. Z. J. Zool. 9:427–436.

Guerin, P.M., W. Baltensweiler, H. Arn, and H.R. Buser. 1984. Host race pheromone polymorphism in the larch budmoth. Experientia, Basel 40:892–894.

Hansson, B.S., M. Tòth, C. Löfstedt, G. Szöcs, M. Subchev, and J. Löfqvist. 1990. Pheromone variation among eastern European and a western Asian population of the turnip moth, *Agrotis segetum*. J. Chem. Ecol. 16:1611–1622.

Harris, M.O., and S.P. Foster. 1991. Wind tunnel studies of sex pheromone-mediated behavior of the Hessian fly (Diptera: Cecidomyiidae). J. Chem. Ecol. 17:2421–2435.

Hildebrand, J.G., and R.A. Montague. 1986. Functional organization of olfactory pathways in the central nervous system of *Manduca sexta*. Pp. 279–285 *in* T.L. Payne, M.C. Birch and C.E.J. Kennedy, eds. Mechanisms in insect olfaction. Clarendon Press, Oxford.

Ho, M.-W., C. Tucker, D. Keeley, and P.T. Saunders. 1983. Effects of successive generations of other treatment on penetrance and expression of the bithorax phenocopy in *Drosophila melanogaster*. J. Exp. Zool. 225:357–368.

Imperato-McGinley, J., L. Guerrero, T. Gauiter, and R.E. Peterson. 1974. Steroid 5 α-reductase deficiency in man: An inherited form of male pseudohemaphroditism. Science 186:1213–1215.

Kingan, T.G., D.B. Teplow, J.M. Phillips, J.P. Riehm, K.R. Rao, J.G. Hildebrand, U. Homberg, A.E. Kammer, and I. Jardine. 1990. A new peptide in the FMRFamide family isolated from the CNS of the hawkmoth *Manduca sexta*. Peptides 11:849–856.

Krasnoff, S.E., and W.L. Roelofs. 1990. Evolutionary trends in the male pheromone systems of arctiid moths: Evidence from studies of courtship in *Phragmatobia fuliginosa* and *Pyrrharctia isabella* (Lepidoptera: Arctiidae). Zool. J. Linn. Soc. 99:319–338.

Krasnoff, S.E., and D.D. Yager. 1988. Acoustic response to a pheromonal cue in the arctiid moth *Pyrrharctia isabella*. Physiol. Entomol. 13:433–440.

Lambert, D.M. 1982. Mate recognition in members of the *Drosophila nasuta* complex. Anim. Behav. 30:438–443.

Lambert, D.M., B. Michaux and C.S. White. 1987. Are species self-defining? Syst. Zool. 36:196–205.

Lambert, D.M., and H.E.H. Paterson. 1982. Morphological resemblance and its relationship to genetic distance measures. Evol. Theor. 5:291–300.

Lambert, D.M., P.M. Stevens, C.S. White, T. Gentle, N.R. Phillips, C.D. Millar, J.R. Barker, and R.D. Newcomb. 1989. Phenocopies, heredity and evolution. Evol. Theor. 8:285–304.

Linn, C.E., L.B. Bjostad, J.-W. Du, and W.L. Roelofs. 1988a. Specificity of male response to multi-component pheromones in noctuid moths *Trichoplusia ni* and *Pseudoplusia includens*. J. Chem. Ecol. 14:47–57.

Linn, C.E., M.G. Campbell, and W.L. Roelofs. 1988b. Temperature modulation of behavioral thresholds controlling male moth sex pheromone response specificity. Physiol. Entomol. 13:59–67.

Linn, C.E., and W.L. Roelofs. 1981. Modification of sex pheromone blend discrimination in male oriental fruit moths by pre-exposure to (*E*)-8-dodecenyl acetate. Physiol. Entomol. 6:421–429.

Löfstedt, C. 1991a. Population variation and genetic control of pheromone communication systems in moths. Entomol. Exp. Appl. 54:199–218.

———. 1991b. Evolution of moth pheromones. Pp. 57–73 *in* I. Hrdy, ed. Insect chemical ecology. Academia, Prague.

Löfstedt, C., J. Löfqvist, B.S. Lanne, J.N.C. Van der Pers, and B.S. Hansson. 1986. Pheromone dialects in European turnip moths *Agrotis segetum*. Oikos 46:250–257.

Löfstedt, C., B.S. Hansson, W.L. Roelofs, and B.O. Bengtsson. 1989. No linkage between genes determining female pheromone production and male olfactory response in the European corn borer *Ostrinia nubilalis*. Genetics 123:553–556.

Margulis, L., and D. Sagan. 1986. Microcosmos: Four billion years of evolution from our microbial ancestors. Summit Books, New York.

Millar, C.D., and D.M. Lambert. 1984. The mating behaviour of individuals of *Drosophila pseudoobscura* from New Zealand. Experientia, Basel 41:950–952.

Miller, J.R., and W.L. Roelofs. 1980. Individual variation in sex pheromone component ratios in two populations of the redbanded leafroller moth, *Argyrotaenia velutinana*. Environ. Entomol. 9:359–363.

Paterson, H.E.H. 1978. More evidence against speciation by reinforcement. S. Afr. J. Sci. 74:369–371.

———. 1981. The continuing search for the unknown and unknowable: A critique of contemporary ideas on speciation. S. Afr. J. Sci. 77:113–119.

———. 1982. Perspective on speciation by reinforcement. S. Afr. J. Sci. 78:53–57.

———. 1985. The recognition concept of species. Pp. 21–30 *in* E.S. Vrba, ed. Species and speciation. Transvaal Museum Monograph No. 4. Pretoria.

Post, D.C., and R.L. Jeanne. 1984. Venom as an interspecific sex pheromone and species recognition by a cuticular pheromone in paper wasps (*Polistes*, Hymenoptera: Vespidae). Physiol. Entomol. 9:65–75.

Priesner, E. 1986. Correlating sensory and behavioral responses in multichemical pheromone systems of Lepidoptera. Pp. 225–233 *in* T.L. Payne, M.C. Birch and C.E.J. Kennedy, eds. Mechanisms in insect olfaction. Clarendon Press, Oxford.

Raina, A.K. 1989. Male-induced termination of sex pheromone production and receptivity in mated females of *Heliothis zea*. J. Insect Physiol. 35:821–826.

Raina, A.K., and J.J. Menn. 1987. Endocrine regulation of pheromone production in Lepidoptera. Pp. 159–174 *in* G.D. Prestwich and G.J. Blomquist, eds. Pheromone biochemistry. Academic Press, New York.

Roelofs, W.L., and A. Comeau. 1969. Sex pheromone specificity: Taxonomic and evolutionary aspects in Lepidoptera. Science 187:398–400.

Roelofs, W.L., J.-W. Du, C.E. Linn, T.J. Glover, and L.B. Bjostad. 1986. The potential for genetic manipulation of the redbanded leafroller moth sex pheromone blend. Pp. 263–272 *in* M.D. Huettel, ed. Evolutionary genetics of invertebrate behavior—progress and prospects. Plenum Press, New York.

Roelofs, W.L., T. Glover, X.-H. Tang, I. Sreng, P. Robbins, C. Eckenrode, C. Löfstedt, B.S. Hansson, and B.O. Bengtsson. 1987. Sex pheromone production and perception in European corn borer moths is determined by both autosomal and sex-linked genes. Proc. Natl. Acad. Sci. 84:7585–7589.

Schneiderman, A.M., S.G. Matsumoto, and J.G. Hildebrand. 1982. Trans-sexually grafted antennae influence development of sexually dimorphic neurones on moth brain. Nature 298:844–846.

Shorey, H.H., and R.J. Bartell. 1970. Role of a volatile female sex pheromone in stimulating male courtship behavior in *Drosophila melanogaster*. Anim. Behav. 18:159–164.

Simon, H.A. 1962. The architecture of complexity. Proc. Am. Philos. Soc. 106:467–482.

Standfuss, M. 1896. Handbuch der pal'darctischen Gross-Schmetlerlinge für Forscher und Sammler. G. Fischer, Jena.

Stenesh, J. 1989. Dictionary of biochemistry and molecular biology. John Wiley, New York.

Subchev, M.A.I., A. Krusteva, and H. Arn. 1986. Bulgarian pheromone "dialect" of

Agrotis segetum Den. & Schiff. (Lepidoptera, Noctuidae). Ecology (Bulg. Acad. Sci.) 19:71–75.
Teal, P.E.A., J.H. Tumlinson, and H. Oberlander. 1990. Endogenous suppression of pheromone production in virgin female moths. Experientia, Basel 46:1047–1050.
Tinbergen, N. 1951. A study of instinct. Clarendon Press, Oxford.
Verrell, P.A. 1988. Stabilizing selection, sexual selection and speciation: A view of specific-mate recognition systems. Syst. Zool. 37:209–215.
Walgebach, C.A., J.K. Phillips, D.L. Faustini, and W.E. Burkholder. 1983. Male produced aggregation pheromone of the maize weevil, *Sitophilus zeamais*, and interspecific attraction between three *Sitophilus* species. J. Chem. Ecol. 9:831–841.
Witzgall, P., and B. Frérot. 1989. Pheromone emission by individual females of the carnation tortrix, *Cacoecimorpha pronubana*. J. Chem. Ecol. 15:707–717.
Wood, D.L. 1982. The role of pheromones, kairomones, and allomones in the host selection and colonization behavior of bark beetles. Annu. Rev. Entomol. 27:411–446.

16

Genetic Variation in Mating Signals and Responses

ROGER BUTLIN

School of Pure and Applied Biology
University of Wales, College of Cardiff

SEXUALLY REPRODUCING ANIMALS have to find mates, and achieve fertilization, in order to pass on their genes to subsequent generations. Finding a mate involves a sequence of signals and responses that leads to copulation or to the synchronized release of gametes. Fertilization may then involve further signal-response interactions between gametes. An enormous diversity of mating signals exists among animal species utilizing visual, tactile, chemical, and acoustic channels or combinations of these channels. While closely related species tend to use similar channels, they almost always have distinct signals, and frequently there is a wide range of signal characteristics within morphologically conservative groups.

There can be little doubt that an *effect* (*sensu* Williams 1966) of mating signal differences between species is to cause assortative mating: partners of the same species are able to complete the signal-response sequence and achieve matings, but the sequence breaks down in interspecific pairs. Thus, signal divergence is an important component of speciation—indeed it may frequently be a primary component of speciation, arising before the evolution of postzygotic barriers to gene flow (Coyne and Orr 1989).

How does signal divergence occur? An answer to this question clearly requires an examination of the selection pressures acting on signals, and this presents an immediate problem. The primary selection pressure acting on a signal arises from the tuning of its receptor. If the receptor is closely tuned to the signal, as might be expected in an efficient signal-response system, then new variant signals will produce lower responses, and selection will tend to remove them. A complementary argument applies to receptors—new variants will tend to be removed by selection.

Author's current address: Department of Genetics, The University of Leeds

Moore (1979) made this problem explicit in a single locus model of assortative mating. Alleles at the assortative mating locus are subject to positive frequency dependent selection—selection acts against them most strongly when they are rare—and so new alleles cannot invade a population. This model has been criticized because it oversimplifies the mating process—a single gene affecting both sexes to produce a major effect on mating pattern seems unlikely. However, it does serve to emphasize the difficulty faced by rare variants, whatever the genetic system. In fact, there is at least one example that fits Moore's model quite closely. In the snail *Partula suturalis* the direction of coiling of the shell (dextral or sinistral) is controlled by a single locus, and the morphs show strong assortative mating (Johnson 1982). In polymorphic populations there is evidence for positive frequency-dependent selection (Murray and Clarke 1980).

More generally, this model reflects a view of mating signals which emphasizes their inherent stability, arising from the need for coordination between mating partners. However, a contrasting viewpoint exists, derived ultimately from Darwin's (1871) ideas about intersexual selection and the development of the theory by Fisher (1930). Here, mating signals are viewed as highly labile because female preferences can exert strong selection pressures, overcoming the stabilizing influence of other components of selection.

Evolutionary change in any character is constrained by the genetic variability available within populations. Conversely, genetic variability within and among contemporary populations reflects the patterns of selection operating on a character. Each character evolves in the context of many other characters, and this is especially true for components of the mating signal system. Nevertheless, focusing attention on variation in individual characters, and its effects on the probability of mating, is more likely to advance understanding than attempts to consider "mate recognition systems" as invariant entities, with only the "black box" of development to account for evolutionary change (White et al., this volume). In this chapter, I will examine published studies of the variability of signals and receptors and relate these data to the two contrasting views of signal evolution outlined above. First, it will be necessary to examine these viewpoints in more detail. I will call them the Recognition and Competition models.

The Recognition Model

Paterson (1985) has placed strong emphasis on the need to maintain an efficient system of signals and responses leading to fertilization within a population. He calls this system the Specific-Mate Recognition System or SMRS. The major selective forces on an SMRS component derive from the need to initiate and maintain exchanges of signals and responses with potential mates. This has two aspects. The signal (or response) must be recognized by the potential mate, and it must be appropriate to the environment in which the animals live. Mutual recognition leads to stabilizing selection exerted by each sex on the signals or responses of the other, since the most efficient signal will be the one corresponding to the mean receptor type and vice versa. Superimposed on this will be the suitability of signals to the environment in terms of their propagation, attractiveness to predators, and so on. The net result will be a narrow distribution of signal and receptor types in a population with a mean determined by adaptation to local environmental conditions. In Paterson's own words, "Such a chain of coadapted stages will, of course, be subject to intense stabilizing selection," but each component "is subject to a small, characteristic variance" (Paterson 1978: 370). Stabilizing selection will reduce genetic variation in components of the SMRS, increasing the stability of the system.

Within a continuous distribution there is likely to be little geographical variation in the SMRS since its evolution is dominated by the stabilizing effects of interactions between the sexes. However, isolated populations may be faced with different environmental conditions which favor distinct optima for the signal components of the SMRS. These populations may evolve toward new equilibria. Thus, the Recognition Model is not incompatible with geographic variation in signal characteristics, but it predicts that this variation will be correlated with environmental parameters.

The generality of the proposed environmental influences may be open to question. The physical environment may favor signals that propagate more efficiently in different environments (Bailey 1991). Possible examples are the anurans *Ranidella riparia* (Odendaal et al. 1986) and *Acris crepitans* (Ryan et al. 1990). However, much of the observed signal diversity among species is in characters that are less likely to be influenced in this way, such as pheromone blend composition or the pulse repetition frequency of a substrate-transmitted vibration. For characters such as these the biotic environment may be more important, for example, through acoustically orienting predators (Tuttle and Ryan 1981) or parasites (Cade 1975) or through interference between sympatric species using the same signal channel (Otte 1989). Paterson emphasizes physical rather than biotic components of the environment (1982:55–56) and predicts that species will remain within their preferred habitat wherever possible. Thus, evolutionary change in the SMRS will tend to be restricted to short time periods in small, isolated populations (Paterson 1982:56).

The Competition Model

The contrasting viewpoint of the Competition Model arises from a distinct approach to signal evolution, originally aimed at understanding the extreme development of secondary sexual characters (primarily mating signals) rather than the nature of the boundaries between species. It focuses on the consequences of intersexual selection: variation in mating success among individuals of one sex due to the behavior of individuals of the other sex. However, it is important to note immediately that the stabilizing selection on signals and receptors described above is, itself, an example of intersexual selection: males with divergent signals obtain few matings because the majority of females have receptors tuned to the "typical" signals.

The contrast between this model and the Recognition Model can best be understood from a consideration of Lande's (1981) influential quantitative genetic model. This has been widely applied to the question of signal evolution (e.g. West-Eberhard 1983). Its conclusions are typical of a class of models dubbed nonadaptive by Kirkpatrick (1987) to denote the lack of a correlation between the equilibrium states of signals and other aspects of the organism or environment. This is in contrast to other models of sexual selection in which signals are related either to the direct benefits a female may receive from her mate, such as parental care, or to genetic benefits for the female's offspring (for review see Bradbury and Andersson 1987).

Lande (1981) considers a male trait and a female preference each determined by a large number of loci. He assumes that, despite selection, significant genetic variation persists for both characters as a result of the input of mutational variation in each generation (Lande 1975). Viability selection acts on the male trait toward an environmentally determined optimum. Sexual selection also acts on the male trait as a result of preferential matings by females based on the value of the male trait. No selection acts directly on the female preference, but a genetic correlation is established between preference and preferred character due to assortative mating. As a result of this correlation, evolutionary changes in the female preference occur whenever changes occur in the male trait.

Lande finds that a line of equilibria exists at which viability selection toward the environmentally determined optimum is exactly balanced by sexual selection away from the optimum. At equilibrium there is net stabilizing selection on the male trait due to the opposition of the two components of selection. Furthermore, if drift or environmental fluctuations displace a population from equilibrium, the genetic correlation between male trait and preference ensures that the population will return to the line of equilibria at a different point. Thus, populations are able to move along the line of equilibria, and geographically isolated populations may diverge rapidly without the need for any environmental change. In some circumstances, when sexual selection is strong relative to viability selection the equilibria may be unstable, and there may be "runaway" evolution of the male trait.

This model predicts extensive geographic variation in mating signals and no necessary relationship between this variation and environmental parameters. The model assumes that genetic variation within populations is a general feature of both signals and receptors. A further distinctive feature of this model is that, in most populations, sexual selection should be directional, rather than stabilizing, and opposed by nonsexual components of selection.

Two major areas of difficulty have been identified for this model. The first concerns the genetical assumptions of the model and in particular the likelihood of maintaining genetic variation and genetic correlations in natural populations: they are not maintained in simulations of finite populations (Nichols and Butlin 1989). Second, the assumption that selection does not act directly on female preferences seems improbable: females with extreme preferences might be expected to suffer some costs due to delayed mating or failure to mate, as would females with very restricted preferences. The effects of female costs have been examined by Pomiankowski (1987), although there are few data relevant to this question. In general, costs of extreme preferences should result in a single equilibrium at the optimum for nonsexual components of selection, a result essentially similar to the Recognition Model. If costs apply more to narrow preferences, they will tend to weaken sexual selection and reduce the genetic correlation between the male signal and female preference (De Jong and Sabelis 1991). This does not remove the line of equilibria but reduces the tendency for populations to move along it.

The details of Lande's model are not critical to the competition viewpoint. As West-Eberhard (1983, 1985) has emphasized, it is intraspecific competition for mates that is at the center of the idea. Whatever its origin, competition can give rise to rapid evolutionary change in unpredictable directions.

Paterson's Recognition Concept leads to the view that "SMRS characters show remarkable stability throughout the geographical ranges of species" (Masters et al. 1987). By contrast West-Eberhard (1983) states that "rapid divergence and speciation can occur between populations with or without ecological differences under selection for success in intraspecific social [including sexual] competition." The evidence is under debate. Coyne et al. (1988) contend that stability is not a feature of mating signals, citing observations by Mayr (1963) and Spieth and Ringo (1983) of geographical variation within species. But the relevance of these data is questioned by Masters and Spencer (1989) for two reasons: that variable characters may not be components of the SMRS and that assortative mating within species may indicate the presence of a complex of cryptic species. They offer a study of geographical populations of *Drosophila melanogaster* (Henderson and Lambert 1982) that failed to detect assortative mating as "the only well studied example to date" (Masters and Spencer 1989:276). What follows is an attempt at a broader review of the available data.

The intention of this survey is not to provide a quantitative distinction between two clearly defined predictions, but rather to assess the weight of the available evidence in relation to two general viewpoints on the evolution of mating signal systems. With respect to variation within populations, Paterson (1978:370) expects the phenotypic variation in all components of the SMRS to be "small" and that, as a result, evolutionary change "must, inevitably, be slow." While he makes no direct statement about genetic variance, the implication is that intense stabilizing selection operating on the components of the SMRS will reduce this relative to the levels typical of less strongly selected quantitative characters. However, Paterson sees the SMRS as one group of characters in a larger set of "adaptive characters under stabilizing selection" due to the tendency of species to "remain in their preferred habitat" (1982:54). He sees change in all of these characters occurring most often when "a small population of conspecific individuals becomes displaced into, and restricted to, a new habitat" (1985:26). This leads to the expectation of geographical stability of SMRS components across a species' range (Masters et al. 1987) and of temporal stasis (Paterson 1982:56). Masters and Spencer (1989:274) suggest that the recognition view predicts "less variation in components of an SMRS than in other phenotypic features of a species" so that the variability of the SMRS could be assessed relative to other characters. However, this is clearly at variance with Paterson's expectation that other adaptive characters are held under stabilizing selection. There is no obvious yardstick for the extent of variability that is consistent with the recognition view.

The competition view relies on the theoretical conclusion that "polygenic mutation . . . can maintain additive genetic variance . . . in spite of selection tending to deplete genetic variation" (Lande 1981:3724). The contrast between this and the recognition view thus has two components: the specific question of the intensity of stabilizing selection on mating signals and preferences and the more general question of the maintenance of genetic variation for quantitative traits. This latter question is largely unresolved (Barton and Turelli 1989) and is beyond the scope of this chapter. With regard to the intensity of stabilizing selection, Lande (1981:3724) assumes that in "polygamous species where males are promiscuous . . . and females have many potential mates, there is no selection directly on female mating preferences." As noted above, this is a critical assumption accounting for much of the difference between the two views. Male traits are subject to strong stabilizing selection resulting from the combination of viability selection and female preferences. Thus, within populations both views predict limited phenotypic variation in male traits, but the competition view predicts that this will not impede rapid evolutionary change because genetic variation is maintained despite stabilizing selection. The competition view predicts less intense selection on female preferences, and thus more variation within populations.

The competition view clearly predicts "rapid indeterminate evolution"

(Lande 1981:3724) of mating signals and preferences, even in the face of some gene flow (Lande 1982). The expectation of extensive geographical variation—greater than expected simply through adaptation to local environmental conditions because of the exaggerating effect of competition—is clearly distinct from the prediction of the recognition view.

The Data

An immediate problem is encountered when searching for data on the variability of mating signals: Which characters are to be included? Paterson (1980) defines the SMRS as "a coadapted signal-response chain" whose raison d'etre "is to ensure effective syngamy within a population occupying its preferred habitat." He emphasizes the use of "specific-mate" to exclude other types of recognition, particularly individual recognition (Paterson 1985). There is evidence in some species (from birds [Emlen 1972] to *Drosophila* [Schilcher 1976]) that the two functions of species recognition and individual recognition are served by separate characters (although this point may not have been adequately demonstrated [Butlin 1986]). However, a character currently used in individual recognition may acquire a species recognition function during evolution.

Paterson (1982) criticized Littlejohn (1981) for equating recognition with selection of a partner. In Paterson's view the SMRS consists of those characters that must be present in order for the sequence of events leading to fertilization to be completed. It excludes additional characters that might be used by one sex or the other to choose mates on the basis of direct or indirect fitness benefits. This is one reason for the divergence of the recognition and competition views of mating signal evolution. But are there really two sets of characters? Suppose a gene for female preference for low-pitched calls enters a population and spreads because males with low-pitched calls happen also to be good fathers (as in the bullfrog *Rana catesbiana* [Howard 1978]). At first the preference and the call character are outside the SMRS (as defined by Paterson)—they are used for selection, not recognition. But once all females in a population have the preference, the call character becomes part of the SMRS since males without the character fail to complete the sequence of behaviors leading to mating. As with individual recognition characters, "selected" characters cannot be distinguished unequivocally from "specific-mate recognition" characters, and so I include in this review all signals and responses involved in obtaining a mate.

In what follows I have generally restricted myself to two types of data: signal characters and their associated responses, and mate choice experiments. Both have their drawbacks. It may be well documented, for example, that the acoustic signals of acridid grasshoppers are a major part of the SMRS in some species (e.g. *Chorthippus biguttulus* [Perdeck 1957]), but this cannot necessarily be extended to all species or to all song characters. In practice there are very few examples of signal variation where it is possible to predict potential levels of assortative mating from observed signal divergence.

An alternative to studying variation in signals is to look for assortative mating between strains or populations of a species, or in hybrid generations. Significant positive assortment is taken as evidence of divergence in the mating signal system. This approach has the advantage that the organisms themselves assess the divergence and so there is no need for the characters to be accessible to our observation or for us to assess their functions. However, there are problems. Laboratory assortative mating experiments typically underestimate premating isolation: for example, the "semispecies" of *Drosophila paulistorum* coexist without gene exchange in the field, but assortative mating is never complete in laboratory tests (Ehrman 1961). On the other hand, a low level of assortment can be obtained in the laboratory simply because the two groups to be tested are reared separately (Spiess and Spiess 1967).

With these caveats in mind, I have surveyed variation within populations and among populations. Variation among populations has been subdivided into geographical variation not associated with other genetic discontinuities and variation among genetically divergent groups, mainly between parapatric races or subspecies. Cases in the first group in which the variation may be explained by overlap of the species distribution with a closely related species, potential examples of reproductive character displacement, are also considered separately.

The study of communication systems is greatly complicated by the potential in some animal groups for learning. Not only does this obscure the genetic basis of variation in signals and responses, but it also opens up an entirely new dimension of variability and the potential for distinct modes of evolution. Passerine birds illustrate the impact of learning very clearly: in the suboscines (about 1000 species) songs are inherited and they are simple and species specific, but in the oscines (more than 4000 species) songs are learned and they are complex and highly variable. I have attempted to avoid this complication by excluding birds and mammals from my survey. The great majority of data, therefore, come from acoustic Anura and insects, especially the Orthoptera, Lepidoptera, and *Drosophila*.

Variation within Populations

Both the Recognition and Competition models predict stabilizing selection on mating signals within populations. In the Recognition Model both sexual and nonsexual components of selection are expected to be stabilizing, whereas in the Competition Model the net stabilizing selection is expected to result in most cases from a balance between opposing forces of sexual and nonsexual selection. However, there have been surprisingly few studies that have considered the possibility of stabilizing selection on signal characters, and none for response characters. There is some direct evidence for stabilizing selection on mating signal system components: acoustic signals in the grasshopper *Chorthippus brunneus* (Butlin et al. 1985) and the frog *Hyla cinerea* (Gerhardt 1987); and indirect evidence for tibial combs in *Drosophila silvestris* (Carson and Teramoto 1984). On the other hand, there is evidence for directional female preferences on various secondary sexual characters that function as mating signals such as tail length in swallows (Møller 1990), bower features in satin bower birds (Borgia 1985), and call pitch in anurans (e.g. Robertson 1986) (reviewed by Searcy and Andersson [1986] for song characters).

In the guppy, *Poecilia reticulata*, there is a balance between sexual selection, due to female preference for conspicuous males, and selection by predators favoring crypsis (Endler 1983). Male color patterns may also function in species recognition although there is evidence that display characteristics are more important (Liley 1966; Endler 1983). Furthermore, populations remain highly polymorphic for male color pattern despite these selection pressures. The guppy, therefore, appears to provide a clear example in support of the competition view. However, Endler (1983) doubts whether the sexual selection acting on male color pattern is actually of the Fisherian type modeled by Lande (1981) and proposed as the major cause of mating signal divergence by West-Eberhard (1983). Instead, he considers male patterns to be indicators of foraging ability and that female choice is aimed at obtaining mates of demonstrated high fitness. This example remains much more consistent with the competition view than with the alternative at the within-population level, but it also demonstrates the potential complexity of selection pressures acting on courtship behavior.

Mating signals, like most quantitative characters, are expected to show some environmental variation within populations. Much more important for signal evolution is the extent of genetic variation. I have excluded from this section populations within hybrid zones that have genetic variation by virtue of the divergence between two parapatric populations, and also variation between sympatric subpopulations identifiable on grounds independent of the mating signal system. Both of these groups will be considered below. I have also excluded variation that leads to disassortative mating, as, for example, in *Panaxia dominula* (Ford and Sheppard 1969), *Drosophila melanogaster* (Averhoff and Richardson 1976), or *Coelopa frigida* (Day and Butlin 1987), since this variation is unlikely to be involved in the evolution of ethological isolation.

Studies of genetic variation in signals or responses, or genetic variation leading to assortative mating, are listed in Table 16.1 for a total of twenty-five species. In all of these species there is evidence of within-population genetic variation in signal or response characters. Only four studies indicated lack of variation: two in *Drosophila melanogaster,* one in the moth *Argyrotaenia velutinana,* and one in the planthopper *Nilaparvata lugens* (Table 16.1). Other studies have given clear evidence of variation in each species. In only two cases (the fish *Xiphophorus nigrensis* [Ryan and Wagner 1987] and the ladybird *Adalia bipunctata* [O'Donald and Majerus (1985) but see Kearns et al. (1992)]) is there evidence for the involvement of a major locus. In the remaining cases the variation can be described as polygenic although in no case has the genetic basis been analyzed in detail. Explicit studies of the variability of the responder side of the signal system are very limited, although it is clear that they have been modified in at least some of the experiments in which artificial selection has increased the assortative mating between strains (e.g. *Drosophila mojavensis,* Koepfer 1987*a,b*).

In several studies, particularly those on pheromone blend, the coefficient of variation for signal characters is low by comparison with other quantitative characters, while only one study reports higher variation (*Poecilia reticulata,* Table 16.1). In *Chorthippus brunneus* the heritability of song characters is low relative to morphological characters. Thus the existence of genetic variation need not imply that rapid evolution is possible. For example, in *Pectinophora gossypiella* genetic variation in pheromone blend is certainly present, but twelve generations of selection only produced a change from 43 percent to 48 percent of the Z, E isomer. The significance of such a shift depends, of course, on the specificity of the responder—in this case very little difference in male response would be expected for this shift in blend (Collins and Cardé 1989*a*).

There appears to be a common thread in these data. Mating signals typically show limited phenotypic variation within populations but this variation probably has a significant genetic component. Thus they are able to respond to selection, but the response is often slow. What little information there is on the range of acceptable signals suggests that the variation present in signal characteristics is of little significance to responders (Haynes and Baker 1988; Butlin 1992). In experiments that have generated an increase in assortative mating it is not usually clear which components of the signal system have been altered and, overall, there is a distinct lack of information on variation among individuals in the responder side of the system. Insufficient studies have considered the possibility of stabilizing selection for an assessment to be made of its prevalence. There is no indication of two distinct classes of characters involved in recognition and competition.

Variation among Populations

Geographical variation in mating signals and responses which is not associated with hybrid zones or with overlap between closely related species is con-

Table 16.1. Genetic variation within populations

Species	Character	Notes	Reference
Reptiles			
Anolis distichus and *brevirostris*	Dewlap coloration	Variation presumed heritable	Case 1990
Amphibians			
Hyla versicolor	Pulse rate of call	Heritable variation	Klump and Gerhardt 1987
Fish			
Poecilia reticulata	Male color pattern	Several polymorphic loci	Endler 1983
Xiphophorus nigrensis	Courtship components	Polymorphic locus on Y	Ryan and Wagner 1987
Insects			
Pectinophora gossypiella	Pheromone titre	$h^2 = 0.41$, $h^2 = 0.71$; 6 generations response to selection	Collins and Cardé 1985 Collins et al. 1990
	Pheromone blend	$h^2 = 0.34$ (C.V. = 5%); 12 generations response to selection (43% to 48% Z,E)	Collins and Cardé 1985, 1989d
		No response to use of pheromone in control	Haynes et al. 1984
	Pheromone response	Responds to selection $h^2 = 0.16$	Collins and Cardé 1989c
	Pheromone preference	$h^2 = 0.39$, own blend h^2 low, 25% or 65% Z, E blends	Collins and Cardé 1989b
Argyrotaenia velutinana	Pheromone blend	Lack of response to selection	Roelofs et al. 1987
	Pheromone blend	$h^2 = 0.4$; genetic correlation with male preference?	Roelofs et al. 1986
	Male blend preference	Heritable variation	Roelofs et al. 1986
Ephestia cautella	Pheromone blend	Variation among females, probably heritable	Barrer et al. 1987

(*continued*)

Table 16.1. (Continued)

Species	Character	Notes	Reference
Colias eurytheme	Pheromone blend (male)	h^2 high in one of two populations	Sappington and Taylor 1990
Chorthippus brunneus	Song parameters	h^2 low	Butlin and Hewitt 1986
	Female preferences	Respond to selection	Charalambous 1990
Gerris odontogaster	Male genitalia	$h^2 = 1.01$	Arnqvist 1989
Nilaparvata lugens	Pulse repetition frequency	$h^2 = 0.25$	Roper and Butlin, unpubl.
		h^2 low	Butlin 1992
	Female preference	h^2 low	Butlin 1992
Ribautodelphax spp.	Pulse repetition frequency	Rapid response to selection	de Winter 1992
Tribolium castaneum	Assortative mating	Generated by disruptive selection	Halliburton and Gall 1981
		Variation among laboratory strains	Graur and Wool 1982
Ips pini	Pheromone blend	Bimodal in some populations, presumed heritable	Miller et al. 1989
Adalia bipunctata	Female preference for melanic	Responds to selection, probably one locus	Majerus et al. 1982; O'Donald and Majerus 1985, but see Kearns et al. 1992
Drosophila pseudoobscura	Assortative mating	Variation between laboratory stocks	Millar and Lambert 1986
	Isolation from persimilis	Increased by selection	Koopman 1950
D. persimilis	Isolation from pseudoobscura	Increased by selection	Koopman 1950
	Mating propensity	Influenced by polymorphic inversions	Spiess and Spiess 1967, 1969
D. melanogaster	Assortative mating	Increases when "hybrid" offspring eliminated	Ehrman 1971, 1973, 1983; Crossley 1974
	Isolation from simulans	Increased by selection	Eoff 1975
	Wing vibration	Responds to selection	McDonald and Crossley 1982

(continued)

Table 16.1. (*Continued*)

Species	Character	Notes	Reference
	Isolation between French and Japanese strains	*Not* increased by selection	Petit et al. 1980
	Assortative mating	Increased by disruptive selection	Thoday and Gibson 1962, 1970
	Assortative mating	*Not* increased by disruptive selection	e.g. Spiess and Wilke 1984
	Discrimination against *yellow* males	Variation among strains (X and autosomal loci)	Heisler 1984
D. simulans	Interpulse interval	Variation among Japanese strains	Kawanishi and Watanabe 1980
	Isolation from *melanogaster*	Responds to selection	Eoff 1975, 1977
	Wing vibration	Weak response to selection	Wood and Ringo 1982
D. silvestris	Tibial comb	$h^2 = 0.50$; responds to selection; polygenic	Carson 1985; Carson and Lande 1984; Carson and Teramoto 1984
	Courtship success	Genetic variation present	Ahearn and Templeton 1989
D. mojavensis	Isolation between Baja and Sonora populations	Responds to selection	Koepfer 1987a,b
D. paulistorum "Transitional"	Assortative mating	variation among strains (autosomal, additive)	Dobzhansky et al. 1969
D. paulistorum "Llanos" and "Interior"	Assortative mating	Increased by selection	Dobzhansky et al. 1976
D. mercatorum	Interpulse interval	Responds to selection; $h^2 = 0.32$ up selection; $h^2 = 0.14$ down selection; polygenic	Ikeda and Maruo 1982
D. immigrans	Assortative mating	Among isofemale lines	Ehrman and Parsons 1981

sidered in this section. In most species, if two populations differ consistently for a signal or response character or show assortative mating in the laboratory it can be assumed that they differ genetically.

Two serious problems face the assessment of geographic variation in mating signals. One is the use of secondary sexual characters in the definition of species. This practice is widespread in classifying animals in some groups such as Orthoptera and Anura, even among taxonomists who do not follow Paterson's Recognition Concept of species. The danger from the point of view of the present survey is that any two populations found to differ consistently in a mating signal will be defined as separate species, and Paterson's prediction that the SMRS will not vary within species will be self-fulfilling. In cases of parapatry or overlap, I have used the criterion of gene exchange in natural situations and thus consider the *Drosophila paulistorum* semispecies to be good species (with the possible exception of the Transitional populations) but *Litoria ewingi* and *Litoria paraewingi* to be conspecific (Table 16.2, Section 3). In allopatry the criterion of gene exchange is not available, and so it has been necessary to accept currently recognized species.

The opposite problem is one of reporting. If authors tend not to report "lack of variation in mating signals" of a species, publications may be biased toward evidence of geographic variation. There appears to be no way of overcoming this problem. However, if the signaling system is really highly stable we would expect to find rather few published examples of variation, especially given the above tendency to define species where variation does occur.

Table 16.2, Section 1, lists sixty-nine species or species groups in which statistically significant geographic variation in signal or response characters, or assortative mating among populations, has been demonstrated. Only eighteen cases lacking variation have been reported (Table 16.2, Section 2), and in some of these there is variation either at hybrid zones (*sensu* Barton and Hewitt 1985) or in zones of overlap. A few species are included in both groups where different characters show different patterns; for example, *Drosophila melanogaster* song characters show some geographic variation, but no assortative mating is observed among populations. About one-third of the examples are in the genus *Drosophila,* and these show a higher proportion of species with variation (twenty-nine out of thirty-four). At least two of the *Drosophila* studies that failed to detect variation were on a very small scale (*gaucha* and *pavani*).

Excluding *Drosophila,* nine studies out of fifty have not reported any variation, unless it is associated with zones of hybridization or overlap. All of these cases are based on measurements of the parameters of acoustic signals and, strictly speaking, only indicate lack of variation in these parameters. In general, the biological significance of variation in a signal character can be decided only by investigating its contribution to assortative mating, either directly or through study of the associated response. Signal parameters found to vary geographically may not function in mate choice, or the variation may be small relative to the range of signals that can elicit responses. Direct demonstrations of a link between signal variation and assortative mating are available in only a few cases (e.g. *Nilaparvata lugens,* Claridge et al. 1985*a,b*). However, as outlined

Table 16.2. Reports of geographic variation

Species	Character	Reference
	1. Continuous variation	
Reptiles		
Anolis distichus complex	Dewlap coloration	Crews and Williams 1977; Case 1990
Anolis brevirostris	Dewlap coloration	Webster and Burns 1973; Case 1990
Anolis marmoratus complex	Head and body colour	Crews and Williams 1977
Amphibians		
Acris crepitans and *gryllus*	Calls and responses	Nevo and Capranica 1985
Bufo woodhousei	Call parameters	Sullivan 1989
Gastrophryne olivacea	Call parameters	Bogert 1960
Eleutherodactylus coqui	Carrier frequency of call	Narins and Smith 1986
Hyla eximia	Call parameters	Bogert 1960
Hyla versicola and *chrysoscelis*	Call parameters	Ralin 1977
Desmognathus ochrophaeus	Assortative mating	Verrell and Arnold 1989; Tilley et al. 1990
Plethodon vehiculum	Pheromones	Ovaska 1989
Fish		
Gasterosteus aculeatus	Red/black throat	McPhail 1969; Moodie 1982
Poecilia reticulata	Color pattern and displays	Endler 1983
	Female preferences	Houde 1988; Houde and Endler 1990
Crustacea		
Uca spp.	Morphology and displays	Crane 1975
Insects		
Agrotis segetum	Preferred pheromone blend	Cardé and Baker 1984
	Pheromone blend	Löfstedt et al 1986; Hansson et al. 1990
Choristoneura occidentalis	Pheromone blends and preferences	Leibhold and Volney 1985
Pectinophora gossypiella	Pheromone blend	Rothschild 1975
Grapholita molesta	Pheromone blend	Rothschild and Minks 1977
Synanthedon pictipes	Pheromone blend	McLaughlin et al. 1977

(*continued*)

Table 16.2. (Continued)

Species	Character	Reference
Spodoptera littoralis	Pheromone blend	Campion et al. 1980
Ephippiger ephippiger	Call parameters	Busnel 1963; Duijm 1990
Pterophylla camellifolia	Call parameters	North and Shaw 1979
Pterophylla spp.	Call parameters	Barrientos-Lozano 1988
Chorthippus brunneus	Call parameters	Perdeck 1957
Chorthippus yersini	Call parameters	Ragge and Reynolds 1988
Chorthippus parallelus	Assortative mating and call parameters	Dagley 1988
Allonemobius socius	Call parameters	Howard 1986
Oecanthus fultoni	Call parameters	Walker 1974
quadripunctatus	Call parameters	Walker 1974
exclamationis	Call parameters	Walker 1974
Gryllus rubens	Call parameters	Walker 1974
Mullerianella brevipennis and *fairmairei*	Call parameters	Booij 1982
Nilaparvata lugens	Male and female calls	Claridge et al. 1985b, 1988
Nephotettix virescens and *nigropictus*	Call parameters	Claridge 1985
Musca domestica	Body size and size preferences	Baldwin and Bryant 1981
Cochliomyia hominovorax	Pheromone titre and response	Hammack 1987
Chrysoperla spp.	Call parameters and preferences	Henry 1985; Henry and Wells 1990; Wells and Henry 1992
Blackburnium spp.	"horns"	Howden 1979
Ips pini	Pheromone blend	Lanier et al. 1980; Miller et al. 1989
Adalia bipunctata	Clinal variation in female preferences	Majerus, pers. comm.
Pteronarcella badia	Call parameters	Stewart et al. 1982
Aedes albopictus	Cuticular hydrocarbons	Kruger et al. 1991
Lipara lucens	Call parameters	Kanmiya 1990
Drosophila equinoxialis	Assortative mating	Ayala et al. 1974
D. pseudoobscura	Isolation from *miranda*	Lakovaara and Saura 1982

(continued)

Table 16.2. (Continued)

Species	Character	Reference
D. miranda	Assortative mating	Lakovaara and Saura 1982
D. melanogaster	Cuticular hydrocarbons	Jallon and David 1987
	Assortative mating	Petit et al. 1976
	Male response to pheromones	Tompkins and Hall 1984
	Interpulse interval and sine song frequency	Cowling 1980
D. simulans	Interpulse interval	Kawanishi and Watanabe 1980
	Cuticular hydrocarbons	Jallon and David 1987
	Assortative mating	Ringo and Wood 1980
D. silvestris	Tibial comb	Carson et al. 1982
	Courtship success	Ahearn and Templeton 1989
D. heteroneura	Courtship success	Ahearn and Templeton 1989
D. grimshawi	Assortative mating	Carson and Yoon 1982
D. pullipes	Assortative mating	Ohta 1978
D. serrata	Assortative mating	Dobzhansky 1962
D. paulistorum "Transitional"	Isolation from other semispecies	Dobzhansky et al. 1969
	Courtship elements	Koref-Santibanez 1972
D. immigrans	Assortative mating	Ehrman 1972; Ehrman and Parsons 1980
D. mercatorum	Interpulse interval	Ikeda and Maruo 1982
	Female response threshold	Ikeda et al. 1981
D. littoralis	Call parameters	Hoikkala 1985; Hoikkala and Lumme 1990
D. bifasciata	Assortative mating	Dobzhansky et al. 1968
D. imaii	Assortative mating	Dobzhansky et al. 1968
D. virilis	Assortative mating	Anderson and Ehrman 1969
D. americana	Assortative mating	Anderson and Ehrman 1969

(continued)

Table 16.2. (*Continued*)

Species	Character	Reference
D. arizonae	Assortative mating	Anderson and Ehrman 1969
D. auraria	Assortative mating	Anderson and Ehrman 1969
D. birchii	Assortative mating	Anderson and Ehrman 1969
D. crocina	Assortative mating	Anderson and Ehrman 1969
D. gasici	Assortative mating	Anderson and Ehrman 1969
D. miranda	Assortative mating	Anderson and Ehrman 1969
D. montana	Assortative mating	Anderson and Ehrman 1969
D. nebulosa	Assortative mating	Anderson and Ehrman 1969
D. peninsularis	Assortative mating	Anderson and Ehrman 1969
D. sturtevanti	Assortative mating	Anderson and Ehrman 1969
D. texana	Assortative mating	Anderson and Ehrman 1969
2. Failures to detect variation		
Amphibians		
Litoria ewingi complex	Call parameters (excluding cases of character displacement)	Littlejohn and Watson 1985
Geocrinia laevis and *victoriana*	Call parameters	Littlejohn and Watson 1985
Insects		
Omocestus spp.	Call parameters	Ragge 1986
Euchorthippus spp.	Call parameters	Ragge and Reynolds 1984
Chorthippus biguttulus group	Call parameters (excluding above)	Ragge and Reynolds 1988
Teleogryllus commodus and *oeceanicus*	Call parameters	Hill et al. 1972

(*continued*)

Table 16.2. (Continued)

Species	Character	Reference
Orocharis saltator and luteolira	Call parameters	Walker 1974
Gryllus integer	Call parameters	Walker 1974
Belocephalus sabalis	Call parameters	Walker 1974
Drosophila pseudoobscura	Assortative mating	Anderson and Ehrman 1969
	(including subspecies bogotana)	Prakash 1972
D. melanogaster	Assortative mating	Henderson and Lambert 1982; Lambert and Henderson 1986
	Interpulse interval	Kawanishi and Watanabe 1980
D. arizonae	Isolation from mojavensis	Wasserman and Koepfer 1977
D. willistoni	Assortative mating	Dobzhansky 1962, 1975
D. gaucha	Assortative mating	Anderson and Ehrman 1969
D. pavani	Assortative mating	Anderson and Ehrman 1969
D. annanassae	Assortative mating	Futch 1973
D. pallidosa	Assortative mating	Futch 1973
3. Possible cases of extra divergence in sympatry		
Reptiles		
Anolis distichus complex	Dewlap coloration	Crews and Williams 1977
Anolis brevirostris	Dewlap coloration	Webster and Burns 1973
Sceloporus undulatus and graciosus	"Push-up" display	Ferguson 1973
Anura		
Pseudacris nigrita/triseriata	Call parameters and preferences	Littlejohn 1960
P. nigrita/feriarum	Call parameters	Fouquette 1975
Scaphiosus bombifrons and hammondi	Pulse rate	Blair 1974
Bufo americanus, terrestris and woodhousei	Call parameters	Blair 1962, 1974
B. microscaphus californicus and punctatus	Call parameters	Bogert 1960
B. compactilis compactilis and c. speciosus	Call parameters	Bogert 1960

(continued)

Table 16.2. (Continued)

Species	Character	Reference
Litoria ewingi/verreauxi	Call parameters	Littlejohn 1965
L. verreauxi/paraewingi	Call parameters	Littlejohn and Watson 1985
Crinia glauerti/insignifera	Call parameters	Blair 1958
Hyla meridionalis/arborea	Call parameters	Bogert 1960
Hyla versicolor/chrysoscelis	Call parameters	Ralin 1977
Crustacea		
Uca rapax/pugnax	Courtship display	Crane 1975
Mollusca		
Partula suturalis/mooreana	Coil direction	Murray and Clarke 1980; Johnson 1982
Insects		
Muellerianella brevipennis and fairmairei	Call parameters	Booij 1982
Archips argyrospilus and mortuanus	Pheromone response	Cardé et al. 1977; Cardé and Baker 1984
Calopteryx maculata and aequabilis	Wing pattern	Waage 1979
Oecanthus rileyi/fultoni	Call parameters	Walker 1974
Barytettix spp.	Genital morphology	Cohn and Cantrall 1974
Allonemobius fasciatus and socius	Call parameters	Benedix and Howard 1991
Laupala spp.	Pulse rate	Otte 1989
Aedes albopictus subgroup	Assortative mating	McLain and Rai 1986
Drosophila paulistorum semispecies	Assortative mating	Ehrman 1965; Dobzhansky et al. 1969
Drosophila mojavensis and arizonae	Assortative mating	Wasserman and Koepfer 1977; Markow 1981

above, signal characters may acquire a recognition function, and so the variability may be relevant even for characters that are not currently significant in mate choice.

According to the recognition view, any geographical variation that does exist in SMRS components is likely to be associated with environmental variables. Rather few studies have looked specifically at this possibility, and yet the examples surveyed do include some cases in which signal variation correlates with environmental variables. In the frogs *Acris crepitans* and *A. gryllus* some parameters of the acoustic signals vary with the environment in a way that may be explained by the need for efficient propagation through different vegetation types (Nevo and Capranica 1985), and propagation differences have been demonstrated directly (Ryan et al. 1990). However, differences between species appear to be qualitatively distinct from intraspecific variation and not linked to the environment. In another frog, *Eleutherodactylus coqui*, call frequency varies with altitude, probably as a result of variation in body size but nevertheless potentially enough to restrict gene flow. Zimmerman (1983) found little evidence of a relationship between habitat and call in a survey of anurans.

The guppy, *Poecilia reticulata*, provides a particularly instructive example in this context also. Male color pattern varies among populations as a result of variation in the intensity of predation which acts in opposition to sexual selection for brightly colored males (Endler 1983). Female preferences may also vary in line with male patterns (Houde 1988; Breden and Stoner 1987; Endler 1988; Houde and Endler 1990). Similarly, stickleback (*Gasterosteus aculeatus*) populations vary in throat color with black throats where predation is high and red throats, which are favored by female choice, where predation is low (McPhail 1969). This variation among populations is more in line with the recognition view of adjustments in the SMRS to suit local environmental conditions than it is with the competition view of random variation among populations in "arbitrary" characters. Measurements of female preferences may help to make this distinction since the recognition view would predict female preference for the type of male that is well suited to its environment, that is, preference for dull males in high-predation populations. The competition view would not predict such a correlation; female preferences balance selection due to predation, but neither male coloration nor female preferences are necessarily limited by environmentally determined optima. At present it appears that neither position adequately explains the data since in both fish species male coloration is correlated with predation pressure, but in all populations there is female preference for bright males, although it varies in intensity (Houde and Endler 1990). This pattern may be better explained by adaptive female preferences, as suggested by Endler (1983), where male coloration is an indicator of male quality.

Extra Divergence in Sympatry

I have excluded from the previous section cases in which the distributions of closely related species overlap and the species show greater divergence in mating signals in the overlap zone. There are at least three possible interpretations of this pattern of variation (Butlin 1989). The pattern may result from reinforcement (as defined by Butlin 1987a,1989) where selection favors divergence in the signaling system to prevent the production of hybrid offspring of reduced fitness and thus to reduce gene flow between incipient species. Second, it may result from reproductive character displacement where selection favors divergence in the signals to avoid wasteful heterospecific matings. A third alternative explanation is that sympatry is possible between species only when they have diverged sufficiently in allopatry to avoid mating interference. There are serious theoretical objections to the process of reinforcement and no convincing examples (Butlin 1987a,1989). The major difficulty with the idea of reproductive character displacement is that it requires coexistence of the two species during the evolution of divergence despite mating interference and possibly ecological similarity (Spencer et al. 1986,1987; Butlin 1987b). This may not be a major problem if the species spread into sympatry from allopatry and can be maintained in an area of overlap by dispersal. Laboratory experiments in which assortative mating increases over generations when hybrid offspring are removed (see Table 16.1) clearly indicate the potential for reproductive character displacement where populations are maintained in sympatry. However, it remains difficult to distinguish reproductive character displacement from overlap of populations after allopatric divergence, and this distinction has not yet been made for any particular example.

The impression of reproductive character displacement may be given by variation within species for other reasons, as in the case of ecological character displacement (Grant 1975). For example, the overlap between *Acris crepitans* and *A. gryllus* has been cited as an example of character displacement, but the very detailed work of Nevo and Capranica (1985) shows that song variation is explained by subspecific differences within each species and by correlations with environmental parameters, not by sympatry versus allopatry.

My survey has revealed twenty-two species or species groups containing possible examples of extra divergence in sympatry (Table 16.2, Section 3). The fact of variation in mating signals within one or both species is well substantiated in these examples, but alternative explanations have not been eliminated even in the best-studied cases. In most cases mapping of variation is insufficient to establish conclusively the link with the presence of the other species. *Litoria ewingi* and *L. verreauxi* probably provide the best example—no environmental parameter explains the signal variation as well as sympatry, but the "overlap after allopatric divergence" explanation remains possible (Littlejohn and Watson 1985). This is also true of two other well-studied examples, *Partula suturalis* and *Drosophila mojavenis/arizonae* (Table 16.2, Section 3). In the *D. mojavensis/arizonae* case a switch in host plant of *mojavensis* may also be implicated. Nevertheless, the rather frequent occurrence of this pattern does lend some support to the idea of reproductive character displacement.

Coyne and Orr's (1989) comparative study of *Drosophila* species pairs also indicates that mating signals are more divergent between sympatric species pairs than between allopatric species pairs that have been diverging for a similar length of time, as judged by electrophoretic distance. Laboratory measures of postmating isolation between these species suggest that it is incomplete, and, therefore, Coyne and Orr consider the possibility that reinforcement is responsible for the divergence in sympatry, although they recognize the theoretical difficulties posed by this model. Reproductive character displacement could be implicated in at least some cases if postmating isolation in the field were complete, or very nearly so. Laboratory measures of postmating isolation are inevitably underestimates since they exclude ecological factors such as competition and predation and, while they incorporate F1 fertility, they rarely consider F1 reproductive success. The comparison made by Coyne and Orr is between allopatric pairs and sympatric pairs of species rather than between allopatric and sympatric populations of a species. This causes some problems since *sympatric* is defined as occurring together in any part of the range. Thus the races of *Drosophila athabasca* are considered sympatric whereas in reality they are parapatric, which means that the vast majority of the populations are actually allopatric. This procedure leaves rather a small group of species in the allopatric category with which to test the "divergence in allopatry" explanation, and these species may be atypical; for example, several of them are Hawaiian species.

Paterson considers the process of reproductive character displacement implausible (Paterson 1978; Spencer et al. 1986). This is because he considers elimination of one species to be the more likely outcome of contact between two species with similar signal systems, but which failed to exchange genes due to postmating barriers. (In fact Paterson would consider such populations conspecific since they share a common recognition system.) However this prediction is based on models of sympatric associations whereas in reality such species would initially be parapatric and form very narrow hybrid zones. Zones in which first-generation hybrids are completely inviable or infertile do exist (e.g. in *Thomomys bottae,* Patton and Smith 1989), but they are rare, perhaps because of reproductive character displacement.

If one views the other species as a component of the environment, then reproductive character displacement is consistent with Paterson's expectation of modification of mating signals to suit environmental conditions. An efficient fertilization system must avoid confusion with other signaling systems operating in the same environment, and closely related species are particularly likely to have similar signals. This view is emphasized by Otte (1989) in connection with the extraordinary variation in songs of Hawaiian *Laupala* crickets, which are morphologically and ecologically very similar and whose signaling environment contains few other acoustic species. However, Paterson (1982:55–56) considers this sort of interaction among species too ephemeral to influence the SMRS.

From the competition viewpoint the presence of another species may provide an initial selective advantage to female preferences, but subsequent evolution is likely to be dominated by the Fisherian process (Lande 1982). This could produce rapid divergence in areas of overlap and could explain continued divergence once hybridization became rare; that is, it could produce complete ethological isolation. This is a problem for other models of reproductive character displacement in which selection for divergence declines as divergence increases, making an explanation of the final separation of the signaling systems difficult.

Variation between Races

Many species are subdivided into genetically divergent races or subspecies that meet at hybrid zones (Barton and Hewitt 1985). In most cases these races probably diverged in allopatry when the species ranges were less extensive and more dissected by geographical barriers. Following range expansion, hybrid zones were formed where they met and are maintained by a balance between selection and dispersal (tension zones, Barton and Hewitt 1985). These races can be viewed as "frozen" on the way toward allopatric speciation, and they are extremely valuable for examination of the extent of divergence and genetic basis of many characters, including those involved in mating.

Barton and Hewitt's survey of more than 150 hybrid zones (in this case, including birds and mammals but not song dialect boundaries in birds) includes thirty-five cases in which parapatric interactions are accompanied either by signal divergence or by assortative mating between "parental" forms. In twenty-three such interactions ethological divergence has been sought but not found. The remaining zones have not been studied from this point of view. The numbers here are somewhat arbitrary since they include several complexes of forms in which many independent interactions exist, such as the *Thomomys bottae* complex in which both types have been observed among the many zones studied. The twenty-three negative interactions include two cases (in orioles [Rising 1983] and flickers [Moore 1987]) in which there is variation in plumage characteristics that potentially function as sexual signals but where there is also evidence of random mating within the hybrid zones. However, this random mating may result from the independent assortment of characters in the zone which, when together in parentals, would be sufficient to produce assortative mating.

Hybrid zones represent a wide range of levels of divergence, reflected by hybrid fitnesses varying from very low to nearly 100 percent. Signal divergence does not appear to correlate closely with general divergence, reinforcing the impression that the signal system evolves independently of postmating isolation. However, in relation to Coyne and Orr's (1989) *Drosophila* data which indicate a more rapid evolution of premating than postmating isolation, it is surprising that the proportion of zones showing signal divergence is not greater. Of course, most of the differences observed at hybrid zones have evolved in allopatry, and so these data may strengthen Coyne and Orr's conclusion that selection in sympatry is responsible for the rapid evolution of premating isolation. The absence of evidence for reinforcement in hybrid zones (Barton and Hewitt 1985; Butlin 1987a,1989) suggests that this is not the way in which selection has operated. Once again, it is possible that, where allopatric divergence in the signal system has gone far enough, hybrid zones do not form following range expansion. Instead range overlap is possible and contributes to the observed link between sympatry and ethological isolation.

In addition to parapatric races some species may be divided into subpopulations by ecological preferences. I include here three examples of mating signal variation between "races" in sympatry; all three are pheromonal systems in moths.

Both *Zeiraphera diniana* (Baltensweiler and Priesner 1988) and *Laspeyrisia pomonella* (Phillips and Barnes 1975) show divergence between host-associated populations, *Zeiraphera* in pheromone blend and *Laspeyrisia* in mating behavior. The controversy over the involvement of host races in speciation is beyond the scope of this chapter, but clearly the coexistence of two distinct signal-response systems within one population requires explanation. In these two cases there are insufficient data available at present to allow a proper assessment of the status of the host-associated populations—they may be shown to be good species. If they are exchanging genes, these examples may fit a recent extension of Lande's model (Lande and Kirkpatrick 1988). Clearly the coexistence of two distinct signal systems is inconsistent with the recognition view.

In the European corn borer, *Ostrinia nubilalis*, two distinct pheromone "races" coexist in parts of Europe and the United States. They differ in pheromone blend and male response, and yet they do hybridize in nature (Löfstedt 1990). The origin and fate of these populations are still an open question, but one possibility is that they are essentially parapatric in Europe and that the sympatry created by introduction into North America will be unstable in the long term. However, a recent study seems to indicate that at least one U.S. population is at some kind of equilibrium (Bengtsson and Löfstedt 1990) with the two signaling systems, whose genetics have been elucidated (Roelofs et al. 1987), continuing to coexist despite gene exchange.

Discussion

Clearly the available data have serious limitations for discriminating among models for the evolution of mating signals. With a very few exceptions genetic analyses of the variation in mating signals have been conducted on a small scale and give only the most basic information on genetic variation at whatever level. Most include only one, or one type of character, which is a serious limitation given that the evolutionary significance of variation in a signal can be assessed only in terms of the specificity of the receiver, and vice versa. Response variation has been particularly neglected relative to signal variation. Nevertheless, the data do show some consistencies and in some respects are sufficient to influence our view of signal evolution.

The few examples that include data at both levels (within and between populations) all tend to reinforce the conclusions from the general survey. *Drosophila* species are heavily overrepresented here, with the information available on the species *silvestris/heteroneura, mojavensis/arizonae,* and *paulistorum* all indicating genetic variation both within and among populations at levels sufficient to influence mating patterns. The best example outside *Drosophila*, the guppy *Poecilia*, shows the same pattern.

Phenotypic variation in signal characters is low in most cases—a result consistent with both views—but has a significant heritable component, which may not be expected if stabilizing selection is intense. The clearest evidence for low signal variation comes from the small number of studies on pheromone systems. This may be a function of the communication channel itself. Presumably because of the biochemistry of pheromone production, blends seem to be influenced by two classes of genes—a few loci capable of causing large shifts in blend (e.g. Roelofs et al. 1987) and numerous polygenes with very minor influence on blend (Collins and Cardé 1989*d*; see Linn and Roelofs, this volume). The predictions of the recognition and competition views are more distinct with respect to preference variation, but in this case there are very few direct studies. However, preference distributions tend to be broad relative to signal distributions, and both selection on preferences directly and selection for assortative mating produce strong responses in most cases. This is not consistent with the recognition view of intense stabilizing selection on all components of the signal-response chain.

In studies of geographic variation in mating signals within species, which is expected to be low on the recognition view, there is a clear excess of examples showing statistically significant variability. There is a need for information on receiver tolerances and variability in many of these cases to assess the evolutionary significance of signal variability. However, those studies involving assortative mating show a similar excess, indicating that preferences as well as signals vary geographically. Some notable exceptions to this pattern, such as assortative mating in *Drosophila melanogaster,* may be due to the recent expansion of the species with limited time for divergence. There is some evidence for a correlation between mating signals and environmental variables, although in cases like *Acris* this variation is apparently distinct from interspecific variation. The strongest evidence for an environmental influence on mating signals is for the effect of sympatry with closely related species. However, the examples of extra divergence in sympatry compared to allopatry may not be due to reproductive character displacement; the extra divergence may have predated sympatry, or other environmental variables may be implicated by more detailed study.

Many of the acoustic and pheromonal characters in question have a clear role in the proper formation of mating pairs and must form part of the SMRS as conceived by Paterson (1985). The studies of assortative mating necessarily involve the SMRS. Thus the observation of variability within and among populations is not simply due to choice of characters outside the SMRS.

The competition view is consistent with both genetic variation within populations and substantial genetic variation among populations. On this basis it is preferable to the Recognition Model. However, some of the examples do not fit comfortably with the more detailed predictions of this model. For example, in guppies signal variation is clearly correlated with predation pressure while Fisherian selection predicts essentially random variation among populations. Variation in bristle number of *Drosophila silvestris* is controlled by a small number of loci, not the large number of loci of additive effect assumed by Lande (1981). A balance between sexual and viability selection seems unlikely for many of the signal components in question which affect only communication. However, these more precise predictions arise from one particular form that reproductive competition might take. In reality, it is likely to take many different forms whose major common feature is the generation of diversity among populations.

An important point on which the two views differ is their treatment of the receiver or preference (usually female) side of the signaling system. Lande (1981) considers female preferences that evolve solely as a correlated response to selection on male signals; that is, they are not subject to direct selection. Tolerances of individual females determine the strength of sexual selection but are not themselves allowed to evolve. Subsequent modifications of the model consider "costs" of female preference by adding direct stabilizing selection on female preference. This is equivalent to assuming that extreme female types are at a disadvantage, perhaps due to delayed mating or increased risk of remaining unmated (Pomiankowski 1987). This modification makes the model much more similar to the recognition view in which stabilizing selection for "efficient" fertilization acts on both males and females. In both cases it tends to stabilize the system at the environmental optimum. However, in reality tolerances will also evolve—responders with narrow tolerances will find it more difficult to obtain a mate than those with broad tolerances relative to the variation in signal types. Broader tolerances will evolve unless this tendency is countered by selection resulting from inappropriate matings—either with signalers of other species or with low-quality males of the same species. Tolerance is also expected to change through an individual female's lifetime, becoming broader as she ages and the risk of failing to mate increases. There are some data suggesting both broad tolerances and increase with age in the planthopper genera *Ribautodelphax* (de Winter and Rollenhagen 1990) and *Nilaparvata* (Butlin 1993); for wide tolerances in *Chorthippus* (Charalambous 1990), *Pectinophora* (Haynes and Baker 1988), and *Grapholita* (Cardé et al. 1976); and for weak selection on interpulse interval in *Drosophila melanogaster* (Cowling 1980), possibly due to broad female tolerances. Selection for broad tolerance in females would lead to a reduction in the stabilizing selection on male signals and might explain the maintenance of intrapopulation genetic variation. This generally looser coordination between male and female components of the signal system would make interpopulation variation in response to nonsexual selection pressures or through drift more comprehensible and may facilitate the fixation of major mutations in some circumstances. Much more experimental work is needed in this area to separate within- and among-individual variation in the mean preferred signal and the range of acceptable signals (Butlin 1993).

The competition view considers two classes of selection on signals: viability selection resulting from interactions with the environment and sexual selection resulting from female preferences. However, for many animal signals nonsexual components of selection may be extremely weak or absent. Thus there is likely to be a cost to female moths associated with pheromone production, but this cost is not likely to vary with the blend of components produced. Similarly, a male grasshopper may use energy and risk attracting predators or parasites by singing, but the cost is unlikely to depend on the pulse rate of the song. There will be a balance between sexual and nonsexual components of selection on signaling activity, but the evolution of signal characteristics will be dominated by sexual selection.

A model by De Jong and Sabelis (1991), mentioned above, goes some way toward this scenario by modifications to Lande's (1981) model. Viability selection on the signaler is excluded, and costs to females of delayed mating are included (the Wallflower Effect). A line of equilibria exists at which mean signal and preference are equal, but significant genetic correlations between preference and signal are unlikely, and so the Fisher runaway process does not operate and drift along the line of equilibria is slow. The model predicts more stable mating signals than does the Lande (1981) model, but since variation in signals is maintained, it allows the possibility of changes in signals in small, isolated populations or in response to external selection pressures.

In the context of mate recognition systems the Fisherian mode of sexual selection has received much more attention than other modes, included here under "external" selection pressures since they do not result exclusively from the signal-response interaction. Three other broad categories of sexual selection are generally recognized: intrasexual selection, preferences based on direct benefits, and preferences based on "good genes" (including both the Zahavi Handicap principle and the Hamilton-Zuk hypothesis) (Bradbury and Andersson 1987). Any of these may cause evolutionary changes in signal characters that have the potential to act as barriers to gene flow. For example, in some anurans male-male competition is mediated by call pitch, producing selection favoring low-pitched calls (Arak 1983). Given the relative flexibility of signal-response interactions that I have argued for above, a response to this selection pressure could be tracked by evolution of female preferences. If the intensity of male-male competition varies between populations for ecological reasons, this process could produce divergence, assortative mating, and perhaps speciation. Similar scenarios could be constructed for any of the other modes of sexual selection. While preferences based on good genes may remain a controversial idea (Kirkpatrick 1987), male-male competition and female preferences based on direct benefits are well established, both theoretically and empirically (Bateson 1983).

Neither the extreme stability of mating signals favored by the recognition view, nor the extreme lability favored by the competition view is necessary or sufficient to explain the data currently available on the variability of these systems. The interaction between signaler and receiver is likely to be maintained in an intermediate state by a balance between selection for the ability to obtain the best possible mate and selection against individuals that mate late or not at all. In this state the system is able to respond to changes in environmental conditions, especially the signaling environment. To understand these selection pressures future attention must be concentrated on the receiver component of the system where current data are very limited, and on an integrated approach to particular cases including both signal and receiver components of the signaling system and all levels of variation.

Summary

Divergence in mating signals and responses is an important step in speciation in sexually reproducing animals and may precede the evolution of complete postmating isolation. Two contrasting views of the evolution of mating signals have been identified: the recognition view emphasizes their stability and adaptation to environmental conditions, while the competition view considers them to be labile and to evolve in essentially arbitrary directions.

I have reviewed data on the variability of mating signals and responses within populations and among populations and related these data to the two opposing views. There is evidence of widespread variation in mating signals both within and among populations with little evidence of a link to environmental variation except for the influence of sympatry with related species. Variation among populations in particular is more compatible with the competition than the recognition view, although more information on preferences is required to assess its significance. The limitations of the Competition Model were discussed, particularly the assumptions regarding the preference side of the system. This crucial component of the signaling system is poorly represented in the data available on variation within and among populations.

Acknowledgments

I am very grateful to Mike Ritchie, Godfrey Hewitt, and Christer Löfstedt for helpful discussions and to Steven Telford and Hamish Spencer for comments on an earlier draft. My research is supported by the Royal Society, National Environment Research Council, and Science and Engineering Research Council.

References

Ahearn, J.N., and A.R. Templeton. 1989. Interspecific hybrids of *Drosophila heteroneura* and *D. silvestris*. I. Courtship success. Evolution 43:347–361.
Anderson, W.W., and L. Ehrman. 1969. Mating choice in crosses between geographic populations of *Drosophila pseudoobscura*. Am. Midl. Nat. 81:47–53.
Andersson, M. 1982. Female choice selects for extreme tail length in a widowbird. Nature 299:818–820.
Arak, A. 1983. Sexual selection by male-male competition in natterjack toad choruses. Nature 306:261–262.
Arnqvist, G. 1989. Sexual selection in a water strider: The function, mechanism of selection, and heritability of a male grasping apparatus. Oikos 56:344–350.
Averhoff, W.W., and R.H. Richardson. 1976. Multiple pheromone system controlling mating in *Drosophila melanogaster*. Proc. Natl. Acad. Sci. 73:591–593.
Ayala, F.J., M.L. Tracey, L.G. Barr, and J.G. Ehrenfeld. 1974. Genetic and reproductive differentiation of the subspecies *Drosophila equinoxialis caribbinensis*. Evolution 28:24–41.
Bailey, W.J. 1991. Acoustic behavior of insects: An evolutionary perspective. Chapman and Hall, London.
Baldwin, F.T., and E.H. Bryant. 1981. Effect of size upon mating performance within geographic strains of the housefly *Musca domestica*. Evolution 35:1134–1141.
Baltensweiler, W., and E. Priesner. 1988. A study of pheromone polymorphism in *Zeiraphera diniana* Gn. (Lepidoptera: Tortricidae). 3. Specificity of attraction to synthetic pheromone sources by different male response types from two host races. J. Appl. Entomol. 106:217–231.
Barrer, P.M., M.J. Lacey, and A. Shani. 1987. Variation in relative quantities of airborne sex pheromone components from individual female *Ephestia cautella* (Lepidoptera: Pyralidae). J. Chem. Ecol. 13:639–653.
Barrientos-Lozano, L. 1988. Acoustic behavior and taxonomy of Mexican *Pterophylla* (Orthoptera: Tettigoniidae: Pseudophyllinae). Ph.D. dissertation, University of Wales, College of Cardiff.

Barton, N.H., and G.M. Hewitt. 1985. Analysis of hybrid zones. Annu. Rev. Ecol. Syst. 16:113–148.
Barton, N.H., and M. Turelli. 1989. Evolutionary quantitative genetics: How little do we know? Annu. Rev. Genet. 23:337–370.
Bateson, P.P.G. 1983. Mate choice. Cambridge University Press, Cambridge.
Benedix, J.H., and D.J. Howard. 1991. Calling song displacement in a zone of overlap and hybridization. Evolution 45:1751–1759.
Bengtsson, B.O., and C. Löfstedt. 1990. No evidence for selection in a pheromonally polymorphic moth population. Am. Nat. 136:722–726.
Blair, W.F. 1958. Mating call in the speciation of anuran amphibians. Am. Nat. 92:27–51.
———. 1962. Non-morphological data in anuran classification. Syst. Zool. 11:72–84.
———. 1974. Character displacement in frogs. Am. Zool. 14:1119–1125.
Bogert, C.M. 1960. The influence of sound on the behavior of amphibians and reptiles. Pp. 137–320 in W.E. Lanyon and W.N. Tavolga, eds. Animal sounds and communication. American Institute of Biological Sciences, Washington, D.C.
Booij, C.J.H. 1982. Biosystematics of the *Muellerianella* complex (Homoptera: Delphacidae), interspecific and geographic variation in acoustic behavior. Z. Tierpsychol. 58:31–52.
Borgia, G. 1985. Bower quality, number of decorations and mating success of male satin bowerbirds (*Ptilonorhynchus violaceus*): An experimental analysis. Anim. Behav. 33:266–271.
Bradbury, J.W., and M. Andersson, eds. 1987. Sexual selection: Testing the alternatives. Wiley, New York.
Breden, F., and G. Stoner. 1987. Male predation risk determines female preference in the Trinidad guppy. Nature 329:831–833.
Busnel, M.-C. 1963. Characterisation acoustique de populations d'*Ephippiger* ecologiquement voisines. Annales des Epiphytes 14:25–34.
Butlin, R.K. 1986. The functions of song components in *Drosophila:* A comment on Ewing and Miyan. Anim. Behav. 35:302–304.
———. 1987a. Speciation by reinforcement. Trends Ecol. Evol. 2:8–13.
———. 1987b. Species, speciation, and reinforcement. Am. Nat. 130:461–464.
———. 1989. Reinforcement of premating isolation. Pp. 158–179 in D. Otte and J.A. Endler, eds. Speciation and its consequences. Sinauer, Sunderland, Mass.
———. 1993. The variability of mating signals and responses in the brown planthopper, *Nilaparvata lugens*. J. Insect Behav. 6:125–140.
Butlin, R.K., and G.M. Hewitt. 1986. Heritability estimates for characters under sexual selection in the grasshopper, *Chorthippus brunneus*. Anim. Behav. 34:1256–1261.
Butlin, R.K., G.M. Hewitt, and S.F. Webb. 1985. Sexual selection for intermediate optimum in *Chorthippus burnneus* (Orthoptera: Acrididae). Anim. Behav. 33:1281–1292.
Cade, W.H. 1975. Acoustically orienting parasitoids: Fly phonotaxis to cricket song. Science 190:1312–1313.
Campion, D.G., P. Hunter-Jones, L.J. McVeigh, D.R. Hall, R. Lester, and B.F. Nesbitt. 1980. Modification of the attractiveness of the primary pheromone component of the Egyptian cotton leafworm, *Spodoptera littoralis* (Boisduval) (Lepidoptera: Noctuidae), by secondary pheromone components and related chemicals. Bull. Ent. Res. 70:417–434.
Cardé, R.T., and T.C. Baker. 1984. Sexual communication with pheromones. Pp. 355–383 in W.J. Bell and R.T. Cardé, eds. Chemical ecology of insects. Chapman and Hall, London.

Cardé, R.T., A.M. Cardé, A.S. Hill, and W.L. Roelofs. 1977. Sex pheromone specificity as a reproductive isolating mechanism among the sibling species *Archips argyrospilus* and *A. mortuanus* and other sympatric tortricine moths (Lepidoptera: Tortricidae). J. Chem. Ecol. 3:71–84.

Carson, H.L. 1985. Genetic variation in a courtship-related male character in *Drosophila silvestris* from a single Hawaiian locality. Evolution 39:678–686.

Carson, H.L., and R. Lande. 1984. Inheritance of a secondary sexual character in *Drosophila silvestris*. Proc. Natl. Acad. Sci. 81:6904–6907.

Carson, H.L., and L.T. Teramoto. 1984. Artificial selection for a secondary sexual character in males of *Drosophila silvestris* from Hawaii. Proc. Natl. Acad. Sci. 81:3915–3917.

Carson, H.L., F.C. Val, C.M. Simon, and J.W. Archie. 1982. Morphometric evidence for incipient speciation in *Drosophila silvestris* from the island of Hawaii. Evolution 36:132–140.

Carson, H.L., and J.S. Yoon. 1982. Genetics and evolution of Hawaiian *Drosophila*. Pp. 297–344 *in* M. Ashburner, H.L. Carson, and J.N. Thompson, eds. Genetics and biology of *Drosophila*. Vol. 3b. Academic Press, London.

Case, S.M. 1990. Dewlap and other variation in the lizards *Anolis distichus* and *A. brevirostris* (Reptilia: Iguanidae). Biol. J. Linn. Soc. 40:373–393.

Charalambous, M. 1990. Genetics of song and female preference in the grasshopper *Chorthippus brunneus* (Orthoptera: Acrididae): Variation in the mate recognition system. Ph.D. dissertation, University of East Anglia, Norwich, England.

Claridge, M.F. 1985. Acoustic signals in the Homoptera: Behaviour, taxonomy and evolution. Annu. Rev. Entomol. 30:297–317.

Claridge, M.F., J. Den Hollander, and J.C. Morgan. 1985a. The status of weed associated populations of the brown planthopper, *Nilaparvata lugens* (Stål)—host race or biological species? Zool. J. Linn. Soc. 84:77–90.

———. 1985b. Variation in courtship signals and hybridization between geographically definable populations of the rice brown planthopper, *Nilaparvata lugens* (Stål). Biol. J. Linn. Soc. 24:35–49.

———. 1988. Variation in host plant relations and courtship signals of weed-associated populations of the brown planthopper, *Nilaparvata lugens* (Stål), from Australia and Asia: A test of the recognition species concept. Biol. J. Linn. Soc. 35:79–93.

Cohn, T.J., and I. Cantrall. 1974. Variation and speciation in the grasshoppers of the Conalcaeini (Orthoptera: Acrididae: Melanoplinae): The lowland forms of Western Mexico, the genus *Barytettix*. San Diego Society of Natural History, Memoir 6.

Collins, R.D., and R.T. Cardé. 1985. Variation in and heritability of aspects of pheromone production in the pink bollworm moth, *Pectinophora gossypiella* (Lepidoptera: Gelechiidae). Ann. Entomol. Soc. Am. 78:229–234.

———. 1989a. Wing fanning as a measure of pheromone response in the pink bollworm moth, *Pectinophora gossypiella* (Lepidoptera: Gelechiidae). J. Chem. Ecol. 15:2635–2645.

———. 1989b. Heritable variation in pheromone response in the pink bollworm moth, *Pectinophora gossypiella* (Lepidoptera: Gelechiidae). J. Chem. Ecol. 15:2647–2659.

———. 1989c. Selection for increased pheromone response in the pink bollworm moth, *Pectinophora gossypiella* (Lepidoptera: Gelechiidae). Behav. Genet. 20:325–331.

———. 1989d. Selection for altered pheromone component ratios in the pink bollworm moth, *Pectinophora gossypiella* (Lepidoptera: Gelechiidae). J. Insect. Behav. 2:609–621.

Collins, R.D., S.L. Rosenblum, and R.T. Cardé. 1990. Selection for increased pheromone titer in the pink bollworm moth, *Pectinophora gossypiella* (Lepidoptera: Gelechiidae). Physiol. Entomol. 15:141–147.

Cowling, D.E. 1980. The genetics of *Drosophila melanogaster* courtship song—diallel analysis. Heredity 45:401–403.

Coyne, J.A., and H.A. Orr. 1989. Patterns of speciation in *Drosophila*. Evolution 43:362–381.

Coyne, J.A., H.A. Orr, and D.J. Futuyma. 1988. Do we need a new species concept? Syst. Zool. 37:190–200.

Crane, J. 1975. Fiddler crabs of the world (Ocypodidae: Genus *Uca*). Princeton University Press, Princeton, N.J.

Crews, D., and E.F. Williams. 1977. Hormones, reproductive behavior, and speciation. Am. Zool. 17:271–286.

Crossley, S.A. 1974. Changes in mating behavior produced by selection for ethological isolation between *ebony* and *vestigial* mutants of *Drosophila melanogaster*. Evolution 28:631–647.

Dagley, J.R. 1988. Population differentiation in the grasshopper, *Chorthippus parallelus* (Orthoptera: Acrididae): A study of the mate recognition system. Ph.D. dissertation, University of East Anglia, Norwich, England.

Darwin, C.R. 1871. The descent of man and selection in relation to sex. John Murray, London.

Day, T.H., and R.K. Butlin. 1987. Non-random mating in natural populations of the seaweed fly, *Coelopa frigida*. Heredity 58:213–220.

De Jong, M.C.M., and M.W. Sabelis. 1991. Limits to runaway sexual selection: The wallflower paradox. J. Evol. Biol. 4:637–656.

De Winter, A.J. 1992. The genetic basis and evolution of acoustic mate recognition signals in a *Ribautodelphax* planthopper (Homoptera: Delphacidae). I. The female call. J. Evol. Biol. 5:249–265.

De Winter, A.J., and T. Rollenhagen. 1990. The importance of male and female behaviour for reproductive isolation in *Ribautodelphax* planthoppers (Homoptera: Delphacidae). Biol. J. Linn. Soc. 40:191–206.

Dobzhansky, T. 1962. Species in *Drosophila*. Proc. Linn. Soc. Lond. 174:1–12.

———. 1975. Analysis of incipient reproductive isolation within a species of *Drosophila*. Proc. Natl. Acad. Sci. 72:3638–3641.

Dobzhansky, T., L. Ehrman, and P.A. Kastritsis. 1968. Ethological isolation between sympatric and allopatric species of the *obscura* group of *Drosophila*. Anim. Behav. 16:79–87.

Dobzhansky, T., O. Pavlovsky, and L. Ehrman. 1969. Transitional populations of *Drosophila paulistorum*. Evolution 23:482–492.

Dobzhansky, T., O. Pavlovsky, and J.R. Powell. 1976. Partially successful attempt to enhance reproductive isolation between semispecies of *Drosophila paulistorum*. Evolution 30:201–212.

Duijm, M. 1990. On some song characteristics in *Ephippiger* (Orthoptera, Tettigonioidea) and their geographic variation. Netherlands J. Zool. 40:428–453.

Ehrman, L. 1961. The genetics of sexual isolation in *Drosophila paulistorum*. Genetics 46:1025–1038.

———. 1965. Direct observation of sexual isolation between allopatric and between sympatric strains of the different *Drosophila paulistorum* races. Evolution 19:459–464.

———. 1971. Natural selection for the origin of reproductive isolation. Am. Nat. 105:479–483.

———. 1972. Rare male advantages and sexual isolation in *Drosophila immigrans*. Behav. Genet. 2:79–84.

———. 1973. More on natural selection for the origin of reproductive isolation. Am. Nat. 107:318–319.

———. 1983. Fourth report on natural selection for the origin of reproductive isolation. Am. Nat. 121:290–293.

Ehrman, L., and P.A. Parsons. 1980. Sexual isolation among widely distributed populations of *Drosophila immigrans*. Behav. Genet. 10:401–407.

———. 1981. Sexual isolation among isofemale strains within a population of *Drosophila immigrans*. Behav. Genet. 11:127–133.

Emlen, S.T. 1972. An experimental analysis of the parameters of bird song eliciting species recognition. Behavior 41:130–171.

Endler, J.A. 1983. Natural and sexual selection on color patterns in poeciliid fishes. Environ. Biol. Fishes 9:173–190.

———. 1988. Sexual selection and predation risk in guppies. Nature 332:593–594.

Eoff, M. 1975. Artificial selection in *Drosophila melanogaster* females for increased and decreased sexual isolation from *Drosophila simulans*. Am. Nat. 109:225–229.

———. 1977. Artificial selection in *Drosophila simulans* males for increased and decreased sexual isolation from *Drosophila melanogaster* females. Am. Nat. 111:259–266.

Ferguson, G.W. 1973. Character displacement of the push-up displays of two partially sympatric species of spiny lizards, *Sceloporus* (Sauria: Iguanidae). Herpetologica 29:281–284.

Fisher, R.A. 1930. The genetical theory of natural selection. Oxford University Press, Oxford.

Ford, E.B., and P.M. Sheppard. 1969. The *medionigra* polymorphism of *Panaxia dominula*. Heredity 24:561–569.

Fouquette, M.J., Jr. 1975. Speciation in chorus frogs. I. Reproductive character displacement in the *Pseudacris nigrita* complex. Syst. Zool. 24:16–22.

Futch, D.G. 1973. On the ethological differentiation of *Drosophila ananassae* and *Drosophila pallidosa* in Samoa. Evolution 27:456–467.

Gerhardt, H.C. 1987. Evolutionary and neurobiological implications of selective phonotaxis in the green treefrog, *Hyla cinerea*. Anim. Behav. 35:1479–1489.

Grant, P.R. 1975. The classical case of character displacement. Evol. Biol. 8:237–337.

Graur, D., and D. Wool. 1982. Dynamics and genetics of mating behavior in *Tribolium castaneum*. Behav. Genet. 12:161–179.

Halliburton, R., and G.A.E. Gall. 1981. Disruptive selection and assortative mating in *Tribolium castaneum*. Evolution 35:829–843.

Hammack, L. 1987. Chemical basis for asymmetric mating isolation between strains of screwworm fly, *Cochliomyia hominovorax*. J. Chem. Ecol. 13:1419–1430.

Hansson, B.S., M. Tòth, C. Löfstedt, G. Szöcs, M. Subchev, and J. Löfqvist. 1990. Pheromone variation among eastern European and a western Asian population of the turnip moth *Agrotis segetum* J.Chem. Ecol. 16:1611–1622.

Haynes, K.F., and T.C. Baker. 1988. Potential for evolution of resistance to pheromones: Worldwide and local variation in chemical communication systems of pink bollworm moth, *Pectinophora gossypiella*. J. Chem. Ecol. 14:1547–1560.

Haynes, K.F., L.K. Gaston, P.M. Mistrot, and T.C. Baker. 1984. Potential for evolution of pheromone resistance: Interindividual and interpopulational variation in chemical communication system of pink bollworm moth. J. Chem. Ecol. 10:1551–1565.

Heisler, I.L. 1984. Inheritance of female mating propensities for *yellow* locus genotypes in *Drosophila melanogaster*. Genet. Res. 44:133-149.

Henderson, N.R., and D.M. Lambert. 1982. No significant deviation from random mating of worldwide populations of *Drosophila melanogaster*. Nature 300:437-440.

Henry, C.S. 1985. Sibling species, call differences and speciation in green lacewings (Neuroptera: Chrysopidae: *Chrysoperla*). Evolution 39:965-984.

Henry, C.S., and M.M. Wells. 1990. Geographical variation in the song of *Chrysoperla plorabunda* (Neuroptera: Chrysopidae) in North America. Ann. Entomol. Soc. Am. 83:317-325.

Hill, K.G., J.J. Loftus-Hills, and D.F. Gartside. 1972. Premating isolation between the Australian field crickets *Teleogryllus commodus* and *T. oceanicus* (Orthoptera: Gryllidae). Aust. J. Zool. 20:153-163.

Hoikkala, A. 1985. Genetic variation in the male courtship sound of *Drosophila littoralis*. Behav. Genet. 15:135-142.

Hoikkala, A., and J. Lumme. 1990. Inheritance of male courtship sound characteristics in *Drosophila littoralis*. Behav. Genet. 20:423-435.

Houde, A.E. 1988. Genetic difference in female choice between two guppy populations. Anim. Behav. 36:510-516.

Houde, A.E., and J.A. Endler. 1990. Correlated evolution of female mating preferences and male color patterns in the guppy *Poecilia reticulata*. Science 248:1405-1408.

Howard, D.J. 1986. A zone of overlap and hybridization between two ground cricket species. Evolution 40:34-43.

Howden, H.F. 1979. A revision of the Australian genus *Blackburnium* Boucomont (Coleoptera: Scarabaeidae: Geotrupinae). Aust. J. Zool. Suppl. Ser. 72:1-88.

Ikeda, H., H. Idoji, and I. Takabatake. 1981. Intraspecific variations in the thresholds of female responsiveness for auditory stimuli emitted by the male in *Drosophila mercatorum*. Zool. Mag. 90:325-332.

Ikeda, H., and O. Maruo. 1982. Directional selection for pulse repetition rate of the courtship sound and correlated responses occurring in several characters in *Drosophila mercatorum*. Japan. J. Genet. 57:241-258.

Jallon, J.-M., and J.R. David. 1987. Variations in cuticular hydrocarbons among the eight species of the *Drosophila melanogaster* subgroup. Evolution 41:294-302.

Johnson, M.S. 1982. Polymorphism for direction of coil in *Partula suturalis*: Behavioral isolation and positive frequency dependent selection. Heredity 49:145-151.

Kanmiya, K. 1990. Acoustic properties and geographic variation in the vibratory courtship signals of the European chloropid fly, *Lipara lucens* Meigen (Diptera: Chloropidae). J. Ethol. 8:105-120.

Kawanishi, M., and T.K. Watanabe. 1980. Genetic variations of courtship song of *Drosophila melanogaster* and *Drosophila simulans*. Japan. J. Genet. 55:235-240.

Kearns, P.W.E., I.P.M. Tomlinson, C.J. Veltman, and P. O'Donald. 1992. Non-random mating in *Adalia bipunctata* (the two-spot ladybird). II. Further tests for female mating preference. Heredity 68:385-390.

Kirkpatrick, M. 1987. Sexual selection by female choice in polygynous animals. Annu. Rev. Ecol. Syst. 18:43-70.

Klump, G.M., and H.C. Gerhardt. 1987. Use of non-arbitrary acoustic criteria in mate choice by female gray tree frogs. Nature 326:286-288.

Koepfer, H.R. 1987a. Selection for sexual isolation between geographic forms of *Drosophila mojavensis*. I. Interactions between the selected forms. Evolution 41:37-48.

———. 1987b. Selection for sexual isolation between geographic forms of *Drosophila mojavensis*. II. Effects of selection of mating preference and propensity. Evolution 41:1409–1412.
Koopman, K.F. 1950. Natural selection for reproductive isolation between *Drosophila pseudoobscura* and *D. persimilis*. Evolution 4:135–148.
Koref-Santibanez, S. 1972. Courtship behaviour in the semispecies of the superspecies *Drosophila paulistorum*. Evolution 26:108–115.
Kruger, E.L., C.D. Pappas, and R.W. Howard. 1991. Cuticular hydrocarbon geographic variation among seven North American populations of *Aedes albopictus* (Diptera: Culicidae). J. Med. Entomol. 28:859–864.
Lakovaara, S., and A. Saura. 1982. Evolution and speciation in the *Drosophila obscura* group. Pp. 1–59 *in* M. Ashburner, H.L. Carson, and J.N. Thompson, eds. Genetics and biology of *Drosophila*. Vol. 3b. Academic Press, London.
Lambert, D.M., and N.R. Henderson. 1986. The stability of the specific-mate recognition system of *Drosophila melanogaster*. Behav. Genet. 16:369–373.
Lande, R. 1975. The maintenance of genetic variability by mutation in a polygenic character with linked loci. Genet. Res. 26:221–235.
———. 1981. Models of speciation by sexual selection on polygenic traits. Proc. Natl. Acad. Sci. 78:3721–3725.
———. 1982. Rapid origin of sexual isolation and character divergence in a cline. Evolution 36:213–223.
Lande, R., and M. Kirkpatrick. 1988. Ecological speciation by sexual selection. J. Theor. Biol. 133:85–98.
Lanier, G.N., A. Classon, T. Stewart, J.J. Piston, and R.M. Silverstein. 1980. *Ips pini:* The basis for interpopulational differences in pheromone biology. J. Chem. Ecol. 6:677–687.
Liebhold, A.M., and W.J.A. Volney. 1985. Effects of attractant composition and release rate on attraction of male *Choristoneura retiniana, C. occidentalis* and *C. carnana* (Lepidoptera: Tortricidae). Can. Entomol. 117:447–457.
Liley, N.R. 1966. Ethological isolating mechanisms in four sympatric species of poeciliid fishes. Behav. Suppl. 13:1–197.
Littlejohn, M.J. 1960. Call discrimination and potential reproductive isolation in *Pseudacris triseriata* females from Oklahoma. Copeia 1960:370–371.
———. 1965. Premating isolation in the *Hyla ewingi* complex (Anura: Hylidae). Evolution 19:234–243.
———. 1981. Reproductive isolation: A critical review. Pp. 298–234 *in* W.R. Atchley and D.S. Woodruff, eds. Evolution and speciation: Essays in honor of M.J.D. White. Cambridge University Press, Cambridge.
Littlejohn, M.J., and G.F. Watson. 1985. Hybrid zones and homogamy in Australian frogs. Annu. Rev. Ecol. Syst. 16:85–112.
Löfstedt, C. 1990. Population variation and genetic control of pheromone communication systems in moths. Entomol. Exp. Appl. 54:199–218.
Löfstedt, C., J. Löfqvist, B.S. Lanne, J.N.C. Van der Pers, and B.S. Hansson. 1986. Pheromone dialects in European turnip moths, *Agrotis segetum*. Oikos 46:250–257.
Majerus, M.E.N., P. O'Donald, and J. Weir. 1982. Female mating preference is genetic. Nature 300:521–523.
Markow, T.A. 1981. Courtship behavior and control of reproductive isolation between *Drosophila mojavensis* and *D. arizonensis*. Evolution 35:1022–1026.

Masters, J. C., R.J. Rayner, I.J. McKay, A.D. Potts, D. Nails, J.W. Ferguson, B.K. Weissenbacher, M. Allsopp, and M.L. Anderson. 1987. The concept of species: Recognition versus isolation. S. Afr. J. Sci. 83:534–537.

Masters, J.C., and H.G. Spencer. 1989. Why we need a new genetic species concept. Syst. Zool. 38:270–279.

Mayr, E. 1963. Animal species and evolution. Harvard University Press, Cambridge, Mass.

McDonald, J., and S. Crossley. 1982. Behavioral analysis of lines selected for wing vibration in *Drosophila melanogaster*. Anim. Behav. 30:802–810.

McLain, D.K., and K.S. Rai. 1986. Reinforcement for ethological isolation in the Southeast Asian *Aedes albopictus* subgroup (Diptera: Culicidae). Evolution 40:1346–1349.

McLaughlin, J.R., J.H. Tumlinson, and J.L. Sharp. 1977. Absence of synergism in the response of Florida lesser peachtree borer males to synthetic pheromone. Fla. Entomol. 60:27–29.

McPhail, J.D. 1969. Predation and the evolution of a stickleback (*Gasterosteus*). J. Fish Res. Bd. Canada 26:3183–3208.

Millar, C.D., and D.M. Lambert. 1986. Laboratory induced changes in the mate recognition system of *Drosophila pseudoobscura*. Behav. Genet. 16:285–294.

Miller, D.R., J.H. Borden, and K.N. Slessor. 1989. Inter- and intrapopulation variation of the pheromone, ipsdienol, produced by male pine engravers, *Ips pini* (Say) (Coleoptera: Scolytidae). J. Chem. Ecol. 15:233–247.

Møller, A.P. 1990. Male tail length and female mate choice in the monogamous swallow *Hirundo rustica*. Anim. Behav. 39:458–465.

Moodie, G.E.E. 1982. Why asymmetric mating preferences may not show the direction of evolution. Evolution 36:1096–1097.

Moore, W.S. 1979. A single locus, mass-action model of assortative mating, with comments on the process of speciation. Heredity 42:173–186.

———. 1987. Random mating in the Northern Flicker hybrid zone: Implications for the evolution of bright and contrasting plumage patterns in birds. Evolution 41:539–546.

Murray, J., and B. Clarke. 1980. The genus *Partula* on Moorea: Speciation in progress. Proc. R. Soc. Lond. B 211:83–117.

Narins, P.M., and S.L. Smith. 1986. Clinal variation in anuran advertisement calls: Basis for acoustic isolation. Behav. Ecol. Sociobiol. 19:135–141.

Nevo, E., and R.R. Capranica. 1985. Evolutionary origin of ethological reproductive isolation in cricket frogs, *Acris*. Evol. Biol. 19:147–214.

Nichols, R.A., and R.K. Butlin. 1989. Does runaway sexual selection work in finite populations? J. Evol. Biol. 2:299–313.

North, R.C., and K.C. Shaw. 1979. Variation in distribution, morphology and calling song of two populations of *Pterophylla camellifolia* (Orthoptera: Tettigoniidae). Psyche 86:363–374.

Odendaal, F.J., C.M. Bull, and S.R. Telford. 1986. Influence of the acoustic environment on the distribution of the frog *Ranidella riparia*. Anim. Behav. 34:1836–1843.

O'Donald, P., and M.E.N. Majerus. 1985. Sexual selection and the evolution of preferential mating in ladybirds. I. Selection for high and low lines of female preference. Heredity 55:401–412.

Ohta, A.T. 1978. Ethological isolation and phylogeny in the *grimshawi* species complex of Hawaiian *Drosophila*. Evolution 32:485–492.

Otte, D. 1989. Speciation in Hawaiian crickets. Pp. 482–526 *in* D. Otte and J.A. Endler, eds. Speciation and its consequences. Sinauer, Sunderland, Mass.

Ovaska, K. 1989. Pheromonal divergence between populations of the salamander *Plethodon vehiculum* in British Columbia. Copeia 1989:770–775.

Paterson, H.E.H. 1978. More evidence against speciation by reinforcement. S. Afr. J. Sci. 74:369–371.

———. 1980. A comment on "mate recognition systems." Evolution 34:330–331.

———. 1982. Perspective on speciation by reinforcement. S. Afr. J. Sci. 78:53–57.

———. 1985. The recognition concept of species. Pp. 21–29 *in* E.S. Vrba, ed. Species and speciation. Transvaal Museum Monograph No. 4. Pretoria.

Patton, J.L., and M.F. Smith. 1989. Population structure and the genetic and morphologic divergence among pocket gopher species (genus *Thomomys*). Pp. 284–304 *in* D. Otte and J.A. Endler, eds. Speciation and its consequences. Sinauer, Sunderland, Mass.

Perdeck, A.C. 1957. The isolating value of specific song patterns in two species of grasshoppers (*Chorthippus brunneus* Thunb. and *Ch. biguttulus* L.). Behavior 12:1–75.

Petit, C., O. Kitagawa, and T. Takamura. 1976. Mating system between Japanese and French geographic strains of *Drosophila melanogaster*. Japan. J. Genet. 51:99–108.

Petit, C., O. Kitagawa, E. Takanashi, and D. Nouad. 1980. The failure to obtain sexual isolation by artificial selection. Genetica 54:213–219.

Phillips, P.A., and M.M. Barnes. 1975. Host race formation among sympatric apple, walnut and plum populations of the codling moth, *Laspeyrisia pomonella*. Ann. Entomol. Soc. Am. 68:1053–1060.

Pomiankowski, A. 1987. The costs of choice in sexual selection. J. Theor. Biol. 128:195–218.

Prakash, S. 1972. Origin of reproductive isolation in the absence of apparent genic differentiation in a geographic isolate of *Drosophila pseudoobscura*. Genetics 72:143–155.

Ragge, D.R. 1986. The songs of the western European grasshoppers of the genus *Omocestus* in relation to their taxonomy (Orthoptera: Acrididae). Bull. Br. Mus. Nat. Hist. Entomol. 53:213–249.

Ragge, D.R., and W.J. Reynolds. 1984. The taxonomy of the western European grasshoppers of the genus *Euchorthippus* with special reference to their songs (Orthoptera: Acrididae). Bull. Br. Mus. Nat. Hist. Entomol. 49:103–151.

———. 1988. The songs and taxonomy of the grasshoppers of the *Chorthippus biguttulus* group in the Iberian Peninsula (Orthoptera: Acrididae). J. Nat. Hist. 22:897–929.

Ralin, D.B. 1977. Evolutionary aspects of mating call variation in a diploid-tetraploid complex of treefrogs (Anura). Evolution 31:721–736.

Ringo, J.M., and D. Wood. 1980. Frequencies of mating among geographic populations of *Drosophila simulans*. Behav. Genet. 10:492.

Rising, J.D. 1983. The progress of oriole hybridization in Kansas. Auk 100:885–897.

Robertson, J.G.M. 1986. Male territoriality, fighting and fighting assessment in the Australian frog *Uperoleia rugosa*. Anim. Behav. 34:763–772.

Roelofs, W.L., J.-W. Du, C. Linn, T.J. Glover, and L.B. Bjostad. 1986. The potential for genetic manipulation of the redbanded leafroller moth sex pheromone blend. Pp. 263–272 *in* M.D. Huettel, ed. Evolutionary genetics of invertebrate behavior. Plenum, New York.

Roelofs, W.L., T.J. Glover, X.-H. Tang, I. Sreng, P. Robbins, C. Eckenrode, C. Löfstedt, B.S. Hansson, and B.O. Bengtsson. 1987. Sex pheromone production and perception in European corn borer moths is determined by both autosomal and sex-linked genes. Proc. Natl. Acad. Sci. 84:7585–7589.

Rothschild, G.H.L. 1975. Attractant for monitoring *Pectinophora scutigera* and related species of Australia. Environ. Entomol. 4:983–985.

Rothschild, G.H.L., and A.K. Minks. 1977. Some factors influencing the performance of pheromone traps for oriental fruit moth in Australia. Ent. Expt. Appl. 22:171–182.

Ryan, M.J., R.B. Cocroft, and W. Wilczynski. 1990. The role of environmental selection in intraspecific divergence of mate recognition signals in the cricket frog, *Acris crepitans*. Evolution 44:1869–1872.

Ryan, M.J., and W.E. Wagner. 1987. Asymmetries in mating preferences between species: Female swordtails prefer heterospecific males. Science 236:595–597.

Sappington, T.W., and O.R. Taylor. 1990. Genetic sources of pheromone variation in *Colias eurytheme* butterflies. J. Chem. Ecol. 16:2755–2770.

Schilcher, F. von. 1976. The functions of pulse song and sine song in the courtship of *Drosophila melanogaster*. Anim. Behav. 24:622–625.

Searcy, W.A., and M. Andersson. 1986. Sexual selection and the evolution of song. Annu. Rev. Ecol. Syst. 17:507–533.

Spencer, H.G., D.M. Lambert, and B.H. McArdle. 1987. Reinforcement, species, and speciation: A reply to Butlin. Am. Nat. 130:958–962.

Spencer, H.G., B.H. McArdle, and D.M. Lambert. 1986. A theoretical investigation of speciation by reinforcement. Am. Nat. 128:241–262.

Spiess, E.B., and L.D. Spiess. 1967. Mating propensity, chromosomal polymorphism and dependent conditions in *Drosophila persimilis*. Evolution 21:672–678.

Spiess, E.B., and C.M. Wilke. 1984. Still another attempt to achieve assortative mating by disruptive selection in *Drosophila*. Evolution 38:505–515.

Spiess, L.D., and E.B. Spiess. 1969. Mating propensity, chromosomal polymorphism and dependent conditions in *Drosophila persimilis*. II. Factors between larvae and between adults. Evolution 23:225–236.

Spieth, H.T., and J.M. Ringo. 1983. Mating behavior and sexual isolation in *Drosophila*. Pp. 223–284 *in* M. Ashburner, H.L. Carson and J.N. Thompson, eds. The genetics and biology of *Drosophila*. Vol. 3c. Academic Press, London.

Stewart, K.W., S.W. Szczytko, and B.P. Stark. 1982. Drumming behavior of four species of North American Pteronarcyidae (Plecoptera): Dialects in Colorado and Alaska *Pteronarcella badia*. Ann. Entomol. Soc. Am. 75:530–533.

Sullivan, B.K. 1989. Interpopulational variation in vocalizations of *Bufo woodhousii*. J. Herpetol. 23:368–373.

Thoday, J.M., and J.B. Gibson. 1962. Isolation by disruptive selection. Nature 193:1164–1166.

―――. 1970. The probability of isolation by disruptive selection. Am. Nat. 104:219–230.

Tilley, S.G., P.A. Verrell, and S.J. Arnold. 1990. Correspondence between sexual isolation and allozyme differentiation: A test in the salamander *Desmognathus ochrophaeus*. Proc. Natl. Acad. Sci. 87:2715–2719.

Tompkins, L., and J.C. Hall. 1984. Sex pheromones enable *Drosophila* males to discriminate between conspecific females from different laboratory stocks. Anim. Behav. 32:349–352.

Tuttle, M.D., and M.J. Ryan. 1981. Bat predation and the evolution of frog vocalizations in the Neotropics. Science 214:677–678.

Verrell, P.A., and S.J. Arnold. 1989. Behavioral observations of sexual isolation among allopatric populations of the mountain dusky salamander, *Desmognathus ochrophaeus*. Evolution 43:745–755.

Waage, J.K. 1979. Reproductive character displacement in *Calopteryx* (Odonata: Calopterygidae). Evolution 33:104–116.

Walker, T.J. 1974. Character displacement and acoustic insects. Am. Zool. 14:1137–1150.

Wasserman, M., and H.R. Koepfer. 1977. Character displacement for sexual isolation between *Drosophila mojavensis* and *D. arizonensis*. Evolution 31:812–823.

Webster, T.P., and J.M. Burns. 1973. Dewlap color variation and electrophoretically detected sibling species in a Haitian lizard, *Anolis brevirostris*. Evolution 27:368–377.

Wells, M.M., and C.S. Henry. 1992. The role of courtship songs in reproductive isolation among populations of green lacewings of the genus *Chrysoperla* (Neuroptera: Chrysopidae). Evolution 46:31–42.

West-Eberhard, M.J. 1983. Sexual selection, social competition and speciation. Q. Rev. Biol. 58:155–183.

———. 1985. Sexual selection, competitive communication and species-specific signals in insects. Pp. 283–324 *in* T. Lewis, ed. Insect communication. Academic Press, London.

Williams, G.C. 1966. Adaptation and natural selection: A critique of some current evolutionary thought. Princeton University Press, Princeton, N.J.

Wood, D., and J.M. Ringo. 1982. Artificial selection for altered male wing display in *Drosophila simulans*. Behav. Genet. 12:449–458.

Wood, D., J.M. Ringo, and L.L. Johnson. 1980. Analysis of the courtship sequences of the hybrids between *Drosophila melanogaster* and *D. simulans*. Behav. Genet. 10:459–466.

Zimmerman, B.L. 1983. A comparison of structural features of calls of open and forest habitat frog species in the Central Amazon. Herpetologica 39:235–246.

Zouros, E. 1981. The chromosomal basis of sexual isolation in two sibling species of *Drosophila—D. mojavensis* and *D. arizonensis*. Genetics 97:703–718.

17

The Recognition Concept of Species Applied in an Analysis of Putative Hybridization in New Zealand Cicadas of the Genus *Kikihia* (Insecta: Hemiptera: Tibicinidae)

DAVID HENRY LANE

Department of Zoology
Victoria University, Wellington

NEW ZEALAND CICADAS of the genus *Kikihia* (Dugdale 1972) include a group of bright green species, which have proved difficult to classify applying the Biological Species Concept (Mayr 1942, 1963), owing to widespread color polymorphism (Fleming 1973). Studies of two morphologically defined species of *Kikihia*, *K. subalpina* (Hudson) and *K. cutora* (Walker), led Fleming to state that "some hybridisation" between the two species was "suspected," in some areas where they are sympatric in the North Island (Fig. 17.1). He added: "This situation is deserving of more study." Central to the problem was the apparent phenomenon of character swapping in the volcanic populations of *Kikihia* on Mount Egmont (Fox 1982) and Mount Ruapehu (see Fig. 17.2), together with the intriguing absence of this situation in the North Island main axial ranges (Fig. 17.3, inset) where the same species occupy the same altitudinal and climatic zones. The presence of a third species of *Kikihia* (morphologically very close to the putative hybrids) sympatric with the other two species is important in the interpretation of the problem. I recognized this third species on the basis of its unique male song, and its distribution was mapped in the North Island on the basis of aural records. Its singing station was described under the epithet "species L" (Fleming 1975a), and its song was first characterized when it was formally described as *Kikihia laneorum* (Fleming 1984).

Fleming (1973) recognized that the green *Kikihia* species "form a gradational series marked by increasing (or decreasing) amounts of black pigmentation." *K. cutora* is represented on the North Island by two parapatric subspecies (or recognizable entities), *K. cutora cutora* (Walker) in the north and *K. c. cumberi*

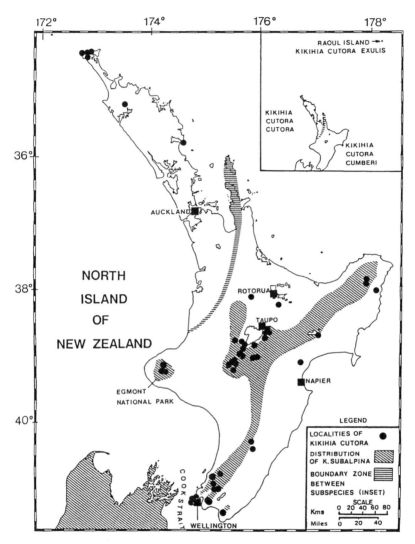

Fig. 17.1. Distribution of *Kikihia cutora* (*sensu lato*) and *Kikihia subalpina*, showing major zones of sympatry. Distribution of three subspecies of *K cutora* shown, including the "boundary" between *K. c. cutora* and *K. c. cumberi* (inset).

Fleming, in the south (Fig. 17.1, inset). The former is more uniformly green, while the latter exhibits increasing amounts of black pigment, and is confined to the central volcanic regions of the North Island, where it extends into the subalpine zone and south to Wellington. It is absent from the South Island. The male calling songs do not differ significantly between the subspecies. *K. subalpina,* present throughout the South Island and Stewart Island, is sym-

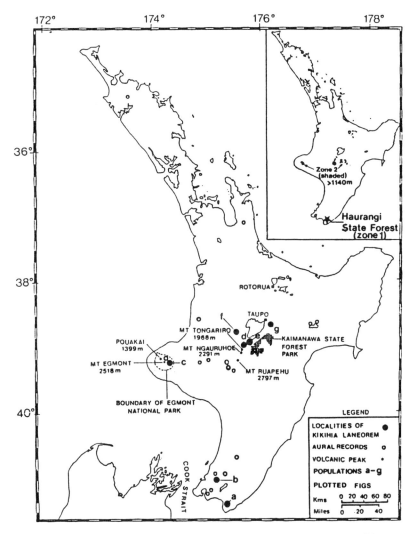

Fig. 17.2. Distribution of *K. laneorum* showing important geographic features. Aural records included. Populations a–g incorporated in Anderson Hybrid Index (Fig. 17.10). Zone 1 populations a = Haurangi State Forest Park (475–700 m), b = Kaitoke (200 m), f = Waituhi Saddle (884 m). Zone 3 populations c = Mount Egmont (below Dawson Falls, 800 m); d = Ketatahi Springs, Mount Tongariro (700 m); e = Lake Rotopounamu (713 m); g = Opepe (640 m).

patric with *K. c. cumberi* over much of the lower North Island (Figs. 17.1 and 17.4) and also extends into the subalpine zone. *K. laneorum,* confined to North Island upland indigenous forests (Fig. 17.2), is rarely found in the subalpine zone, unlike the other two species which dominate this biotype. In forested

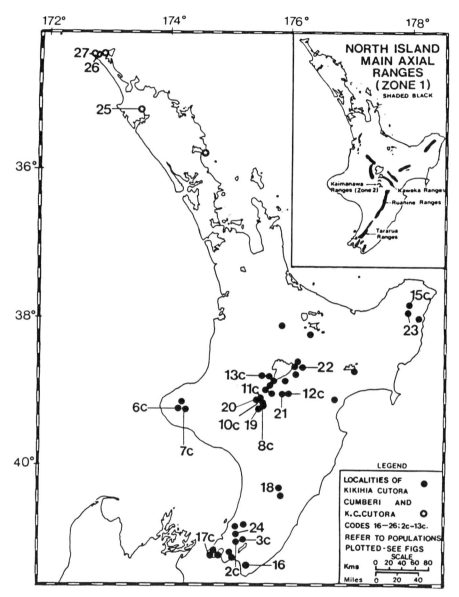

Fig. 17.3. Distribution of *K. c. cumberi* and *K. c. cutora* showing all sites where specimens were collected for study. All populations numbered with the exception of 23 (Ruatahunga, 790 m, zone 1) and 15c (Mount Hikurangi, 1219 m, zone 1) were used to calculate mean "hybrid index scores" (Fig. 17.10). Ranges and frequencies of scores (including populations 23 and 15c) are shown in Figs. 17.11–17.13. Zone 1 populations (main axial ranges) include: 16. Haurangi State Forest Park (300–475 m). 18. Manawatu Gorge (200 m). 24. Cloustonville, Akatarawa

areas all three species generally sing high in the canopy and are extremely difficult to capture.

Fleming (1973) suggested that *K. cutora* and *K. subalpina* are "sibling species [or sister species] derived from a common ancestor, undoubtedly closely related in both morphology and acoustic behaviour." Later he incorporated *K. laneorum*, postulating that *K. subalpina* and *K. laneorum* are sister species derived from a common ancestor and that *K. cutora* is "less closely related [to *K. subalpina* and *K. laneorum*], being, perhaps, a sister species of their common ancestor" (Fleming 1984), based on examination of more *laneorum*. Fleming (1984) continued to support the case for *cutora/subalpina* hybridization and noted that Myers (1929a) had also suggested that hybridization may occur. In this chapter I present detailed evidence which suggests that hybridization is not occurring and that coloration can be an incomplete predictor of species identity.

The Case for Hybridization

Fleming (1973) suggested that *K. cutora cumberi* and *K. subalpina* were hybridizing based on specimens which appeared intermediate in color pattern and that they were closely related phylogenetically. He stated: "The most strongly pigmented individuals of *K. cutora*—those superficially resembling *K. subalpina*—are restricted to the areas adjacent to the subalpine regions of Taranaki and Tongariro where *cutora* and *subalpina* have the most intimate relationship, as if the effects of introgression decreased away from these areas." *Hybridization* has been defined by Mayr (1963:110) as "the crossing of individuals belonging to two different species," and the term *introgression* as "the incorporation of genes of one species into the gene pool of another species, as a result of successful hybridisation." Fleming (1973) used intermediacy of pig-

---

(240 m). 2c. Kaitoke (200 m). 3c. Rimutaka Ranges (555 m). 13c. Waituhi Saddle (945 m). 17c. Johnsons Hill, Karori (320 m). Zone 2 populations (zone of suspected hybridization) include 6c. Mount Egmont (1140–1450 m). 8c. Ohakune Mt. Rd., Mount Ruapehu (1360–1440 m). 10c. Bruce Rd. (above Silica Springs), Mount Ruapehu (1200–1400 m). 12c. Umukarikari Range, Kaimanawa Mountains (1250–1300 m). Zone 3 populations (Pleistocene volcanic areas below bushline) include: 19. Ohakune Mt. Rd, Mount Ruapehu (700 m). 20. Whakapapaiti Stream, Mount Ruapehu (860 m). 21. Kaimanawa Rd. (660 m). 22. Opepe (640 m). 7c. Dawson Falls, Mount Egmont (905 m). 11c. Chateau, Mount Ruapehu (1050 m). Populations of *K. c. cutora* include 25. Maungataniwha Range (300 m). 26. Tapotupotu Bay. 27. Cape Reinga (183 m). Populations of *K. c. cumberi* sympatric with *K. subalpina* populations (Fig. 17.4) indicated by suffix c (e.g. population 6c, Fig. 17.3, is sympatric with population 6, Fig. 17.4).

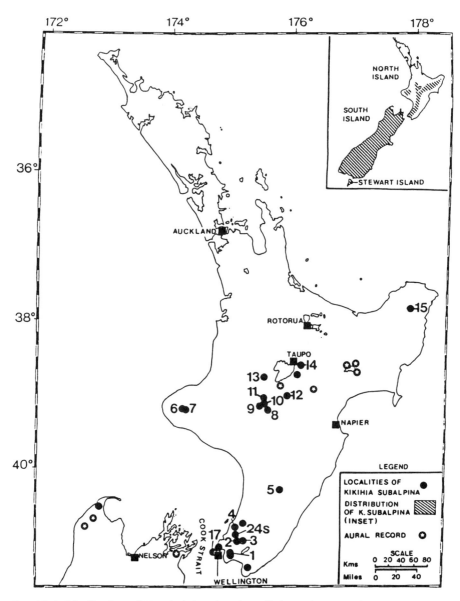

Fig. 17.4. Distribution of *K. subalpina* showing all sites where specimens were collected for study. All populations numbered (except 17) were used to calculate mean "hybrid index scores" (Fig. 17.10). Ranges and frequencies of scores are shown in Figs. 17.11–17.13. Zone 1 populations (main axial ranges) include: 1. Whakanui Track summit (550 m). 2. Kaitoke (200 m). 3. Rimutaka (555 m). 4. Akatarawa Saddle (457 m). 5. Wharite, Ruahine Ranges (800 m). 13. Waituhi Saddle (884–945 m). 15. Mount Hikurangi (1219 m). 17. Johnsons Hill, Karori (320

mentation as prima facie evidence for hybridization, in the absence of field or laboratory observation of interbreeding between the two species. Identical reasoning has been used in numerous studies of "hybridization," by systematists employing the Biological Species Concept (e.g. Gibbs 1968). Fleming (1984) highlighted the problem, as he saw it, in obtaining empirical evidence: "Generally no mixed matings have been observed, but their recognition is virtually impossible because females of related species are so similar."

The study of laboratory hybrids (often showing similar intermediate characters) may strengthen the case for the "hybrid" status of field populations by demonstrating that hybridization is possible (e.g. Hawaiian *Drosophila* studies, Kaneshiro and Val 1977). Molecular studies have been employed to confirm that hybrids do indeed form in nature and, moreover, that these hybrids can and do backcross, leading to limited introgression (e.g. DeSalle and Templeton 1987). Under the Biological Species Concept, species are defined on the basis of reproductive isolation in the field rather than in the laboratory. Applying this concept in laboratory and field studies of two fruit fly species, Gibbs (1968) assumed that true hybrids form in nature. He concluded that "if hybrids are formed in natural populations the direct demonstration of this turns out to be a somewhat intractable problem." His "evidence" that hybridization occurred in field populations was rejected by a researcher who assessed the phenotypic variability of field populations more thoroughly and did not rely on a single morphological character (which overlaps widely between the two species) to assess species status (Wolda 1967a,b).

Fleming (1973) found that the two species of *Kikihia* remained "quite distinct [in song], in no way bridged by intermediates" in the areas of "suspected hybridisation." He also found that all male specimens singing like *subalpina* were found to possess only two long tymbal bars (Fig. 17.5d,e) and those singing like *cutora* had three long tymbal bars (Fig. 17.5f). Dugdale (pers. comm.), however, reported the swapping of these tymbal characters in subalpine populations of the two species on Mounts Egmont and Ruapehu. He also reported on the existence of "hybrid songs" based on his own field work and supported Fleming's case for hybridization. Such "hybrid songs" have

---

m). 24s. Cloustonville, Akatarawa (240 m). The single specimen of *K. subalpina* from this site was classified as a "morphological hybrid" (see Figs. 17.5a–c, 17.11a,j). Zone 2 populations (zones of suspected hybridization) include 6. Mount Egmont (1140 m–1450 m). 8. Ohakune Mt. Rd, Mount Ruapehu (1360–1440 m). 9. Hauhungatahi (1160–1521 m). 10. Bruce Rd. (above Silica Springs), Mount Ruapehu (1200–1400 m). 12. Umukarikari Range, Kaimanawa Mountains (1250–1300 m). Zone 3 populations (Pleistocene volcanic areas below bushline) include: 7. Dawson Falls, Mount Egmont (905 m). 11. Chateau, Mount Ruapehu (1050 m). 14. Mount Tauhara (1088 m).

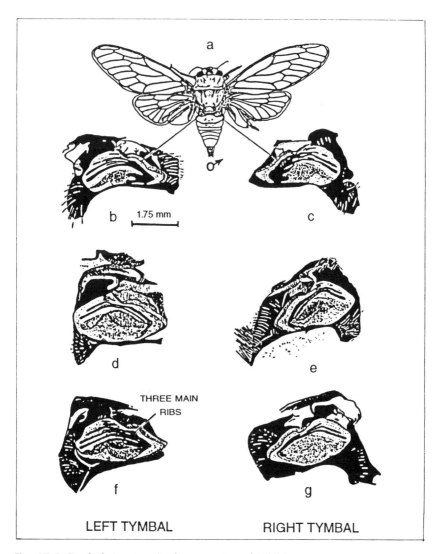

Fig. 17.5. Tymbal structure in three species of *Kikihia*.
- a. "Phenetic hybrid" *K. subalpina* × *K. cutora cumberi* (?). Song typical of *K. subalpina*, but tymbal like *K. c. cumberi*. Compare with illustration *f*. This specimen scored 24 on hybrid index (see Fig. 17.11a).
- b,c. Tymbals—third main rib interrupted differing slightly from typical *K. c. cumberi*.
- d,e. Typical tymbal *K. subalpina* male. Two main ribs.
- f. Typical tymbal *K. c. cumberi* male. (*K. c. cutora* male is identical.)
- g. Typical tymbal of *K. laneorum*. Two main ribs.

been produced in laboratory crosses in other insects (e.g. Bentley and Hoy 1974; Yusof 1982; Claridge 1985; Heady 1987). Having considered the data on which this present chapter is based (Lane 1984, 1985, 1990a, 1993a), Dugdale (pers. comm.) has now rejected the hybridization hypothesis.

Fleming's case for hybridization was in part based on his model of speciation for *K. cutora* and *K. subalpina*, which was derived from the species' distributions. He argued that their partial sympatry in the mountains of the southern North Island was a zone of secondary contact that arose following allopatric speciation (Fleming 1973). Selection against hybrids in such zones of overlap, or hybridization, is seen to reinforce species separation (Dobzhansky 1937, 1940) or lead to reproductive character displacement (Lack 1945, 1947; Walker 1974; Littlejohn 1981; Otte 1989). Character displacement (or divergence, Grant 1972) in components of male courtship song has been reported in insects (Wasserman and Koepfer 1977; Waage 1979; Howard and Furth 1986; Butlin 1985, 1987a,b, 1989; Benedix and Howard 1991) and in a number of vertebrate taxa (e.g. *Cnemidophorus* lizards, Case 1979; Taper and Case 1992). Futuyma (1986:32, 34) cites studies by Brown and Wilson (1956) and Lack (1947) as examples of this phenomenon, but acknowledges that "it has been generally agreed" that "rather few cases . . . have been convincingly demonstrated" (p. 243). Grant (1984) maintains that "recent observational and experimental results" support Lack's (1947) reasoning about natural selection operating upon incipient species in sympatry (see Ratcliffe and Grant 1983). The most widely known cases reported involve amphibians (e.g. Littlejohn 1965; Loftus-Hills and Littlejohn 1971), and such cases have been shown to be highly dubious (Lambert and Paterson 1984). Harrison (1990:114) in his review, concluded that "in none of the cases is there any direct evidence that selection against hybrids has been the driving force producing the pattern of variation." Walker (1974) found no convincing case for reinforcement in acoustic insects. Mayr (1963) argued against the reinforcement model, based on his observations that some hybrid zones persisted for long periods without any apparent development of sexual selection. Like many biologists today, he does not consider reinforcement a necessary part of allopatric speciation.

Fleming advanced his model of speciation, involving the differentiation of *K. cutora* and *K. subalpina* on either side of Cook Strait (see Figs. 17.1 and 17.4) from a common stock, when he still considered them "sibling species" (cf. Fleming 1971:468–71). He later abandoned this model and applied it instead to "explain" the origin of *K. subalpina* and *K. laneorum* (Fleming 1984).

The Case for Species-Specific Songs as "Isolating Mechanisms"

Fleming (1973, 1984) in applying the Biological Species Concept actively sought "isolating mechanisms preventing interbreeding." This approach has also been adopted by Ewart (1989a:290) in his studies of Australian cicadas

(*Pauropsalta*). Fleming (1975a) regarded "song differences . . . as of *primary* importance as isolating mechanisms" (italics added), listing it with four other such "mechanisms," which he believed maintained genetic isolation, namely, distribution and altitude, singing station, seasonal isolation, and host plant preferences. Fleming (1984) viewed song as a mechanism to "deter interbreeding, loss of genetic identity, and the wastage that results from hybrids with generally reduced fertility." In an attempt to "explain" the existence of two sibling species (*K. laneorum* and *K. subalpina*), he stated that "when a behavioural isolating mechanism [male song] is efficient, there is no need for divergence in morphology to discourage interbreeding." Fleming (1975a, 1984) assumed that cicada song structure (the "primary isolating mechanism") was perfected through character displacement, for the function of efficiently isolating one species from another, when they became sympatric, but he provided no data to support this view.

Mayr (1963:551) saw no role for natural selection in the "fine tuning" of *"primary* isolating mechanisms." He stated: "Natural selection does play a role in the improvement of some of the isolating mechanisms, only it concerns subsidiary isolating mechanisms. The primary, basic one must be fully efficient when contact is first established." Selection was held only to maintain the efficiency of isolating mechanisms and possibly "add to their perfection" (p. 109). In recent years, the question of whether signal differences develop while populations are evolving independently of one another, or emerge subsequently as a result of selection against hybridization, has been debated (Littlejohn 1981; Hölldobler 1984).

The first theory maintains that the species isolation properties of song do not evolve as isolating mechanisms *between* populations; rather they arise as by-products of "adaptive divergence" (Darwin 1859; Muller 1939, 1942; Templeton 1981) or as incidental effects of selection (in allopatry) for mechanisms that ensure the effective union of gametes *within* populations (Paterson 1978, 1981). Paterson (1985) attributes these properties to adaptations to physical features of the "preferred" habitats of the species, rather than as solely nonadaptive by-products of divergence (e.g. by pleiotropic effects [Muller 1942] or genetic revolutions or transiliences [Carson and Templeton 1984]). Others contend that differences in sexual selection pressures are expected to result in rapid divergence in premating signals in allopatry (Carson 1982; Lande 1981, 1982; West-Eberhard 1983, 1984; Thornhill and Alcock 1983). Another model to account for the gradual divergence of these signals relies entirely on genetic drift (Nei et al. 1983).

The second theory, reinforcement theory, contends that differences in these signaling systems actually evolved to prevent mispairing or mismating (Dobzhansky 1937; Ayala 1982), which, historically, resulted in a selective disadvantage (due to time and energy costs in responding to heterospecific males or in the production of hybrid offspring with decreased fitness). Differences in premating displays, and the response to them by mates, were reinforced as the incipient species became sympatric, and thus mating display differences resulted only in conspecific pairings (Dobzhansky 1940; Fisher 1958; Littlejohn 1969). Models of speciation in which selection favors characters for their isolating effects have been heavily criticized (Futuyma and Mayer 1980; Templeton 1981, 1987; Butlin 1985, 1987, 1992). Recent studies on acoustic insects claim some support for this model (e.g. Harrison and Rand 1989; Howard and Waring 1991).

A number of studies support both theories (Gwynne and Morris 1986), while a number of serious theoretical criticisms have been raised against the reinforcement theory (e.g. Paterson 1978; Lambert et al. 1984; Spencer et al. 1986, 1987; Sanderson 1989). Paterson and colleagues argue that a logical nexus exists between the Biological Species Concept and reinforcement theory (Paterson 1981; Lambert and Paterson 1984). Phelan and Baker (1987) note that the theory has been rejected due to "both theoretical difficulties and a paucity of examples of the predicted pattern of reproductive character displacement." Barton and Hewitt (1985) argue that many of the well-studied hybrid zones contain the "sort of variation that might form the basis of reinforcement." Barton et al. (1988) note that "reinforcement of isolating mechanisms in response to selection against hybrids seems to be a less important phase of speciation than was once thought." Weaknesses in the theory (process) are seen to undermine the concept (Lambert et al. 1987). A number of authors, however, while conceding serious deficiencies in both, reject the view that a logical nexus exists between the two (Chandler and Gromko 1989) and defend the Biological Species Concept (Coyne et al. 1988; Verrell 1988). These objections have been dealt with (Masters and Spencer 1989; White et al. 1990).

Fleming (1984) stressed the "isolation function" of cicada song thus: "However achieved, a one-to-one relationship between male song and female response must deter or prevent a female from mating with a male that is producing the wrong song for her species. In general, this seems likely to apply . . . [However] interspecific matings . . . *must* occasionally occur in nature" (italics added). Fleming's assertion that hybridization "*must* occasionally occur in nature" between *K. cutora cumberi* and *K. subalpina* (despite the lack of field evidence of mixed matings) is explained by his commitment to the reinforcement theory. He quoted Walker (1974), an advocate of the theory: "Theoretically, 'Character displacement in areas of sympatry should encourage divergence of songs of related species' . . . The two New Zealand sibling species [*K. subalpina, K. laneorum*] . . . are evidence that this has occurred" (Fleming 1984:205). The temporal patterns of cicada calling songs are generally regarded as sufficient in isolating sympatric species by most researchers (Jiang 1985; Nakao and Kanmiya 1988; Villet 1988). When related sympatric species share an environment, selection is believed to minimize signal differences within a species and maximize differences between species by reinforcement (Fleming 1984).

The Case for Song in Mate Recognition

A mate recognition/fertilization concept, ("interbreeding natural populations") is an integral part of the Biological Species Concept (Mayr 1963, 1970, 1988). Fleming's work emphasizes this aspect: "The complex and stereotyped ethology of cicadas, in particular the specificity of the males' songs and of the females' response to them, play a notable role in speciation . . . a species-diagnostic song expedite[s] the approach of the sexes for mating; since the females respond best (if not solely) to their own species' song" (Fleming 1975a:299; see Dugdale and Fleming 1969:955). He clearly indicated some equivocation as to the degree of specificity of the male song. This doubt was raised again in his 1984 paper. He stated, "Song, like all other characters, shows individual variation; how finely tuned is the female's discriminatory power?" Fleming noted that experimental evidence from North American periodical cicadas (genus *Magicicada*) emphasizes the biological importance of distinctive songs, since the auditory response of males (and probably females) is "keyed in" to the species-diagnostic song pattern of males of the same species (Simmons et al. 1971; Huber 1984). The critical interspecies differences in mate recognition are the varying pulse structures and temporal variations of cicada songs, rather than frequency differences (Doolan and Young 1989; Huber 1984; Huber et al. 1990). The specificity of male cicada song and female response have been established from quantitative fieldwork experiments using species of *Magicicada* (Alexander and Moore 1958).

Fleming's doubt over the degree of specificity of male song was based on the following evidence:

1. His surveys of well-distributed species (based mainly on aural assessments, not sonograms) suggested to him that "song may vary geographically in wholly or partially disjunct populations" (Fleming 1975a), as "for instance between undescribed races of *K. rosea*" (Fleming 1984:204)

2. The observation of a single intergeneric mating in the field (Fleming 1984:204)

3. Interspecific matings of cicadas confined to cages (Fleming 1984)

4. Putative hybridization in *Kikihia* (Fleming 1973, 1984) and "hybrid song zones" involving *K. muta* (Fabricius) and an undescribed species of *Kikihia* (Fleming 1975a:299)

5. The apparent overlap in sound spectra and auditory sensitivity to the frequency component of male song in the studies of *Magicicada* (Simmons

et al. 1971, cited in Fleming 1984). The auditory sensitivity (in both sexes of *Magicicada*) peaks at correspondingly different frequencies in the two sympatric species. (In many cicadas there is no clear overlap between sound spectra and optimal hearing sensitivity [Popov et al. 1985; Huber et al. 1990; Huber pers. comm.]).

Ewart (1989a) has raised similar doubts about the specificity of male cicada song. He has documented variation in the gross song structure of calling songs (used in long-range communication between the sexes) of some Australian cicada species. He states that this "illustrates the problem of using song as a diagnostic taxonomic character [and] raises the question (in the absence of direct experimental data) of the acoustic characteristics by which cicada[s] can recognize [their] own song pattern." However, Ewart did not apply acoustic analysis to courtship songs used by cicadas in close-range communication between the sexes.

Stemming perhaps from doubts over the degree of song specificity, populations of *Kikihia* have been defined by Fleming as subspecies, despite the fact that they are almost exclusively allopatric and differ in song (Fleming 1975a). Populations in the genus *Maoricicada* have been defined as subspecies, yet they differ in both song and morphology (Dugdale and Fleming 1978:328–33). Lambert (1982) suggests that such inconsistencies indicate that either their taxonomic "status is a matter for arbitrary designation, or that isolating mechanisms are *not necessary* since the populations are allopatric."

Theoretical Considerations

Problems Involved in Investigating Hybridization

The definition of *hybridization* by Mayr (1963) as "the crossing of individuals belonging to two different species" is formulated within the context of the Biological Species Concept (cf. Mayr 1942:258–259). The keystone of this concept is reproductive isolation (Mayr 1970; White 1978). The incorporation of interbreeding criteria into the definition precludes any real investigation into hybridization. It has led to circularity, since the decision whether to include two populations in the same, or two different, species depends on the occurrence of hybridization. What is the subject of empirical test (evidence of interbreeding) cannot be part of the definition of the term. The term *hybridization* so defined cannot be used by definition, until the presence of two different species can be assumed (Bigelow 1965). This excludes the term from the very context in which it is most useful. In assuming the existence of two "biological species," one is forced to accept that they are reproductively isolated by definition. Mayr (1970) adjusted his species definition, weakening the criterion of reproductive isolation in the case where "isolating mechanisms" broke down, thereby allowing for limited introgression between species. Hybridization is therefore seen as a breakdown or failure of these "isolating mechanisms." Mayr (1992) has described the "incompleteness of isolating mechanisms" in the following terms: "Very few . . . are all-or-none devices (see Mayr 1963, ch. 5). They are built up step by step, and most isolating mechanisms of an incipient species are imperfect and incomplete. Species level is reached when the process of speciation has become irreversible, even if some of the secondary isolating mechanisms have not yet reached perfection (see Mayr 1963, ch. 17)."

"Evidence" that isolating mechanisms have broken down is generally based on the intermediacy of morphology of the putative hybrids (as discussed in the section above on the case for hybridization). However, there is obvious circularity of argument in assuming that certain morphological characters are those of hybrids and then using them to prove the hybrid nature of the organisms bearing them (Schueler and Rising 1976). Much of the difficulty involved in studies of suspected hybridization arises because of the failure to distinguish clearly between two distinct concepts, taxonomic species and genetical species. Woodruff (1973) noted that "the concept of hybridization (essentially a genetic phenomenon)" has for too long been embedded in "the matrix of taxonomy" from which it needs to be extracted. Woodruff's definition of the term *hybridization* (accepted by Harrison [1990:70]) does not restrict it to interbreeding between individuals assigned to different taxa (as Woodruff considers taxonomic decisions to be arbitrary, especially at subspecies level). Harrison's modified version defines "natural hybridization" as "the interbreeding of [individuals from] two populations, or groups of populations, which are distinguished on the basis of one or more [heritable] characters (Woodruff 1973, p. 214)." The central concern of any genetic species concept is summed up by Carson's (1957) succinct phrase, "the species as a field for gene recombination." Few would contest Carson's definition according to Paterson (1988), yet disagreement exists over how the field for gene recombination is delineated. The genetical species concept used must be conceptually sound. It must clarify (1) how the limits for the field for gene recombination are set and (2) how such limits arise (Paterson 1985). According to the Biological Species Concept, it is the "isolating mechanisms" that delimit the field for gene recombination, and since "sympatry is the final test of species status" (Grant 1957), such "mechanisms" are seen to become more efficient in sympatry via character displacement (e.g. Bigelow 1964:102; Dunning et al. 1979:1087–88).

Although a noted critic of the Reinforcement Model of speciation, Mayr made statements consistent with this model. For example, he stated: "When no other closely related species occur all courtship signals can afford to be general, non-specific and variable" (Mayr 1963:109). Such a view is based on Mayr's "population thinking," which stresses variation in all phenotypic characters of a species and is diametrically opposed to essentialist philosophies

(Mayr 1970). No one or set of individual species characters is viewed as the species essence. There is an insistence on variation in mating systems in nature (Mayr 1963; Coyne et al. 1988) since natural selection can operate only where sufficient variation exists to give rise to new species. Zones of hybridization are interpreted as zones of secondary contact (Brown and Wilson 1956) where "general, non-specific and variable" mating systems are perfected (Bock 1979). More recently, Bock (1986) has viewed the criterion of reproductive (i.e., genetic) isolation as consistent with processes other than reinforcement. Mayr and Provine (1980) argue that "the major intrinsic [intraspecific] attribute characterizing a species is a set of isolating mechanisms that keeps it distinct from other species." Clearly, the Biological Species Concept is a relational one, since species "are more succinctly defined by isolation from non-conspecific populations than by the relation of conspecific individuals to each other" (Mayr 1957). Futuyma (1986:219) considers the evolution of new species to be equivalent to the evolution of "isolating mechanisms" between populations. These "mechanisms," he suggests, are "the biological characteristics that cause sympatric species to exist—to maintain distinct gene pools" (p. 112).

The Recognition Concept of Species

Paterson (1978) provoked a far-reaching reappraisal of the Biological Species Concept of Mayr, arguing that it should be termed the *Isolation Concept*. He advanced an alternative "genetical species concept," which he terms the *Recognition Concept*, and argues that it is conceptually superior to the Isolation Concept (see Templeton [1987, 1992] and Coyne et al. [1988] for discussion of this claim). The Recognition Concept regards species "as the most inclusive population of individual biparental organisms which share a common fertilization system" (Paterson 1985). Members of a species consequently mate positively assortatively, through the operation of their shared Specific-Mate Recognition System (SMRS), which is a subset of the fertilization mechanisms of the species. Paterson (1988) has argued that under the Biological Species Concept outlined earlier, the influence of the "isolating mechanisms," on the exchange of genes between species, should be attributed to incidental consequences (i.e., effects) of their true functions (viz., specific-mate recognition). He has shown that these "mechanisms" cannot be described in terms of biological interactions between individuals of different species. Isolation is thus an effect, not a function, of the characters, and the term *isolating mechanisms* is better avoided (Butlin 1987a,b).

Contrary to suggestions that courtship functions to preserve species integrity (Dobzhansky 1937), some species-pairs exhibit asymmetries in individual mate preference. Although most common among flies in the genus *Drosophila* (reviewed by Ehrman and Wasserman 1987), this phenomenon occurs in other insects (e.g., Heady et al. 1989) and other taxa including fish (*Poeciliidae;* Ryan and Wagner 1987). These mating asymmetries have been interpreted as indicating that while male mating behavior has diverged between the species, there has been a lack of divergence in female preference. Insect studies demonstrate that species recognition need not be restricted to one sensory mode (e.g. Schneider 1969). Paterson's (1985) choice of the term *specific-mate* instead of simply *mate* recognition is to avoid confusing the SMRS with "individual mate recognition."

The Recognition Concept Applied to Cicadas

Ewing (1984) has stated that "the empirical evidence for the *isolating function* of song [in insects] is limited and essentially of two kinds. First, there are female phonotactic responses, and second, female acoustic behaviour can be elicited in response to male calling in some cases." In phonotactic experiments females are given a choice between the calling songs, usually recorded, of a conspecific and heterospecific male (Claridge 1985; Heady 1987). Investigators have been able to determine which features of cicada song confer specificity, by altering the acoustic parameters of artificial song. Such model calling songs regularly elicit flight phonotaxis and courtship behavior, in the female of at least one species of cicada (Doolan and Young 1989). "Isolating mechanisms" are usually identified by analysis of the activities of individuals within each species, most of which are entirely directed toward conspecific mates. Individuals perceive courtship, but cannot perceive isolation. The Isolation Concept does not therefore convey any real biological property that could form the basis of species definition (Lambert et al. 1987). Despite claims to the contrary (e.g. Templeton 1987, 1992; Coyne et al. 1988) the Recognition Concept (unlike the Biological Species Concept) is nonrelational, in terms of the biological process of reproduction (White et al. 1990).

This study of *Kikihia* attempts to test the heuristic value of the Recognition Concept, as it applies to studies of hybridization. According to Paterson (pers. comm.) under this concept "hybridization occurs when characters of the fertilization systems of two distinct populations are shared to some extent (see Lorenz 1958)." Hybridization is not viewed as a failure of "isolating mechanisms." "Fertilization has been achieved, and this after all is the function of the fertilization system! What happens then is up to natural selection." He considers "the different view of hybridization" his concept generates, to be "one of its major advantages" (Paterson 1985, pers. comm.). Since the central argument of the Recognition Concept is that selection operates to secure syngamy rather than hinder hybridization, it is not surprising that SMRSs sometimes have similar features that may be misconstrued (Masters and Spencer 1989).

Field observations of cicada courtship and mating throughout New Zealand (1963–1994), convinced the writer that the coadaptation in terms of signaler and receiver is a two-way dynamic communication network—"communication" being defined as "the passing of information coded in a form of signals, in one or other of the sensory modes, to which the receiver is sensitive" (Carthy 1966). Both sexes are in receipt of auditory information during close-range contacts in the courtship sequence (Doolan 1981; Gwynne 1987). Recent speciation studies have focused on the extent to which these types of male and female components of communication are genetically coupled (Butlin and Ritchie 1989). The species-specific character of male song in cicadas worldwide (Myers 1929b; Schremmer 1960; Duffels 1988:75; Moore 1993; Ewart, Moore and Popov pers. comms.) does suggest that the female central nervous system is capable of differentiating her own species song (Huber et al. 1990) or, via an auditory recognizer network, is specifically "keyed in" to the temporal pattern of conspecific song only (Doolan and Young 1989). Some cicadas have been found to possess approximately 1500–2000 auditory receptor cells in each ear (Young 1975; Young and Hill 1977; Huber et al. 1979, 1980; Huber 1984) in both sexes, and electrophysiological studies have shown that these cells are capable of copying individual sound pulses up to rates of 200 cycles per second (and very likely higher rates [Huber et al. 1980; Huber pers. comm.]).

The female cicadas (lacking tymbals) of some species have been observed to indicate their receptivity to a singing male by a series of faint wing clicks (Dugdale and Fleming 1969:955; Doolan 1981; Lane 1984, 1990a,b; Gwynne 1987), which are barely audible to the human ear at close range. Pringle (1954:557) noted that, in some cicada species, the female emits sound by the "wing-clacking mechanism" which *may* act as a stimulus to the male prior to mating. I have observed that these signals can be detected by males at distances of up to ten meters in species of *Kikihia*. They greatly stimulate the singing male and adjacent conspecifics to actively seek out the receptive female (Figs. 17.6a,b and 17.7a–e). The female may initially fly toward a singing male, but closer-range contact is expedited by this wing-flicking signal while she is stationary (Doolan 1981). The female's wing clicking in *Kikihia* species is initially "keyed in" to one particular male song, even though a number of conspecifics may be singing within close range. In *K. scutellaris*, a number of males will synchronize their call songs when in close range to a signaling female. The series of female wing-clicking signals not only indicates the female's receptivity for mating, but assists the singing male in locating her and stimulates the male into the final phases of courtship leading to copulation and fertilization (Lane, unpub. data).

An efficient communication system facilitating mate location is important, enabling cicadas to reduce mortality from acoustically orienting predators (Soper et al. 1976; Gwynne 1987). The cicada call-answer system of pair-formation parallels the acoustic signaling of certain phaneropterine katydids (Spooner 1968) and lacewings (Henry 1980a; Wells and Henry 1992) and flash communication between the sexes of *Photinus* fireflies (Lloyd 1966). These

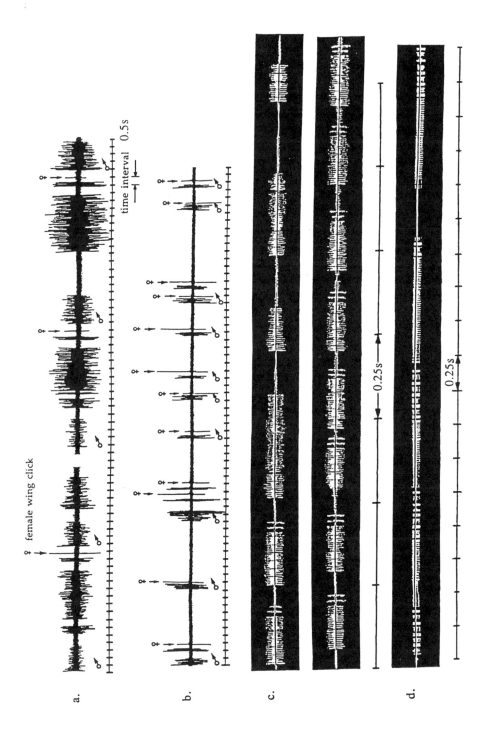

Fig. 17.6
a. Courtship sequence *Kikihia scutellaris*. Female wing click indicated by arrow. Wing click response always inserted between same section of tymbal bursts in male song.
b. Courtship sequence *Kikihia muta muta*. Male tymbal bursts procede click response.
a,b. Both sequences analyzed using a Strobes Acquisition Unit 901 A in conjunction with a Macintosh Apple Computer.
c. Male courtship song of *Kikihia laneorum*.
d. Male courtship song of *Kikihia cutora cumberi*.
c,d. Both analyses made using a Tektronix Dual Beam Storage Oscilloscope (model 5113). Note: different time bases for C and D.

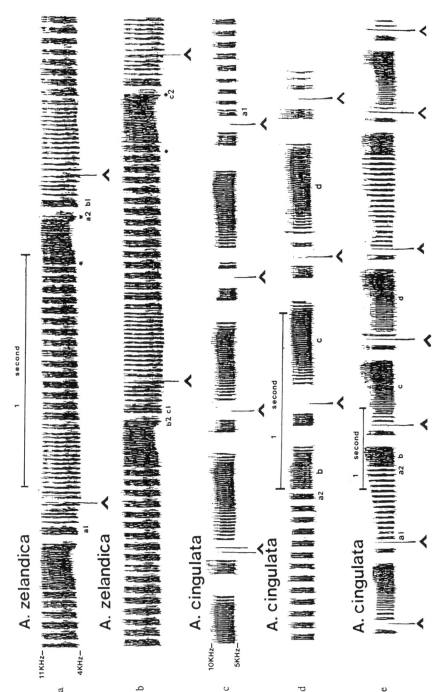

Fig. 17.7.

Fig. 17.7.

a,b. Courtship sequence involving male/female *Amphisalta zelandica*. Male tymbal song continuous from a to b includes three complete song cycles (a1-a2,b1-b2,c1-c2). Species diagnostic frequency modulation notable between points marked by asterisk (*). Male tymbal song is normally accompanied by rhythmic wing clicks (Dugdale and Fleming 1969), but these are omitted during close-range courtship, allowing female to signal receptivity by wing clicking exhibiting a species-specific temporal pattern (see lower frequency trace—female clicks indicated by arrows). Recording made in Hataitai, Wellington, March 1991.

c,d. Courtship sequence *A. cingulata* (specimen 1) continuous from C–D. One complete song cycle shown between a1–d represented as follows "dididididididididi$^{(a1-a2)}$ . . . eéka$^{(b)}$ (click), eéka$^{(c)}$ (click), eéka$^{(d)}$ (click)." The male normally produces the click in free (tymbal) song. In close-range courtship the male omits the click allowing the female to signal its receptivity (female clicks shown by arrows). Female click is identical with male click in terms of temporal pattern relative to tymbal song. Recording made in Hataitai, Wellington, 11 March 1991.

e. Courtship sequence involving second pair of *A. cingulata*. One complete song cycle shown between a1 and d. Arrows show female click response. Recorded at Hataitai, Wellington, 11 March 1991.

Analyses a–e made using DSP Sona-Graph (model 5500), frequency range 0–16 kHz, sonograms reduced by factor of 4.

female wing-click signals have been recorded in a number of species belonging to all five genera of New Zealand cicadas (Lane, unpub. data). No evidence of female signaling has been documented in North American cicadas or in studies of cicadas in the USSR (Heath, Moore, Popov, pers. comms.; Moore has observed such signaling in *Tibicen linei*). I failed to find evidence for it in my studies of three species of Brood XIV *Magicicada* in the 1991 emergence (Lane 1993b). Dunning et al. (1979) observed female wing flicking sometimes during the courtship of *Magicicada* species, but concluded that it signaled the female's *lack* of receptivity. Artificially produced signals have been used on numerous occasions to attract males of *K. scutellaris* from distances of up to ten meters from vegetation to the sound source where they continue to sing (Lane, unpub. data). These artificial signals attract the males only if they precisely mimic the temporal pattern of the female *scutellaris* response (see Fig. 17.6a). This technique has not yet proved successful with the three species of *Kikihia* under investigation in this study.

The Recognition Concept defines species in terms of the very characters by which mating partners recognize each other in their typical habitat, with the SMRS being the central defining property of species. Applying this concept, hybridization occurs when characters of the fertilization systems of two distinct cicada populations are shared to some extent, and is not due to a "breakdown of isolating mechanisms." Cross-attraction (if it ever occurs in nature) due to shared components of the SMRS does not necessarily lead to interbreeding in nature. In insect pheromone systems, (see Linn and Roelofs, this volume) cross-attraction has been reported (e.g. Ganyard and Brady 1971,1972; Phelan and Baker 1986). In a survey of the literature Lambert et al. (1987) found that such studies gave no details on field or laboratory observations of interbreeding. They concluded that there was little relationship between the frequency of observed cross-attractions and levels of taxonomic separation. The Recognition Concept predicts that courtship signals in cicada species will be equally specific in allopatry and in sympatry, because of the structure of the SMRS. Any individual that is deviant with respect to its signal or receiver is less likely to be recognized by, or recognize, a conspecific mate (Lambert and Paterson 1984).

Methods

The Recognition Concept Applied to Kikihia

The summer emergence of adults of the three species of *Kikihia* is within the period mid-December to mid-April, but numbers are sufficiently high to warrant major field work only during mid-January to late March. Initial field work and analysis (Lane 1984) was undertaken during three summers (February

1980 to April 1983), and specimens were collected from within and beyond the zones of suspected hybridization. Further field work was continued each summer to 1991. Field recordings were made of as many singing males as possible, using a Report L 4000 Uher magnetic tape recorder and a dynamic hypercardioid directional microphone (AKG D900). Because the three species are extremely wary, taking to flight at the slightest disturbances, and their singing stations are generally high in the forest canopy, adequate samples were hard to obtain at many sites. Males sing only in fine weather. Specimens were collected by hand or using a sweep net. Analyses of all data additional to the 1980–83 period are included in this chapter.

The only way reports of "hybrid songs," produced by morphological male "hybrids" (Dugdale, pers. comm.) could be verified was by carefully matching song recordings with recorded males. Only male specimens for which adequate acoustic records had been made were set aside for analysis of morphological characters. The decision as to species status of males rested solely on an assessment of the SMRS, in particular the male courtship song. The term *intermediate* was given only to individual males for which the acoustic parameters, after analysis, were judged to be "intermediate" to any extent between any of the taxa involved in the suspected introgression. This could be done only after gaining an appreciation of the variation in the respective species' songs across their entire geographic ranges, which had been done during field work prior to 1980.

The application of the Recognition Concept was difficult in the case of female specimens, since they do not sing and are so rarely caught in copulation with males. It became apparent that there would be problems in areas of sympatry, particularly in the subalpine zones, where males and females sometimes congregate in proximity, with both sexes of the other species under investigation. The method employed was based on the fact that the responsive female replies to the male courtship song by wing clickings which are species-specific in temporal pattern (e.g. Figs. 17.6a,b and 17.7a–e). These signals can just be detected at close range, by the trained ear, and some females can be identified in the field, after careful scrutiny, to be actively engaged in some stage of this signal-response chain. Mere proximity of a female to a male in the field is not an adequate basis for decisions on specific status. All females used in the analysis of morphology were those for which sufficient data relating to the SMRS were available. The interaction of signalers and receivers was observed as far as possible along this coadapted signal-response chain, in many cases through to copulation (cf. Gwynne 1987) to gain knowledge of the operation of the signaling system.

A female was labeled an "intermediate" only if it had been observed to be involved in a signal-response chain involving a male judged to be an intermediate, based solely on acoustic character. (On rare occasions male "intermediates" have been collected and recorded between other species pairs of *Kikihia* [Lane, unpub. data].) The Recognition Concept suggests that such males share components of the SMRS of both species. The view was taken that this is indicative of a population with a different genetic structure (Hoy et al. 1977). Close similarity in terms of acoustic structure in widely separated populations is viewed as indicative of close genetic relationship (i.e., the populations are conspecific or share a common genetic heritage). Critical to such decisions was the high-resolution analysis of the temporal parameters of male courtship song, inaudible to the human ear, but capable of being detected by the female cicada (Huber 1984; Huber et al. 1990) and central to conspecific mate recognition at close range (Doolan and Young 1989).

Acoustic Analysis

Sound recordings of cicadas were analyzed using a Sonograph (model 7030, Kay Electric Co.). This model is an audio-frequency spectrum analyzer which produces permanent graphic recordings of sound waves 5–16 kilocycles/second (sonograms). In the most detailed analyses of pulse structure, only tape intervals of 1.2–2.4 seconds could be analyzed and recorded onto one sonogram. Recordings were also analyzed using a Tektronix Dual Beam Storage Oscilloscope (model 5113), which displays an amplitude versus time graph. Such displays were photographed and recorded on videotape for analysis, to build up a complete picture of one song sequence. A Gould Digital Storage Oscilloscope (model 4030) was used to calculate frequencies from period analysis. Tape recordings were monitored at reduced speeds for analysis of pulse structure (see Fleming 1975*b*). This method allowed the species-diagnostic pulse structures to be analyzed in the field. A Strobes Acquisition Unit 901A was used in conjunction with a Macintosh Apple Computer and a Kay DSP Sona-Graph (model 5500), which were used to analyze recordings of courtship sequences.

Analysis of Morphology

Since species status of the *Kikihia* specimens had been determined a priori using components of the SMRS, there was a need to assess species status, independent of acoustic character, to test whether or not the two classifications correlated. The technique of Anderson (1949, 1954) was used to construct a "hybrid index." The application of this technique was seen as imperative, since it is central to the current use of the term *introgressive hybridization* (Mayr 1963:111; Heiser 1973; Futuyma 1986:115; Harrison 1990:82). This term, first used by Anderson and Hubricht (1938) to describe the repeated backcrossing of hybrids with parental types, is currently used to refer to "gene leakage" into the range of each "semispecies." This term and Anderson's Hybrid Index (or modified versions of it) have received wide usage in the study of plant hybrids (e.g. Anderson and Hubricht [1938], discussed in Harrison [1990:82, 90–95]; Riley [1938], discussed in Futuyma [1986]), and the index has been widely used in similar studies in animal groups (e.g. Hubbs 1959; Littlejohn and Martin 1964; Brown 1974; Woodruff 1977; McDonnell et al. 1978; Wake et al. 1989). The technique involves sampling a number of populations and assessing the states of morphological characters diagnostic of "pure" species (i.e., unaffected by introgression). By using a large number of characters, the average phenotypic character of populations can be compared. This is done quantitatively by assigning numerical values in a standardized manner to character states and constructing composite index values for each specimen.

Howard and Waring (1991), in their electrophoretic studies of a hybrid zone involving two species of ground crickets, calculated character index scores for each individual based on its genotype at four loci they surveyed. They stated: "Character index profiles provide a good way to visualize the genotypic composition of a population and hence the level of isolation between two taxa . . . The character index score of an individual [ranging from -4 to $+6$] represented the sum of its scores over all loci" (p. 1123; cf. Wake et al. 1989). In its simplest form, only three grades are scored and standardized for all characters; the values are: 0, 1, 2. The scores are set such that resemblance to species A is always high (2), and resemblance to species B is always low (0). The intermediate state is assigned the value 1. Sixteen characters diagnostic of each "pure" species (i.e., from sites outside the zones of suspected introgression) were chosen for males, and thirteen for females (Lane 1984:44–67; Lane 1993a). The hybrid index is based on the belief that the "average" hybrid lies midway between the "average" member of the two populations (Vogt and McPherson 1972). Anderson (1954:106) stressed that species do not differ at random, but that "species are differentiated by combinations of characters more certainly than by single characters."

Defining Populations and Geographic Zones

The total sample of 2428 specimens (collected from 1980 to 1991) was divided into three zones based on locality. Three zones were defined:

Zone 1: Nonvolcanic areas, including main axial ranges (and their plains) (Fig. 17.3,inset) and the nonvolcanic hill country between Egmont and Ruapehu, but excluding the pumice-coated greywacke Kaimanawa Ranges (see Figs. 17.2 and 17.3)

Zone 2: Pleistocene volcanic areas above the present bushline (Egmont, Pouakai Range, Egmont National Park, Ruapehu, Ngauruhoe, Tongariro, Tongariro National Park, alpine ridges of Kaimanawa Range)

Zone 3: Pleistocene volcanic areas below the bushline, either on dark ash (Taranaki) or pumice (Tongariro National Park, Taupo-Rotorua)

"Pure" populations of *K. cutora* and *K. subalpina* were classed as Zone 1, while putative hybrids reported by Fleming and Dugdale were from Zone 2. All specimens were pinned (wings spread) and stored in 21- × 34-cm entomological boxes (designated Lane collection).

Summary of Results

The three taxa, *K. cutora*, *K. subalpina*, and *K. laneorum*, each proved to have a unique SMRS throughout their respective distributions and exhibited no variation in terms of the unique sound pulse patterns underlying male song (Figs. 17.6c–d, 17.8, and 17.9). The total song structure was unique to each species, in terms of the patterning of clusters of pulses. The acoustic motifs audible to the human ear are unique, and no "intermediate" songs were ever heard or recorded in the field, including Zone 2, from where they had been reported. The uniqueness of the female component of the SMRS to each species was unable to be documented, due to the difficulties in obtaining suitable field recordings. Adults of these species did not prove amenable to laboratory behavioral studies. However, in two other species in the genus *Kikihia*, *K. muta muta* and *K. scutellaris* (Figs. 17.6a,b), and in two species in the genus *Amphisalta* (*A. cingulata* and *A. zelandica*, Figs. 17.7a–e), it was shown that the temporal patterning of female clicks, in courtship sequences, was unique to each species. I inferred from these data that the same holds true for the other species of *Kikihia* based on aural records and observations of such interactions. The intraspecies variation in acoustic character is related to the length of phrases and the gaps between them, (Lane 1984) and was generally related to temperature effects (as in Josephson and Young 1979), time of singing, and motivational state (presence or absence of responsive females). *K. cutora* (*sensu lato*) exhibited the greatest variability throughout the North Island ranges of the species, while the songs of *K. subalpina* and *K. laneorum* exhibited much greater consistency in gross temporal structure (Lane 1984). The three species overlap in the dominant frequency of male song, with peak amplitudes lying between 14 and 16 kHz.

In the assessment of species status, using only morphological criteria, no single character could be totally relied on in either sex, in populations where at least two of these species of *Kikihia* are sympatric (cf. Gibbs 1968). With the presence of *K. laneorum* (morphologically almost identical to *K. subalpina* and

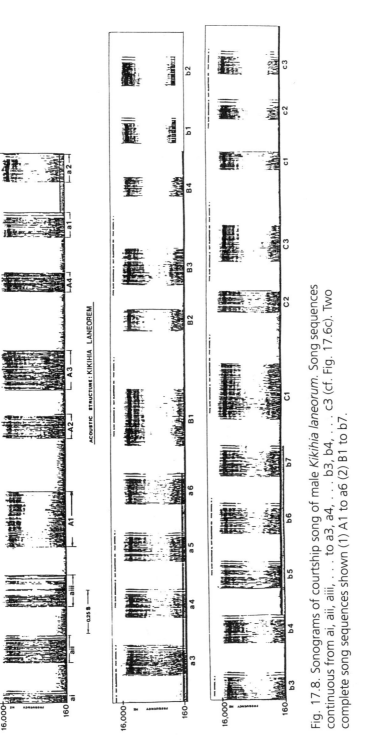

Fig. 17.8. Sonograms of courtship song of male *Kikihia laneorum*. Song sequences continuous from ai, aii, aiii, . . . to a3, a4, . . . b3, b4, . . . c3 (cf. Fig. 17.6c). Two complete song sequences shown (1) A1 to a6 (2) B1 to b7.

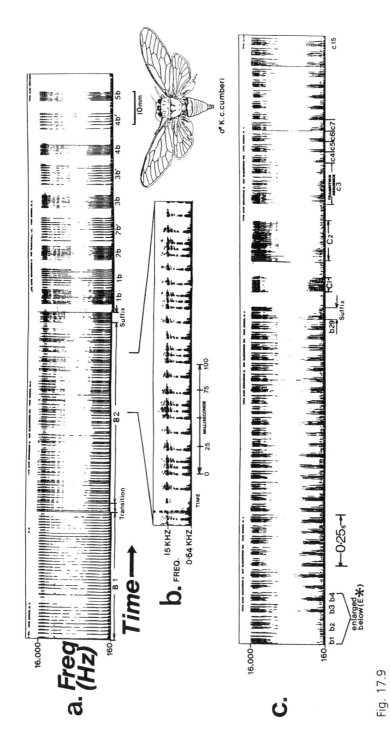

Fig. 17.9

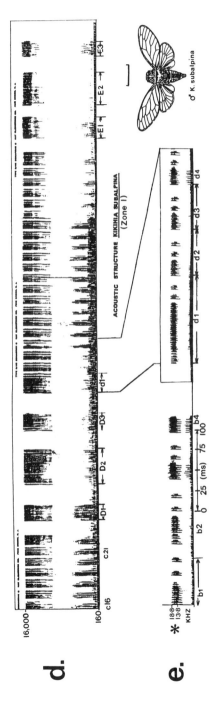

Fig. 17.9
a,b. Sonograms of courtship song of male *Kikihia cutora cumberi*, recorded from Wellington (zone 1) (character score 31.8, Anderson Hybrid Index). a. One complete song sequence consisting of sequence of pulses (B1) which varies in duration from 0.16 s (11 doublets) to 1.27 s (86 doublets) and varies according to temperature. A complex pulse group cluster underlies section B2 of the main phrase, which similarly varies in duration. An alternating sequence of faster and slower pulses forms the set of pulse groupings: 1b, 1b', 2b, 2b'–5b. Analysis made using 40-kHz frequency range (sonograph) and ¼ tape speed giving a sonogram frequency range 0.16–16 kHz. (160–16,000 Hz).

b. Analysis of species-specific pulse group clusters. Sonogram frequency range 0.64–15 kHz achieved using 16-kHz range (sonograph) and ⅛ tape speed.

c,d. Sonograms of courtship song of male *Kikihia subalpina*. Three complete song units illustrated. Song unit commences with distinctive pulse group clusters b1, b2, b3 ... to b29, followed by a "suffix" consisting of 10–13 doublet pulses. The rapid sequences of pulses comprising C1, C2; D1, D2; E1, E2, E3 generally show little variability in duration or number (sonograph analysis identical to sonogram *a*, i.e., same time base).

e. Analysis of species-specific pulse group clusters. Sonograph analysis identical to sonogram *b* (i.e., same time base). Note distinctive pulse group clusters labeled d2, d3, d4 (8:2:2 pulses).
All figures from Lane (1984).

the putative hybrids) in sympatry in Zone 3 populations, the use of single characters, such as tymbal ridging (effective in distinguishing *K. cutora* and *K. subalpina*), had to be abandoned (see Fig. 17.5). The Anderson Hybrid Index employed, using sixteen characters to distinguish males (thirteen in female), proved a powerful tool in determining species status. Mean male scores plotted against mean female scores from each population of all three species showed that *K. subalpina* and *K. cutora* formed two distinct phenetic clusters (Fig. 17.10). More important, these nonoverlapping clusters existed in all Zone 2 (volcanic subalpine) populations where hybridization had been reported. The frequencies of individual specimen scores were plotted and compared for all populations (Figs. 17.11–17.17). Again, it is evident that the two taxa involved in putative hybridization are distinctive in Zone 2 populations (Figs. 17.13m–q; 17.16m1, 17.17n1,o1). It is noteworthy that a large number of Zone 2 and 3 specimens of *K. c. cumberi* of both sexes received scores close to midway between the mean population scores of *K. subalpina* and *K. c. cumberi*, respectively (Figs. 17.13m–q; 17.14–15, s–x; 17.16–17,m1,n1,t1), and would be classed as morphological "hybrids" under Anderson Hybrid Index procedures. One such "hybrid" (see Figs. 17.5a–c, 17.11a,k) was found from Zone 1 populations (nonvolcanic zone). When mean scores for male *K. laneorum* (Zones 1, 3) are added to the index (Fig. 17.10), these population mean values form a discrete cluster distinct from *K. subalpina* and most Zone 1 populations of *K. c. cumberi*. However, *K. laneorum* scores (Figs. 17.12i,l and 17.14r) do overlap with Zone 2 and 3 populations of *K. c. cumberi* (e.g. Zone 2, Fig. 17.12q). Only one specimen, representing 0.04 percent of the total sample of *K. cutora/K. subalpina*, could not be classified correctly using only morphological criteria (Fig. 17.11). This specimen scored 24 on the hybrid index, and was therefore close to the scores for "pure" *K. cutora* male (32), rather than "pure" *K. subalpina* (0). However, its song was that of *K. subalpina* despite the fact it possessed *K. cutora*–type tymbals (Fig. 17.5a–c). There was broad overlap in morphology between *K. laneorum* and *K. c. cumberi* in one Zone 1 site (Haurangi State Forest; cf. Figs. 17.12, h and i). The "swapping of characters" here was reminiscent of the putative hybridization involving *K. c. cumberi* and *K. subalpina*. (*K. subalpina* is extremely rare in the Haurangis).

Discussion

The uniqueness of the pulse patterns of the male song of each species of *Kikihia* and the absence of "intermediate" songs in the field are strong evidence against the hypothesis of hybridization advanced by Fleming (1973, 1984). Dugdale (following Fleming 1975a) argued that, if introgression was occurring, it must lead to a breakdown of the components of male song, and an eventual merging of the two species, into a continuous "hybrid swarm" (Mayr

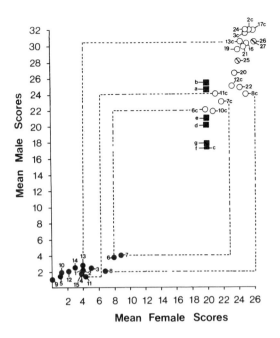

Summary of data used for construction of Hybrid Index (Fig. 17.10)

K. subalpina (●)

	Male			Female		
Pop*	N	X̄	Range	N	X̄	Range
1	51	2.0	0-8	40	4.0	0-8
2	49	2.0	1-7	5	4.0	0-9
3	425	2.4	0-12	74	5.1	1-11
4	116	2.4	0-24	17	4.0	2-8
5	11	1.5	0-3	2	1.0	0-2
6	45	3.7	0-7	4	7.8	4-10
7	34	4.0	0-7	3	8.7	8-9
8	35	2.1	0-6	7	6.7	0-2
9	12	1.3	0-3	1	0	0
10	81	2.0	0-6	5	1.2	0-2
11	29	1.6	0-6	4	4.3	3-6
12	26	2.1	0-6	9	2.2	1-7
13	20	2.8	0-5	1	4.0	4
14	4	2.5	1-5	1	3.0	3
15	3	2.3	1-3	1	4.0	4

K. laneorum (■)

	Male			Female		
Pop‡	N	X̄	Range	N	X̄	Range
a	48	24.5	19-30			
b	4	25.3	25-27			
c	1	17.0	17			
d	2	20.0	19-21			
e	4	21.0	20-25			
f	1	17	17			
g	8	17.0	15-20	1	20	20

K.c. cumberi (O)

	Male			Female		
Pop**	N	X̄	Range	N	X̄	Range
2c	63	31.8	29-32	11	25.6	24-26
3c	50	31.3	29-32	7	25.0	24-26
4c	180	31.8	28-32	9	24.8	24-26
6c	96	22.0	14.30	13	19.8	16-23
7c	68	22.9	17-31	5	21.8	19-24
8c	8	23.9	18-32	2	25.0	24-26
10c	122	21.9	14-30	13	20.7	14-24
11c	60	24.1	14-32	9	20.9	14-26
12c	8	24.8	19-29	3	23.3	23-24
13c	7	30.4	30-31	4	24.3	24-25
16	69	30.3	26-32	11	25.2	23-26
17c	54	31.8	30-32	3	26	26
18	17	31.7	30-32	1	26	26
19	19	29.5	18-32	6	24	24
20	46	26.4	15-32	5	23.7	20-25
21	27	29.7	22-32	3	24.7	23-26
22	54	24.7	14-31	3	24.3	23-25
24	180	31.8	28-32	9	24.8	24-26

K.c.cutora(◉)

	Male			Female		
Pop**	N	X̄	Range	N	X̄	Range
25	14	28	21-31	3	24.0	22-25
26	43	30.4	26-32	1	26.0	26
27	17	30.1	26-32	1	26.0	26

KEY
☐ Zone 1 populations
☐ Zone 2 populations
☐ Zone 3 populations

* Geographic location and altitude of all populations shown on Fig. 17.4
** Geographic location and altitude of all populations shown on Fig. 17.3
‡ Geographic location and altitude of all populations shown on Fig. 17.2
Populations of K. c. cumberi 2c-17c are sympatric with corresponding populations of K. subalpina (2-17). Five such relationships are shown on Fig. 17.10 by use of dotted lines: e.g. population 6 (K. subalpina) is sympatric with population 6c (K. c. cumberi) (cf. Figs. 17.4 and 17.3, respectively).

Fig. 17.10. Anderson Hybrid Index illustrating mean male character scores plotted against mean female character scores (sixteen male and thirteen female morphological characters assessed).

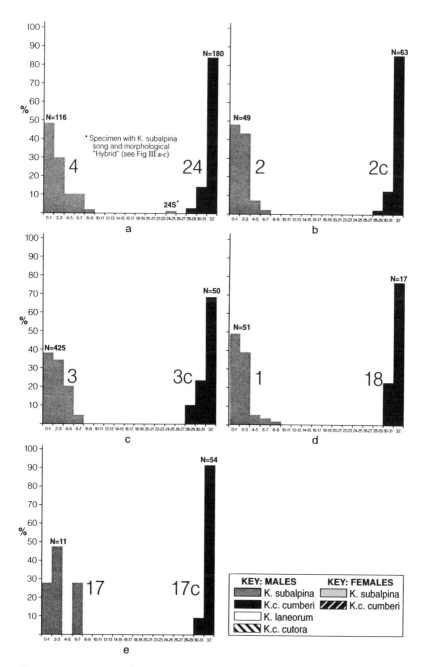

Figs. 17.11–17.17. Character score frequencies compared between populations of *K. c. cumberi, K. c. cutora, K. subalpina* and *K. laneorum*. Graphs a–z1 (males), a1–t1 (females). Sample size (*N*) and codes for sample localities are shown on graphs (refer to Figs. 17.3–17.5); e.g. graph b represents two sympatric populations (2,2c) from zone 1, main axial ranges (Kaitoke, 200 m). *N** = sample size of all populations from one zone, e.g. *N*(1*) = combined sample from all zone 1 populations shown in Figs. 17.11k,l (zone 1), 17.12r (zones 1 and 2), 17.12r (zone 2), 17.13y (zone 3), 17.13r1 (zone 1, females). Graphs a–l all involve zone 1 populations (main axial ranges).

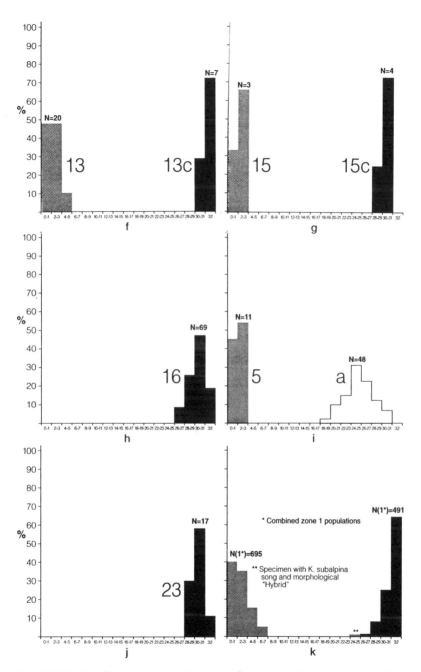

Fig. 17.12. Graphs m to p based on data from zone 2 sympatric populations (zone 2 = area of suspected hybridization). m = Mt. Ruapehu (Silica Springs 1340m); n = Mt Egmont (above 1140m); o = Kaimanawa Ranges (1340m); p = Ohakune Mt Rd, Mt Ruapehu (1360–1440m) (see figs. II, IV). Graphs s to x based on data from zone 3 (pleistocene volcanic areas below bush line).

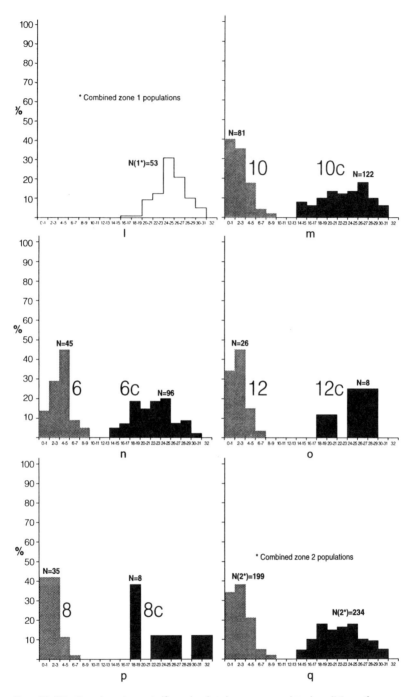

Fig. 17.13. Graphs a1 to t1 (female data) correspond to localities of male data, graphs a–t (fig. XI). a1 to k1 involve sympatric zone 1 populations. m1 to o1 involve zone 2 populations. t1 = zone 3 population (Chateau, Mt Ruapehu 1050m).

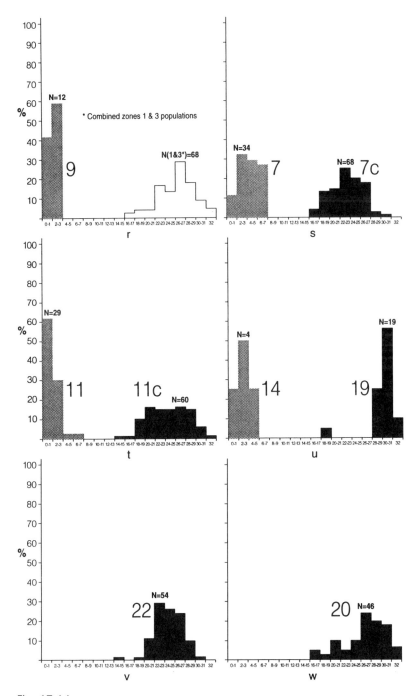

Fig. 17.14.

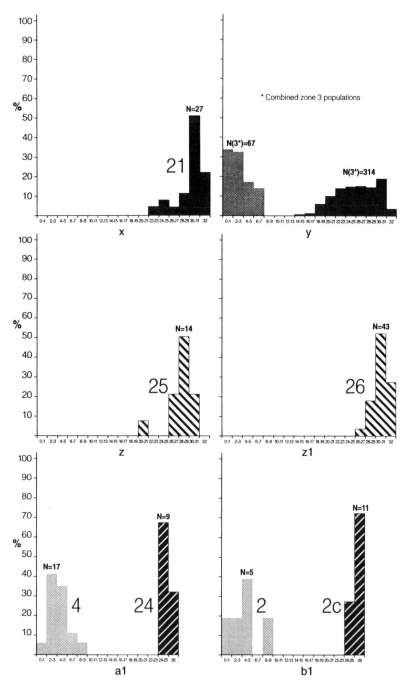

Fig. 17.15.

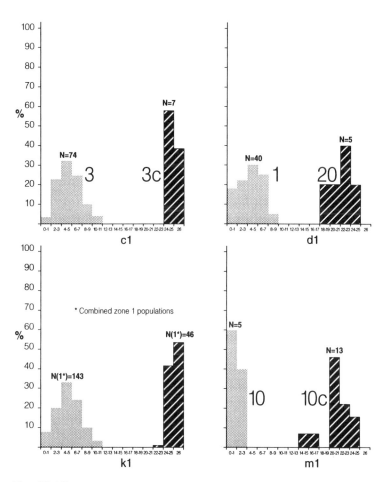

Fig. 17.16.

1963:132). Harrison (1990:72) proposed an "operational definition" for "hybrid zones" as "interactions between genetically distinct groups of individuals resulting in at least some offspring of mixed ancestry. Pure populations of the two genetically distinct groups are found outside of the zone of interaction." He also noted that "hybrid zones may be modified . . . [by] introgressive hybridization [which] results in fusion, the formation of a single homogeneous (perhaps polymorphic) population" (p. 89). Dugdale reported on a large number of "recombinants" in terms of acoustic and morphological characters, in Zone 2 populations. He considered this a parallel case with a "narrow zone of introgression (i.e. intermediate songs)" reported by Fleming (1975a:299) involving two other taxa within the genus *Kikihia*. Dugdale considered the Zone 2 populations to be a case of "peripheral sympatric hybridization" (see Wood-

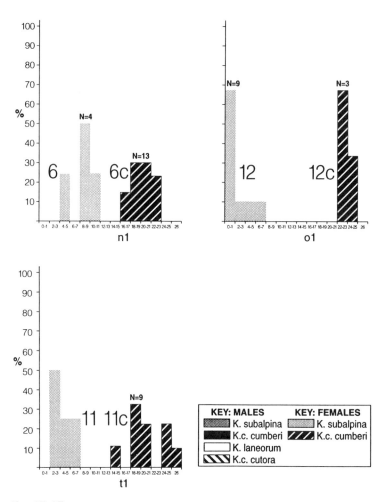

Fig. 17.17.

ruff 1977:216), which is assumed to arise when a population gradually extends its range into another closely related population (see examples in Short 1969).

The widespread occurrence of stable hybrid zones (see review in Barton and Hewitt 1985) has led even the originators of the Biological Species Concept to qualify their reliance on gene flow. For example, Mayr (1982:285) states that "the two species in such a case seem to remain 'reproductively isolated' in the sense that they do not fuse into a single population." If introgression has occurred between any of the *Kikihia* in the past, then why have all three species remained discrete in terms of their acoustic structure, over their entire geographic distributions? The stability of the SMRS of each species (cf. Moore 1993) is a logical prediction of the Recognition Concept (Lambert and Paterson 1984). The significance of the uniqueness of the songs of the three species of *Kikihia* is admittedly of little consequence, if the female exhibits limited or no discriminatory power. This, however, is not the case as evidenced by the absence of acoustic "recombinants" in the field.

Physiological studies of sound reception in other cicada genera have established that sound pulses generated at rates of up to 200 per second are perceived as separate entities and that the patterns of electrical discharge within the tympanal nerve of the female cicada correspond to each incoming pulse of sound. The grouping of these pulses in male song, according to Huber (1984:29), is "the crucial parameter for species-specific sound recognition and sound location" and "the auditory sense cells would be expected to copy the pulses in the activity pattern" (Huber et al. 1979, 1980, 1990; Popov 1981). The fact that receptive females of *Kikihia* have the ability to, and do in fact, synchronize their "click response" to the temporal patterning of conspecific male courtship song (Figs. 17.6a–b and 17.7a–e) is strong evidence in favor of the specificity of male song (Gwynne 1987). Doolan and Young (1989) have demonstrated that the temporal component of the male calling song is central to mate recognition in the cicada *Cystosoma saundersii* (Westwood). Courtship responses were elicited only by model songs with temporal parameters similar to those of the natural song. In contrast, a wide range of carrier frequencies were found to elicit courtship provided that the temporal parameters were similar to those of the control.

The results of biometric analysis summarized in Fig. 17.10, using the methods of Anderson (1949, 1954), provided strong evidence that the taxa *K. cutora* and *K. subalpina* are discrete genetic communities of reproduction; since all specimens, except one (Figs. 17.5a–c and 17.11a, 17.12k), could be assigned to the correct species, using a combination of thirteen to sixteen morphological characters (without knowledge of the SMRS). The overall phenotypic pattern of the two species (with the exception of one specimen), did not overlap in terms of hybrid index scores. If hybridization was occurring in the Zone 2 populations, one would expect significant overlap of the two phenotypic clusters, due to numerous recombinations and backcrosses. The apparent intermediacy in morphology, observed in the putative hybrids, is therefore superficial and can be accounted for under other hypotheses besides hybridization (see Lane 1984:217–20). The one specimen with a "hybrid" score was from Zone 1, where no case for hybridization has been advanced. Its song was that of *K. subalpina* rather than "intermediate," so it could be interpreted as an extreme color variant of the species *K. subalpina* (Lane 1984:106–108,113). The close correlation between the classification of specimens of *K. cutora* and *K.*

subalpina, based on the SMRS, and the independent classification based on combinations of morphological characters, is convincing evidence for dismissing the hypothesis of hybridization (Lane 1993a).

Mayr (1963:130) stated: "If hybrids between two species were to form a third species coexisting with the two parental species, this would be a process of speciation through hybridization. Such a process has . . . never been unequivocally established." More recently, speciation by hybridization has been claimed to have played a major role in plant evolution (e.g. Grant 1981). Harrison (1990:91) stated: " 'hybrid swarms' or populations of stable introgessants often appear to dominate in certain habitats . . . suggest[ing] that introgression may have a significant role in the evolution of some groups of plants." However, frequently cited cases (e.g. Lewis 1953, 1966; Lewis and Raven 1958) turn out to be highly dubious, since they are based on the old fallacy of inferring causality from a correlation (Paterson 1988). A number of examples of speciation by hybridization have been suggested in animal taxa (e.g. Moritz et al. 1992), but the number of strongly documented cases of introgression are very low (Heiser 1973). Sneath (1975:360) concluded that reticulate evolution "may have occurred more than is commonly thought in some invertebrates." Harrison (1990:116) suggests that reproductive character displacement may lead "to the origin of an entirely new species, distinct from either of the two parental types." He believes that such a scenario would be facilitated if the male and female components of the mate recognition system are genetically coupled (see Butlin and Ritchie 1989). He also notes that "in several of the best examples of reproductive character displacement (reinforcement)," (Littlejohn 1965; Wasserman and Koepfer 1977), the data support his scenario.

While examples of hybridization in the field have been reported in insects (see review in White 1978), the present data do not support a case for the hybrid origin of *K. laneorum*. Its distinctive SMRS, lacking combinations of acoustic characters derived from *K. cutora* and *K. subalpina* (Lane 1984), rules it out as being of hybrid origin. It remained distinctive even at a site where *K. laneorum* occurred in sympatry with *K. c. cumberi* and where there was broad overlap in morphology between the two species (see Fig. 17.12h,i). If a co-adapted signal-response system, consisting of a male mating call and female auditory system acting as a finely tuned receiver, is operating as the basis of conspecific mate recognition, there may be no gene flow between the species *K. cutora* and *K. subalpina*, despite marked overlap in morphological characters. The same is true of the genetic relationship of these two species to *K. laneorum*.

Greater variability in character states (reflected in score frequencies and ranges, for example, *K. c. cumberi* (Fig. 17.12k; cf. Figs. 17.12q and 17.15y) between populations presumed to be of "hybrid origin," and homospecific populations are generally taken as prima facie evidence for hybridization (Anderson 1949; Mayr 1963). Endler (1977) has shown that, despite longstanding claims, it is invalid to use levels of morphological variation to distinguish zones of primary hybridization (involving races), from zones of secondary hybridization. There are many cases where "hybrid" populations are no more variable than homospecific ones (Sokal and Rohlf 1969). It has now been established that increased variability in some populations can be explained without taking recourse in hypotheses of hybridization. These include the development of morph-ratio step-clines in populations (Clarke 1966) and the spread of newly evolved characteristics through the range of the species. Rosen (1979:278) notes that there is no a priori reason why changes in intrapopulational variability (and apparent intermediacy) might not be the result of primary causes (rather than secondary hybridization) affecting different parts of a species' range. Species can develop geographic or ecophenotypic morphs (e.g. *Tibicen lyricen* [De Geer], see Moore [1993]), resulting in a situation suggestive of secondary hybridization. This appears to be the case in the extreme plasticity of pigmentation pattern expression in *K. cutora* (*sensu lato*) compared with *K. subalpina*, and similar variability exists in at least one Zone 1 population of *K. laneorum* ($N = 48$; see Fig. 17.14i) when compared with *K. c. cumberi* ($N = 11$; see Fig. 17.11h) from the same site (Haurangi State Forest Park).

The marked overlap in the dominant frequencies in male songs between sympatric cicada species, a puzzle to other researchers (e.g. Hagiwara and Ogura 1960) is confirmed in this study of *Kikihia*. Katsuki and Suga (1960) concluded that frequency discrimination in insects is a possibility, but that cicadas may "recognise their group by discriminating sounds by means of the temporal and spatial patterns of impulses" (p. 289) and that the high frequencies employed aids directionality in long-range communication. Reliance on higher frequencies for the transmission of airborne songs, is one of the constraints upon insects of small body size (Michelson 1985). Pringle (1954:555) in his classic physiological studies of cicada song, noted that "the sound frequency is that of a carrier wave, and the significant information for the insect is transmitted in the pulse pattern" (based on Pumphrey 1940). The songs of some cicadas lack a pulse pattern, and their free song resembles an almost continuous tone (Moulds 1975). Doolan and Young (1989) have demonstrated experimentally that long-range phonotaxis is dependent on a species-specific carrier frequency of male calling song in the cicada *Cystosoma saundersii*. The temporal patterns of male cicada courtship song have been shown to be the basis of mate recognition in close-range communication (Joerman and Schneider 1987). Temporal patterning of pulses has been suggested as a means of avoiding acoustic interference in some insect groups (e.g. Latimer 1981). Three suggestions have been made as to how acoustic interference can be avoided in animals employing sound signals as the basis of conspecific mate recognition (Littlejohn and Martin 1964): (1) frequency (or spectral) separation (utilization of different frequency bands for transmission), (2) spatial separation by aggregation into species-specific choruses, (3) temporal separation, whereby two species synchronize calling rates or times to reduce overlap. All these behavioral characters have been demonstrated in cicadas (e.g. Alexander and Moore 1958; Doolan 1981; Doolan and McNally 1981; Ewart 1989*b*; Williams and Smith 1991). Sanborn (1990) has shown how endothermy may reduce acoustic interference in some cicada species. Singing involves intense activity of the tymbal muscles, which apparently can function effectively only over a relatively narrow range of temperature (Josephson and Young 1979,1981; Young and Josephson 1983). Some species can regulate thoracic temperature within fairly narrow limits by utilizing behavioral mechanisms and evaporative cooling. Individuals of two sympatric species have been found to be active in thermally distinct microhabitats (Hastings and Toolson 1991). The courtship behaviors of the three species of *Kikihia* studied, each employing a unique SMRS, are not disrupted by acoustic "interference," since the songs of males consist of unique temporal patterns (Figs. 17.6,17.8, and 17.9). Other components of song in cicadas, such as frequency modulation (see Fig. 17.7a,b), have been shown to be species specific (Young 1972; Popov 1981; Popov et al. 1986; Weber et al. 1988; Ewart 1989*a,b*; Jiang et al. 1991*a*) and may also be important in species recognition (Popov 1981). The production of varitoned complex songs are not only related to abdominal movement, but are mainly dependent on the vibration property of the tymbal rib structures and contraction properties of the tymbal muscles (Jiang et al. 1991*b*).

The striking feature of the machinery of sound production in the cicada, evident from this study of the genus *Kikihia* and from previous studies, is the diversity of patterns of coordination produced by such little modification of the basic anatomical "design" (Pringle 1954, 1957; Young 1972, 1980; Josephson and Young 1981, 1985; Young and Josephson 1983; Ewart 1990*a,b*; Popov 1990). Pringle (1954:554) noted that it "is unusual in the animal kingdom . . . that such a diversity of patterns of coordination should have evolved within one family with so little modification of the basic anatomical plan" (or "design"; cf. "design features" [Paterson 1988:69] and "well integrated design" [Coyne et al. 1988:197]). As noted by Hartley and Stephen (1992:359) in reference to bush crickets (Tettigoniidae), the sound-producing apparatus in cicadas "is an elaborately constructed device with *design features* peculiar to each species" (italics added). Commenting on biological adaptations in general, Gould (1976) stated: "Certain [examples] should be superior *a priori* as designs for living in new environments. These traits confer fitness by an engineer's criterion of good design." Indeed, the highly efficient two-way acoustic communication system in cicadas reveals an "engineering design" of remarkable complexity. Are the patterns evident from detailed acoustic analysis mere mechanical products ground out to a predetermined pattern by the structure of the sound organ? The answer is *No* (Myers 1929b). The interneural control of patterned movements of the tymbal (by specific sets of command) has been well illustrated by the experimental reproduction of typical sound reflexes in the cicada (Hagiwara and Watanabe 1956; cf. Simmons and Young 1978), suggesting that song specificity is under genetic control. Research has provided evidence that hybridization involving insect taxa employing unique acoustic signals does result in the production of new and intermediate acoustic patterns (Bentley and Hoy 1974; Yusof 1982; Claridge 1985; Heady 1987). Hybrid song patterns in cicadas have been recognized in the *K. muta* complex (Fleming 1975a) as involving acoustic components of both parental species, based on sonogram analyses (Lane, unpub. data). The fact that the form of the elements is like that of either one or the other of the parents is consistent with the hypothesis that these parameters are inherited independently and are not controlled by a group of closely linked loci. Any new and intermediate characters might therefore be due to polygenic inheritance (Leroy 1963). Playback experiments have shown that hybrid field crickets can respond equally well to signals of both parents and indicate a preferential response to hybrids (Bentley and Hoy 1974; cf. Rao and DeBach 1969). The lengthy life cycle of *Kikihia* species (at least four years [Cumber 1952]), and the difficulties involved in raising adults from eggs, meant that no experimental studies of the acoustic characters of laboratory-produced hybrids of *Kikihia* were attempted in the initial inquiry (Lane 1984). More recent experimental work has produced large numbers of interspecific matings within the genus *Kikihia* by manipulating the acoustic environment during courtship sequences (Lane, unpub. data). No crosses were achieved between *K. c. cumberi* and *K. subalpina* nor between *K. laneorum* and the other two species. (Similar methods have achieved interspecific matings between the two species of *Amphisalta*, whose courtship sequences are illustrated in Fig. 17.7). As discrimination trials (Lane, unpub. data) have demonstrated, the temporal patterns underlying close-range acoustic communication are an important component of the SMRS in species of the genus *Kikihia;* therefore, interspecific matings between *K. cutora cumberi* and *K. subalpina* are predicted to be very unlikely in the field.

Conclusions

1. The absence of "hybrid songs" in the zones of suspected hybridization between *K. cutora cumberi* and *K. subalpina* remains the strongest evidence against the hypothesis of hybridization, advanced by Fleming (1973, 1984).

2. The unique SMRS of each of the three species of *Kikihia* studied indicates that each of the taxa represents a distinct community of reproduction, since it has been established that the temporal parameters of male court-

ship song form the basis of conspecific mate recognition and the female auditory system acts as a finely tuned receiver (Doolan and Young 1989; Huber 1984; Huber et al. 1980, 1990).

3. The Recognition Concept of species (Paterson 1985, 1988) is applicable to investigations of suspected hybridization and as a concept has significant heuristic value in such studies. It has some advantages (see Templeton 1992) over the Biological Species Concept (Mayr 1942, 1963), which treats hybridization as a failure, a breakdown of the "isolating mechanisms" of an incipient species. These "mechanisms" or "devices" are considered to be still in an "imperfect and incomplete" stage of evolution (Mayr 1992). According to the Recognition Concept, hybridization occurs when characters of the fertilization systems of two populations are shared to some extent.

Summary

The aim of this chapter has been to investigate the heuristic value of the Recognition Concept of species (Paterson 1981, 1985), as it applies to studies of hybridization. A study of suspected hybridization (Lane 1984, 1985, 1990*a,b*) employing this concept, rather than the Biological Species Concept (Mayr, 1942, 1963) has been briefly reviewed.

Fleming (1973, 1984) suggested that two New Zealand species of cicada, *K. subalpina* and *K. cutora cumberi,* are hybridizing in the subalpine zones of the North Island's volcanic region. These studies employing the Biological Species Concept suggested that the morphological intermediacy of the putative hybrids, their apparent close phylogenetic relationship, and a speciation model for their origin, involving reinforcement, or perfection of "isolating mechanisms" in sympatry, provided the best support for hybridization.

Two aspects of the study of suspected hybridization have been reviewed: acoustic analysis and morphological analysis. The species *K. laneorum* was also investigated, because of its morphological similarity to *K. subalpina* and the putative hybrids and its sympatry with the other two species.

Consistent with the Recognition Concept, species status of all specimens was determined a priori, using acoustic character (in males), as this is a central component of the SMRS (Doolan 1981; Doolan and Young 1989; Huber 1984; Huber et al. 1990).

The response of females to male song was used to assess the species status of females. I collected 2428 specimens between 1980 and 1991 and demonstrated that all three species possess a unique SMRS in terms of the temporal patterning of sound pulses in male courtship song and female receptivity. Moreover, these patterns were remarkably stable throughout the geographical ranges of the respective species.

The application of the Anderson Hybrid Index enabled an independent assessment of species status to be made based only on morphological criteria. All specimens except one (0.04 percent) could be assigned to the "correct" species (consistent with species determination based on the SMRS) using combinations of morphological characters (sixteen in males, thirteen in females). The close correlation between the two classifications proved strong evidence against the hypothesis of hybridization.

Conceptual difficulties with the Biological Species Concept have been discussed. The application of the Recognition Concept to this inquiry demonstrated its heuristic value in studies of hybridization.

Acknowledgments

I wish to express my special thanks to Mr. John Dugdale for making available to me for examination a number of specimens of *Kikihia* from the New Zealand Arthropod Collection, Mount Albert Research Centre, Auckland. He provided helpful advice and encouragement during the early stages of the study.

I thank Dr. Peter Jenkins and Dr. Andy Edgar for assistance with the acoustic analysis. Dr. David Lambert has given valuable advice on the theoretical aspects of species concepts and has read and commented on an early draft of this chapter.

The research formed part of an M.Sc. (Hons.) thesis, submitted at Victoria University, Wellington, in April 1984. I am grateful to a large number of academics worldwide who have taken considerable interest in the work since its submission and provided helpful comments on aspects of the completed thesis. In particular, I thank Professors M.F. Claridge, J.E. Heath, M.S. Heath, F. Huber, M. Lloyd, T.E. Moore, H.E.H. Paterson, A.V. Popov, and H. Strübing; Drs. M. Boulard, P.W.F. de Vrijer, S. Drosopoulos, J.P. Duffels, A. Ewart, D. Gwynne, K. Hill, J. den Hollander, M.J. Littlejohn, J. Moss, J.A. Quartau, R. Rowe, C.M. Simon, J.W. Schedl, V. Thompson, B. Whittaker, and D. Young; and Mr. A.C. Harris and Mr. M.S. Moulds.

It is a pleasure to acknowledge my parents Rex and Ruth Lane for their support and encouragement in the course of this study. I am indebted to the late Sir Charles Fleming for the loan of specimens of *Kikihia laneorum* (Fleming) and for stimulating my interests in cicadas in my formative years. He provided no other assistance in the course of the thesis study, an arrangement he insisted upon (and to which I agreed), to ensure that we arrived at independent conclusions in our studies of *Kikihia*.

References

Alexander, R.D., and T.E. Moore. 1958. Studies on the acoustic behavior of seventeen-year cicadas (Homoptera: Cicadidae: *Magicicada*). Ohio J. Sci. 58:107–127.

Anderson, E. 1949. Introgressive hybridization. John Wiley, New York.

———. 1954. Efficient and inefficient methods of measuring specific differences. Pp. 93–106 *in* O. Kempthorne, T.A. Bancroft, J.W. Gowen, and J.L. Lush, eds. Statistics and mathematics in biology. Iowa State College Press, Ames.

Anderson, E., and L. Hubricht. 1938. Hybridization in *Tradescantia*. III. The evidence for introgressive hybridization. Am. J. Bot. 25:396–402.

Ayala, F.J. 1982. Gradualism versus punctualism in speciation: Reproductive isolation, morphology, genetics. Pp. 51–66 *in* C. Barigozzi, ed. Mechanisms of speciation. Proceedings from the international meeting on mechanisms of speciation May 4–8, 1981, Rome, Italy. A.R. Liss, New York.

Barton, N.H., and G.M. Hewitt. 1985. Analysis of hybrid zones. Annu. Rev. Ecol. Syst. 16:113–148.

Barton, N.H., J.S. Jones, and J. Mallet. 1988. No barriers to speciation. Nature 336:13–14.

Benedix, J.H., and D.J. Howard. 1991. Calling song displacement in a zone of overlap and hybridization. Evolution 45:1751–1759.

Bentley, D., and R.R. Hoy. 1974. The neurobiology of cricket song. Sci. Am. 231(2):34–44.

Bigelow, R.S. 1964. Song differences in closely related cricket species and their significance. Aust. J. Sci. 27:99–102.

———. 1965. Hybrid zones and reproductive isolation. Evolution 19:449–458.

Bock, W.J. 1979. The synthetic explanation of macroevolutionary change—a reductionistic approach. Pp. 20–69 *in* J.H. Schwartz and H.B. Rollins, eds. Models and methodologies in evolutionary theory. Bull. Carnegie Mus. Nat. Hist. 13:20–69.

———. 1986. Species concepts, speciation, and macroevolution. Pp. 31–57 *in* K. Iwatsuki, P.H. Raven and W.J. Bock, eds. Modern aspects of species. University of Tokyo Press, Tokyo.

Brown, C.W. 1974. Hybridization among the subspecies of plethodontid salamander *Ensantina eschscholtzi*. Univ. Calif. Pub. Zool. 94:1–57.

Brown, J.L. 1975. The evolution of behavior. Norton and Co. Inc., New York.

Brown, W.L., and E.O. Wilson. 1956. Character displacement. Syst. Zool. 5:49–64.

Butlin, R.K. 1985. Speciation by reinforcement. Pp. 84–113 *in* J. Gosalvez, C. Lopez-Fernandez and C. Garcia de la Vega, eds. Orthoptera 1. Fundación Ramón Areces, Madrid.

———. 1987a. Species, speciation and reinforcement. Am. Nat. 130:461–464.

———. 1987b. Speciation by reinforcement. Trends Ecol. Evol. 2:8–13.

———. 1989. Reinforcement of premating isolation. Pp. 158–179 *in* D. Otte and J.A. Endler, eds. Speciation and its consequences. Sinauer, Sunderland, Mass.

Butlin, R.K., and M.G. Ritchie. 1989. Genetic coupling in mate recognition systems: What is the evidence? Biol. J. Linn. Soc. 37:237–246.

Carson, H.L. 1957. The species as a field for gene recombination. Pp. 23–38 *in* E. Mayr, ed. The species problem. Symposium of the American Association for the Advancement of Science. Atlanta. December 28–29, 1955. Am. Assoc. Adv. Sci., Washington, D.C.

———. 1982. Evolution of *Drosophila* on the newer Hawaiian volcanoes. Heredity 48:3–25.

Carson, H.L., and A.R. Templeton. 1984. Genetic revolutions in relation to speciation phenomena: The founding of new populations. Annu. Rev. Ecol. Syst. 15:97–131.
Carthy, J.D. 1966. Insect communication. Pp. 69–80 in P.T. Haskell, ed. Insect behavior. Symp. R. Entomol. Soc. Lond. 3:69–80.
Case, T.J. 1979. Character displacement and co-evolution in some *Cnemidophorus* lizards. Fortschr. Zool. 25:235–282.
Chandler, C.R., and M.H. Gromko. 1989. On the relationship between species concepts and speciation process. Syst. Zool. 38:116–125.
Claridge, M.F. 1985. Acoustic signals in the Homoptera: Behavior, taxonomy, and evolution. Annu. Rev. Entomol. 30:297–317.
Clarke, B.C. 1966. The evolution of morph-ratio clines. Am. Nat. 100:389–402.
Coyne, J.A., H.A. Orr, and D.J. Futuyma. 1988. Do we need a new species concept? Syst. Zool. 37:190–200.
Cumber, R.A. 1952. Notes on the biology of *Melampsalta cruentata* Fabricius (Hemiptera-Homoptera: Cicadidae), with special reference to the nymphal stages. Trans. R. Entomol. Soc. London 103:219–238.
Darwin, C.R. 1859. On the origin of species. John Murray, London.
De Salle, R., and A.R. Templeton. 1987. Comments on "the significance of asymmetrical sexual isolation." Evol. Biol. 21:21–27.
Dobzhansky, T. 1937. Genetics and the origin of species. 1st ed. Columbia University Press, New York.
———. 1940. Speciation as a stage in evolutionary divergence. Am. Nat. 74:312–321.
———. 1970. Genetics of the evolutionary process. Columbia University Press. New York.
Doolan, J.M. 1981. Male spacing and the influence of female courtship behavior in the bladder cicadas *Cystosoma saundersii,* Westwood. Behav. Ecol. Sociobiol. 9:269–276.
Doolan, J.M., and R.C. McNally. 1981. Spatial dynamics and breeding ecology of the cicada, *Cystosoma saundersii*: The interaction between distributions of resources and intraspecific behavior. J. Anim. Ecol. 50:925–940.
Doolan, J.M., and D. Young. 1989 Relative importance of song parameters during flight phonotaxis and courtship in the bladder cicada *Cystosoma saundersii.* J. Exp. Biol. 141:113–131.
Duffels, J.P. 1988. The cicadas of the Fiji, Samoa and Tonga Islands, their taxonomy and biogeography (Homoptera, Cicadoidea). Brill and Scandanavian Science Press, Leiden (Entomonograph. Volume 10 ed. L. Lyneborg).
Dugdale, J.S. 1972. Genera of New Zealand Cicadidae (Homoptera). N. Z. J. Sci. 14:856–882.
Dugdale, J.S., and C.A. Fleming. 1969. Two New Zealand cicadas collected on Cook's Endeavour voyage, with description of a new genus. N. Z. J. Sci. 12:929–957.
———. 1978. New Zealand cicadas of the genus *Maoricicada* (Homoptera, Tibicinidae). N. Z. J. Zool. 5:295–340.
Dunning, D.C., J.A. Byers, and C.D. Zanger. 1979. Courtship in two species of periodical cicadas, *Magicicada septendecim* and *Magicicada cassini.* Anim. Behav. 27:1073–1090.
Ehrman, L., and M. Wasserman. 1987. The significance of asymmetrical sexual isolation. Evol. Biol. 21:1–20.
Endler, J.A. 1977. Geographic variation, speciation and clines. Princeton University Press, Princeton, N.J.

Ewart, A. 1989a. Revisionary notes on the genus *Pauropsalta*, Goding and Froggatt (Homoptera: Cicadidae) with special reference to Queensland. Mem. Qld. Mus. 27:289–375.

———. 1989b. Cicada songs: Song production, structures, variation and uniqueness within species. News Bull. Ent. Soc. Qld. 17:75–82.

Ewing, A.W. 1984. Acoustic signals in insect sexual behavior. Pp. 223–240 *in* T. Lewis, ed. Insect communication. Twelfth Symp. Roy. Entomol. Soc. Lond. 7–9 Sept. 1983. Academic Press, London.

Fisher, R.A. 1958. The genetical theory of natural selection. 2d ed. Dover Publications, New York.

Fleming, C.A. 1971. A new species of cicada from rock fans in southern Wellington, with a review of three species with similar songs and habitat. N. Z. J. Sci. 14:443–479.

———. 1973. The Kermadec Islands cicada and its relatives (Hemiptera: Homoptera). N. Z. J. Sci. 16:315–332.

———. 1975a. Adaptive radiation in New Zealand cicadas. Proc. Am. Philos. Soc. 119:298–306.

———. 1975b. Acoustic behavior as a generic character in New Zealand cicadas. J. R. Soc. N. Z. 5:47–64.

———. 1984. The cicada genus *Kikihia* Dugdale (Hemiptera, Homoptera). 1. The New Zealand green foliage cicadas. Nat. Mus. N.Z. Records. 2:191–206.

Fox, K.J. 1982. Entomology of the Egmont National Park. N.Z. Entomol. 7:286–289.

Futuyma, D.J. 1986. Evolutionary biology. 2d ed. Sinauer, Sunderland, Mass.

Futuyma, D.J., and G.C. Mayer. 1980. Non-allopatric speciation in animals. Syst. Zool. 29:254–271.

Ganyard, M.C., and U.E. Brady. 1971. Inhibition of attraction and cross-attraction by interspecific sex pheromone communication in Lepidoptera. Nature 234:415–416.

———. 1972. Interspecific attraction in Lepidoptera in the field. Ann. Entomol. Soc. Am. 65:1279–1282.

Gibbs, G.W. 1968. The frequency of interbreeding between two sibling species of *Dacus* (Diptera) in wild populations. Evolution. 22:667–683.

Gould, S.J. 1976. This view of life: Darwin's untimely burial. Nat. Hist. 85(8):24–30.

Grant, P.R. 1972. Convergent and divergent character displacement. Biol. J. Linn. Soc. 4:39–68.

———. 1984. Recent research on the evolution of land birds on the Galapagos. Biol. J. Linn. Soc. 21:113–136.

Grant, V. 1957. The plant species in theory and practice. Pp. 39–80 *in* E. Mayr, ed. The species problem. Symposium of the American Association for the Advancement of Science, Atlanta. December 28–29, 1955. American Association for the Advancement of Science, Washington, D.C.

———. 1981. Plant speciation. 2d ed. Columbia University Press, New York.

Gwynne, D.T. 1987. Sex-biased predation and risky mate-locating behavior of male tick-tock cicadas (Homoptera: Cicadidae). Anim. Behav. 35:571–576.

Gwynne, D.T., and G.K. Morris. 1986. Heterospecific recognition and behavioural isolation in acoustic Orthoptera (Insecta). Evol. Theor. 8:33–38.

Hagiwara, S., and K. Ogura. 1960. Analysis of songs of Japanese cicadas. J. Insect Physiol. 5:259–263.

Hagiwara, S., and A. Watanabe. 1956. Discharges in motoneurons of cicada. J. Cell Comp. Physiol. 47:415–428.

Harrison, R.G. 1990. Hybrid zones: Windows on evolutionary process. Pp. 69–128 *in* D.J. Futuyma and J. Antonovics, eds. Oxford surveys in evolutionary biology. Vol. 7. Oxford University Press.

Harrison, R.G., and D.M. Rand. 1989. Mosaic hybrid zones and the nature of species boundaries. Pp. 111–133 *in* D. Otte and J.A. Endler, eds. Speciation and its consequences. Sinauer, Sunderland, Mass.

Hartley, J.C., and R.O. Stephen. 1992. A paradoxical problem in insect communication: Can bush crickets discriminate frequency? J. Exp. Biol. 163:359–365.

Hastings, J.M., and E.C. Toolson. 1991. Thermoregulation and activity patterns of two syntopic cicadas *Tibicen chiricahua* and *Tibicen duryi* (Homoptera: Cicadidae), in central New Mexico [USA]. Oecologia 85:513–520.

Heady, S.E. 1987. Acoustic communication and phylogeny of *Dalbulus* leafhoppers. Pp. 189–202 *in* M.R. Wilson and L.R. Nault, eds. Proc. Second Int. Workshop. Leafhoppers Planthoppers Econ. Import., CIE London.

Heady, S.E., L.V. Madden, and L.R. Nault. 1989. Courtship behavior and experimental asymmetrical hybridization in *Dalbulus* leafhoppers (Homoptera: Cicadellidae) with evolutionary inferences. Ann. Entomol. Soc. Am. 82:535–543.

Heiser, C.B. 1973. Introgression re-examined. Bot. Rev. 39:347–366.

Henry, C.S. 1980. The courtship call of *Chrysopa downsei* Banks (Neuroptera: Chrysopidae): Its evolutionary significance. Psyche 86:291–297.

Hölldobler, B. 1984. Evolution of insect communication. Pp. 349–377 *in* T. Lewis, ed. Insect communication. Twelfth Symp. Roy. Entomol. Soc. Lond. 7–9 Sept. 1983. Academic Press, London.

Howard, D.J., and D.G. Furth. 1986. Review of the *Allonemobius fasciatus* (Orthoptera: Gryllidae) complex with the description of two new species separated by electrophoresis, songs and morphometrics. Ann. Entomol. Soc. Am. 79:472–481.

Howard, D.J., and G.L. Waring. 1991. Topographic diversity, zone width, and the strength of reproductive isolation in a zone of overlap and hybridization. Evolution 45:1120–1135.

Hoy, R.R., J. Hahn, and R.C. Paul. 1977. Hybrid cricket auditory behavior: Evidence for genetic coupling in animal communication. Science 195:82–84.

Hubbs, C. 1959. Population analysis of a "hybrid swarm" between *Gambusia affinis* and *G. heterochir*. Evolution. 13:236–246.

Huber, F. 1984. The world of insects: Periodical cicadas and their behavior. Alexander von Humboldt Stiftung Mitteilungen. 43:24–31.

Huber, F., H.-U. Kleindienst, T.E. Moore, K. Schildberger, and T. Weber. 1990. Acoustic communication in periodical cicadas: neuronal responses to songs of sympatric species. Pp. 217–228 *in* F.G. Gribakin, K. Weise and A.V. Popov, eds. Sensory systems and communication in arthropods. Advances in life sciences. Birkhäuser Verlag, Basel.

Huber, F., D.W. Wohlers, and T.E. Moore. 1980. Auditory nerve and interneurone responses to natural sounds in several species of cicadas. Physiol. Entomol. 5:25–45.

Huber, F., D.W. Wohlers, J.L.D. Williams, and T.E. Moore. 1979. Structure and function of the auditory pathway in cicadas (Homoptera: Cicadidae). Verh. Dtsch. Zool. Ges., 212, Gustav Fischer Verlag, Stuttgart.

Jiang, J.-C. 1985. A study of the song characteristics in cicadas at Jinghong in Yunnan Province (China). (In Chinese.) Acta Entomol. Sinica. 28:257–265.

Jiang, J.-C., Y.-B. Wang, M.-L. Xu, Z. Hong, and C. Hao. 1991a. Frequency conversion property and dynamic process in songs of Mingming cicada (*Oncotympana maculaticollis* Motsch). Sci. China Ser. B. Chem. Life Sci. Earth Sci. 34:576–586.

Jiang, J.-C., M.-L. Xu, and L. Yan. 1991b. The vari-toned complex songs in cicadas and an analysis of the mechanism of their sound production. Acta Entomol. Sinica 34:159–165. [In Chinese with Chinese and English summaries.]

Joermann, G., and H. Schneider. 1987. The songs of four species of cicada in Yugoslavia (Homoptera: Cicadidae). Zool. Anzeiger 219:283–296.

Josephson, R.K., and D. Young. 1979. Body temperature and singing in the bladder cicada *Cystosoma saundersii*. J. Exp. Biol. 80:69–81.

———. 1981. Synchronous and asynchronous muscles in cicadas. J. Exp. Biol. 91:219–237.

———. 1985. A synchronous insect muscle with an operating frequency greater than 500 Hz. J. Exp. Biol. 118:185–208.

Kaneshiro, K., and F.C. Val. 1977. Natural hybridization between a sympatric pair of Hawaiian *Drosophila*. Am. Nat. 111:897–902.

Katsuki, Y., and N. Suga. 1958. Electrophysiological studies on hearing in common insects in Japan. Proc. Japan. Acad. 34:633–638.

———. 1960. Neural mechanism of hearing in insects. J. Exp. Biol. 37:279–290.

Lack, D. 1945. The Galapagos finches (Geospizinae): A study of variation. Occasional papers of the California Academy of Sciences 21:1–159.

———. 1947. Darwin's finches. Cambridge University Press, Cambridge.

Lambert, D.M. 1982. Darwinian views and patterns of speciation in New Zealand; or, what's adaptive about radiation? Tuatara 25:41–47.

Lambert, D.M., M.R. Centner, and H.E.H. Paterson. 1984. Simulation of the conditions necessary for the evolution of species by reinforcement. S. Afr. J. Sci. 80:308–311.

Lambert, D.M., B. Michaux, and C.S. White. 1987. Are species self-defining? Syst. Zool. 36:196–205.

Lambert, D.M., and H.E.H. Paterson. 1984. On "bridging the gap between race and species": The isolation concept and an alternative. Proc. Linn. Soc. N.S.W. 107:501–514.

Lande R. 1981. Models of speciation by sexual selection on polygenic traits. Proc. Natl. Acad. Sci. 78:3721–3725.

———. 1982. Rapid origin of sexual isolation and character divergence in a cline. Evolution 36:213–23.

Lane, D.H. 1984. An inquiry into suspected hybridisation in zones of overlap involving species of the genus *Kikihia* (Homoptera: Tibicinidae). M.Sc. dissertation. Victoria University, Wellington.

———. 1985. Bioacoustic studies on New Zealand cicadas demonstrating a new approach to the species concept. Invited paper to symposium What's a species?. Thirty-fourth Annual Conference Entomol. Soc. N.Z. 21–23 May. Auckland University, Auckland.

———. 1990a. An inquiry into suspected hybridization in the genus *Kikihia*. Two papers. Thirty-ninth Annual Conference Entomol. Soc. N. Z. 7–9 May. Lincoln University, New Zealand.

———. 1990b. The specific mate recognition system in cicadas of the genus *Kikihia*. Paper and poster presentation. Seventh International Auchenorrhyncha Congress. 13–17 August. Ohio State University, Wooster.

———. 1993a. Symposium "Variation in Insects." Intraspecific variation in the green foliage cicadas of the genus *Kikihia* (Homoptera: Tibicinidae) Forty-second Annual Conference Entomol. Soc. N.Z. 10–12 May. Canterbury University, Christchurch, New Zealand.

———. 1993b. Symposium "Variation in Insects." The 13- and 17-year North American periodical cicadas (*Magicicada*) of Brood XIV Forty-second Annual Conference Entomol. Soc. N.Z. 10–12 May. Canterbury University, Christchurch, New Zealand.

Latimer, W. 1981. Acoustic competition in bush crickets. Ecol. Entomol. 6:35–45.

Leroy, Y. 1963. Étude du chant de deux espèces de Grillons et de leur hybride (*Gryllus commodus* Walker, *Gryllus oceanicus* Le Guillou, Orthoptères). C.R. Acad. Sci. Paris 256:268–270.

Lewis, H. 1953. The mechanism of evolution in the genus *Clarkia*. Evolution 7:1–20.

———. 1966. Speciation in flowering plants. Science 152:167–172.

Lewis, H., and P.H. Raven. 1958. Rapid evolution in *Clarkia*. Evolution 12:319–336.

Littlejohn, M.J. 1965. Premating isolation in the *Hyla ewingi* complex (Anura: Hylidae). Evolution 19:234–243.

———. 1969. The systematic significance of isolating mechanisms. Pp. 459–482 *in* Systematic biology: Proceedings of an International Conference. Natl. Acad. Sci., Washington, D.C.

———. 1981. Reproductive isolation: A critical review. Pp. 298–334 *in* W.R. Atchley and D.S. Woodruff, eds. Evolution and speciation: Essays in honour of M.J.D. White. Cambridge University Press, Cambridge.

Littlejohn, M.J., and A.A. Martin. 1964. The *Crinia laevis* complex (Anura, Leptodactylidae) in south-eastern Australia. Aust. J. Zool. 12:70–83.

Lloyd, J. 1966. Studies on the flash communication system in *Photinus* fireflies. Misc. Publ. Mus. Zool. Univ. Mich. 130:1–95.

Loftus-Hills, J.J. 1975. The evidence for reproductive character displacement between the toads *Bufo americanus* and *B. woodhouseii fowleri*. Evolution 22:368–369.

Loftus-Hills, J.J., and M.J. Littlejohn. 1971. Pulse repetition rate as the basis for mating call discrimination by two sympatric species of *Hyla*. Copeia. 154–156.

Lorenz, K.Z. 1958. The evolution of behavior. Sci. Am. 199:67–78.

Masters, J.C., and H.G. Spencer. 1989. Why we need a new genetic species concept. Syst. Zool. 38:270–279.

Mayr, E. 1942. Systematics and the origin of species. Columbia University Press, New York.

———. 1949. Speciation and systematics. Pp. 281–298 *in* G.L. Jepsen, E. Mayr, and G.G. Simpson, eds. Genetics, paleontology and evolution. Princeton University Press, Princeton, N.J.

———. 1957. Species concepts and definitions. Difficulties and importance of the biological species concept. Pp. 1–22, 371–388 *in* E. Mayr, ed. The species problem. Symposium of the American Association for the Advancement of Science. Atlanta. December 28–29, 1955. Am. Assoc. Adv. Sci., Washington, D.C.

———. 1963. Animal species and evolution. Harvard University Press, Cambridge, Mass.

———. 1970. Population, species and evolution. Harvard University Press, Cambridge, Mass.

———. 1976. Evolution and diversity of life. Harvard University Press, Cambridge, Mass.

———. 1982. The growth of biological thought: Diversity, evolution and inheritance. Harvard University Press, Cambridge, Mass.

———. 1988. The why and how of species. Biol. Philos. 3:431–441.

———. 1992. Species concepts and their application. Pp. 15–25 *in* M. Ereshefsky, ed. The units of evolution: Essays on the nature of species. Massachusetts Institute of Technology Press, Cambridge, Mass.
Mayr, E., and W.B. Provine. 1980. The evolutionary synthesis. Harvard University Press, Cambridge, Mass.
McDonnell, L.J., D.F. Gartside, and M.J. Littlejohn. 1978. Analysis of a narrow hybrid zone between two species of *Pseudophryne* (Anura Leptodactylidae) in southeastern Australia. Evolution 32:602–612.
Michelsen, A. 1985. Environmental aspects of sound communication in insects. Pp. 1–9 *in* K. Kalmring and N. Elsner, eds. Acoustic and vibrational communication in insects. Parey, Berlin.
Moore, T.E. 1993. Acoustic signals and speciation in cicadas (Insecta: Homoptera: Cicadidae). Pp. 269–284 *in* D.R. Lees and D. Edwards, eds. Evolutionary Patterns and Processes. Academic Press, London.
Moritz, C., J.W. Wright, and W.M. Brown. 1992. Mitochondrial DNA analyses and the origin and relative age of parthenogenetic *Cnemidophorus*: Phylogenetic constraints on hybrid origins. Evolution 46:184–192.
Moulds, M.S. 1975. The song of the cicada *Lembeja brunneosa* (Homoptera:Cicadidae) with notes on the behavior and distribution of the species. J. Aust. Entomol. Soc. 14:251–254.
Muller, H.J. 1939. Reversibility in evolution considered from the standpoint of genetics. Biol. Rev. Camb. Philos. Soc. 14:261–280.
———. 1942. Isolating mechanisms, evolution and temperature. Biol. Symp. 6:71–125.
Myers, J.G. 1929a. The taxonomy, phylogeny and distribution of New Zealand cicadas (Homoptera). Trans. Roy. Entomol. Soc. Lond. 77:29–60.
———. 1929b. Insect singers: A natural history of the cicadas. Routledge, London.
Nakao, S., and K. Kanmiya. 1988. Acoustic analysis of the calling songs of a cicada, *Meimuna kuroiwae*, Matsumura in Nansei Islands, Japan. I. Physical properties of the sound. Kumume Uni. J. 37:25–46.
Nei, M., T. Maruyama, and C.-I. Wu. 1983. Models of evolution of reproductive isolation. Genetics 103:557–579.
Otte, D. 1989. Speciation in Hawaiian crickets. Pp. 482–526 *in* D. Otte and J.A. Endler, eds. Speciation and its consequences. Sinauer, Sunderland, Mass.
Paterson, H.E.H. 1978. More evidence against speciation by reinforcement. S. Afr. J. Sci. 74:369–371.
———. 1981. The continuing search for the unknown and unknowable: A critique of contemporary ideas on speciation. S. Afr. J. Sci. 77:113–119.
———. 1985. The recognition concept of species. Pp. 21–29 *in* E.S. Vrba, ed. Species and speciation. Transvaal Museum Monograph No. 4. Pretoria.
———. 1988. On defining species in terms of sterility: Problems and alternatives. Pacific Sci. 42:65–71.
Phelan, P.L., and T.C. Baker. 1986. Cross attraction of five species of stored-product Phycitinae (Lepidoptera: Pyralidae) in a wind tunnel. Environ. Entomol. 15:369–372.
———. 1987. Evolution of male pheromones in moths: reproductive isolation through sexual selection. Science 235:205–207.
Popov, A.V. 1981. Sound production and hearing in the cicada *Cicadetta sinuatipennis* Osh. (Homoptera, Cicadidae). J. Comp. Physiol. 142:271–280.
———. 1990. Species of singing cicadas revealed on the basis of peculiarities of acoustic

behavior. I. *Cicadatra cataphractica* Popov (ex Gr. *Querula*) (Homoptera, Cicadidae). [Eng. transl. Entomol. Obozr. No 2(1989):291-307]. Entomol. Rev. 68:62-78.
Popov, A.V., I.B. Aronov, and M.V. Sergeeva. 1985. Calling songs and hearing in cicadas from Soviet Central Asia. [In Russian] Z. Evol. Biokh. Fiziol. 21:451-462. (Translated in J. Evol. Biochem. Physiol. 21:288-97)
Pringle, J.W.S. 1954. A physiological analysis of cicada song. J. Exp. Biol. 31:525-560.
———. 1957. The structure and evolution of the organs of sound-production in cicadas. Proc. Linn. Soc. Lond. 167:144-159.
Pumphrey, R.J. 1940. Hearing in insects. Biol. Rev. 15:107-132.
Rao, S.V., and P. DeBach. 1969. Experimental studies on hybridization and sexual isolation between some *Aphytis* species (Hymenoptera: Aphelinidae). III. The significance of reproductive isolation between interspecific hybrids and parental species. Evolution 23:525-533.
Ratcliffe, L.M., and P.R. Grant. 1983 Species recognition in Darwin's finches (*Geospiza*, Gould). II. Geographic variation in mate preference. Anim. Behav. 31:1154-65.
Riley, H.P. 1938. A character analysis of colonies of *Iris fulva, Iris hexagona* var. *giganticaerulea* and natural hybrids. Am. J. Bot. 25:727-38.
Rosen, D.E. 1979. Fishes from the uplands and intermontane basins of Guatemala: Revisionary studies and comparative geography. Bull. Am. Mus. Nat. Hist. 162:267-376.
Ryan, M.J., and W.E. Wagner. 1987. Asymmetries in mating preferences between species: Female swordtails prefer heterospecific mates. Science 236:595-597.
Sanborn, A.F. 1990. Endothermy in cicadas (Homoptera: Cicadidae). Ph.D. dissertation, University of Florida.
Sanderson, N. 1989. Can gene flow prevent reinforcement? Evolution 43:1223-1235.
Schneider, D. 1969. Insect olfaction: Deciphering system for chemical messages. Science 163:1031-1036.
Schremmer, F. 1960. Über die Bedeutung des Gesanges der Singzikadenmännchen Anzeiger osterr. Akad. Wiss. Wien Math. Nat. Kl., 97:83-86.
Schueler, F.W., and J.D. Rising. 1976. Phenetic evidence of natural hybridization. Syst. Zool. 25:283-289.
Simmons, J.A., E.G. Wever, and J.M. Pylka. 1971. Periodical cicada: Sound production and hearing. Science 171:212-213.
Simmons, P., and D. Young. 1978. The tymbal mechanism and song patterns of the bladder cicada *Cystosoma saundersii*. J. Exp. Biol. 76:27-45.
Short, L.L. 1969. Taxonomic aspects of avian hybridization. Auk. 86:84-105.
Sokal, R.R., and F.J. Rohlf. 1969. Biometry, W.H. Freeman, San Francisco.
Sneath, P.H.A. 1975. Cladistic representation of reticulate evolution. Syst. Zool. 24:360-368.
Spencer, H.G., D.M. Lambert, and B.H. McArdle. 1987. Reinforcement, species and speciation: A reply to Butlin. Am. Nat. 130:958-962.
Spencer, H.G., B.H. McArdle, and D.M. Lambert. 1986. A theoretical investigation of speciation by reinforcement. Am. Nat. 128:241-262.
Spooner, J.D. 1968. Pair-forming acoustic systems of phaneropterine katydids (Orthoptera: Tettigoniidae). Anim. Behav. 16:197-212.
Soper, R.S., G.E. Shewell, and D. Tyrell. 1976. *Colcondamyia auditrix* nov. sp. (Diptera: Sarcophagidae), a parasite which is attracted by the mating song of its host, *Okanagana rimosa* (Homoptera: Cicadidae). Can. Entomol. 108:61-68.

Taper, M.L., and T.J. Case. 1992. Models of character displacement and the theoretical robustness of taxon cycles. Evolution 46:317–333.

Templeton, A.R. 1981. Mechanisms of speciation—a population genetic approach. Annu. Rev. Ecol. Syst. 12:23–48.

———. 1987. Species and speciation. Evolution 41:233–235.

———. 1992. The meaning of species and speciation: A genetic perspective. Pp. 159–183 in M. Ereshefsky, ed. The units of evolution: Essays on the nature of species. Massachusetts Institute of Technology Press, Cambridge, Mass.

Thornhill, R., and J. Alcock. 1983. The evolution of insect mating systems. Harvard University Press, Cambridge, Mass.

Verrell, P.A. 1988. Stabilizing selection, sexual selection and speciation: A view of specific-mate recognition systems. Syst. Zool. 37:209–215.

Villet, M. 1988. Calling songs of some South African cicadas (Homoptera: Cicadidae). S. Afr. J. Zool. 23:71–77.

Vogt, W.G., and D.G. McPherson. 1972. The weighted separation index: A multivariate technique for separating members of closely related species using qualitative differences. Syst. Zool. 21:187–198.

Waage, J.K. 1979. Reproductive character displacement in *Calopteryx* (Odonata: Calopterygidae). Evolution 33:104–116.

Wake, D.B., K.P. Yanev, and M.M. Frelow. 1989. Sympatry and hybridization in a "ring species": The plethodontid salamander *Ensatina eschscholtzii*. Pp. 134–157 in D. Otte and J.A. Endler, eds. Speciation and its consequences. Sinauer, Sunderland, Mass.

Walker, T.J. 1974. Character displacement and acoustic insects. Am. Zool. 14:1137–1150.

Wasserman, M., and H.R. Koepfer. 1977. Character displacement for sexual isolation between *Drosophila mojavensis* and *D. arizonensis*. Evolution 31:812–823.

Weber, T., T.E. Moore, F. Huber, and U. Klein. 1988. Sound production in periodical cicadas (Homoptera: Cicadidae: *Magicicada septendecim, M. cassini*). Pp. 329–336 in C. Vidano and A. Arzone, eds. Proc. Sixth Auchenorrhyncha Meeting, Sept. 7–11, 1987. Turin, Italy. CNR-IPRA.

Wells, M.M., and C.S. Henry. 1992. The role of courtship songs in reproductive isolation among populations of green lacewings of the genus *Chrysoperla* (Neuroptera: Chrysopidae). Evolution 46:31–42.

West-Eberhard, M.J. 1983. Sexual selection, social competition, and speciation. Q. Rev. Biol. 58:155–183.

———. 1984. Sexual selection, competitive communication and species-specific signals in insects. Pp. 283–324 in T. Lewis, ed. Insect communication. Academic Press, London.

White, M.J.D. 1978. Modes of speciation. W.H. Freeman, San Francisco.

White, C.S., B. Michaux, and D.M. Lambert. 1990. Species and neo-Darwinism. Syst. Zool. 39:399–413.

Williams, K.S., and K.G. Smith. 1991. Dynamics of periodical cicada chorus centers (Homoptera: Cicadidae: *Magicicada*). J. Insect Behav. 4:275–292.

Wolda, H. 1967a. Reproductive isolation between two closely related species of the Queensland fruit fly, *Dacus tryoni* (Frogg) and *D. neohumeralis*, Hardy (Diptera, Tephritideae). I. Variation in humeral callus pattern and the occurrence of intermediate color forms in the wild. Aust. J. Zool. 15:501–513.

———. 1967b. Reproductive isolation between two closely related species of the Queensland fruit fly, *Dacus tryoni* (Frogg) and *D. neohumeralis*, Hardy (Diptera, Tephritideae). II. Genetic variation in humeral callus pattern in each species as compared with laboratory-bred hybrids. Aust. J. Zool. 15:515–539.

Woodruff, D.S. 1973. Natural hybridization and hybrid zones. Syst. Zool. 22:213–218.

———. 1977. Hybridization between two species of *Pseudophryne* (Anura, Leptodactylidae) in Sydney basin, Australia. Proc. Linn. Soc. N.S.W. 102:131–147.

Young, D. 1972. Analysis of songs in some Australian cicadas (Homoptera, Cicadidae). J. Aust. Entomol. Soc. 11:237–243.

———. 1975. Chordotonal organs associated with the sound producing apparatus of cicadas (Insecta, Homoptera). Z. Morph. Tiere. 81:111–135.

———. 1980. The calling song of the bladder cicada *Cystosoma saundersii*: A computer analysis. J. Exp. Biol. 88:407–11.

Young, D., and K.G. Hill. 1977. Structure and function of the auditory system of the cicada *Cystosoma saundersii*. J. Comp. Physiol. A. 117:23–45.

Young, D., and R.K. Josephson. 1983. Mechanisms of sound production and muscle contraction kinetics in cicadas. J. Comp. Physiol. A. 152:183–195.

Yusof, O.M. 1982. Biological and taxonomic studies on some leafhopper pests of rice in South East Asia. Ph.D. dissertation, University of Wales.

18

Intraspecific Variability in SMRS Signals: Some Causes and Implications in Acoustic Signaling Systems

MARTIN VILLET

Department of Zoology
University of the Witwatersrand

SEXUAL REPRODUCTION requires the union of two gamete cells for the development of a new organism. The parent organisms either may release gametes into the environment and allow them to meet suitable partners by chance or may exhibit behavioral patterns, including signaling, that increase the likelihood of such meetings. When two organisms or their gametes meet, signals pass between them that serve to lead to fertilization (Robertson and Paterson 1982). The entire system is termed a fertilization mechanism (Paterson 1982a,b, 1985), and the suite of signals that enables the organisms and their gametes to mate and fuse, respectively, has been called the Specific-Mate Recognition System or SMRS (Paterson 1978, 1985).

Paterson (1978, 1985) has pointed out that there must be some degree of congruence between the signal structure of the emitter and the perceptual abilities of the intended receiver of these SMRS signals and that coadaptation between the sexes in this regard would have important implications for the evolution of SMRSs (Lambert and Paterson 1984; Lambert 1984; Robertson and Paterson 1982). In particular, signalers or receivers that deviate from the norm of their gene pool may be less likely to succeed in finding mates, and their deviant SMRS traits should be eliminated from the population (Lambert and Paterson 1984; Paterson 1978, 1985), inasmuch as they have a genetic basis.

Coadaptation between the sexes is postulated to result in close congruence between the traits of a signal and the perceptual abilities of receivers. Physi-

---

Author's current address: Department of Zoology, Rhodes University, Grahamstown, 6140 South Africa

ological and anatomical investigations have shown that the ears of birds (Dooling 1982, 1985), frogs (Ryan and Wilczynski 1988; Wilczynski and Capranica 1984), and insects (Engler et al. 1969; Katsuki and Suga 1960; Nocke 1972; Simmons et al. 1971; Young and Hill 1977) are generally well tuned to the reception of the sound frequencies produced by their own species, providing empirical support for this line of reasoning. Exceptions are the frogs *Acris crepitans* (Ryan et al. 1992) and *Hyla* spp. (Wilczynski et al. 1993).

In attempting to formalize testable and unique predictions of Paterson's concept of species, it was assumed from the above arguments and supporting evidence that receivers would be *finely* tuned to conspecific signals. It was therefore inferred that "characters of calls which exhibit high variability are unlikely to be those involved in mate recognition . . . we must look for characters of rather low variability" (Lambert and Paterson 1984:507). The exact degree of variation that could be expected was not stated.

Important corollaries of coadaptation between the sexes are that SMRS signals are expected to show small and stable amounts of variation across time and space (Lambert and Paterson 1984) and that changes in the SMRS during speciation are expected to follow a stepwise pattern of small changes interspersed with periods when coadaptation is reestablished (Paterson 1978).

Examples of apparent variation in SMRS signals within and between conspecific populations have been reported (e.g. Gayou 1984; Claridge et al. 1988; Ryan and Wilczynski 1988, 1991; Villet 1988; Sullivan and Wagner 1988; Bretagnolle 1989; Gillham 1992; Ryan et al. 1992.). Either the SMRS theory is logically flawed or its predictions do not take into account some aspect of SMRSs that conserve variation despite coadaptation between the sexes.

Hugh Paterson has a broad knowledge of African bird life and often draws on the biology of birds for examples to support his points. Thus, there is a measure of appropriateness in turning to the field of acoustic signaling to examine the subject of this chapter: the variability of SMRS signals and the associated tolerances of the receivers, and the implications of such variation for change in SMRSs and speciation.

Variation within Populations

Signal Production

In some cases, variation in SMRS signals between individuals may be statistically significant, but not biologically meaningful. Furthermore, indicators of variation, such as variance or standard deviation, have to be interpreted within the context of the statistical distribution of the trait they describe. If variation in some aspect of a signal is evenly distributed within a small, sharply defined range, or if it is very skewed (e.g. Claridge et al. 1988), most statistical measures will give an exaggerated impression of the variability. Since such information is rarely provided in descriptions of animal calls, one might automatically (but not necessarily correctly) assume a normal distribution in an intuitive assessment of variation.

Although moments of skewness and kurtosis can be calculated for any distribution, there is as yet no rigorous method of assessing the implications of their absolute magnitudes (Sokal and Rohlf 1981) and thus no quantitative way of relating them to variance. The coefficient of variation provides a unitless measure of variation in relation to the mean (Sokal and Rohlf 1981), but it is based on the standard deviation, and thus shares its limitations. Perhaps the simplest solution is to examine a histogram of the distribution of the trait of interest.

It is also important to have a benchmark against which to compare statistical variation in the signals being analyzed in order to assess its *biological* significance. Indirect experimental controls may involve comparisons with acoustic and other signals that have different contexts, or with other, noncommunicatory characters such as those listed by Henderson and Lambert (1982). Of course, the significance of variation is better tested directly, by recording the responses of conspecifics to it. When variation in the signals is biologically significant, and not merely a result of conceptual limitations in its measurement and interpretation, it may still not be important for several reasons.

The characteristics of an SMRS signal may not be under precise, direct genetic control. Epigenetic effects may alter the expression of underlying genetic mechanisms of signal production, and the mechanisms determining some parameters may provide more precise control than those of others (Villet 1988). Neural mechanisms govern many temporal features of calls, and can be markedly susceptible to the effects of temperature (Gayou 1984; Gerhardt 1982; Villet 1986; Bauer and von Helversen 1987; McClelland and Wilczynski 1989). By contrast, spectral properties are often less variable (Gerhardt 1982; Villet 1988; Masters 1991) because the biophysical properties of the sound-producing organs (Michelsen 1978; McClelland and Wilczynski 1989; Wilczynski et al. 1993) govern them more directly.

Pleiotropic effects can also be important. The sound frequency of animal calls can be constrained by body size (Michelsen 1978; Rand 1985; Ryan and Brenowitz 1985), both within species (Dyson 1989; Narins and Smith 1986; Ramer et al. 1983) and within higher taxa (Michelsen and Nocke 1974; Ryan and Brenowitz 1985; Villet 1986; Wallschlager 1980). If body size is, in turn, pleiotropically constrained by, or adapted to, some other function, such as predator avoidance or feeding (Lloyd and Dybas 1966), a trade-off may explain the variation in song frequencies. It has also been reported that the quality of the nymphal host plants has an effect on the body size of adult cicadas (Fleming and Scott 1970), and this might affect the sound frequency of their songs (Villet 1986).

Narrow variation may be compromised by the organism in the interests of other aspects of signaling. Some signal emitters, such as chorusing frogs, actively vary aspects of their signal to increase its distinctness against background noise or the signals of other organisms. Some of the affected signal features have been implicated in signal recognition (Wells 1988), and intended receivers must have a way of compensating for this, such as responding to a suite of features rather than only one. This last point applies to chemical SMRS signals too, at least in moths (Linn and Roelofs, this volume).

In all of these instances, variation inherent in the production of the signal cannot be eliminated because it does not have a direct genetic basis. However, some variance may clearly be genetic (e.g. de Winter 1992).

Signal Transmission

It may still be argued that the production of some signals is under direct genetic control, and therefore open to stabilizing selection and coadaptation. Even under these circumstances, the signal that reaches a receiver may contain variation that cannot be removed by selection because it arises during transmission. Such environmentally induced variation can occur in both temporal and spectral properties of acoustic signals, especially over long ranges.

The sound frequency spectrum of a call can be distorted during transmission in four ways. Higher frequencies attenuate faster in air than do lower ones, so that the range of sound frequencies becomes increasingly restricted to lower notes as transmission continues (Bass et al. 1984). Next, below a height of two or three meters, lower sound frequencies are attenuated more than higher frequencies by ground absorption caused by the porosity of the substrate. This, combined with atmospheric attenuation, gives rise to a "sound window" of least distortion that lies between about 1 and 3 kHz (Michelsen 1978; Michelsen and Larsen 1983; Morton 1975; Waser and Brown 1984; Wiley and Richards 1982). Third, differential attenuation can be produced by absorption and scattering of sounds by objects such as vegetation. This depends on both the frequency of the sound and the dimensions of the objects (Wiley and Richards 1982). Finally, there is a rare case involving movement of the emitter: males of the New Zealand cicada, *Maoricicada otagoensis,* call in flight, causing a noticeable Doppler shift in the sound frequency of their songs as they pass a receiver (Dugdale and Fleming 1978).

Distortion of temporal patterns results from amplitude fluctuations and reverberations produced in the call by the environment; both are due to scattering and are therefore also frequency dependent. Reverberations are due to the scattering of sound from objects, such as vegetation, which causes some sound waves to reach the receiver after the direct wave. This "smudges" fast, brief amplitude modulations (Wiley and Richards 1982) and increases with the ratio of the objects' dimensions to the wavelength of the sound (Richards and Wiley 1980; Wiley and Richards 1982). Patterns of amplitude modulation are affected by amplitude fluctuations caused by scattering in heterogeneities in the transmitting medium, such as air turbulence due to heating (Wiley and Richards 1978, 1982; Richards and Wiley 1980; Michelsen 1978; Michelsen and Larsen 1983).

Finally, background noises, including acoustic signals produced by other members of the community, can also contribute to the random spectral and temporal distortion of an acoustic signal (Wiley and Richards 1982; Ryan and Brenowitz 1985).

If calls are sampled by a biologist at some distance from their source, statistical analyses of variation may reflect these environmental effects more than they do the intrinsic variability of the call. At larger distances, discrimination by intended receivers would not result in effective coevolutionary selection. On the other hand, over short distances the distorting effects of transmission may be negligible, in which case genetically controlled signals would still be open to selection. Which features of a signal constitute parts of a SMRS, and thus characterize a species, may depend on their relative robustness during signal transmission.

In this context, the habitat specificity of a species is important. The environment an individual inhabits can be quite heterogeneous from the point of view of acoustic ecology, especially if individuals are quite vagile. However, the range of environmental contexts in which SMRS signals are emitted (i.e., contextual parts of the fertilization mechanism) may be more limited, decreasing the influence of this source of variability in acoustic signals. For instance, some tree crickets call from acoustic baffles built from leaves, while cicadas (Villet 1986), frogs (Littlejohn 1977), and birds call from characteristic microhabitats or song posts within their environment. Cicadas (Young 1981; Villet 1986) and birds (Henwood and Fabrick 1979) often show crepuscular peaks in their calling activity, when environmental conditions for the transmission of acoustic signals are generally at their best (Henwood and Fabrick 1979). This last habit has the disadvantage that, if shared by many species, it can increase the level of background noise, which affects a receiver's ability to perceive signals.

Signal Reception

Variability may persist, even in a genetically controlled, cleanly transmitted signal, if the receiver either tolerates or does not perceive it. The auditory physiology of cicadas provides a good example.

The ears of cicadas serve as frequency filters, selecting only a certain range of sound frequencies from which they then record amplitude modulations (Huber et al. 1980; Popov 1981). Such frequency filtering is imposed, in part, by the biophysical properties and structure of the auditory membrane and its associated chordotonal organs (Doolan and Young 1981). The ears do not show any differentiation into areas specialized for the reception of different sound wavelengths (Young and Hill 1977).

The auditory receptors are able to perceive the rhythm of sounds falling within their frequency range. Temporal patterns of sound can be recorded accurately by the ears of orthopterans and cicadas, but pulse repetition rates above about 100 cycles per second may not be perceived because of the physiological limitations (i.e., the latencies) of the nerves serving the receptors (Elsner and Popov 1978). Grasshoppers' ears may perceive pulse repetition rate as a spectral property of the call, rather than as a temporal character (Michelsen 1978), in a way similar to that by which the human ear perceives cicada songs (Pringle 1954). However, the pulse repetition rate of cicada calls is often an order of magnitude smaller than the frequency range to which their ears will respond.

Receivers may perceive call features under some conditions but not under others. Female gray tree-frogs were found to respond to a larger range of variation in aspects of male calls as temperatures rose (Gerhardt 1982). Generally, an increase in temperature lowers the auditory thresholds of poikilotherms (Gerhardt 1982), presumably by increasing the excitability of neurons. On the other hand, ears are often specifically tuned, by logarithmic *physiological responses,* to receive conspecific signals, concentrating their "attention" on these stimuli (e.g. Dooling 1982). They may therefore perceive more variation than can be measured using linear statistics (McGregor 1991).

These investigations show which features of acoustic signals the insects are capable of perceiving and which *could* therefore be recognition cues; it does not follow that they *are.* For example, although an organism may perceive a trait in an acoustic signal, variation in that trait may remain opaque if the physiological mechanisms generating the signal in one sex are coupled to the perception and recognition processes of the other sex. Coupling implies that the tolerance of a receiver will match the variation in its conspecific signal. Models of coupling can invoke genetic or biophysical mechanisms.

Physiological coupling arises if the neural signal-pattern-generating mechanism in the emitter is also the pattern recognition mechanism in the receiver. Such coupling could theoretically also occur in nonacoustic signaling systems where signal patterns are neuronally generated and decoded, for example, electrical, behavioral, or some tactile signals. In crickets, two neural mechanisms for recognizing temporal acoustic patterns have been postulated: feature detectors whose activation requires particular patterns of auditory input; and auditory templates against which auditory input is compared (Hoy et al. 1977; Doherty and Hoy 1985). It was postulated that the neurons responsible for the generation of the rhythms in the calls of males might also be involved in providing the template against which acoustic stimuli are compared in females.

Evidence of physiological coupling in crickets is difficult to interpret because the signal and the response are in different individuals, but the model has been discredited for grasshoppers (Bauer and van Helversen 1987), where emission and perception occur in the same individual. It has not been more widely tested. Alternative explanations of the available evidence postulate a common gene governing each physiological system, or genetic linkage of two different physiological mechanisms, rather than coupling by a single gene or physiological mechanism (Elsner and Popov 1978; Butlin and Ritchie 1989; Boake 1991). The genetic linkage model predicts coupling of signal and response, but not coupling of variance in signal emission and reception. Because of its implications for change in signals during speciation (Paterson 1978, 1981, 1982b, 1985), genetic coupling needs further investigation (Boake 1991). Butlin and Ritchie (1989) reviewed evidence for genetic coupling and did not find support for it.

In cicadas and frogs, biophysical effects of body size on calling and hearing may cause coupling of spectral parameters between the sexes. Cicada song is produced by the sudden and repeated distortion of a roughly circular, ribbed, and resilient membrane, the tymbal (Myers 1928, 1929). The ear of a cicada consists of a bundle of scolopales attached to another membrane, the tympanum, of similar dimensions to the tymbal but thinner and lacking ribs (Doolan and Young 1981; Young and Hill 1977). These ears act as passive peripheral frequency filters (Huber et al. 1980). The sound frequency emitted best by an oscillator such as the tymbal membrane depends in part on the dimensions of its axes of vibration (Michelsen 1978). Conversely, the sound frequency that best excites an oscillator depends (in the same way) on the dimensions of the latter. The diameters of conspecific tettigonian tymbal and typanic membranes are thus linked by a biophysical constraint. Their exact size in any individual is probably largely dependent on individual body size. Similar effects have been shown in some frogs (Wilczynski et al. 1993).

These relationships also remain fairly constant between species within a clade (Villet 1986; Wilczynski et al. 1993), so that evolutionary changes in the body size of a species will affect the spectral structure of its signal; provided that no sexual dimorphism in body size develops, the sexes should remain co-adapted.

Convincing experimental evidence of physiological or genetic coupling in acoustic signaling systems is very scarce (Butlin and Ritchie 1989; Boake 1991), leaving the impression that it is unlikely to occur universally, even in signaling systems that could support it. If variation in parameters of a signal is perceptible to receivers, they may yet be compelled to tolerate it to some extent.

Substantial receiver tolerance to variation in long-range signals may be necessary to allow recognition of signals despite their being scrambled during transmission. Although finely tuned receivers could theoretically compensate for differential atmospheric attenuation of sound frequencies, this would require them to be able to measure, in some way, their distance from the source. Since the precise effects of other frequency-distorting influences are not predictable in time or space, there is little chance of complete success in recovering the undistorted signal, although it may help to make it more recognizable. The unpredictability of temporal distortions means that, even if an organism can perceive the fine temporal characteristics of a call, it cannot afford to be too finely tuned because such traits can be unreliable in the recognition of a call over larger distances in natural settings. Receivers are virtually forced to have broad tolerances to both temporal and spectral variation in long-range communication.

One manner in which such leeway can be accommodated is if the receiver assesses a suite of parameters, particularly if it includes additional, nonacoustic signals too. Such analytical trade-offs have been reported in crickets (Doherty 1985; Doherty and Hoy 1985; Doolan and Pollack 1985), fish (Crapon de Caprona and Ryan 1990), frogs (Gerhardt and Doherty 1988; Rand 1985; Schwartz 1987), and birds (Nelson 1988; Date et al. 1991). Cicadas use their loud songs as long-distance attractants, and subsequent steps in the chain of signals that make up their SMRSs may include distinct courtship songs (Villet 1988) and behavioral patterns used over shorter ranges (Villet 1986). Song alone is enough to evoke a response indicating recognition of a conspecific in birds (Becker 1982), but here distinctive features are important in the suite, irrespective of their variability (Nelson 1989).

A second manner of compensating for environmental distortion is related to the scale of these effects. If the scale of "meaningful" pattern in a signal is more than an order of magnitude greater than the scale of extraneous distortion, the signal-to-noise ratio will be greatly improved. A receiver may preferentially assess signal parameters that have gross structures that are beyond the effects of environmental scrambling. This does not mean that variation in fine characteristics of a signal are necessarily unimportant to a receiver.

If their neural systems are sufficiently complex, animals may have additional, positive reasons for tolerating variation in a signal, including using them as cues for range-finding (McGregor and Krebs 1984; Richards and Wiley 1980; McGregor 1991), individual identification (Emlen 1972; Falls 1982; McGregor 1991), and mate selection (Emlen 1972; Searcy 1982; Wells 1988; West-Eberhard 1983). However, features of calls that contain the type or degree of variation needed for these purposes may not be involved in the SMRS. Song components of indigo buntings *(Passerina cyanea)* show a hierarchy of variability: features characterizing the species showed little variation; those involved in individual recognition, a great deal; and those relating to mate choice, an intermediate level (Emlen 1972). During the search for a mate, these secondary and tertiary parameters are excluded from the primary set of characteristics or stimuli that produce specific-mate recognition (e.g. Ryan 1983). Because of this hierarchical structure, variation related to sexual selection may be limited by the needs of the SMRS (Lambert et al. 1982; Rand 1985).

The questions of learning, individuality, and sexual selection greatly complicate the scientific analysis of variation in SMRS signals and will not be addressed further in this discussion. The important point here is that, while it is biologically significant, such variation may not affect the SMRS, and must then be excluded from tests of Paterson's (1985) predictions.

Variation between Populations

The preceding discussion has emphasized sources of variation that might be important to receivers within populations. Attention may now be turned to patterns of geographic variation between populations and their causes. Evidence of geographical variation in acoustic SMRS signals has been found in four species of leafhopper (Booij 1982; Claridge et al. 1988; Gillham 1992), *Drosophila mercatorum* (Ikeda et al. 1980), the frogs *Eleuthrodactylus coqui* (Narins and Smith 1986) and *Acris crepitans* (Ryan and Wilczynski 1988), and a bird, Wilson's storm petrel (Bretagnolle 1989). It was negligible in several species of cicada (Villet 1988) and in *Drosophila melanogaster*, which was studied quite thoroughly (Henderson and Lambert 1982).

The leafhopper and bird studies were not specifically designed to test the prediction of low variability in SMRS features, and thus fail to control for many of the factors discussed above. In *E. coqui* variation was related to body size differences at different altitudes (Narins and Smith 1986). The study of *A. crepitans* (Ryan and Wilczynski 1988) shows coadaptation of signaler and receiver in different populations and the independence of geographical variation from body size. In all cases except the last, it is not clear whether the variation has any significance to the animals. Work on great tits suggests that sound frequency is the parameter most important to these birds in classifying songs (Weary 1990). If this is the case in storm petrels too, then it is significant that Wilson's storm petrel shows least geographic variation in the spectral features of its mating call (Bretagnolle 1989).

Factors that could affect the degree of geographic variation in an SMRS include: gene flow and the vagility of individuals between populations; adaptation of the SMRS to local environmental conditions; pleiotropic changes in the SMRS caused by local adaptation of other traits; epigenetic influences (e.g.

of altitude and latitude), especially on body size; and anything that alters polygenic balances during the establishment of new populations. For instance, SMRS variation in the leafhopper *Nilaparvata lugens* (Claridge et al. 1988) could have resulted from founder effects as the species established new populations through the Indonesian archipelagoes (de Winter 1992).

But how can these changes take place in the face of coadaptation between signaler and receiver? One answer is that the emitter and receiver systems are coupled by neurological, genetic, or biophysical mechanisms, but rigorous evidence of this remains to be collected (Butlin and Ritchie 1989; Boake 1991).

Second, the examination of sources of variation in SMRS signals that are not open to selection shows that receivers may be forced to tolerate some variation. Thus, *organisms do not respond to ideal SMRS signals, but to the best reasonable approximation available* (e.g. Claridge et al. 1988; Crapon de Caprona and Ryan 1990; Gerhardt and Doherty 1988). In a large population, individuals are likely to encounter stimuli that are near the ideal, because variation will be almost continuous and because so many individuals are present. In small populations, the number of mates an organism will meet is much smaller, and the choice of stimuli more limited. The genetic variation encountered in small populations need not be any less than in large populations (Kemp 1985:68). Under these conditions, organisms still respond to the best available stimulus, but this may differ markedly from the ideal.

An important implication of this model is that in very small, newly founded populations, although variation may have a genetic basis, the absence of choices may disrupt coadaptation. Instead, the SMRS may alter because of mutations and changes in polygenic balances due to genetic drift, inbreeding, and founder effects (cf. Spencer, this volume). Mutations are more likely to be fixed in small populations than in large ones. This may be particularly relevant to the extreme case of captive breeding programs, and perhaps also the planning of conservation breeding programs. Temporal variation in the SMRS of very small, established populations is predicted to occur through the same mechanism and should be sought as supporting evidence.

Implications for SMRS Theory

Despite the stabilizing effects of coadaptation between the sexes, parameters of acoustic signals that may be important to their recognition by conspecifics can retain a measure of variability that may even be accommodated within the signaling system. A degree of geographic variation can also be expected, even in genetically heritable traits that are essential to signal recognition. The exact magnitude of this variability is still not predictable, but must depend at least partly on its cause. There is also no way of telling a priori from measures of variability which parameters will be important to the organism;

one can, however, rule out those that the organism is not capable of perceiving. These conclusions are likely to apply to other communication channels too.

Thus, the predictions concerning signal variability put forward by Lambert and Paterson (1984) can be tested only after considering a labyrinth of confounding factors; this discussion is not a disproof of these predictions of Paterson's theory and should not be interpreted as such. However, designing adequately controlled experiments to gather convincing data for exploring these predictions rigorously will not be an enviable task. For instance, signals must be shown (and not assumed) to be important to the recognition process. The selection of controls for comparisons of statistical variation is also a vital aspect of such testing, since it has never been postulated that SMRS signals are invariable, but only that they show less variation than other characters. The magnitude of variation in SMRS signals caused by the factors identified above also remains to be examined to see if it accounts for the variability found in empirical studies.

Rather than undermining Paterson's theory, the recognition of the causes of variation discussed above may contribute to the theory's refinement. In particular, some of the factors preserving variation in signals may be very important in providing ways of circumventing the problem of change in a coadapted SMRS during speciation. Paterson made at least four assumptions in his model of speciation: (1) that it occurs in small, isolated populations; (2) that it involves a process of stepwise changes to the fertilization mechanism caused by an alternation between changes to the signal and coadaptation of signaler and receiver (Paterson 1978, 1985); (3) that it occurs in a new habitat (Paterson 1978, 1981, 1982a,b, 1985, 1986); and (4) that adaptation to the new habitat was the major cause of change in the SMRS (Paterson 1985, 1986).

As argued in the discussion of geographic variation in SMRS signals, the smaller a population, the more a change in the SMRS should be ameliorated by the smaller number of potential mates emitting signals, because organisms should respond to the closest stimulus to their SMRS available and should tolerate some imperfection *in the absence of a better stimulus*. This implies that the degree of discrimination shown by the limiting sex in epigenic selection may be density dependent and have little importance in directing changes in the SMRS during speciation (cf. Kaneshiro 1989). It also implies that changes in the SMRS need not follow a gradualistic, step-by-step course, but may undergo a single quantum change to the limit of the tolerance of the receiver (cf. Kauffman 1969, 1984, 1990, 1993), followed by the reestablishment of coadaptation. When several properties of a signal are used in its recognition, some may even alter beyond the tolerances of the receiver. Once the population starts growing, coadaptation, genetic equilibrium, and gene flow can be reestablished between the sexes, resulting in the stabilization of a new SMRS and thus the evolution of a new species. If neurological or biophysical coupling exists, it may even obviate the need for a phase of coadaptation, at least with respect to some parts of the signal (Paterson 1978, 1981, 1982b, 1985).

Founder effects, genetic drift, and inbreeding, especially in combination with mutations in regulatory or metabolic genes (see also Linn and Roelofs, this volume; Spencer, this volume) may produce large, random changes in polygenic balances affecting SMRS signals. However, these processes will probably not result in directed changes in the SMRS. The direct and pleiotropic effects of adaptation on the signaling system (see Rand 1985; Endler 1992) in a new environment may serve as a directing force to reshape variation at such times. Sexual selection under these circumstances might also cause directed selection (Rand 1985; Lande 1981; Lande and Kirkpatrick 1988), perhaps through pleiotropy. The existence of other species competing for the same signaling channel can be interpreted as biotic components of environmental noise, which can promote directional selection or selection for decreased variance (McGregor 1991), but it need not be distinguished from abiotic sources of background noise, at least from the point of view of speciation (cf. Linn and Roelofs, this volume). The importance of illegitimate receivers (e.g. parasitoids) also needs to be explored.

The robustness to environmentally induced variation conferred on any acoustic signal by scale, redundancy, and tolerance seems to make it less susceptible to direct selection, and many such signals are in fact not optimally suited to the emitter's habitat (Toms 1985; Villet 1986; Ryan and Brenowitz 1985; Masters 1991). The examples given by Paterson (1985) of adaptation in SMRSs are generalized rather than specific. This adaptationist aspect of Paterson's theory seems to vary in its importance, depending partly on the signaling channel (cf., e.g. chemical signals; Linn and Roelofs, this volume) and the range over which the signal is required to work. In other words, SMRS signals are generally not closely adapted to the constraints of the transmission environment and are unlikely to come under direct selection during speciation, even in a new habitat. Although he placed a heavy emphasis on such adaptation (Paterson 1986), Paterson (1985:26) specifically denied pan-selectionism, but did not develop the ramifications of his comment for his theory.

If the requirement for adaptation of the SMRS is omitted from the core of the theory, the need for the newly isolated population to occupy a new habitat (Paterson 1985, 1986) is less compelling, but may still be needed as a source of selective forces that bring about pleiotropic changes in the SMRS.

Thus, Paterson's model of speciation can increase its generality by abandoning its assumptions (listed above) of (2) gradualistic change in the SMRS during speciation and (4) direct adaptation as the major force reshaping SMRSs. The requirement for (3) a change of habitat during speciation may be necessary even when there is no direct selection on the SMRS. The crucial assumption is

(1) that the speciating population must be small in order to fix changes to the genome (Paterson 1981; Kemp 1985; Spencer, this volume) and overcome coadaptation between signaler and receiver.

These refinements to the original model simplify and strengthen it and do not affect Paterson's way of looking at and defining species. The essential difference between intraspecific variation and speciation lies not in their mechanisms, but in whether the changes to the SMRS are sufficiently large to incidentally cause the closure of a new gene pool and thereby the origin of a new species.

Summary

It can be expected that the system of signals (SMRS) that leads to syngamy must show coadaptation between the sexes. Lambert and Paterson (1984) inferred that this would lead to close tuning between the signaler and the receiver, and to a diminution in the variability of features of a signal that are important to its recognition, especially compared to variation in other characters. Variation within and between populations over large areas and time periods was not expected. Although the degree of variation to be expected was not stated, empirical studies of acoustic components of SMRSs suggest that these predictions do not hold true.

Detailed examination of the production, transmission, and perception of acoustic signals in SMRSs identified several factors that may counteract close coadaptation. Many of these factors can be generalized to include other signaling channels. Tests of the prediction of limited variability of SMRS signals, using either direct experimental approaches or the interpretation of variation in SMRS components within or between populations, are greatly complicated by the need to take these factors into account.

It is concluded that the original prediction was based on an incomplete assessment of sources of variation; that the predictions may still hold once the newly identified sources have been taken into account; and that these sources have an important role in conserving variation that may contribute to speciation in small, isolated populations. Paterson's model of how SMRS signals change at speciation is refined to take into account the effects of population size on the occurrence of coadaptation between the sexes and the implications of environmentally induced signal variability for adaptation to new habitats. Direct adaptation of SMRSs to new habitats is probably not generally important in speciation, although indirect effects may have a more frequent role in that process.

Acknowledgments

This work is a product of my interactions with Hugh Paterson, who was a cosupervisor of my M.Sc. project, and I am grateful for the time he spent discussing the subtleties of evolutionary biology with me. Mark Claridge, Sarah Davies, Murray Littlejohn, Lydia Maltby, Judith Masters, and Susan Risi provided thoughtful and welcome comments on the manuscript. Charles Linn and Wendel Roelofs kindly provided copies of their very useful manuscript, as did Hamish Spencer. This work was supported by the Department of Zoology, University of the Witwatersrand.

References

Bass, H., L. Sutherland, J. Piercy, and L. Evans. 1984. Absorption of sound by the atmosphere. Phys. Acoustics 17:145–232.

Bauer, M., and O. von Helversen. 1987. Separate location of sound recognizing and sound producing neural mechanisms in a grasshopper. J. Comp. Physiol. A. 161:95–101.

Becker, P.H. 1982. The coding of species-specific characteristics in bird sounds. Pp. 214–252 in D. Kroodsma and E. Miller, eds. Evolution and ecology of acoustic communication in birds. Vol.1. Academic Press, New York.

Boake, C. 1991. Coevolution of senders and receivers of sexual signals: Genetic coupling and genetic correlation. Trends Ecol. Evol. 6:225–227.

Booij, C.J.H. 1982. Biosystematics of the *Muellerianella* complex (Homoptera, Delphacidae), interspecific and geographic variation in acoustic behavior. Z. Teirpsychol. 58:31–52.

Bretagnolle, V. 1989. Calls of Wilson's storm petrel: Functions, individual and sexual recognitions, and geographic variation. Behavior 111:98–112.

Butlin, R.K., and M.G. Ritchie. 1989. Genetic coupling in mate recognition systems: What is the evidence? Biol. J. Linn. Soc. 37:237–246.

Claridge, M.F., J. den Hollander, and J.C. Morgan. 1988. Variation in hostplant relations and courtship signals of weed-associated populations of the brown planthopper, *Nilaparvata lugens* (Stål), from Australia and Asia: A test of the recognition species concept. Biol. J. Linn. Soc. 35:79–93.

Crapon de Caprona, M.-D., and M.J. Ryan. 1990. Conspecific mate recognition in swordtails, *Xiphophorus nigrensis* and *X. pygmaeus* (Poeciliidae): Olfactory and visual cues. Anim. Behav. 39:290–296.

Date, E.M., R.E. Lemon, D.M. Weary, and A.K. Richter. 1991. Species identity by birdsong: Discrete or additive information? Anim. Behav. 41:111–120.

de Winter, A.J. 1992. Genetic studies on acoustic differentiation in *Ribautodelphax* planthoppers (Homoptera, Delphacidae). Proc. Exper. Appl. Entomol., N. E. V. Amsterdam 3:116–120.

Doherty, J.A. 1985. Trade-off phenomena in calling song recognition and phonotaxis in the cricket, *Gryllus bimaculatus* (Orthoptera, Gryllidae). J. Comp. Physiol. A 156:787–801.

Doherty, J., and R. Hoy. 1985. Communication in insects. III. The auditory behavior of crickets: Some views of genetic coupling, song recognition and predator detection. Q. Rev. Biol. 60:457–471.

Doolan, J.M., and G.S. Pollack. 1985. Phonotactic specificity of the cricket *Teleogryllus oceanicus:* Intensity-dependent selectivity for temporal parameters of the stimulus. J. Comp. Physiol. A 157:223–233.

Doolan, J.M., and D. Young. 1981. The organisation of the auditory organ of the bladder cicada, *Cystosoma saundersii.* Philos. Trans. R. Soc. Lond. B 291:525–540.

Dooling, R.J. 1982. Auditory perception in birds. Pp. 95–131 *in* D. Kroodsma and E. Miller, eds. Evolution and ecology of acoustic communication in birds. Vol.1. Academic Press, New York.

———. 1985. Hearing and balance. Pp. 276–277 *in* B. Campbell and E. Lack, eds. A dictionary of birds. Poyser, Calton, England.

Dugdale, J.S., and C.A. Fleming. 1978. New Zealand cicadas of the genus *Maoricicada* (Homoptera: Tibicinidae). N. Z. J. Zool. 5:295–340.

Dyson, M. 1989. Aspects of social behavior and mate choice in a caged population of the painted reed frog, *Hyperolius marmoratus.* Ph.D. dissertation, University of the Witwatersrand, Johannesburg.

Elsner, N., and A.V. Popov. 1978. Neuroethology of acoustic communication. Adv. Insect Physiol. 13:229–355.

Emlen, S.T. 1972. An experimental analysis of the parameters of bird song eliciting species recognition. Behavior 41:130–171.

Endler, J.A. 1992. Signals, signal conditions, and the direction of evolution. Am. Nat. 139:S125–S153.

Enger, P.S., D.J. Aidley, and T. Szabo. 1969. Sound reception in the Brazilian cicada *Fidicina rana* Wlk. J. Exp. Biol. 51:339–345.

Falls, J. 1982. Individual recognition by sounds. Pp. 237–278 *in* D. Kroodsma and E. Miller, eds. Evolution and ecology of acoustic communication in birds. Vol.1. Academic Press, New York.

Fleming, C.A., and G.H. Scott. 1970. Size differences in cicadas from different plant communities. N. Z. Entomol. 4:38–42.

Gayou, D.C. 1984. Effects of temperature on the mating call of *Hyla versicolor.* Copeia 1984:733–738.

Gerhardt, H.C. 1982. Sound pattern recognition in some North American treefrogs (Anura: Hylidae): Implications for mate choice. Am. Zool. 22:581–595.

Gerhardt, H.C., and J.A. Doherty. 1988. Acoustic communication in the gray treefrog, *Hyla versicolor:* Evolutionary and neurobiological implications. J. Comp. Physiol. A 162:261–278.

Gillham, M.C. 1992. Variation in acoustic signals within and among leafhopper species of the genus *Alebra* (Homoptera, Cicadellidae). Biol. J. Linn. Soc. 45:1–15.

Henderson, N., and D.M. Lambert. 1982. No significant deviation from random mating of worldwide populations of *Drosophila melanogaster.* Nature 300:437–440.

Henwood, K., and A. Fabrick. 1979. A quantitative analysis of the dawn chorus: Temporal selection for communicatory optimisation. Am. Nat. 114:260–274.

Hoy, R.R., J. Hahn, and R.C. Paul. 1977. Hybrid cricket behavior: Evidence for genetic coupling in animal communication. Science 195:82–84.

Huber, F., D. Wholers, and T. Moore. 1980. Auditory nerve and interneurone responses to natural sounds in several species of cicada. Physiol. Entomol. 5:25–45.

Ikeda, H., I. Takabatake, and N. Sadawa. 1980. Variation in courtship sounds among three geographical strains of *Drosophila mercatorum.* Behav. Genet. 10:361–375.

Kaneshiro, K.Y. 1989. The dynamics of sexual selection and founder effects in species formation. Pp. 279-296 *in* L.V. Giddings, K.Y. Kaneshiro and W.W. Anderson, eds. Genetics, speciation and the founder principle. Oxford University Press, Oxford.

Katsuki, Y., and N. Suga. 1960. Neural mechanism of hearing in insects. J. Exp. Biol. 37:279-290.

Kauffman, S.A. 1969. Metabolic stability and epigenesis in randomly constructed genetic nets. J. Theor. Biol. 22:437-467.

―――. 1984. Emergent properties in random complex automata. Physica D 10:145-156.

―――. 1990. Requirements for evolvability in complex systems: Orderly dynamics and frozen components. Physica D 42:135-152.

―――. 1993. Origins of order: Self-organization and selection in evolution. Oxford University Press, Oxford.

Kemp, A.C. 1985. Individual and population expectations related to species and speciation: The case of raptorial birds. Pp. 59-70 *in* E.S. Vrba, ed. Species and speciation. Transvaal Museum Monograph No. 4. Pretoria.

Lambert, D.M. 1984. Specific-mate recognition systems, phylogenies and asymmetrical evolution. J. Theor. Biol. 109:147-156.

Lambert, D.M., and H.E.H. Paterson. 1984. On "Bridging the gap between race and species": The Isolation Concept and an alternative. Proc. Linn. Soc. N.S.W. 107:501-514.

Lambert, D.M., P.D. Kingett, and E. Slooten. 1982. Intersexual selection: The problem and a discussion of the evidence. Evol. Theor. 6:67-78.

Lande, R. 1981. Models of speciation by sexual selection on polygenic traits. Proc. Natl. Acad. Sci. 78:3721-3725.

Lande, R., and M. Kirkpatrick. 1988. Ecological speciation by sexual selection. J. Theor. Biol. 133:85-98.

Littlejohn, M.J. 1977. Long-range communication in anurans: An integrated and evolutionary approach. Pp. 263-294 *in* D.H. Taylor and S.I. Guttman, eds. The reproductive biology of amphibians. Plenum, New York.

Lloyd, M., and H.S. Dybas. 1966. The periodical cicada problem. II. Evolution. Evolution 20:466-505.

Loftus-Hills, J.J. 1973. Comparative aspects of auditory function in Australian anurans. Aust. J. Zool. 21:353-367.

Masters, J.C. 1991. Loud calls of *Galago crassicaudatus* and *G. garnettii* and their relation to habitat structure. Primates 32:153-167.

McClelland, B., and W. Wilczynski. 1989. Sexually dimorphic laryngeal morphology in *Rana pipiens*. J. Morphol. 201:293-299.

McGregor, P.K. 1991. The singer and the song: On the receiving end of bird song. Biol. Rev. 66:57-81.

McGregor, P.K., and J.R. Krebs. 1984. Sound degradation as a distance cue in great tit *(Parus major)* song. Behav. Ecol. Sociobiol. 16:49-56.

Michelsen, A. 1978. Sound reception in different environments. Pp. 345-373 *in* M. Ali, ed. Perspectives in sensory ecology. Plenum, New York.

Michelsen, A., and O. Larsen. 1983. Strategies for acoustic communication in complex environments. Pp. 321-331 *in* F. Huber and H. Markl, eds. Neuroethology and behavioral physiology: Roots and growing points. Springer-Verlag, Berlin.

Michelsen, A., and H. Nocke. 1974. Biophysical aspects of sound communication in insects. Adv. Insect Physiol. 10:247-296.

Morton, E.S. 1975. Ecological sources of selection on avian sounds. Am. Nat. 109:17–34.
Myers, J. 1928. The morphology of the Cicadidae. Proc. Zool. Soc. Lond. 1928:365–472.
———. 1929. Insect singers: A natural history of the cicadas. Routledge, London.
Narins, P., and S. Smith. 1986. Clinal variation in anuran advertisement calls: Basis for acoustic isolation? Behav. Ecol. Sociobiol. 19:135–141.
Nelson, D. 1988. Feature weighting in species song recognition by the field sparrow (*Spizella pusilla*). Behavior 106:158–182.
———. 1989. The importance of invariate and distinctive features in species recognition in bird song. Condor 91:120–130.
Nocke, H. 1972. Physiological aspects of sound communication in crickets (*Gryllus campestris*). J. Comp. Physiol. A 80:141–162.
Paterson, H.E.H. 1978. More evidence against speciation by reinforcement. S. Afr. J. Sci. 74:369–371.
———. 1981. The continuing search for the unknown and unknowable: A critique of contemporary ideas on speciation. S. Afr. J. Sci. 77:113–119.
———. 1982a. Perspective on speciation by reinforcement. S. Afr. J. Sci. 78:53–57.
———. 1982b. Darwin and the origin of species. S. Afr. J. Sci. 78:272–275.
———. 1985. The recognition concept of species. Pp. 21–30 *in* E.S. Vrba, ed. Species and speciation. Transvaal Museum Monograph No. 4. Pretoria.
———. 1986. Environment and species. S. Afr. J. Sci. 82:62–65.
Penna, M., C. Palazzi, P. Paolinelli, and R. Solis. 1990. Midbrain auditory sensitivity in toads of the genus *Bufo* (Amphibia—Bufonidae) with different vocal repertoires. J. Comp. Physiol. A 167:673–681.
Popov, A. 1981. Sound production and hearing in the cicada *Cicadetta sinuatipennis* Osh. (Homoptera, Cicadidae). J. Comp. Physiol. A 142:271–280.
Pringle, J.W.S. 1954. A physiological analysis of cicada song J. Exp. Biol. 31:525–560.
Ramer, J.D., T.A. Janssen, and C.J. Hurst. 1983. Size-related variation in the advertisement call of *Rana calmitans* (Anura: Ranidae), and its effects on conspecific males. Copeia 1983:141–155.
Rand, S. 1985. Tradeoffs in the evolution of frog calls. Proc. Ind. Acad. Sci. (Anim. Sci.) 94:623–637.
Richards, D.G., and R.H. Wiley. 1980. Reverberations and amplitude fluctuation in the propagation of sound in a forest: Implications for animal communication. Am. Nat. 115:381–399.
Robertson, H.M., and H.E.H. Paterson. 1982. Mate recognition and mechanical isolation in *Enallagma* damselflies (Odonata, Coenagrionidae). Evolution 36:243–250.
Ryan, M.J. 1983. Frequency modulated calls and species recognition in a neotropical frog. J. Comp. Physiol. A 150:217–221.
Ryan, M.J., and E.A. Brenowitz. 1985. The role of body size, phylogeny and ambient noise in the evolution of bird song. Am. Nat. 126:87–100.
Ryan, M.J., S.A. Perrill, and W. Wilczynski. 1992. Auditory tuning and call frequency predict population-based mating preferences in the cricket frog, *Acris crepitans*. Am. Nat. 139:1370–1383.
Ryan, M.J., and W. Wilczynski. 1988. Coevolution of sender and receiver: Effect on local mate preference in cricket frogs. Science 240:1786–1788.
———. 1991. Evolution of intraspecific variation in the advertisement call of a cricket frog (*Acris crepitans*, Hylidae). Biol. J. Linn. Soc. 44:249–271.
Schwartz, J.J. 1987. The importance of spectral and temporal features in species and call recognition in a neotropical frog with a complex vocal repertoire. Anim. Behav. 35:340–347.

Searcy, W.A. 1982. The evolutionary effects of mate selection. Annu. Rev. Ecol. Syst. 13:57–85.
Simmons, J.A., E.G. Wever, and J.M. Pylka. 1971. Periodical cicada: Sound production and hearing. Science 171:212–213.
Sokal, R.R., and F.J. Rohlf. 1981. Biometry. 2d ed. W.H. Freeman, San Francisco.
Sullivan, B.K., and W.E. Wagner. 1988. Variation in advertisement and release calls, and social influences on calling behavior in the Gulf Coast toad *(Bufo valliceps)*. Copeia 1988:1014–1020.
Toms, R.B. 1985. Speciation in tree crickets (Gryllidae Oecanthinae). Pp. 109–114 *in* E.S. Vrba, ed. Species and speciation. Transvaal Museum Monograph No. 4. Pretoria.
Villet, M. 1986. Aspects of calling song in some African cicadas. M.Sc. dissertation, University of the Witwatersrand, Johannesburg.
———. 1988. Calling songs of some South African cicadas (Homoptera: Cicadidae). S. Afr. J. Zool. 23:71–77.
Wallschlager, D. 1980. Correlation of song frequency and body weight in passerine birds. Experientia, Basel 36:412.
Waser, P.M., and C.H. Brown. 1984. Is there a "sound window" for primate communication? Behav. Ecol. Sociobiol. 15:73–76.
Weary, D.M. 1990. Categorization of song notes in great tits: Which acoustic features are used and why? Anim. Behav. 39:450–457.
Wells, K. 1988. The effects of social interactions on anuran vocal behavior. Pp. 433–454 *in* B. Fritzsch, M. Ryan, W. Wilczynski, E. Hetherington and W. Walkowiack, eds. The evolution of the anuran auditory system. John Wiley, New York.
West-Eberhard, M.J. 1983. Sexual selection, social competition and speciation. Q. Rev. Biol. 58:155–183.
Wilczynski, W., and R. Capranica. 1984. The auditory system of anuran amphibians. Prog. Neurobiol. 22:1–38.
Wilczynski, W., B.E. McClelland, and A.S. Rand. 1993. Acoustic, auditory and morphological divergence in three species of neotropical frog. J. Comp. Physiol. A 172:425–438.
Wiley, R.H., and D.G. Richards. 1978. Physical constraints on acoustic communication in the atmosphere: Implications for the evolution of animal vocalizations. Behav. Ecol. Sociobiol. 3:69–94.
———. 1982. Adaptations for acoustic communication in birds: Sound transmission and signal detection. Pp. 131–181 *in* D. Kroodsma and E. Miller, eds. Acoustic communication in birds. Vol.1. Academic Press, New York.
Young, A.M. 1981. Temporal selection for communicatory optimization: The dawn-dusk chorus as an adaptation in tropical cicadas. Am. Nat. 117:826–829.
Young, D., and K.G. Hill. 1977. Structure and function of the auditory system of the cicada *Cystosoma saundersii*. J. Comp. Physiol. A 117:23–45.

19

Acoustic Signals as Specific-Mate Recognition Signals in Leafhoppers (Cicadellidae) and Planthoppers (Delphacidae) (Homoptera: Auchenorrhyncha)

JERONE DEN HOLLANDER
School of Pure and Applied Biology
University of Wales, College of Cardiff

IN SEXUAL REPRODUCTION the gametes must be brought together to achieve fertilization. This may be achieved by the evolution of such simple mechanisms as chemicals which act as attractants and guide the gametes to each other, the coordinated release of gametes only at specific times, the rendezvous of gametes at particular environmental features, or the release of gametes in a restricted area. These are termed *fertilization mechanisms* (Paterson 1985).

In mobile organisms with separate sexes, the fertilization system can become more complex: it now needs to bring sexually receptive, conspecific members of opposite sexes to the same place at the same time. Individuals must then be able to recognize conspecific members of the opposite sex and assess their reproductive condition once they have been brought together. This has been termed the *Specific-Mate Recognition System* (SMRS) by Paterson (1980, 1985). If differences in mate recognition systems between two populations are sufficiently large, the sexes will not recognize those of the other population as potential mates. Each population thus has an SMRS which results in positive assortative mating and sets the limits to the field of genetic recombination, and thus to the species. Species of mobile, bisexually reproducing organisms are thus generally characterized by behavior patterns that serve to bring conspecific males and females together for mating.

This Recognition Concept of species differs from the Biological Species Concept (Mayr 1942) as the maintenance of the genetic integrity of the species is an incidental result of positive assortative mating due to a shared SMRS and not to isolating mechanisms (Dobzhansky 1951) selected for the function of preventing interbreeding between species to preserve coadapted gene complexes. Mating with conspecifics is a necessity; therefore SMRSs are necessary, irrespective of the presence of other species, closely related or otherwise. Although other species may have an effect on the SMRS, they are not the reason for the SMRS and cannot initiate the system (Paterson 1985).

The Recognition Concept of species also explains a number of anomalies of the Biological Species Concept, such as how crosses between some species can produce apparently normal, viable, fertile hybrids; how some species may have no close relatives yet have elaborate courtship; how "isolating mechanisms" are characteristic over the entire range of a species even when it is in contact with different species in different parts of its range; or how speciation can occur in total allopatry without reinforcement.

Many of the problems in taxonomy involve the elucidation of the status of different populations that are morphologically very similar or identical but behave as biologically distinct species. As morphology can be used with difficulty, or not at all, differences in host preferences, ecology, electrophoresis of enzymes, cytology, or other characteristics may be used as markers of specific status. However, the crucial difference between species is the SMRS. The best way of establishing the specific status of populations is to use the criteria the organisms themselves use to identify conspecifics; there can be no finer criteria.

The array of signals used as SMRSs is vast and includes chemical, visual, tactile, and acoustic signals or combinations of these. These signals may, of course, be involved in other functions such as in signaling alarm, in aggressive behavior, or in spacing of individuals within a population. Acoustic signals are used to various degrees as SMRSs by mammals, birds, frogs, and insects. This chapter deals only with acoustic signals of planthoppers (Delphacidae) and leafhoppers (Cicadellidae) (Homoptera: Auchenorrhyncha).

Acoustic Signals in Leafhoppers and Planthoppers

Ossisannilsson (1946, 1949) was the first to demonstrate the production of sounds by many Auchenorrhyncha, but his and most other early studies assumed the sounds were always airborne. Later studies revealed the acoustic signals of some groups, for example, planthoppers, were substrate transmitted (Ichikawa 1976, 1979, 1982; Ichikawa and Ishii 1974; Ichikawa et al. 1975). It is now clear that most, if not all, families of Auchenorrhyncha use acoustic signals in communication and that for all, except Cicadidae, substrate-transmitted sounds are dominant (Claridge 1985*a*,*b*, 1988). The signals consist of trains of damped pulses, repeated at characteristic rates in distinctive temporal patterns (Claridge 1985*a*,*b*). Several authors have studied the substrate-transmitted acoustic signals of different planthopper and leafhopper species, including *Euscelus* (Strubing 1970, 1978), *Empoasca* (Shaw et al. 1974), *Macrosteles* (Purcel and Loher 1976), *Nephotettix* (Inoue 1982), *Nilaparvata* (Claridge et al. 1984b), *Ribautodelphax* (den Bieman 1986), *Javasella* (de Vrijer 1986), and *Dalbulus* (Heady et al. 1986). Although some species may have a number of types of acoustic signals, certain signals, termed *calling signals* (Claridge 1985b), function in pair formation. The calling signals are unique to each species and are the principal signals responsible for mate recognition. The pattern and pulse repetition frequency (PRF) of the calling signals are suggested as the important features to which conspecifics respond (Claridge 1985a,b). Generally, mature males and sometimes also females call spontaneously on their food plants. Responsive insects reply by emitting their own signals, and an exchange occurs during which the males normally move and approach the females. The exchange usually ends when the male makes contact with a female (Claridge 1985a).

Auchenorrhyncha are all plant feeders, many species showing extreme host specificity, yet often morphological differences between populations from different hosts are slight. This has resulted in much confusion over the specific status of various populations. The SMRSs can be used to separate the species and to test the validity of morphological taxonomy.

Sibling Species in *Nilaparvata*

The brown planthopper *Nilaparvata lugens* (Stål) is a distinct morphospecies separated from other *Nilaparvata* species by characters of the male genitalia. It is specific to rice and a major pest of rice in southeast Asia, causing damage through feeding on the rice and transmitting viruses. The entire life cycle is associated with the rice plant, which it uses for feeding and oviposition and as a substrate for acoustic communication between the sexes. It has been shown to be very plastic in its ability to adapt to different resistant strains of rice (Claridge and den Hollander 1982; Claridge et al. 1982), but it will survive only on rice.

Recently, a population of brown planthoppers was discovered on *Leersia hexandra* (Swartz), a weed grass which often grows near rice paddies, in ditches and swampy ground (Medrano and Heinrichs 1982; Domingo et al. 1983). Planthoppers from rice would not feed and did not survive on *Leersia*, and conversely those from *Leersia* would not feed or survive on rice (Claridge et al. 1985a). A clear difference was also seen in oviposition preference, with each population showing a strong tendency to oviposit on its own host (Claridge et al. 1985a). Detailed examination of their morphology, including the male genitalia and abdominal apodemes, did not reveal any characters distinguishing the populations. A morphometric analysis using ten distance measures (Claridge et al. 1984a) also failed to separate the populations clearly (Fig. 19.1).

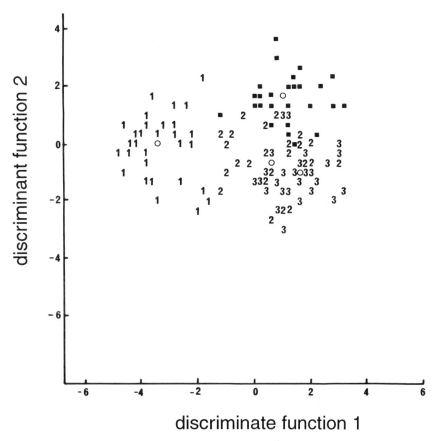

Fig. 19.1. Plots of discriminant function 1 against 2 for canonical analysis of morphometric characters of the IRRI biotypes of *Nilaparvata lugens* and the population from *Leersia*. Biotype 1 on TN1 (1), biotype 2 on Mudgo (2), biotype 3 on ASD7 (3), and *Leersia* (■). ○ indicates the population centroids.

The populations must be considered morphologically virtually indistinguishable. Saxena et al. (1983) and Saxena and Barrion (1985) claimed morphometric differences between the rice and *Leersia* inhabiting populations but considered these as of no greater significance than differences between the biotypes of the rice-inhabiting population. They regarded the *Leersia*-inhabiting population as merely another "biotype" of *N. lugens*. Laboratory studies have revealed that the two populations could readily interbreed and produce apparently normal, viable, and fertile offspring (Claridge et al. 1985a). However, when individuals were given a choice of mates from either host in mate choice experiments, they showed total positive assortative mating (Claridge et al. 1985a). Calling signals of populations from rice and *Leersia* were recorded and compared (Fig. 19.2). The signals were similar in structure in both populations, but both sexes differed significantly between populations and did not overlap in PRF (Fig. 19.3). The PRFs of F1 and F2 hybrid signals were also distinct and intermediate to the parental generations' (Fig. 19.3), which strongly suggests a polygenic inheritance. No such intermediate signals were ever recorded from field-collected specimens, even those collected from areas where *Leersia* and rice grew intermingled. A laboratory rearing experiment enclosing fifty nymphs from both populations in a single small cage containing both host plants produced only one F1 individual with a signal PRF within the hybrid range (but also within the known range of the rice population) (Fig. 19.4). Playback of acoustic signals clearly showed that individuals respond only to signals from the opposite sex derived from the same host plant (Claridge et al. 1985a). An exchange of signals never occurred between males and females from different host plants.

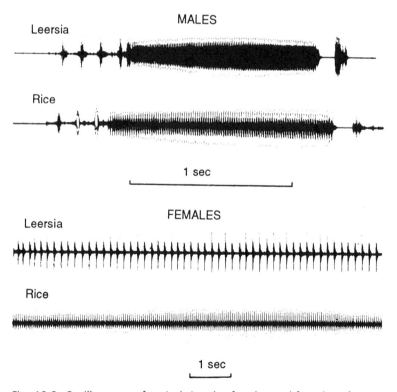

Fig. 19.2. Oscillograms of typical signals of males and females of *Nilaparvata* from rice-inhabiting populations and *Leersia*-inhabiting populations from the Philippines (after Claridge et al. 1988).

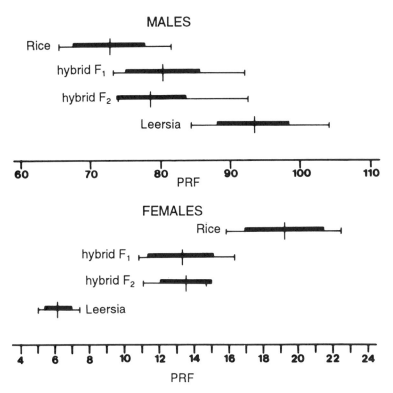

Fig. 19.3. Pulse repetition frequencies (pulses per second) (PRF) of male and female signals from rice-inhabiting and *Leersia*-inhabiting populations of *Nilaparvata* and F1 and F2 hybrids between them. Vertical line represents the mean for each population, thick bar one standard deviation on either side of the mean, and the line the total range (after Claridge et al. 1988).

The evidence is unequivocal that these two populations are separate species. The difference in their calling signals could not be maintained in the field if there was not positive assortative mating. Because there is no evidence of postmating iolation, the calling signals must function to promote positive assortative mating rather than to prevent the production of inferior hybrids as expected under the Biological Species Concept.

The recognition of the two populations as separate species has important implications on many aspects of brown planthopper control. Surveillance systems or insecticide resistance tests that cannot assign insects to the correct host must be treated with caution. The *Leersia*-inhabiting population may also act as a reservoir of predators, parasites, and pathogens of the rice-feeding species (Heinrichs and Medrano 1984).

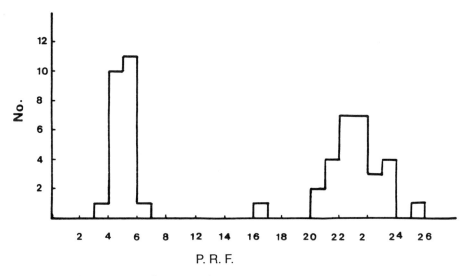

Fig. 19.4. Pulse repetition frequency (pulses per second) (PRF) of a sample of first-generation nymphs produced from a mixture of fifty first instar nymphs of rice-inhabiting and fifty of *Leersia*-inhabiting *Nilaparvata* enclosed in the same cage containing both host plants.

Claridge et al. (1984*b*, 1985*b*) recorded populations of *N. lugens* from rice from nineteen locations over southeast Asia under standard conditions and found significant geographic variation in PRFs. Laboratory crosses between *N. lugens* populations from different geographic areas with different PRFs indicated that the greater the difference in PRF the fewer successful matings are achieved (Claridge et al. 1984*b*). Those males successful in crosses between populations with different PRFs had PRFs significantly different from their own population mean and closer to the range of the other population (Fig. 19.5). The hybrids obtained from the crosses gave apparently normal F1 and F2 generations. Thus it is clear that divergence in PRF of the calling signals (SMRS) can be achieved without the development of postmating isolating mechanisms and suggests that changes in SMRS need not be the result of major genetic changes. It also shows that divergence of SMRS can occur in allopatry and does not require a period of sympatry.

Possible Hybridization between *Nephotettix* Species

Nephotettix virescens (Distant) and *N. nigropictus* (Stål) commonly known as green leafhoppers, occur sympatrically throughout much of tropical Asia. *N. virescens* occurs on rice and *N. nigropictus* chiefly on weeds but also on rice. Both species are vectors of rice tungro virus. The two species are very similar mor-

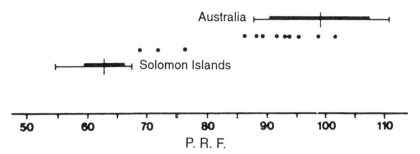

Fig. 19.5. Pulse repetition frequency (pulses per second) (PRF) of samples of *Nilaparvata lugens* males on rice from Australia and Solomon Islands. Solid dots represent PRF of individuals that were successful in crosses with females of the other population. Conventions as in Fig. 19.3 (after Claridge et al. 1988).

phologically and are separated primarily on the shape of the vertex and the extent and arrangement of the color pattern on the vertex, pronotum, and tegmen of the males (Ramakrishnam and Ghauri 1979; Cruz 1974). However, some confusion may occur as all these external characters show considerable variation. In the field, individuals occur that are intermediate in the appearance of one or more of these characters. These have been regarded as naturally occurring interspecific hybrids (Ramakrishnam and Ghauri 1979). Conflicting reports have been published as to whether the two species will (Ling 1968; Inoue 1983) or will not (Cruz 1974) hybridize under laboratory conditions.

Species determination in *Nephotettix* has been considered as confirmed only by the use of male genitalia characters, particularly the number of spines on the aedeagus (Ghauri 1971). The external female genitalia are extremely similar, and no differences have been recorded. However, examination of a number of male genital characters, including the number of spines on the aedeagus, the number of spines on the subgenital plate, shape and size of the median paraphyses, shape of the subgenital plates, and shape of the pygofers revealed the variation in all these characters overlapped between the two species (Yusof 1982; Haslam 1984).

A canonical discriminant analysis using male genital characters separated field-collected individuals into two groups (Fig. 19.6). The interesting point was that field-caught intermediates were assigned to one or other species, with most being attributed to *N. virescens* (Fig. 19.6). Laboratory-produced hybrids were intermediate in color pattern, but unlike field intermediates, they were also intermediate in genital characters. In the discriminant analysis they formed a distinct group truly intermediate between the two parental species (Fig. 19.6).

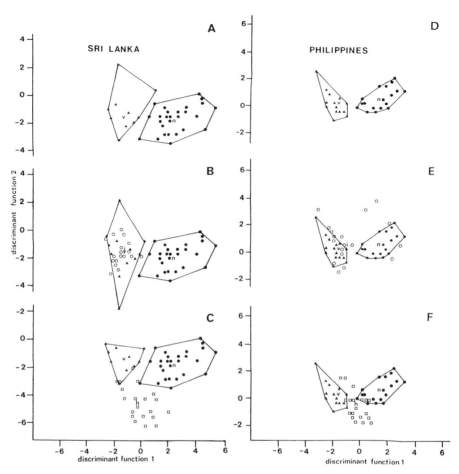

Fig. 19.6. Plots of canonical discriminant function 1 against 2 for *Nephotettix virescens* (▲), *N. nigropictus* (●), field-caught intermediates (○) and laboratory-produced F1 hybrids between *N. virescens* and *N. nigropictus* (□), from Sri Lanka (A, B, and C) and the Philippines (D, E, and F). Letters indicate group centroids.

The acoustic signals of males of the two species are clearly different (Fig. 19.7) (Haslam 1984; Inoue 1983; Yusof 1982), even to the ear. Several field-caught intermediates were recorded and could be unambiguously assigned to one or other species by their calls. Female calls, although simpler in structure than male calls, were also significantly different (Fig. 19.7). Signals of laboratory-produced hybrids of both sexes were different from either parental population with no clear repeatable pattern (Fig. 19.7). To the ear they sounded weak and scratchy. Hybrid signals varied remarkably between individuals and even in the same individual over time.

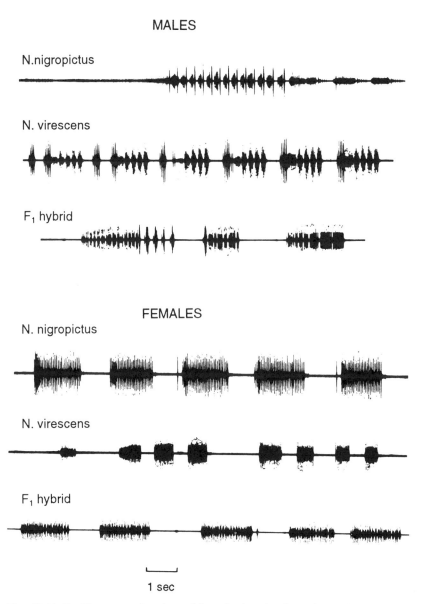

Fig. 19.7. Oscillograms of male and female signals of *Nephotettix virescens* and *N. nigropictus* and of F1 hybrids between them (after Claridge 1985b).

Mate choice experiments between the two species showed total positive assortative mating (Table 19.1). Mate choice experiments using hybrids revealed that the hybrids rarely accepted either species or other hybrids as mates (Table 19.1). This shows that the hybrid signals were recognized by neither

Table 19.1. Results of mate choice between *Nephotettix virescens* (V) and *N. nigropictus* (N) and their hybrids (H)

| | | | Mates with which females mated (N) | | |
Females	Males	Trials	N	V	H
N	N V	22	22	0	—
V	N V	23	0	23	—
N	N H	10	10	—	0
V	V H	10	—	10	0
H	N V H	18	0	1	2
H	N V H	18	0	0	0

species nor by other hybrids, which suggests there is no genetic coupling between the signal-receiver system in these species (Hoy et al. 1977; Hoy and Paul 1973).

Sibling Species in *Oncopsis*

Six species of leafhopper in the genus *Oncopsis* are currently recognized in Britain (Claridge et al. 1977; Claridge and Nixon 1981). The species are extremely difficult to separate as the characters usually used in leafhopper taxonomy, the male genitalia, are extremely similar in all species. The characters are also variable and often not reliable. To add to the difficulty, most species are polymorphic for nymphal and adult color patterns.

In south Wales one of the species, *Oncopsis flavicollis* (L.) occurs on the birch trees *Betula pendula* Roth and *B. pubescens* Ehrh. Nymphs have been reported as being able to feed and survive equally well on, and to show no preference for either, host tree (Claridge et al. 1977). However, females tended to prefer to oviposit on the species of *Betula* from which they were collected (Claridge et al. 1977). This difference correlated with a chromosome difference in the males, with 10 II + X0 males occurring on *B. pubescens* and 9 II + XY males occurring on *B. pendula* (John and Claridge 1974). The situation was explained as one species with both a chromosome polymorphism and a host preference polymorphism for oviposition. As a host preference polymorphism is the first requirement in some models of sympatric speciation, this example has been seized upon by advocates of this mode of speciation (Diehl and Bush 1984).

More detailed work on the chromosomes of the two populations, including the females, has provided strong evidence that no interbreeding occurs between the two populations, even in areas where the two host trees grow in mixed stands. Females were either $2n = 22$ on *B. pubescens* or $2n = 20$ on *B. pendula*. No evidence of the expected category of $2n = 21$ females resulting

Table 19.2 Sex chromosome complement of Oncopsis males and the chromosome complement of Oncopsis females on Betula pendula and B. pubescens at four localities in Wales (for females; 20 = 2n = 9 II + XX; 22 = 2n = 10 II + XX)

	Males				Females			
	B. pendula		B. pubescens		B. pendula		B. pubescens	
	XY	X0	XY	X0	20	22	20	22
Merthyr Mawr	12	1	0	18	10	0	0	12
Coed-y-Wenallt	40	0	3	17	8	0	0	9
Cym-Llwch			0	15				
Blaen-y-Glyn			0	8			0	8
Total	52	1	3	58	18	0	0	29

from interbreeding occurred (Table 19.2). A discriminant analysis of morphological characters, including male abdominal apodemes, also showed significant differences between the two populations, although no completely diagnostic characters were found (Claridge and Nixon 1986) (Fig. 19.8).

The insects are difficult to maintain in the laboratory; therefore, crosses and mate choice experiments, to assess possible interbreeding, could not be carried out. However, when the acoustic signals of both populations were recorded, they were found to be specific to, and very different between, the populations on the different trees (Claridge and Nixon 1986) (Fig. 19.9). The studies of Claridge and Nixon (1986) also revealed that the situation may be even more

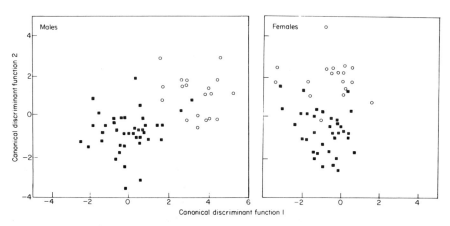

Fig. 19.8. Plot of canonical discriminant function 1 against 2 for morphometric analysis of Oncopsis "flavicollis" from Betula pubescens (○) and B. pendula (■).

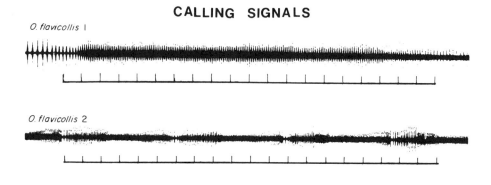

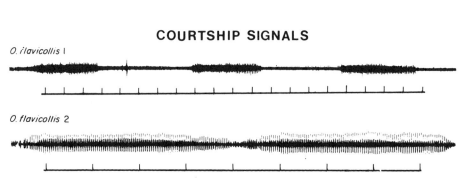

Fig. 19.9. Representative oscillograms of calling songs and courtship songs of *Oncopsis* "flavicollis" 1 from *Betula pubescens* and *O.* "flavicollis" 2 from *B. pendula* in south Wales. Not all to the same scale but time marks at 0.25-s intervals (after Claridge and Nixon 1986).

complex, with a third species present in eastern England. Certainly, in south Wales, the populations on the different birch trees do not represent morphs of one species but two sibling species, each associated with a particular host tree.

The two populations do not represent an intermediate stage (either a host polymorphism or sympatric host races) in the process of sympatric speciation. How the speciation occurred is difficult to establish. Although the two birches are sympatric over most of their range, *B. pendula* is a lowland species while *B. pubescens* has a greater altitudinal range, and both species often occur alone (Forbes and Kenworthy 1973). In the past, part of the *Oncopsis* population could have become isolated with only one host available and eventually developed a preference for it. If it also acquired changes in its SMRS during this period, on subsequent reestablishment of contact the two forms would be able to coexist as distinct species.

Sibling Species in *Alebra*

A number of studies have been made on British tree-feeding Typhlocybine leafhoppers (Claridge and Wilson [1981] and references therein). Most species have been found to be associated with either one taxonomically distinct host species or a few closely related ones; very few are polyphagous. Because species determination is based on morphological criteria, the possibility of sibling species within those currently regarded as polyphagous still exists.

Three species of the genus *Alebra* have been recognized in Britain: *A. coryli* (Le Quesne), *A. wahlbergi* (Boheman), and *A. albostriella* (Fallen). They are extremely similar morphologically and can be reliably separated only by using characters of the male abdominal apodemes (Le Quesne 1976; Le Quesne and Payne 1981). Two of the species, *A. wahlbergi* and *A. albostriella*, show a color pattern polymorphism in the adults (Ribaut 1936), and both have more than one host plant species (Claridge and Wilson 1981).

A discriminant analysis on morphometric characters by Wilson (1979) separated *A. coryli* on *Corylus avellana* (hazel) and *A. albostriella* on *Quercus robur* (oak) from *A. wahlbergi* on other hosts, seemingly confirming the presence of three species. However, a more detailed study by Gillham (1989) revealed significant differences in the morph frequencies of *A. albostriella* from different *Quercus* species (Table 19.3). He also found the male acoustic signals of the three species gave four song types (Fig. 19.10) (Gillham 1992). One song type was restricted to individuals from *C. avellana*; another to those from *Tilia* spp., *Acer campestre* and *Aesculus hippocastanum*; a third to *Q. robur*; and the fourth to *Q. petraea* and *Q. cerris*. The two song types on the oaks corresponded to two morphs, *viridis* on *Q. petraea* and *Q. cerris* and *typica* on *Q. robur*. The conclusion drawn was that there are four species within *Alebra*: *A. coryli*, *A. wahlbergi*, and two sibling species within *A. albostriella*, one on *Q. robur* (species A) and the other on *Q. petraea* and *Q. cerris* (species B), not three as previously thought. Examination of male abdominal apodemes has revealed slight but consistent differences between the two taxa on oak (Gillham 1991). No such discontinuities have been found among populations of *A. wahlbergi* from different

Table 19.3. Number and percentages of the different adult morphs of *A. albostriella* collected from species of oak in south Wales

	Q. robur		Q. cerris		Q. petraea	
	N	%	N	%	N	%
typica	618	89	1	1	5	5
viridis	61	9	121	99	95	95
discicollis	17	2	0	0	0	0

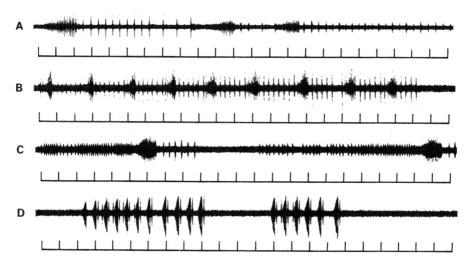

Fig. 19.10. Representative oscillograms of calling signals of *Alebra* males collected as nymphs from known host plants. Time marks at 0.25-s intervals. A: *Corylus avellana;* B: *Tilia cordata, Acer campestre,* and *Aesculus hippocastanum;* C: *Quercus petraea* and *Q. cerris;* D: *Q. robur* (after Gillham 1989, 1992).

host plant species. Therefore, *A. wahlbergi* is regarded as a genuinely polyphagous species.

Acoustic studies on the three species of *Alebra* found on *Castanea sativa* Miller in Kasanitsa, southern Greece, showed them to correspond in male calling song to *A. wahlbergi* and the two sibling species of *A. albostriella* (species A and species B) in Britain. No morphological differences were found between populations of species A from Britain and Greece, but populations of species B from Greece had more distinctive markings on the vertex. A canonical discriminant analysis on variables of the male calls of species B separated the populations from Britain and Greece, but failed to separate populations from the two host plants in Britain (Fig. 19.11). The evidence suggests divergence in SMRS is occurring among allopatric rather than between sympatric host-associated populations (Gillham 1989).

Acoustics of *Alnetoidea*

Six species are recorded in the leafhopper genus *Alnetoidea,* but only one, *A. alneti* (Dahlbom), occurs in Britain where it has been reared from at least seventeen species of trees and shrubs (Claridge and Wilson 1981). Ribaut (1936) reported color and size differences between populations from *Alnus glutinosa* and *Corylus avellana* and considered them different species, but they were later synonymized by China (1943). Differences in size and host prefer-

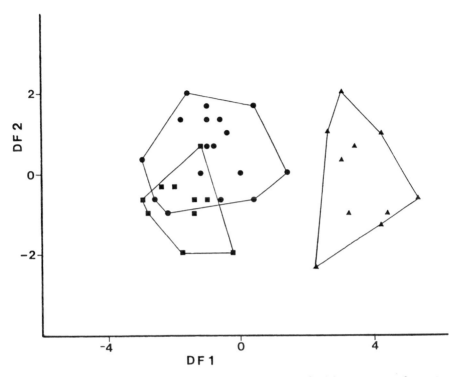

Fig. 19.11. Plot of canonical variables of the male call of *Alebra* species B from *C. sativa* in Kastanitsa, Greece (▲), *Q. petraea* (■) and *Q. cerris* (●) in Cardiff, U.K. (after Gillham 1989).

ences of these populations were also reported by Claridge and Wilson (1976). A morphometric study (Gillham 1989) of *A. alneti* from thirteen different host trees revealed a series of overlapping ranges of variation, with those from *A. glutinosa* and *C. avellana* forming extreme groups (Fig. 19.12). Transfer experiments indicated that these differences were host plant induced (Fig. 19.12).

Gillham (1989) recorded male signals (Fig. 19.13) of individuals from *A. glutinosa, Tilia cordata,* and *C. avellana* in south Wales and found no differences in the structure of the acoustic signals of populations from the different hosts. He concluded that only one species occurred that was truly polyphagous.

Discussion

As each species has a unique SMRS, the process of speciation essentially consists of the development of distinct SMRSs in populations of what was once one species. How this comes about is the subject of much controversy (Dobzhansky 1951, 1970; Mayr 1942, 1963, 1970; Bush 1975; White 1978; Paterson

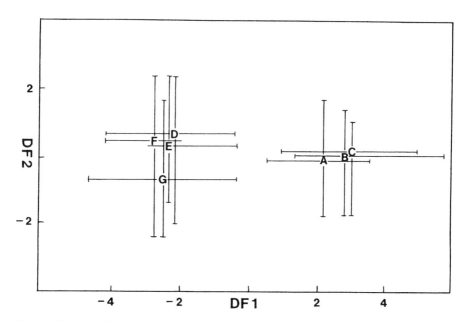

Fig. 19.12. Plot of canonical discriminant function 1 and 2 for male *Alnetoidia alneti* collected as first and second instar nymphs from alder and reared on alder (A, B, and C); from hazel and reared on hazel (D, E, and F); and from alder and reared on hazel (G) (after Gillham 1989).

1978, 1985). The concept of speciation promoted by Dobzhansky and Mayr requires a period of sympatry between the diverging populations during which natural selection can act to build up "isolating mechanisms" to prevent hybridization from breaking up coadapted gene complexes and producing recombinants of lower fitness. Mayr, although a strong supporter of allopatric speciation, also views species in relation to other species, each depending on "isolating mechanisms" for their integrity.

Paterson (1978, 1985) lists many criticisms of the above models and states that the only incontrovertible model of speciation is that of change in total allopatry without reference to other species. Therefore, a concept for speciation must be developed that lacks the necessity for a period of sympatry. Part of

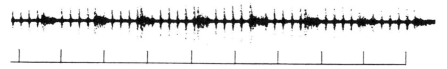

Fig. 19.13. Representative oscillogram of the calling song of male *Alnetoidia alneti*. Time marks at 0.25-s intervals.

the problem in developing such a model is that little is known about the evolution of SMRSs.

Planthoppers and leafhoppers may be very suitable subjects for the study of speciation and the evolution of SMRSs. They are common, there are a large number of species, including numerous species complexes, and many species cover a wide geographic range. Many species are easy to rear in the laboratory and have a short generation time, and the acoustic signals (SMRS) can be quickly and easily recorded from large numbers of individuals and are amenable to rigorous statistical analysis.

Paterson's model predicts that SMRSs should be stable. To function as a recognition system it is imperative that the SMRSs are stereotyped and stable over the entire species' distribution, during an individual's lifetime and under a range of environmental conditions (Bentley and Hoy 1972). This seems largely true for acoustic signals in planthoppers and leafhoppers (Claridge 1985*a;* den Bieman 1986).

The mutual tuning or coadaptation between the sexes provides a degree of stability to the system through negative feedback. Although variation may exist, if either the transmitter or receiver varies outside a certain range it may not be recognized as a potential mate or be less efficient in procuring a mate. Natural selection will thus act constantly to fine-tune the system and keep both signal and receiver near the optimum. Each component of the system is under rigid genetic control, and additional stability to the system may be achieved through polygenic and multichromosomal determination of the elements of the system buffering it against change by mutations (Ewing 1969; Bentley and Hoy 1972, 1974; Cowling and Burnett 1981; Zouros 1981). Although each component of the SMRS is under genetic control, components will exhibit some variation. Selection will act so the SMRS operates most efficiently in the normal environment of the species.

This does not mean the system cannot change, since it manifestly has, but it can normally do so only in a series of small steps with coadaption between the sexes being reestablished after each change. Change can occur through pleiotropic effects of genes selected for increased fitness (Bastock 1956; Manning 1966), or selection may act directly on the mate recognition system to increase its efficiency in a new environment.

In a large population all the stabilizing conditions apply, and any deviant individuals will be selected against as they will not be as effective in securing a mate. Changes in SMRS are thus unlikely to develop in large populations. In a small isolate, however, such as a founder population, which may be established from one mated female, most likely from the margins of the species range and thus possibly unrepresentative of the species as a whole, the situation is different. Selection pressures as a result of new environmental conditions may be different. Through action directly upon the SMRS or on other genes which have pleiotropic effects, the SMRS may change. Lack of choice in mates may enable the success of aberrant SMRSs. Selection should reestablish the coadaptation but perhaps at new values. The new SMRS established will remain fixed as the population expands. If the SMRS is sufficiently different from those of the original population, and individuals from each population no longer recognize those of the other population as potential mates, then speciation has occurred. There is little information on the length of time required for this to occur, but it could be fairly rapid.

Few studies have been made on variation in SMRSs. Individual variation does occur in planthopper and leafhopper acoustic signals (de Vrijer 1986; Gillham 1992), but little work has been carried out linking variation in acoustic signals with mating success.

Some geographic variation has also been found in SMRSs. Geographic variation for a number of song variables has been shown by Booij (1982) for two species of *Muellerianella*, by den Bieman (1986) for some species of *Ribautodelphax* over western Europe, and by Claridge et al. (1985a, 1988) for *N. lugens* in southeast Asia. However, it is not clear whether the variation is in components that are important in mate recognition. Care must also be taken to avoid a circular argument that because characters show variation they are not important in mate recognition. There is no clear pattern in the variation found so far and no information on the gene flow or history of the populations concerned. It is not known whether this variation occurs in large panmictic populations or only between a series of more or less isolated populations, each originating from a small founder population. It must be remembered that variation in SMRS is not inconsistent with the Recognition Concept of species. One theory is that sexual selection may be an important factor in promoting differences in mating signals between allopatric populations (West-Eberhard 1983, 1984). Despite the geographic variation in acoustic signals in *N. lugens*, there is little evidence of any hybrid inviabiliy between populations, suggesting changes in SMRS may not require major genetic differences. The evidence from acoustic signals in leafhoppers and planthoppers is most consistent with specific divergence occurring in total allopatry.

Summary

All sexual reproducing organisms have developed methods of promoting fertilization. In motile organisms this requires bringing the sexes together, at the same place and time, as well as a system of communication to enable the recognition of the opposite sex for the purposes of mating. The mate recognition system consists of a signal-response system and may utilize visual, olfactory, tactile, or acoustic signals, or various combinations of these. Those signals used in the mate recognition system are highly stereotyped and unique to each species. The mate recognition system determines the limits to the species, since all members of a species share a common mate recognition system, the SMRS (Paterson 1985). The SMRS can therefore be used to resolve the specific status of populations where other taxonomic methods cannot be used or have given conflicting results. This chapter has shown how the substrate-transmitted acoustic signals that function as SMRSs in leafhoppers and planthoppers have been used to resolve a number of such situations.

Acknowledgments

I wish to thank my colleagues in the rice research group at Cardiff for their help, discussions, and suggestions, in particular, Prof. M. F. Claridge, who read a draft of the paper; Dr. M. C. Gillham, who allowed me to incorporate some of his results; Mr. G. A. Nixon; and Mr. J. C. Morgan. The work on rice pests at Cardiff has been largely funded by the Overseas Natural Resources Institute (London) and the European Economic Community DG XII Programme for Tropical Agriculture.

References

Bastock, M. 1956. A gene mutation which changes a behavior pattern. Evolution 10:421–439.
Bentley, D.R., and R.R. Hoy. 1972. Genetic control of the neuronal network generating cricket *(Teleogryllus gryllus)* song patterns. Anim. Behav. 20:478–492.
———. 1974. The neurobiology of cricket song. Sci. Am. 231:34–44.
den Bieman, C.F.M. 1986. Acoustic differentiation and variation in planthoppers of the genus *Ribautodelphax* (Homoptera: Delphacidae). Netherlands J. Zool. 36:461–480.
Booij, C.J.H. 1982. Biosystematics of the *Muellerianella* complex (Homoptera: Delphacidae), interspecific and geographic variation in acoustic behavior. Z. Tierpsychol. 58:31–52.
Bush, G.L. 1975. Modes of animal speciation. Annu. Rev. Ecol. Syst. 6:339–364.
China, W.E. 1943. New and little known species of British Typhlocybidae (Homoptera), with keys to the genera *Typhlocyba, Erythroneura, Dikraneura, Notus, Empoasca* and *Alebra*. Trans. Soc. Br. Entomol. 8:111–153.
Claridge, M.F. 1985a. Acoustic signals in the Homoptera: Behavior, taxonomy, and evolution. Annu. Rev. Entomol. 30:297–317.
———. 1985b. Acoustic behavior of leafhoppers and planthoppers: Species problems and speciation. Pp. 103–125 *in* L.R. Nault and J.G. Rodriguez, eds. The leafhoppers and planthoppers. John Wiley, New York.
———. 1988. Species, speciation and acoustical signals in Auchenorrhyncha. Pp. 233–243 *in* C. Vidano and A. Arzone, eds. Proc. Sixth Auchenorrhyncha Meeting, Turin, Italy, 7–11 September, 1987, Inst. Agr. Entomol. Apiculture, University of Turin.

Claridge, M.F., and J. den Hollander. 1982. Virulence to rice cultivars and selection for virulence in populations of the brown planthopper *Nilaparvata lugens*. Entomol. Experi. Appl. 32:213-221.

Claridge, M.F., J. den Hollander, and I. Furet. 1982. Adaptations of brown planthopper *(Nilaparvata lugens)* populations to rice varieties in Sri Lanka. Entomol. Experi. Appl. 32:222-226.

Claridge, M.F., J. den Hollander, and D. Haslam. 1984a. The significance of morphometric and fecundity differences between the "biotypes" of the brown planthopper, *Nilaparvata lugens*. Entomol. Experi. Appl. 36:107-114.

Claridge, M.F., J. den Hollander, and J.C. Morgan. 1984b. Specificity of acoustic signals and mate choice in the brown planthopper *Nilaparvata lugens*. Entomol. Experi. Appl. 35:221-226.

———. 1985a. The status of weed-associated populations of the brown planthopper, *Nilaparvata lugens* (Stål)—host race or biological species? Zool. J. Linn. Soc. 84:77-90.

———. 1985b. Variation in courtship signals and hybridization between geographically definable populations of the rice brown planthopper, *Nilaparvata lugens* (Stål). Biol. J. Linn. Soc. 24:35-49.

———. 1988. Variation in host plant relations and courtship signals of weed-associated populations of the brown planthopper, *Nilaparvata lugens* (Stål), from Australia and Asia: A test of the recognition species concept. Biol. J. Linn. Soc. 35:79-93.

Claridge, M.F., and G.A. Nixon. 1981. *Oncopsis* leafhoppers on British trees: Polymorphism in adult *O. flavicollis* (L.). Acta Entomol. Fennica 38:15-19.

———. 1986. *Oncopsis flavicollis* (L.) associated with tree birches *(Betula)*: A complex of biological species or a host plant polymorphism? Biol. J. Linn. Soc. 27:381-397.

Claridge, M.F., W.J. Reynolds, and M.R. Wilson. 1977. Oviposition behavior and food plant discrimination in leafhoppers of the genus *Oncopsis*. Ecol. Entomol. 2:19-25.

Claridge, M.F., and M.R. Wilson. 1976. Diversity and distribution patterns of some mesophyll-feeding leafhoppers of temperate woodland canopy. Ecol. Entomol. 1:231-250.

———. 1981. Host plant associations, diversity and species area relationships of mesophyll-feeding leafhoppers of shrubs and trees in Britain. Ecol. Entomol. 6:217-238.

Cowling, D.E., and B. Burnett. 1981. Courtship songs and genetic control of their acoustic characters in sibling species of the *Drosophila melanogaster* subgroup. Anim. Behav. 29:924-935.

Cruz, Y.P. 1974. Reproductive isolation between *Nephotettix virescens* (Distant) and *N. nigropictus* (Stål) (Euscelidae: Homoptera). M.Sc. Dissertation, University of the Philippines, Los Banos.

Diehl, S.R., and G.L. Bush. 1984. An evolutionary and applied perspective of insect biotypes. Annu. Rev. Entomol. 29:471-504.

Dobzhansky, T. 1951. Genetics and the origin of species. 3d ed. Columbia University Press, New York.

———. 1970. Genetics of the evolutionary process. Columbia University Press, New York.

Domingo, I.T., E.A. Heinrichs, and R.C. Saxena. 1983. Occurrence of brown planthopper on *Leersia hexandra* in the Philippines. Int. Rice Res. News. 18:17.

Ewing, A.W. 1969. The genetic basis of sound production in *Drosophila pseudoobscura* and *Drosophila persimilis*. Anim. Behav. 17:556-560.

Forbes, J.C., and J.B. Kenworthy. 1973. Distribution of two birch forming stands on Deeside. Trans. Bot. Soc. Edinburgh 42:101–110.

Gillham, M.C. 1989. Biosystematic studies on some mesophyll-feeding leafhoppers associated with trees and shrubs. Ph.D. dissertation, University of Wales, Cardiff.

———. 1991. Polymorphism, taxonomy and host plant associations in *Alebra* leafhoppers (Homoptera: Cicadellidae: Typhlocybinae). J. Nat. Hist. 25:233–255.

———. 1992. Variation in acoustic signals within and among leafhopper species of the genus *Alebra* (Homoptera: Cicadellidae). Biol. J. Linn. Soc. 45:1–15.

Ghauri, M.S.K. 1971. Revision of the genus *Nephotettix* Matsumura (Homoptera: Cicadelloidea: Euscelidae) based on the type material. Bull. Entomol. Res. 60:481–512.

Haslam, D. 1984. Taxonomic studies on some planthopper and leafhopper pests of rice. M.Sc. dissertation, University of Wales, Cardiff.

Heady, S.E., L.R. Nault, G.F. Shambaugh, and L. Fairchild. 1986. Acoustic and mating behavior of *Dalbulus* leafhoppers (Homoptera: Cicadellidae). Ann. Entomol. Soc. Am. 79:727–736.

Heinrichs, E.A., and F.G. Medrano. 1984. *Leersia hexandra*, a weed host of the rice brown planthopper, *Nilaparvata lugens* (Stål). Crop Prot. 3:77–85.

Hoy, R.R., J. Hahn, and R.C. Paul. 1977. Hybrid cricket auditory behavior: Evidence for genetic coupling in animal communication. Science 195:82–84.

Hoy, R.R., and R.C. Paul. 1973. Genetic control of song specificity in crickets. Science 180:82–83.

Ichikawa, T. 1976. Mutual communication by substrate vibration in the mating behavior of planthoppers (Homoptera: Delphacidae). App. Entomol. Zool. 11:8–23.

———. 1979. Studies on the mating behavior of four species of Auchenorrhynchous Homoptera which attack the rice plant. Mem. Fac. Agr. Kagawa Univ. 34:1–60.

———. 1982. Density related changes in male-male competitive behavior in the rice brown planthopper, *Nilaparvata lugens* (Stål) (Homoptera: Delphacidae). App. Entomol. Zool. 17:439–452.

Ichikawa, T., and S. Ishii. 1974. Mating signals of the brown planthopper, *Nilaparvata lugens* (Stål) (Homoptera: Delphacidae): Vibration of the substrate. App. Entomol. Zool. 9:196–198.

Ichikawa, T., M. Sakuma, and S. Ishii. 1975. Substrate vibrations: Mating signals of three species of planthoppers which attack the rice plant. App. Entomol. Zool. 10:162–171.

Inoue, H. 1982. Species-specific calling sounds as a reproductive isolating mechanism in *Nephotettix* spp. (Hemiptera: Cicadellidae). App. Entomol. Zool. 17:253–262.

———. 1983. Reproductive isolation and hybridisation of *Nephotettix* spp. Pp. 339–349 *in* W.J. Knight, N.C. Pant, T.S. Robertson, and M.R. Wilson, eds. First Int. Workshop Leafhoppers Planthoppers Econ. Import. Commonwealth Institute of Entomology, London.

John, B., and M.F. Claridge. 1974. Chromosome variation in British populations of *Oncopsis* (Homoptera: Cicadellidae). Chromosoma 46:77–89.

Le Quesne, W.J. 1976. A new species of *Alebra* Fieber (Hemiptera: Cicadellidae). Entomol. Month. Mag. 112:49–52.

Le Quesne, W.J., and K. Payne. 1981. Hemiptera Cicadellidae (Typhlocybinae). Handbooks for the identification of British insects 2(2). Royal Entomological Society, London.

Ling, K.C. 1968. Hybrids of *Nephotettix impicticeps* (Ish.) and *N. apicalis* (Motsch.) and their ability to transmit the tungro virus of rice. Bull. Entomol. Res. 58:393–398.

Manning, A. 1966. Sexual behavior. Pp. 59–68 *in* P.T. Haskell, ed. Sexual behavior. Symposium Royal Society, London.
Mayr, E. 1942. Systematics and the origin of species. Columbia University Press, New York.
———. 1963. Animal species and evolution. Harvard University Press, Cambridge, Mass.
———. 1970. Populations, species and evolution. Harvard University Press, Cambridge, Mass.
Medrano, F.G., and E.A. Heinrichs. 1982. *Leersia hexandra* as a weed host for the brown planthopper. Int. Rice Res. News. 7:15–16.
Ossiannilsson, F. 1946. On sound-production and the sound-producing organ in Swedish Auchenorrhyncha. (A preliminary note). Opus. Entomol. Sup. 11:82–84.
———. 1949. Insect drummers: A study on the morphology and function of the sound-producing organ of the Swedish Homoptera Auchenorrhyncha with notes on their sound-production. Opus. Entomol. Sup. 10:1–145.
Paterson, H.E.H. 1978. More evidence against speciation by reinforcement. S. Afr. J. Sci. 74:369–371.
———. 1980. A comment on "mate recognition systems." Evolution 34:330–331.
———. 1985. The recognition concept of species. Pp. 21–29 *in* E.S. Vrba, ed. Species and speciation. Transvaal Museum Monograph No. 4. Pretoria.
Purcell, A.H., and W. Loher. 1976. Acoustical and mating behavior of two taxa of the *Macrosteles fascifrons* species complex. Ann. Entomol. Soc. Am. 69:513–518.
Ramakrishnan, U., and M.S.K. Ghauri. 1979. Probable natural hybrids of *Nephotettix virescens* (Distant) and *N. nigropictus* (Stål) (Hemiptera: Cicadellidae) from Sabah, Malaysia. Bull. Entomol. Res. 69:357–361.
Ribaut, H. 1936. Faune de France 31: Homopteres Auchennorrhynques. 1 (Typhlocybidae). Paul Lechevalier, Paris.
Saxena, R.C., and A.A. Barrion. 1985. Biotypes of the brown planthopper *Nilaparvata lugens* (Stål) and strategies in deployment of host plant resistance. Insect Sci. Appl. 3:193–210.
Saxena, R.C., M.V. Velasco, and A.A. Barrion. 1983. Morphological variations between brown planthopper biotypes on *Leersia hexandra* and rice in the Philippines. Int. Rice Res. News. 8:3.
Shaw, K.C., A. Vargo, and O.V. Carlson. 1974. Sounds and associated behavior of some species of *Empoasca* (Homoptera: Cicadellidae). J. Kansas Entomol. Soc. 47:284–307.
Strubing, H. 1970. Zur artberechtigung von *Euscelis alsius* Ribaut gegenuber *Euscelis plebejus* Fall. (Homoptera: Cicadina). Ein bietrag zur neunen systematik. Zool. Beitr. 16:441–478.
———. 1978. *Euscelis lineolatus* Brulle 1832 und *Euscelis ononidis* Remane 1967. 1. Ein okologischer, morphologischer und bioakustischer vergleich. Zool. Beitr. 24:123–154.
de Vrijer, P.W. 1986. Species distinctiveness and variability of acoustic calling signals in the planthopper genus *Javesella* (Homoptera: Delphacidae). Netherlands J. Zool. 36:162–175.
West-Eberhard, M.J. 1983. Sexual selection, social competition and speciation. Q. Rev. Biol. 58:155–183.
———. 1984. Sexual selection, competitive communication and species-specific signals in insects. Pp. 283–324 *in* T. Lewis, ed. Insect communication. Academic Press, London.
White, M.J.D. 1978. Modes of speciation, W.H. Freeman, San Francisco.

Wilson, M.R. 1979. Studies on the taxonomy and ecology of some mesophyll-feeding leafhoppers. Ph.D. dissertation, University of Wales, Cardiff.

Yusof, O.M. 1982. Biological and taxonomic studies on some leafhopper pests of rice in south east Asia. Ph.D. dissertation, University of Wales, Cardiff.

Zouros, E. 1981. The chromosomal basis of sexual isolation in two sibling species of *Drosophila: D. arizonensis* and *D. mojavensis*. Genetics 97:703–718.

20

Species, Selection, and Paterson's Concept of the Specific-Mate Recognition System

NILES ELDREDGE

Department of Invertebrates
The American Museum of Natural History

WHAT ARE SPECIES? Perhaps no other issue in comparative or evolutionary biology has provoked quite so much disparate opinion as this simple question. A fundamental ontological issue, the nature of species stands near the very center of recent attempts to clarify the nature of biological entities generally (e.g. Ghiselin 1974; Hull 1980; Eldredge 1985; also Otte and Endler 1989; and the entire April 1987 issue of *Biology and Philosophy* as entry points to this literature). It is my conviction (Eldredge 1989) that Hugh Paterson's concept of the *Specific-Mate Recognition System* (SMRS; see especially Paterson 1978, 1980, 1981, 1982, 1985) has contributed more to the difficult task of sorting competing claims on the nature of species—and related issues—than any other formulation that has appeared since the origins of the "modern synthesis" in the late 1930s and early 1940s. Following Paterson, I will use SMRS in its most general sense: in sexually reproducing organisms (including hermaphrodites), it follows that there must be a "fertilization system" facilitating successful mating; each such system in sexual nature constitutes an SMRS. It will be my purpose here to state the nature of the problem of species fully, to explore the implications of the SMRS concept to the issue, and to connect Paterson's notion with the distinction between economic and reproductive organismic attributes and consequent concepts of biotic hierarchy and levels of selection.

The SMRS in Intellectual Context

Paterson (e.g. 1985) generally presents his notion of SMRS as a contrast with what he calls the "isolation" species concept of the modern synthesis. The Isolation Concept was first fully articulated (at least in modern terms) in the writings of Mayr (1942) and, especially, Dobzhansky (1937a,b). It is worthwhile to inquire, first, what these two biologists actually said about species—and why—before considering the similarities and differences between the Isolation and Recognition concepts of species.

Dobzhansky (1937b), followed closely by Mayr (1942), added the notion of *discontinuity* as a second theme to be explained by the general notion of evolution. Darwin's (1859) original problem was to explain *diversity*—meaning *phenotypic (including morphological, physiological, and behavioral) differences between organisms*. Natural selection was the original mechanism promulgated to explain the modification of organismic phenotypic properties through time; it remains the prime causal mechanism invoked today. The problem, as Dobzhansky and Mayr were concerned to formulate it, is that natural selection would be expected to produce a smoothly continuous gradient, a spectrum of anatomical variation interconnecting all forms of life. But organisms seem to come in more or less discrete packages—groups whose members resemble each other rather closely, as a rule, and are generally quite distinct from even their closest relatives.

To be sure, it had been appreciated even before Darwin that lineages had multiplied; Darwin and many of his successors simply postulated extinction of intermediate forms to produce the discontinuity among members of the recent biota. Dobzhansky and Mayr argued that additional causal theory was required: a process that could initiate divergence among lineages, setting lineages on independent evolutionary courses and fragmenting the spectrum of adaptive continuity into discrete packages of phenotypic properties. The causal theory they adduced emphasized disruption of reproductive continuity as a forerunner of the general adaptive discontinuities they were concerned to explain.

There has been an uneasy duality in conceptions of species over the past two hundred years, a duality reflected directly in the opening passages of *Genesis*, where an organism begets offspring "after his kind." Species are variously seen as collections of similar organisms and as collections of interbreeding organisms. Often a species definition entails both concepts (in modern terms, a reproductive community is understood to share a pool of genetic information—and the two notions *must* be intimately linked). But it is nonetheless true that species concepts, almost inevitably, stress one aspect over the other (as will be seen still to be true in the most recent discussions; see below).

Because of their concern with explaining discontinuity, superposed as it is on the spectrum of adaptive continuity, and because their explanation of adaptive discontinuity entailed the establishment of reproductive discontinuity first, *Dobzhansky and Mayr came to see species first and foremost as reproductive communities*. If all of biology has not embraced the conceptualization of species as primarily reproductive communities (systematics, for example, has remained generally recalcitrant; see below), certainly evolutionary biology (study of causal mechanisms of evolution) has done so. And this is precisely the context in which Paterson has developed his notion of the SMRS. First, Dobzhansky and Mayr had to establish the primacy of reproduction as the *sine qua non* of species; it was left to Paterson to refine the conceptualization of the nature of those reproductive communities.

Isolation versus Recognition Concepts

Paterson (e.g. 1985) contrasts his SMRS ("recognition") with the "isolation" ontological conceptualization of species. It is important to stress at the outset the features that these competing views of species actually share. Thus, species in both views are reproductive communities, that is, for example, collectivities of sexual organisms among whom mating occurs; mating is rare or absent with organisms of other such collectivities. Moreover, where mating does not occur because of simple geographic distance (separation), both concepts nonetheless attribute organisms to the same species provided that nonmating is exclusively a function of geography, that is, that mating would occur were the organisms in proximity. Vrba (1985) is, however, perfectly correct in noting that Paterson's notion of the SMRS provides far sharper criteria than the older "isolation" concept for reaching judgment of species membership in allopatric situations; this point becomes clearer immediately below.

Most of Paterson's distinctions between the two concepts actually concern a debate of *how species arise*, rather than what species are per se (see Chandler and Gromko [1989] on this point). Dobzhansky (1937b) saw discontinuity between species as arising as a function of the evolutionary process. Discontinuity, in other words (and in the customary rhetoric of evolutionary theory), must be "for" something; selection, Dobzhansky presumed, was responsible as much for discontinuity between species as it was for the anatomical differences between component organisms of different species. In his model, Dobzhansky saw isolation as a means of simultaneously eliminating inharmonious gene complexes and of focusing a species on an adaptive peak: too much variation diffuses the economic adaptations of organisms. Much later, Dobzhansky (in Dobzhansky et al. 1977:168) simply wrote: "The question of why there should be species can thus be answered: because there are many adaptive peaks."

Mayr (1942) was rather a different case—and to some extent Mayr's (1988) complaint that Paterson incorrectly links him with Dobzhansky on the matter of isolation is justified. Dobzhansky's model of speciation by reinforcement suffered the potentially fatal objection that a species, originating in allopatry, might well never become sympatric at any subsequent stage with the "parental" or "sister" species from which it became separated. Less elaborate than Dobzhansky's notion of reinforcement, Mayr's model simply saw geographic isolation leading to the accumulation of sufficient differences that mating would not, or could not, occur should secondary sympatry ever be established. This, of course, has become the most general statement of the canonical model of speciation in contemporary evolutionary theory.

It should, however, also be noted that Mayr (1942, 1963) did retain the expression "isolating mechanisms" coined by Dobzhansky (though his classification of isolating mechanisms in 1942 differed significantly from that of Dobzhansky [1937b]; Dobzhansky [e.g. 1951] later modified his classification to resemble Mayr's). And even though there are passages in Mayr (e.g. 1942:254) where he expressly acknowledges that isolating mechanisms could as easily be seen as means for conspecifics to "recognize" one another as they could be construed as means of keeping species apart, there is no doubt that Mayr's point of view was primarily isolationist. At any rate, it has not been until Paterson's thinking became generally available that there has been any attempt to follow the full implications of "recognition."

There are two major points of difference between Paterson's and the received view of speciation. The first, of course, is the aforementioned debate between "reinforcement" (where isolation is actually selected) and "recognition." The second point is closely related, but potentially of even greater interest; it contrasts the Mayrian view with a more precise formulation: where Mayr simply invoked (genetically based) allopatric differentiation *of whatever kind* eventually leading to "reproductive isolation," Paterson sees speciation as fundamentally entailing modification of the SMRS. This insight, in my opinion, is a logical outgrowth of seeing species primarily as reproductive communities—and is a great conceptual step forward in our understanding of the organization of biotic nature.

The success of Paterson's argument hinges on a distinction between *economic* and *reproductive* characteristics of (sexual) organisms. Those phenotypic attributes (in the widest sense, including behavior and physiology in addition to morphology) pertaining to reproduction constitute the SMRS; these include, of course, primary sexual organs, but also all somatic and behavioral elements brought to bear in the reproductive process. (As Darwin may have been the first to note, the distinction between "secondary sexual characteristics" [and other features that may be involved in reproduction] and general economic features is not in every case a straightforward matter. At the very least, it is clear that features may serve multiple functions, including some economic and some reproductive, the mammalian penis being an excellent example).

In contrast with the view that speciation entails simple accrual of any manner of genetically based change, Paterson's view sees modification of the SMRS as both necessary and sufficient for speciation to occur. In other words, a great deal of economic change can accrue within a polytypic species (whether through selection, genetic drift, or any other mechanism of genetic change) without reproductive isolation necessarily following. The converse is also true:

as we know from numerous documented examples of "sibling species," reproductive isolation can exist between two closely related species that are hardly to be distinguished on the basis of external, economic phenotypic attributes. Economic and SMRS adaptive change need not go hand in hand—though substantial economic change in general seems not to accumulate in phylogeny *without* SMRS change. As I have recently discussed in detail (Eldredge 1989:ch. 4), although theoretically economic and SMRS change are quasi-independent (though see Paterson [1985] for a somewhat different view), it seems to be the case that SMRS change (speciation) can occur without substantial economic change (i.e., standard morphological transformation of non-SMRS features), but the converse is not true. In particular, it appears that (as Darwin himself pointed out), economic differentiation *within* species (i.e., without accompanying SMRS alteration, or "speciation") is apt to be lost, that is, not incorporated into the "phylogenetic mainstream." Conversely, SMRS differentiation without appreciable concomitant economic modification results in "sibling species"—the antithesis of accumulation of (economic) adaptive change in evolution. Only when the two go hand in hand is significant adaptive change in general likely to accumulate phylogenetically.* Such a manner of stating this characteristic "punctuated" pattern of evolutionary change hinges on a distinction between economic and reproductive features, a distinction that lies at the heart of Paterson's formulation of the SMRS.

Even though Paterson has developed his notion of the SMRS and the Recognition Concept of species primarily in the context of theories of the speciation process itself, there are still further implications of the SMRS concept for the ongoing debate on the very nature of species, as well as related issues on the nature of, and levels of, selection—issues to which I now turn.

The SMRS and Species: The Collision between Function and History

One need look no further than the first two of the papers recently published in the very useful collection edited by Otte and Endler (1989) to see that the fundamental dichotomy in species conceptualizations is still very much with us, albeit in subtly modified form. Templeton (1989), hinging his analysis on cohesion, reviews three species concepts (Evolutionary, Isolation, and Recognition) and attempts to integrate the better features of all three into his Cohesion Species Concept. According to Templeton, cohesion mechanisms include gene flow, genetic drift, and natural selection, all of which clearly presuppose sexual reproduction.

*I thank Dr. Frederick M. Cohan, Wesleyan University, for suggesting this succinct summation of my views on the relation between SMRS and general economic adaptive change in phylogenesis.

On the other hand, Cracraft (1989), while recognizing that any species concept cannot ignore the "reproductive community" aspect of the Biological Species Concept, articulates the growing concern among systematists that *economic* (i.e., nonreproductive) differentiation within species defined in purely reproductive terms often leads to anomalies: within well-differentiated "biological species," a systematist may well find that one portion of a species is more closely related phylogenetically to a second (reproductively isolated) species than it is other portions of its own reproductive community (see Fig. 20.1). Such a conclusion would arise if a part of a species were to be found to share one or more synapomorphies with the second species—synapomorphies not found in the other populations of the first species. Thus, cladists are fond of pointing out that interbreeding (reproductive cohesion) in fact may constitute a symplesiomorphy and thus not constitute positive evidence that a phylogenetic entity—a taxon, in this case, a species—exists in any meaningful sense.

This is not a trivial objection. Systematists rightly ask why species should be seen as taxa different in kind from taxa at any other rank. Both reproductive

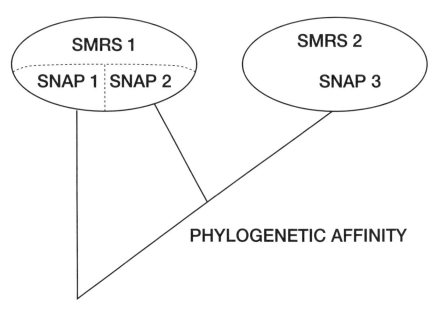

Fig. 20.1. Diagram illustrating the difference between species defined as reproductive communities ($SMRS_i$), and species recognized in conjunction with a phylogenetic analysis based on the distribution of synapomorphies ($SNAP_i$). Such conflicts arise only when reproductive attributes are distributed more widely than economic attributes—rendering the SMRS symplesiomorphous.

and economic characters are used wherever relevant to define higher-level taxa (hair and placentas in mammals, for example). It is only at the level of species that problems arise, specifically when economic differentiation is more finely grained than reproductive differentiation and the SMRS is thereby rendered symplesiomorphous. That many botanists have long opposed adoption of the Biological Species Concept is ample testimony that, in some groups of sexually reproducing organisms, economic differentiation regularly outstrips reproductive adaptive change. North American oaks constitute but one example of a cluster of well-differentiated "species" that hybridize freely.

Systematics is the search for phylogenetic history, and it is indeed an anomaly when details of phylogenetic resolution are ignored in favor of recognition of species based on reproductive cohesion, when the two sets of criteria lead to different conclusions on species identities. As de Queiroz and Donoghue (1988) and Eldredge (1989) have pointed out, both conceptualizations of species are valid in their own contexts. Often, specifically when reproductive (SMRS) differentiation is equal to or greater than economic differentiation, the species groups recognized by both phylogeneticists and evolutionists relying on reproductive cohesion will coincide. But the fact that there is a definite set of circumstances in which species identities do not coincide in the two systems suffices to establish that there are, minimally, two different and valid conceptualizations of species.

Nor is the existence of two partly contradictory species concepts, each valid in its own right, to be viewed most fruitfully as a conflict in search of the usual sort of resolution: that is, in the form of a hybrid or compromise concept, blending or resolving the differences into a single definition. Still less is it either rational or desirable to expect that one concept will drive the other one out. The Phylogenetic Concept arises from the straightforward conviction that, from the time of Linnaeus on, species are to be regarded as taxa in fully the same sense as are genera, families, and so forth. Species are merely the lowest ranked level in a series of categories; taxa assigned to these levels do not otherwise differ in nature (Nelson 1989). (However, de Queiroz and Donoghue [1988] present a useful review and commentary on the range of opinions of phylogeneticists on the reconciliation of the Biological Species Concept with the logical structure of phylogenetic systematics.)

Biologists focused on causal processes in evolution—especially within-population processes of drift and selection—have (for reasons reviewed above) come to see species primarily as reproductive entities. Instead of the phylogenetic perspective that sees species simply as the lowest of a many-tiered array of systematic categories, species to a causal evolutionary theorist arise as a consequence of organismic activity, specifically, reproductive activity. Species are built from the ground up, through the direct activities of component organisms. The Phylogenetic (systematics) Concept is closest to the original use of the term *species* (albeit originally in an a-phylogenetic sense in original usage). Yet, because evolution is, at base, the maintenance and modification of genetically based information and that information is manifestly packaged in a nested set of stable entities (organisms, demes, species, and monophyletic taxa [see Eldredge 1985]), species as reproductively coherent entities are an essential component of a "genealogical hierarchy" that forms as a simple consequence of reproductive behavior.

Thus we have two conflicting and equally valid and important concepts of species. Moreover, each is important in a distinctly evolutionary sense: one for its insistence on genealogical precision, the other for its recognition of the formation of stable entities that arise as a consequence of (sexual) reproductive behavior. Perhaps the more recent usage—the reproductive cohesion sense of *species*—should be renamed. At the very least, causal evolutionary theory should acknowledge whatever may be known about phylogenetic structure that cuts across the grain of a cluster of species defined on reproductive criteria. And phylogeneticists should likewise remain aware that, should any given SMRS system be shown to constitute a symplesiomorphy, that symplesiomorphy nonetheless constitutes a functional system. As botanists are only too keenly aware, hybridization has a way of erasing, or at least strongly blurring, phylogenetic signals. Today's well-differentiated features, the synapomorphies linking a subset of one reproductive community with a second reproductive community, may well be impermanent on the phylogenetic landscape, engulfed in amoeba fashion by the parental ("ancestral") reproductive community. Such is the power of ongoing reproductive cohesion. As I concluded in the previous section (and have discussed at greater length in Eldredge [1989, especially ch. 4; see also Futuyma 1987]), disruption of the SMRS seems vital to the lasting phylogenetic accumulation of economic adaptive change.

The SMRS and the Nature, and Levels, of Selection

The SMRS, as a quintessentially functional system directly concerned with reproductive success, is unquestionably adaptive in the classic sense. The SMRS, therefore, arises and is maintained by a form of selection. This raises several interesting issues: What form of selection is involved in the development and maintenance of the SMRS? And what are the implications, if any, of such forms of selection for the evaluation of the nature (even the existence) of selection at other levels—specifically, the level of the species itself ("species selection")?

To avoid confusion, it is important to stress at this juncture that my use of the terms *natural selection* and *sexual selection*, while lying within the range of historical usage, differs in important respects from the meanings commonly ascribed to these terms in contemporary evolutionary biology. Natural selection can be characterized as *the relative effect of differential economic success (of organisms within a local population) on relative reproductive success:* in other words, on the average, organisms that fare better economically tend to leave a disproportionately greater number of alleles pertaining to that economic success to the next generation than do their less economically successful peers. Thus do we account for the origin, maintenance, and modification of economic adaptive attributes.

In contrast, Darwin (1871:256) defined *sexual selection* as arising from the *"advantage which certain individuals have over other individuals of the same sex and species, in exclusive relation to reproduction,"* a definition more general than his earlier (1859) use of the term. Sexual selection, in my terminology, rather than constituting a vector from the economic to the reproductive side of organismic activity, pertains directly to reproduction: the term, in my usage, embraces fecundity, efficiency in mate finding, care of young, and all other specifiable aspects of organismic reproduction. I am contrasting, in other words, the resultant differential reproductive success arising from, for example, variation in predatory behavior (*natural* selection) from that which arises purely through variation in fecundity, mate recognition, and the like (*sexual* selection). The distinction is based, ultimately, on classes of adaptations: economic versus reproductive. The SMRS is that adaptive complex of structure, physiology, and function pertaining to reproduction in sexual organisms. It seems a reasonable and logical conclusion that the SMRS arises, is maintained, and is further modified through *sexual* selection. Paterson's focus on the resultant, species-specific reproductive adaptive complex further underscores the only recently revived appreciation of the very real distinction Darwin originally drew between natural and sexual selection.

Sexual selection is most commonly viewed in conjunction with (and is at times confounded with) *mate* selection. But it is clear that differential reproductive success arising strictly through variation in structures, behaviors, and physiological functions pertaining directly to reproduction (i.e., not to economic matters per se) can be expected to occur within any sexually reproducing population. There can be within-population variation in mate recognition signals as well as geographic variation (Ryan and Wilczynski 1988). As Paterson and his colleagues and students have documented so assiduously, selection will act in a population to maintain levels of successful reproduction—a sort of within-population stabilizing sexual selection where deviants from the SMRS norm will experience reduced fitness. Modern evolutionary biology (especially sociobiology) tends to view sexual selection as a special aspect of the overall competitive race among organisms to leave relatively larger shares of their genes to the next generation; such an image conjures the familiar picture of constant change: "directional sexual selection." The SMRS perspective, however, provides a welcome counterbalance: one would expect within-population stability in mate recognition systems as the major generation-to-generation pattern within populations and species. Such differing expectations of directional versus stabilizing sexual selection mirror precisely the overall conflict in expectation between evolutionary biologists who see competition for resources leading to inevitable, more or less linear change through time (regardless of whether the environment is changing or remaining stable: see Eldredge [1989] for extensive discussion) and those who have come to expect relative stability in economic adaptation within species over time. The empirical evidence of the fossil record, plus the expectation that habitat tracking in the face of environmental change can still lead to stability, implies that economic adaptive systems are far more stable than simple selection theory extrapolated through time characteristically predicts.

Yet selection for continued mate recognition can lead to divergence in SMRS among allopatric populations; this amounts to a form of among-deme "drift," meaning that selection within populations is simply for continued mating, and *not* for "isolation." Such divergence in any case does not represent adaptive "improvement." No SMRS system could ever be viewed as superior to another; the point is that each one works, and when they differ to the point where organisms of one reproductive community cannot "recognize" those of another sufficiently to allow successful mating, the two constitute separate species.

Thus sexual selection—in the more general sense propounded by Darwin (1871)—is both necessary and sufficient for speciation to occur. As we have already seen, the standard speciation model makes no distinction between economic and reproductive adaptations: accumulated change of either sort could lead, eventually, to "reproductive isolation." That sexual selection (simply to maintain successful mating in allopatry) is minimally necessary for speciation to occur is demonstrated by the existence of sibling species—reproductively isolated, but hardly, if at all, economically differentiated.

Finally, the SMRS concept has great implications for the debate over "species selection" (see Vrba and Eldredge [1984] for review of various conceptualizations of "species selection"). Species, as entities with internal coherence and finite spatio-temporal distributions, arise and are maintained as a consequence of organismic reproductive behavior. Despite assertions to the contrary (most recently, Mayr 1982), I have argued at length that species do not constitute parts of economic (ecological) systems (Eldredge 1985: ch. 6, 1986, 1989: ch. 5). If it is so that species *are* "more-makers" (new species arise from old), but not "interactors" (not being parts of economic systems), then, whatever species selection might be, it cannot constitute an exact analogue of natural selection. Natural selection entails effects of economic success on reproductive success, and occurs because organisms are both (economic) "interactors" *and* reproducers possessing genic "replicators"—to use the terms Hull (1980) maintains are necessary for any selective process to occur.

Thus selection at the level of species must be much more nearly like "sexual" than "natural" selection at the organismic level. Apparent differences in rates of speciation and extinction among subclades within monophyletic taxa constitute the sorts of patterns that a hypothesis of species selection would be adduced to explain. In sexual selection, organisms remain interactors, though in a reproductive rather than economic sense. It is rather difficult to imagine species *as entire entities* "interacting" with other species. Indeed, speciation is most generally a pattern of fission rather than fusion—much more analogous to asexual than sexual reproduction. Thus selection at the species level is not fully comparable with sexual selection at the organismic level, either.

Vrba (e.g. 1984) has argued (convincingly, I believe) that selection at any specified level must also entail properties emergent at that level. Economic and reproductive phenotypic properties are "emergent" from the genome at the organismic level. The point I wish to raise here, in concluding this survey of the important implications of the SMRS concept to current debates in evolutionary biology, is that the SMRS constitutes perhaps the only true intrinsic species-level property yet to be posited (Eldredge 1989). Until recently, I was convinced that there simply are no species-level properties (apart from spatio-temporal distributions, which are extrinsic, rather than intrinsic, features of species). However, it seems clear that the SMRS is an adaptive device distributed among two classes (males and females) within a species; it is specieswide and species-specific.

And, as Paterson (1985) and others have already begun to explore, some mate recognition systems are more readily fragmented, or simply subject to change, than are others. In other words, characteristic stabilities of the SMRS are likely to vary among a series of species, providing a bias to rates of origination of new reproductive communities. Some lineages have species that show naturally higher rates of increase (speciation) than others—obviously an ingredient of "species selection."

No one to my knowledge has yet argued that species extinction arises from reproductive failure caused by some sort of flaw in the breeding system itself; should such a model be developed, that would constitute, of course, another component of true "species selection." But it seems far more reasonable to continue to postulate that species extinction results from the economic difficulties (deaths, or difficulties leading to failure to reproduce) of its constituent organisms. There is economic property comparable to the SMRS that can be cited as a specieswide, species-specific, species-*level* property, affording internal cohesion and biasing species birth and death rates. Most patterns of differential birth and death of species within monophyletic taxa continue to appear to me more simply and reasonably attributable to Vrba's (1980) "effect hypothesis" than to "species selection" (see Eldredge [1989] for further discussion).

Hugh Paterson's notion of species mate recognition, then, in my view underscores the necessity of distinguishing between reproductive and economic features of organisms—a distinction that has many direct implications for our views of the structure of biotic nature and of the causal interactions among biotic elements. The concept will undoubtedly continue to shed light on further elements of evolutionary theory.

Summary

Hugh Paterson's notion of the SMRS represents a significant advance in the analysis and clarification of various issues pertaining to the nature of species. I have argued here that Dobzhansky (1937a,b) and Mayr (1942) came to the view that species are primarily reproductive communities as a consequence of their recognition that organismic morphological diversity does not form a smoothly continuous array, but is, rather, discontinuous. Natural selection would be expected to produce continuity—hence the conclusion that reproductive discontinuity underlies disruption of morphological continuity generally.

Paterson's Recognition Concept of species shares with the earlier Dobzhansky-Mayr Biological Species Concept the notion that species are reproductive communities. The two concepts differ in two important ways: (1) Dobzhansky's "reinforcement" model envisioned selection *for* reproductive isolation (though Mayr never embraced this model closely), while Paterson's point is that selection is for continued mate recognition in allopatry, and speciation is an accidental by-product, an *effect,* of such selection.

Perhaps more significantly, (2) Paterson's notion of the SMRS triggers recognition of the distinction between organismic economic and reproductive features. Mayr's basic model of speciation is that any genetically based change in allopatry may lead to speciation ("reproductive isolation"). Paterson's notion, in contrast, sees change in the reproductive adaptations of a species—the SMRS—as both necessary and sufficient for speciation to occur. Paterson's Recognition Concept of species opens the door to a more general consideration of the relation between reproductive and economic stasis and change in phylogenetic history (Eldredge 1989).

The distinction between economic and reproductive organismic features has further implications for several current issues in evolutionary biology. When reproductive adaptations (i.e., the SMRS) are more widely distributed than economic features, phylogenetic systematists see the SMRS as symplesiomorphous and thus of no use in the delineation of ("species-level") taxa. In contrast, evolutionary biologists see species as arising from the sexual reproductive activities of organisms; species are self-defining through the reproductive activities of their component organisms via their shared SMRS. Both views are "evolutionary," and each has merit: there really are two conflicting ways of describing biotic nature in such instances.

Sexual selection is a process distinct from natural selection: it is differential reproductive success based on variation of reproductive attributes. I have argued in this chapter that the SMRS concept expands appreciation of the nature of sexual selection, including the recognition of a distinction between directional and stabilizing sexual selection. Finally, though the analogy is not exact, species selection seems much more nearly similar to sexual selection than to natural selection in the strict sense.

Acknowledgments

I thank Joel Cracraft and Michael Ghiselin for bringing their special perspectives to bear in reviewing this manuscript. My respect for the points of view of each of them is great, though I reserve the right, of course, to differ. I thank, too, the editors of this volume for providing me with the opportunity to explain the nature of the intellectual debt I owe to Hugh Paterson.

References

Chandler, C.R., and M.H. Gromko. 1989. On the relationship between species concepts and speciation processes. Syst. Zool. 38:116–125.

Cracraft, J. 1989. Speciation and its ontology: The empirical consequences of alternative species concepts for understanding patterns and processes of differentiation. Pp. 28–59 *in* D. Otte and J.A. Endler, eds. Speciation and its consequences. Sinauer, Sunderland, Mass.

Darwin, C. 1859. On the origin of species. John Murray, London.

———. 1871. The descent of man, and selection in relation to sex. John Murray, London.

de Queiroz, K., and M.J. Donoghue. 1988. Phylogenetic systematics and the species problem. Cladistics 4:317–338.

Dobzhansky, T. 1937a. Genetic nature of species differences. Am. Nat. 71:404–420.

———. 1937b. Genetics and the origin of species. Reprint ed., 1982. Columbia University Press, New York.

———. 1951. Genetics and the origin of species. 3d ed. Columbia University Press, New York.

———. 1977. Species and their origins. Pp. 165–194 *in* T. Dobzhansky, F. Ayala, J.W. Valentine, and G.L. Stebbins. Evolution. W.H. Freeman, San Francisco.

Eldredge, N. 1985. Unfinished synthesis: Biological hierarchies and modern evolutionary thought. Oxford University Press, New York.

———. 1986. Information, economics and evolution. Annu. Rev. Ecol. Syst. 17:351–369.

———. 1989. Macroevolutionary dynamics. Species, niches and adaptive peaks. McGraw-Hill, New York.

Futuyma, D.J. 1987. On the role of species in anagenesis. Am. Nat. 130:465–473.
Ghiselin, M.T. 1974. A radical solution to the species problem. Syst. Zool. 23:536–544.
Hull, D.L. 1980. Individuality and selection. Annu. Rev. Ecol. Syst. 11:311–332.
Mayr, E. 1942. Systematics and the origin of species. Reprint ed., 1982. Columbia University Press, New York.
———. 1963. Animal species and evolution. Harvard University Press, Cambridge, Mass.
———. 1982. The growth of biological thought. Harvard University Press, Cambridge, Mass.
———. 1988. The why and how of species. Biol. Philos. 3:431–441.
Nelson, G. 1989. Species and taxa: Systematics and evolution. Pp. 60–81 *in* D. Otte and J.A. Endler, eds. Speciation and its consequences. Sinauer, Sunderland, Mass.
Otte, D., and J.A. Endler, eds. 1989. Speciation and its consequences. Sinauer, Sunderland, Mass.
Paterson, H.E.H. 1978. More evidence against speciation by reinforcement. S. Afr. J. Sci. 74:369–371.
———. 1980. A comment on "mate recognition systems." Evolution 34:330–331.
———. 1981. The continuing search for the unknown and the unknowable: A critique of contemporary ideas on speciation. S. Afr. J. Sci. 77:113–119.
———. 1982. Perspectives on speciation by reinforcement. S. Afr. J. Sci. 78:53–57.
———. 1985. The recognition concept of species. Pp. 21–30 *in* E.S. Vrba, ed. Species and speciation. Transvaal Museum Monograph No. 4 Pretoria.
Ryan, M.J., and W. Wilczynski. 1988. Coevolution of sender and receiver: Effect on local mate preference in cricket frogs. Science 240:1786–1788.
Templeton, A.R. 1989. The meaning of species and speciation: A genetic perspective. Pp. 3–27 *in* D. Otte and J.A. Endler, eds. Speciation and its consequences. Sinauer, Sunderland, Mass.
Vrba, E.S. 1980. Evolution, species and fossils: How does life evolve? S. Afr. J. Sci. 76:61–84.
———. 1984. What is species selection? Syst. Zool. 33:318–328.
———. 1985. Introductory comments on species and speciation. Pp. ix–xviii *in* E.S. Vrba, ed. Species and speciation. Transvaal Museum Monograph No. 4 Pretoria.
Vrba, E.S., and N. Eldredge. 1984. Individuals, hierarchies and processes: Towards a more complete evolutionary theory. Paleobiology 10:146–171.

Author Index

Aarssen, L. W., 195, 218
Abrams, P. A., 201, 219
Adak, T., 101, 237
Ahmad, A., 295
Aidley, D., 436
Aigaki, T., 323
Ajioka, J. W., 234, 235
Akers, R. P., 289, 292
Alberch, P., 41
Albon, S. D., 135
Alcock, J., 124, 134, 137, 263, 267, 299, 376, 420
Alexander, R. D., 412
Allan, S. A., 299
Allsopp, M., 173, 188, 363
Ambrose, W. G., 217, 221
Anderson, D. J., 223
Anderson, E., 390, 391, 405, 406, 412
Anderson, M. L., 173, 188, 363
Anderson, W. W., 134, 154–56, 344, 345, 356, 437
Andersson, M., 356, 357, 365
Andre, R. G., 236
Andrewartha, H. G., 201, 207, 211, 212, 216, 219
Angerilli, N. P. D., 299
Anglade, P., 275, 292
Annett, H. E., 101
Anxolabehere, D., 234, 237
Appel, T. A., 239–41, 257
Appelgren, M., 297
Arber, A., 240, 257
Archie, J. W., 358
Arkell, W. J., 59, 63, 67
Arn, H., 273, 292, 296, 314, 322, 323, 325
Arnold, S., 11, 41
Arnold, S. J., 341, 365, 366
Aronov, I. B., 419
Ashburner, M., 137, 154, 173, 227, 236, 358, 362, 365
Asquith, P. J., 72, 87
Austen, E. E., 101
Averhoff, W. W., 56, 134, 356
Avise, J. C., 31, 35, 41

Ayala, F. J., 13, 31, 36, 41, 62, 67, 69, 188, 220, 257, 277, 292, 342, 356, 376, 412, 476

Backman, J., 43
Baimai, V., 95, 98, 229, 236
Baker, B. S., 318, 322
Baker, H. G., 155
Baker, T. C., 264–69, 271, 273, 278–82, 289, 292, 293, 296, 299, 306, 307, 322, 336, 341, 346, 354, 357, 360, 377, 388, 418
Baldauf, J. G., 43
Balmer, J., 323
Balsubramanian, M., 295
Baltensweiler, W., 322, 323, 356
Barber, M. A., 92, 98, 99
Barker, J. R., 324
Barnes, M. M., 351, 364
Barosa, S., 295
Barr, H. J., 234, 235, 356
Barral, J., 295
Barrer, P. M., 268, 292, 337, 356
Barrion, A. A., 443, 462
Bartell, R. J., 325
Barton, N. H., 62, 67, 144, 147, 148, 150, 151, 153, 154, 159, 162, 167, 168, 172, 272, 277, 286, 292, 332, 340, 350, 351, 357, 377, 404, 412
Bass, H., 425, 435
Bastock, M., 457, 459
Bateman, A. J., 130, 134
Bateson, P. P. G., 127, 134, 355, 357
Bauer, M., 424, 428, 435
Becker, P. H., 283, 292, 429, 435
Beerbower, J. R., 58, 67
Beiles, A., 169, 172
Bekker, P., 101
Belkin, J. A., 90, 98
Bell, G., 8–10, 15, 22, 35, 41
Bell, W. J., 266, 292, 293, 322, 357
Belofs, W. L., 358
Belote, J. M., 318, 322
Benedix, J. H., Jr., 346, 357, 375, 412

479

Bengtsson, B. O., 298, 324, 325, 357, 364
Bentley, D. R., 131, 134, 375, 409, 412
Bently, D., 457, 459
Berger, A., 18, 41
Bergstrom, G., 297
Besansky, N. J., 234, 235
Bienz, M., 323
Bigelow, R. S., 379, 380, 412
Bingham, P. M., 233–35
Birch, L. C., 201, 207, 211, 212, 216, 219
Birch, M. C., 265, 266, 293, 300, 306, 322, 324, 325
Birgersson, G., 298
Birnstiel, M. L., 236
Bjostad, L. B., 266, 269, 275, 281, 284, 287, 290, 293, 297, 298, 306, 322, 324, 325, 364
Blacher, R. W., 298
Blair, W. F., 189, 345, 346, 357
Blum, M. S., 124, 134, 263, 293
Blum, N. A., 124, 134, 263, 293
Boake, C. R. B., 132, 136, 285, 293, 428, 431, 435
Bock, W. J., 59, 67, 381, 412
Bogert, C. M., 341, 345, 346, 357
Böhlen, P., 323
Bonde, N., 25, 41
Bonnell, M. L., 128, 134
Booij, C. J. H., 314, 322, 342, 346, 357, 430, 435, 458, 459
Borden, J. H., 299, 363
Borgia, G., 335, 357
Bosiger, E., 130, 134
Bourgeron, P., 137
Bourgois, M., 136
Boxshall, G. A., 42
Bradbury, J. W., 357
Brady, R. H., 201, 219
Brady, U. E., 388, 414
Breden, F., 347, 357
Brenowitz, E., 438
Bretagnolle, V., 435
Bridgman, P. W., 105, 122
Broecker, W., 41
Brophy, J. A., 18, 41
Brothers, D. J., 183–86, 188, 189
Brown, C. H., 425, 439
Brown, C. W., 391, 412
Brown, J. H., 195, 201, 219
Brown, J. L., 412
Brown, J. Y., 98
Brown, R. L., 266, 278, 298

Brown, W. L., 375, 381, 412
Brown, W. M., 418
Bruce-Chwatt, L. J., 92, 93, 98
Bruford, M. W., 257
Bryan, J. H., 95, 96, 98
Bryant, E. H., 127, 129, 134–36, 142, 154, 342, 356
Bryant, P. J., 132, 135
Bues, M., 292, 322
Bull, C. M., 55, 88, 89, 98–102, 135, 293, 357, 363, 364, 412, 414, 419
Burke, T., 249, 250, 252, 257
Burkholder, W. E., 326
Burnett, B., 457, 460
Burns, J. M., 341, 345, 366
Buser, H. R., 296, 323
Bush, G. L., 60, 67, 450, 455, 459, 460
Bush, M. B., 18, 41
Busnel, M.-C., 342, 357
Butlin, R. K., 13, 24, 28, 41, 62, 67, 155, 158–60, 163, 164, 167, 169, 171–74, 189, 277, 283, 285, 293, 298, 309, 310, 327, 331, 333, 335, 336, 338, 347, 348, 351, 354, 357, 359, 363, 375, 377, 381, 383, 406, 412, 419, 428, 431, 435
Byers, J. A., 413

Cade, W. H., 132, 135, 329, 357
Caffrey, D. J., 275, 293
Caithness, N., 15, 41
Campbell, B., 436
Campbell, B. G., 59, 67
Campbell, M. G., 295, 297, 324
Campbell, N. A., 104, 122
Campero, J., 295
Campion, D. G., 342, 357
Cancrini, G., 235
Cann, R. L., 97, 98
Cannon, H. G., 253, 257
Cantrall, I. J., 346, 358
Capranica, R. R., 341, 348, 363, 439
Cardé, A. M., 266, 268, 290, 293, 322, 358
Cardé, R. T., 264, 266, 267, 268, 270, 278, 284, 290, 292, 294, 298, 307, 311, 314, 322, 336, 337, 341, 346, 352, 354, 357, 358
Carlson, D. A., 323
Carlson, O. V., 462
Carroll, R. L., 58, 65, 67
Carson, H. L., 13, 24, 28, 31, 41, 62, 67, 123, 124, 128–32, 135, 137, 142–45, 147, 149–54, 177, 183, 184, 188, 277, 286,

293, 309, 335, 339, 343, 358, 362, 365, 376, 380, 412, 413
Carthy, J. D., 383, 413
Case, A. C., 67
Case, S. M., 337, 341, 358
Case, T. J., 219, 223, 375, 413, 420
Caswell, H., 4, 10, 41
Causey, O. R., 92, 98
Centner, M. R., 173, 188, 258, 416
Chakraborty, R., 155
Chamberlain, A., 66, 69
Chandler, C. R., 67, 413, 476
Chapin, F. S., 205, 212, 213, 219, 223
Chapman, P. J., 278, 293, 322, 356, 357
Charalambous, M., 358
Charlesworth, B., 62, 67, 147, 150, 154, 172, 219, 292
Charlton, R. E., 298, 299
Charmillot, P. J., 296
Cheetham, A. H., 46, 55
Chen, P. S., 119, 307, 323
China, W. E., 454, 459
Chomsky, N., 81, 88
Christensen, T. A., 294, 300, 323
Christophers, S. R., 98
Chubb, S. H., 60, 68
Churcher, C. S., 67
Claridge, M. F., 358, 413, 423, 435, 441, 442, 450, 451, 453, 454, 457, 459–61
Clark, P. F., 42, 224
Clarke, B., 328, 346, 363
Clarke, B. C., 407, 413
Clarke, C. A., 170, 172, 318, 323
Clarke, F. M., 170, 172
Classon, A., 362
Clearwater, J. R., 294, 295, 323
Clutton-Brock, T. H., 127, 135, 136
Coates, D. J., 237
Cocroft, R. B., 365
Cody, M. L., 198, 219
Coetzee, M., 90, 94, 95, 97–100, 175, 226, 236, 244
Cohen, M., 234, 235
Cohn, T. J., 346, 358
Coleman, W., 253, 257
Colinvaux, P., 18, 41
Collins, R. D., 268, 284, 294, 336, 337, 352, 358
Coluzzi, M., 95, 96, 98, 99, 102, 227, 235
Colwell, R. K., 194, 198, 199, 201–3, 219
Combs, L. M., 134, 154, 335
Comeau, A., 312, 325

Connell, J. H., 201, 207, 212, 213, 219, 289, 292, 295
Coope, C. R., 18, 41, 49, 55, 211, 212, 219
Corradetti, A., 92, 98
Cowling, D. E., 343, 354, 359, 457, 460
Coyne, J. A., 62, 67, 68, 125, 135, 142, 154, 158, 170, 172, 181, 188, 263, 277, 294, 308, 314, 317, 323, 327, 331, 348–51, 359, 377, 381, 382, 408, 413
Coz, J., 94, 99
Cracraft, J., 3, 11, 35, 37, 41, 59, 63, 68, 469, 476
Craig, A. J. F. K., 220, 224
Crane, J., 341, 346, 359
Crapon de Caprona, M., 435
Crews, D., 341, 345, 359
Crompton, R. H., 66, 68
Cronquist, A., 122
Crossley, S. A., 338, 359, 363
Crow, J. F., 147, 154
Crowe, T. M., 67, 68, 101, 157, 174
Cruz, Y. P., 447, 460
Csanyi, V., 41
Cumber, R. A., 409, 413
Cusson, M., 306, 323

Dagley, J. R., 342, 359
Darlington, C. D., 257
Darwin, C. R., 15, 41, 61, 181–83, 188, 216, 219, 328, 359, 376, 413, 472, 473, 476
Date, E., 13, 30, 66, 112, 114, 169, 331, 429, 435
Davey, T. H., 99
David, J. R., 137, 343, 361
Davidson, D. W., 219
Davidson, G., 93, 95, 99
Davies, N. B., 250, 257
Davis, M. B., 211, 219
Davis, N. T., 180, 188
Dawkins, R., 41, 242, 257
Day, M. F., 300
Day, T. H., 335, 359
Dayton, P. K., 216, 217, 219
De Buck, A., 90, 91, 101
De Jong, M. C. M., 331, 354, 359
de la Rosa, M. E., 134
de Meillon, B., 92, 99, 226, 236
d'Entremont, C. J., 170, 174
De Queiroz, K., 41, 476
De Salle, R., 33, 44, 373, 413
De Vries, H., 91, 99

de Vrijer, P. W., 442, 458, 462
De Winter, A. J., 338, 354, 359, 425, 435
Deane, L. M., 98
Deane, M. P., 98
DeBach, P., 214, 219, 409, 419
Demoulin, M. L., 92, 101
den Bieman, C. F. M., 442, 457-59
den Boer, P. J., 219
Den Hollander, J., 358, 435, 440, 442, 460
Denley, E. J., 217, 223
Dennet, N., 32, 42
Denton, G. H., 18, 19, 30, 41
Desprairies, A., 43
Di Deco, M., 98, 235
Diamond, J., 192, 197, 199, 201, 202, 207, 216, 219, 223
Diehl, S. R., 450, 460
Dilger, W. C., 250, 257
Dobkin, D. S., 212, 219
Dobzhansky, T., 13, 42, 95, 99, 154, 172, 177, 182, 188, 202, 220, 257, 294, 345, 359, 413, 440, 455, 460, 464-67, 475, 476
Doherty, J. A., 281, 283, 294, 428, 429, 431, 435, 436
Domingo, I. T., 442, 460
Donald, P. O., 127, 136, 336, 338, 361-63
Donitz, W., 92, 99
Donoghue, M. J., 25, 31, 41, 42, 44, 470, 476
Doolan, J. M., 378, 382, 383, 390, 405, 407, 408, 410, 413, 427-29, 436
Dooling, R., 423, 427, 436
Drake, J. A., 191, 197, 198, 203, 220
Drickamer, L. C., 258
Du, J.-W., 269, 293, 294, 297, 298, 311, 323-25, 364
Duffels, J. P., 383, 413
Dugdale, J. S., 273, 274, 294, 295, 313, 323, 367, 373, 375, 378, 379, 383, 389, 392, 396, 403, 413, 414, 425, 436
Duijm, M., 342, 359
Dunkelblum, E., 294
Dunning, D. C., 380, 388, 413
Dunsmuir, P., 237
Dybas, H., 424, 437
Dyson, M., 424, 436

East, T. M., 173
Eberhard, W. G., 125, 135, 137, 294, 331, 335, 366, 420, 458, 462
Eckenrode, C. J., 295, 298, 325, 364

Edwards, F. W., 91, 92, 99, 219-21
Ehrlich, P. R., 219
Ehrman, L., 127, 135, 316, 323, 334, 338, 339, 343-46, 356, 359, 360, 381, 413
Eisner, T., 266, 294
Eldredge, N., 13, 42, 45, 55, 58, 61-63, 66, 68, 70, 286, 294, 464, 468, 470, 471, 473-77
Ellison, J. R., 56, 234, 235
Elmfors, A., 297
Elton, C., 197, 202, 220
Emerson, S., 41
Emlen, S. T., 333, 360, 429, 430, 436
Endler, J. A., 3, 41, 43, 125, 135, 136, 153, 154, 159, 172, 173, 193, 194, 220, 222, 335, 337, 341, 347, 357, 360, 361, 364, 407, 412, 413, 415, 418, 420, 464, 468, 476, 477
Engels, W. R., 129, 135, 233-35
Enger, P. S., 423, 436
Enger, R., 80, 88
Eoff, M., 338, 339, 360
Esbjerg, R., 292, 322
Evans, A. M., 91, 92, 99
Evans, L., 435
Ewart, A., 375, 379, 408, 414
Ewing, A. W., 357, 382, 414, 457, 460

Fabian, B., 286, 294
Fabrick, A., 426, 436
Fairchild, L., 461
Fairweather, P. G., 223
Fales, H. M., 298
Falleroni, D., 91, 99
Falls, J., 429, 436
Farris, J. S., 227, 236
Faustini, D. L., 326
Felsenstein, J., 172
Fenster, G., 132, 135
Ferguson, G. W., 360
Ferguson, J. W. H., 173, 178, 184, 188, 363
Finnegan, D. J., 234, 236
Fisher, R. A., 127, 135, 294, 328, 354, 360, 377, 414
Fitzpatrick, J. W., 109, 112, 122
Fleming, C. A., 48, 55, 367, 371, 373, 375-79, 383, 390, 392, 396, 403, 409, 410, 413, 414, 425, 436
Flint, H. M., 268, 273, 295
Forbes, J. C., 452, 461
Ford, E. B., 155, 318, 323, 335, 360
Forey, D. E., 295

Foss, L., 248, 258
Foster, S. P., 273–75, 294, 296, 301, 306, 307, 311–13, 323, 324
Foucault, M., 77, 88
Fouquette, M. J., Jr., 360
Fox, K. J., 122, 367, 414
Franklin, I. R., 149, 154
Frelow, M. M., 420
Frérot, B., 270, 300, 311, 326
Fritzsch, B., 41, 439
Fryer, G., 31, 42
Fullard, J. H., 278, 298
Furet, I., 460
Furth, D. G., 375, 415
Futch, D. G., 345, 360
Futuyma, D. J., 13, 24, 28, 42, 68, 141, 154, 172, 188, 192, 194–96, 220, 263, 269, 277, 294, 295, 323, 359, 375, 377, 381, 390, 391, 413–15, 471, 477

Gall, G. A. E., 338, 360
Ganyard, M. C., 388, 414
Gaso, M. I., 134
Gaston, K. J., 205, 221
Gaston, L. K., 296, 360
Gatesy, J. E., 17, 33, 44
Gaud, W. S., 219, 222, 224
Gauiter, T., 324
Gaunce, A. P., 299
Gayou, D., 423, 424, 436
Gentle, T., 324
Geoffrion, S. C., 323
George, T. N., 59, 68, 418
Gerhardt, H. C., 337, 360, 361, 436
Ghauri, M. S. K., 447, 461, 462
Ghiselin, M. T., 8, 9, 13, 20, 27, 28, 42, 55, 464, 477
Gibbins, E. G., 92, 99
Gibbs, G. W., 373, 392, 414
Gibson, J. B., 339, 365
Giebultowicz, J. M., 307, 323
Giles, G. M., 55, 90, 92, 99, 101, 236
Gillett, J. D., 100
Gillham, M. C., 453, 455, 458, 459, 461
Gillies, M. T., 90, 94, 99, 226, 236
Gingerich, P. D., 62, 68
Glenn-Lewin, D. C., 220
Glover, T. J., 270, 275, 280, 281, 283, 284, 287–89, 295, 298, 299, 314, 325, 364
Gockel, C. W., 92, 98
Goldschmidt, R., 317, 318, 323

Goodnight, C. J., 135
Goodwin, B. C., 241, 253, 258, 259
Gordon, D. H., 32, 42
Gordon, R. M., 92, 99
Gosalvez, J., 173, 412
Gothilf, S., 294
Gould, S. J., 31, 42, 45, 50, 55, 61, 66, 68, 249, 258, 286, 294, 316, 323, 408, 414
Graham, R. W., 211, 212, 220
Grant, A. J., 282, 295
Grant, P. R., 207, 220, 348, 360, 375, 414, 419
Grant, V., 380, 406, 414
Graur, D., 338, 360
Green, C. A., 95, 99–101, 226, 227, 231–33, 236
Green, C. J., 273, 295, 313, 323
Green, M. B., 297
Greenacre, M. J., 42
Greenfield, M. D., 295
Grimaldi, D. A., 135
Grimm, E. C., 211, 212, 220
Grine, F. E., 207, 220
Gromko, M. H., 62, 67, 377, 413, 466, 476
Gross, M. R., 17, 45, 132, 136, 226, 379, 392, 429
Grubb, P. J., 208, 220
Grula, J. W., 285, 296
Guerin, P. M., 269, 292, 296, 313, 322, 323
Guerrero, J. L., 324
Guiness, F. E., 135
Guzman, J., 134
Gwynne, D. T., 298, 377, 383, 389, 405, 414

Hackett, L. W., 91, 99
Haddow, A. J., 92, 100
Hagiwara, S., 414
Hahn, J., 415, 436, 461
Hailman, J. P., 104, 105, 106, 122
Haldane, J. B. S., 174, 210, 220
Hall, B. G., 75, 88
Hall, J. C., 132, 136, 169, 173, 343, 365
Hall, M., 43
Halliburton, R., 338, 360
Halstead, B., 71, 74, 88
Hammack, L., 342, 360
Hammond, A. M., 295, 297
Hamon, J., 94, 99, 100
Hansson, B. S., 273, 285, 288, 292, 296–98, 314, 323–25, 341, 360, 362, 364
Hao, C., 416

Harbach, R. E., 98
Harpaz, I., 294
Harper, A. A., 154, 219, 222, 223, 277, 296
Harris, M. O., 276, 306, 307, 323, 324
Harrison, B. A., 90, 100, 226, 236
Harrison, G., 91, 100
Harrison, R. G., 160, 173, 375, 403, 406, 415
Hartl, D. L., 234, 235
Hartley, J. C., 408, 415
Haslam, D., 447, 448, 460, 461
Hassell, M. P., 208, 211, 220
Hastings, J. M., 415
Hatchwell, B. J., 257
Hayes, D. K., 298
Haynes, K. F., 268, 273, 287, 296, 299, 336, 337, 354, 360
Hays, J., 41
Hays, J. D., 18, 42
Heady, S. E., 375, 382, 409, 415, 442, 461
Hecht, M. K., 62, 68, 220
Heck, K. L., 201, 204, 220
Hedgecock, D., 41, 292
Heinrichs, E. A., 442, 460–62
Heiser, C. B., Jr., 390, 406, 415
Heisler, I. L., 339, 361
Heller, H. C., 122
Henderson, N. R., 345, 361, 362, 430, 436
Hendrichs, H., 77, 88
Hendriske, A., 296
Hengeveld, R., 196, 220
Henry, C. S., 221, 253, 342, 361, 366, 383, 415, 420
Henwood, K., 426, 436
Herbert, T., 18, 43
Heth, G., 172
Hewitt, G. M., 129, 136, 157, 159, 160, 162, 167, 168, 172–74, 272, 277, 292, 296, 298, 310, 338, 340, 350, 351, 357, 377, 404, 412
Heyneman, D., 174
Hicks, E. P., 99
Highton, R. B., 100
Hilborn, R., 203–5, 220
Hildebrand, J. G., 294, 300, 319, 324, 325
Hill, A. S., 276, 293, 296, 299, 322, 358
Hill, K. G., 344, 361, 383, 421, 423, 427, 428, 439
Himmelfarb, G., 247, 248, 258
Hochberg, M. E., 195, 220
Hoffman, A., 62, 68
Hoikkala, A., 136, 343, 361

Hölldobler, B., 415
Homberg, U., 324
Homrighausen, R., 43
Hong, Z., 416
Hopkins, G. H. E., 91, 100
Horder, T., 41
Horn, B. R., 295
Houde, A. E., 341, 347, 361
Howard, D. J., 342, 346, 357, 361, 375, 377, 391, 412, 415
Howard, R. W., 362
Howden, H. F., 342, 361
Hoy, R. R., 131, 132, 134, 136, 150, 154, 375, 390, 409, 412, 415, 428, 429, 436, 450, 457, 459, 461
Hubbs, C., 391, 415
Huber, F., 378, 379, 383, 390, 405, 410, 415, 420, 427, 428, 436, 437
Hubricht, L., 390, 412
Huddlestun, P., 43
Hughes, A. J., 201, 221, 253, 258, 296
Hull, D. L., 13, 42, 464, 474, 477
Hulley, P. E., 193, 201, 204–7, 216, 221, 224
Humpage, E. A., 173
Hunt, R. E., 273, 287, 296
Hunt, R. H., 100, 226, 227, 236, 244
Hunter-Jones, P., 357
Huntley, B., 18, 42
Hurst, C., 438, 439
Hutchinson, G. E., 199, 204, 221
Hutter, P., 170, 173
Huxley, J., 57, 68, 155, 250, 253, 258
Huyton, P., 323

Ichikawa, T., 441, 461
Idoji, H., 361
Ikeda, H., 339, 343, 361, 430, 436
Ikeda, K., 72–74, 76, 80–83, 85–88
Iles, T. D., 31, 42
Imanishi, K., 71–74, 76, 78, 80, 81, 84, 88, 89
Imbrie, J., 41, 42
Imperato-McGinley, J., 318, 324
Inoue, H., 442, 447, 448, 461
Ishii, S., 441, 461
Ito, Y., 72, 88

Jackson, E., 93, 99
Jackson, J. B. C., 46, 55
Jaffe, H., 298
Jallon, J.-M., 343, 361

James, S., 31, 43
Jansen, E. J., 189
Janssen, T., 438, 439
Jantsch, E., 253, 258
Jardine, I., 324
Jeanne, R. L., 315, 325
Jiang, J.-C., 377, 408, 415, 416
Joerman, G., 407, 416
John, B., 185, 188, 450, 461
Johnson, C. R., 214, 221
Johnson, L. L., 366
Johnson, M. S., 328, 346, 361
Jones, J. S., 224, 357, 412
Josephson, R. K., 416, 421
Joysey, K. A., 59, 68
Jurenka, R. A., 287, 299

Kaltenbach, A. J., 43
Kammer, A. E., 324
Kaneshiro, K. Y., 133, 136, 141, 142, 151, 153, 154, 416, 437
Kanmiya, K., 342, 361, 377, 418
Karandinos, M. G., 295
Kastritsis, P. A., 359
Katsuki, Y., 407, 416, 423, 437
Kauffman, S., 432, 437
Kawanishi, M., 339, 343, 345, 361
Kearns, P. W. E., 336, 338, 361
Kedes, L. H., 236
Keene, J. B., 43
Keim, P., 298
Kemp, A. C., 263, 277, 279, 286, 296, 431, 434, 437
Kemp, T. S., 68
Kempoe, T. G., 298
Kent, D. V., 43
Kenworthy, J. B., 452, 461
Keough, M. J., 217, 221
Key, K. H. L., 159, 173, 181, 182, 188
Khail, A. F., 295
Kijchalao, U., 98
Kilburn, R. N., 47, 55
Kimbel, W. H., 3, 32, 42
Kimura, M., 147, 154
King, G. G. S., 299
King, M. C., 31, 42
King, R. C., 188
Kingan, T. G., 319, 324
Kingett, P., 437
Kirkpatrick, M., 330, 361, 362, 437
Kitagawa, O., 364
Klein, T., 236

Klein, U., 420
Kleindienst, H.-U., 415
Klump, G. M., 337, 361
Klun, J. A., 275, 296, 298
Knodel, J. J., 295
Koenig, H. E., 41
Koepfer, H. R., 336, 339, 345, 346, 361, 362, 366, 375, 406, 420
Koopman, K. F., 338, 362
Koref-Santibanez, S., 343, 362
Kortmulder, K., 78, 88
Krasnoff, S. B., 265, 266, 296, 306, 307, 324
Krebs, J., 40, 429, 437
Kruger, E. L., 342, 362
Krumsieck, K. A. O., 43
Krusteva, A., 325
Kuhlow, F., 93, 100
Kukla, G., 41
Kurten, B., 58, 59, 68
Kyriacou, C. P., 173

Lacey, M. J., 292, 356
Lachaise, D., 137
Lack, D., 375, 416
Lack, E., 436
Lakovaara, S., 342, 343, 362
Lambert, D. M., 28, 42, 46, 55, 95–97, 100, 103, 121, 125, 127, 136, 155, 157–59, 173, 174, 178, 188, 189, 201, 221, 244, 246, 249, 253, 256, 258, 259, 263, 264, 267, 277, 296, 297, 299, 303, 304, 309, 313, 315, 317, 324, 331, 338, 345, 361, 362, 363, 365, 375, 377, 379, 382, 388, 405, 411, 416, 419, 420, 422, 423, 424, 430, 432, 434, 436, 437
Lande, R., 13, 24, 28, 42, 128, 131, 133, 135, 136, 144, 145, 155, 219, 330–33, 335, 339, 350, 351, 353, 354, 358, 362, 376, 416, 433, 437
Lane, D. H., 367, 375, 383, 388, 390–92, 405, 406, 409–10, 416, 417
Langley, P. A., 323
Lanier, G. N., 342, 362
Lanne, B. S., 208, 324, 362
Larsen, O., 425, 426, 437
Latimer, W., 407, 417
Lawton, J. H., 195, 197, 198, 201, 205, 220, 221
Le Boeuf, B. L., 136
Lee, H., 217, 221
Lehman, J. T., 215, 218, 221
Lemeunier, F., 236

Lemon, R., 435
Le Quesne, W. J., 453, 461
Leroy, Y., 409, 417
Lester, R., 357
Le Sueur, D., 101
Levey, B., 28, 42, 267, 296
Levine, L., 134
Levinton, J. S., 52, 55
Lewis, H., 406, 417
Lewis, K. R., 185, 188
Lewis, T., 137, 366, 414, 415, 420, 462
Lewontin, R. C., 154, 155
Li, C. C., 181, 188
Liebhold, A. M., 362
Lienk, S. E., 278, 293
Liley, N. R., 335, 362
Lim, K.-C., 174
Lima-de-Faria, A., 256, 258
Lincoln, R. I., 40, 42, 416
Ling, K. C., 447, 461
Linn, C. E., 266, 268–71, 278, 280–82, 293, 295, 297, 298, 304, 306, 311, 314–16, 324, 325, 364, 388, 425, 433
Lints, F. A., 129, 136
Littlejohn, M. J., 333, 344, 346, 362, 417, 418, 437
Lloyd, E. A., 155
Lloyd, J., 383, 417
Lloyd, M., 424, 437
Loew, H., 92, 100
Loewenberg, B. J., 248, 258
Löfqvist, J., 292, 294, 296, 297, 324, 360, 362
Löfstedt, C., 269, 292, 294, 296–98, 300, 323–24, 357, 360, 362, 364
Loftus-Hills, J. J., 361, 417, 437
Loher, W., 442, 462
Lomnicki, A., 221
Lopez-Fernandez, C., 173
Lorenz, K. Z., 382, 417
Løvtrup, S., 245, 258
Lumme, J., 131, 136, 343, 361

MacArthur, R. H., 198, 221
MacFadyen, A., 221
Madden, L. V., 415
Madsen, H. F., 200
Mahon, R. J., 95, 100
Maini, S., 275, 296
Majerus, M. E. N., 336, 339, 342, 362, 363
Makela, M. E., 46, 55
Mallet, J., 412

Mann, K. H., 214, 221
Manning, A., 457, 462
Margulis, L., 325
Markow, T. A., 346, 362
Marks, R. W., 145, 146, 155
Marler, P., 250, 258
Martin, A. A., 391, 407, 417
Martin, L. B., 3, 42
Maruo, O., 339, 343, 361
Maruyama, T., 418
Mason, G. F., 99
Mastbaum, O., 92, 100
Masters, J. C., 67, 68, 121, 141, 151, 155, 158, 173, 178, 185, 188, 196, 221, 331, 332, 363, 377, 382, 417, 424, 433, 435, 437, 439
Matsumoto, S. G., 325
Matthew, W. D., 60, 68
Matthews, R. W., 298
Mattingly, P. F., 93, 94, 100
Maturana, H. R., 258
May, R. M., 192, 194, 198, 199, 208, 211, 219, 220, 221, 222, 223
Mayer, G. C., 13, 24, 28, 42, 194, 220, 377, 414
Maynard Smith, J., 8, 9, 11, 35, 41, 42, 62, 64, 68
Mayr, E., 3, 13, 24, 28, 38, 39, 42, 43, 57–59, 61–63, 68, 90, 94, 95, 100, 104, 106, 108, 120, 122, 125, 136, 141–46, 151, 152, 155, 175, 177, 178, 180–84, 186–89, 194–96, 202, 208, 220, 222, 263, 264, 269, 272, 277, 298, 331, 363, 367, 371, 375, 376, 378–81, 390, 396, 404, 406, 407, 410, 412, 414, 417, 440, 455, 456, 462, 464–67, 473, 475, 477
McArdle, B. H., 155, 174, 189, 258, 299, 365, 419
McClelland, B., 437
McCommas, S. A., 136, 154
McDonald, J., 338, 363
McDonald, J. F., 129, 136
McDonnell, L. J., 418
McGregor, I., 98
McGregor, P. K., 427, 429, 433, 437
Mchunu, Z. M., 181, 189
McIntosh, R. P., 221
McKay, I. J., 173, 188, 363
McKenzie, J. A., 172
McLain, D. K., 346, 363
McLaughlin, J. R., 363
McNally, R. C., 4–9, 413

McNeil, J. N., 266, 209, 323
McPhail, J. D., 341, 347, 363
McPherson, D. G., 420
McVeigh, L. J., 357
Mecham, J. S., 177, 189
Medrano, F. G., 442, 445, 461, 462
Meffert, L. M., 129, 134, 154
Meinwald, J., 266, 294
Menge, B. A., 192, 221
Menken, S. B. J., 279, 298
Menn, J. J., 266, 298, 306, 319, 325
Mercot, H., 137
Meyer, W. L., 88, 292
Michaux, B., 29, 46–48, 55, 57, 65, 249, 258, 259, 297, 416, 420
Michelsen, A., 418, 427, 437
Micks, D. W., 99
Miethke, P., 100
Miles, S. J., 95, 99, 100, 110, 227, 236
Millar, C. D., 125, 136, 248, 249, 258, 309, 324, 338, 363
Miller, D. R., 338, 342, 363
Miller, E. H., 281, 283, 292, 298, 300, 435, 436, 439
Miller, J. G., 20, 22, 43
Miller, J. R., 267, 298, 311, 325
Minks, A. K., 341, 365
Mistrot, M. P., 360
Möller, A. P., 335, 363
Montagu, A., 201, 221
Montague, R. A., 319, 324
Moodie, G. E. E., 341, 363
Moore, T. E., 378, 408, 412, 415, 418, 420, 436
Moore, W. S., 328, 350, 363
Moran, C., 186, 189
Morgan, J. C., 358, 435, 459, 460
Moritz, C., 406, 418
Morris, D., 10, 22, 23, 43
Morris, G. K., 278, 298, 377, 414
Morton, A. C., 43
Morton, E. S., 425, 438
Morton, N. E., 154
Moulds, M. S., 407, 418
Muggleston, S. J., 294, 295, 323
Muirhead-Thomson, R. C., 92–94, 100, 101
Muller, H. J., 376, 418
Murray, B. G., 195, 203, 222
Murray, J., 159, 173, 328, 346, 363
Murray, J. W., 43
Myers, J. G., 371, 383, 409, 418, 428, 438

Nails, D., 173, 188, 363
Nakao, S., 377, 418
Narins, P. M., 341, 363, 424, 430, 438
Nault, L. R., 415, 459, 461
Nei, M., 96, 97, 144, 145, 155, 376, 418
Nelson, D., 429, 438
Nelson, G., 4, 21, 25, 35, 39, 43, 470, 477
Nesbitt, B. F., 357
Nevers, P., 233, 236
Nevo, E., 135, 172, 341, 347, 348, 363
Newberry, K., 177, 180, 181, 183–86, 188–90
Newcomb, R. D., 324
Nichols, R. A., 161, 168, 173, 331, 363
Nix, H., 17, 43
Nixon, G. A., 450, 451, 460
Nixon, K. C., 26, 44
Noble, I. R., 213, 219, 222
Nocke, H., 423, 424, 437, 438
North, R. C., 342, 363
Nouad, D., 364

Oberlander, H., 299, 326
O'Connell, R. J., 289, 292, 295
Odendaal, F. J., 329, 363
O'Donald, P., 136, 338, 361–63
Ogura, K., 407, 414
Ogushi, R., 71, 72, 88
Ohta, A. T., 343, 363
Olivera, O., 134
Olivieri, I., 210
Olsen, A. M., 221
Olsen, P. E., 18, 43
Olson, O. P., 47, 55, 192
Omori, N., 175, 179, 181, 183, 185, 186, 188, 189
Ono, T., 269, 298
Orians, G. H., 122, 208, 222
Orr, H. A., 68, 125, 135, 142, 154, 172, 277, 294, 323, 327, 348–51, 359, 413
Ossiannilsson, F., 462
Otte, D., 3, 41, 43, 125, 135, 136, 153, 154, 172, 173, 193, 194, 220, 222, 329, 346, 349, 357, 364, 375, 412, 415, 418, 420, 464, 468, 476, 477
Ovaska, K., 341, 364

Paine, R. T., 198, 216, 217, 222
Palazzi, C., 438
Pantazidis, A. C., 173
Paolinelli, P., 438
Pappas, C. D., 362

Pardue, M. L., 234, 236
Parent, M., 92, 101, 102
Park, J., 18, 43, 392, 414
Parsons, P. A., 127, 129, 135, 136, 339, 343, 360
Paterson, H. E. H., 3, 9, 12, 13, 22, 24, 28, 31, 37, 39, 43, 46, 49, 55, 56, 63, 64, 67, 69, 77, 88, 93, 94, 95, 96, 97, 100, 104, 122, 123, 124, 125, 136, 141, 150, 155, 158, 159, 171, 173, 174, 177, 178, 181, 182, 185, 188, 189, 193, 196, 202, 204, 206, 207, 209, 210, 211, 212, 216, 221, 222, 226, 237, 244, 246, 258, 263, 264, 267, 277, 279, 286, 297, 298, 303, 313, 317, 324, 325, 328, 329, 332, 333, 349, 353, 364, 375, 376, 377, 380, 381, 382, 388, 405, 406, 408, 410, 416, 418, 422, 423, 428, 432, 433, 437, 438, 440, 441, 455, 456, 459, 462, 464, 466, 468, 474, 477
Patton, J. L., 67, 349, 364
Paul, R. C., 131, 132, 136, 415, 436, 450, 461
Pavlovsky, O., 359
Payne, K., 300, 324, 325, 453, 461
Penna, M., 438
Perdeck, A. C., 334, 342, 364
Perez, N., 295
Peteet, D., 18, 43
Peterson, R. E., 324
Petit, C. P., 130, 137, 339, 343, 364
Petrarca, V., 98, 102
Phelan, P. L., 299, 377, 388, 418
Phillips, J. K., 326
Phillips, J. M., 324
Phillips, N. R., 268
Phillips, P. A., 364
Pianka, E. R., 201, 222
Piercy, J., 435
Piston, J. J., 362
Platnick, N., 43
Pollack, G., 429, 436
Pomiankowski, A., 331, 364
Poore, M. E. D., 197, 222
Pope, M. M., 296
Popov, A. V., 379, 405, 408, 415, 418, 419, 427, 428, 436, 438
Poppy, G. M., 293
Post, D. C., 315, 325
Potts, A. D., 173, 188, 363
Powell, J. R., 134, 142, 155, 359
Prakash, S., 345, 364

Preston, C. R., 233, 235
Price, P. W., 27, 36, 199, 201–3, 219, 222, 224
Priesner, E., 294, 313, 316, 322, 325, 351, 356
Pringle, J., 383, 407, 408, 419, 427, 438
Probber, J., 127, 135
Provine, W. B., 381, 418
Pumphrey, R. J., 407, 419
Purcell, A. H., 462
Purves, W. K., 104, 122
Pylka, J. M., 419, 439

Quann, A. G., 189

Rafatjah, H., 189
Ragge, D. R., 342, 344, 364
Rai, K. S., 236, 346, 363
Raina, A. K., 266, 298, 306, 307, 319, 323, 325
Rak, Y., 32, 42
Ralin, D. B., 341, 346, 364
Ramakrishnan, U., 447, 462
Ramer, J., 424, 438, 439
Rand, D. M., 160, 173, 377, 415
Rand, S., 424, 429, 430, 433, 438
Rao, K. R., 324
Rao, S. V., 409, 419
Rao, T. R., 226, 237
Ratcliffe, L. M., 419
Rattanarithikul, R., 95, 101
Raubenheimer, D., 68, 101, 174
Rauscher, S., 292
Raven, P. H., 67, 406, 412, 417
Rayner, R. J., 173, 188, 363
Reichman, O. J., 219
Reid, J. A., 226, 237
Reiter, J., 128, 136
Resh, J. A., 41
Reynolds, W. J., 342, 344, 364, 460
Rhodes, F. H. T., 59, 68, 422
Ribaut, H., 453, 454, 462
Ribbands, C. R., 93, 101
Rice, J. B., 98
Richards, D. G., 300, 438, 439
Richardson, M. L., 67
Richardson, R. H., 46, 55, 56, 134, 356
Richmond, R. C., 41, 292
Richter, A., 435
Ricklefs, R. E., 222
Ridgway, R. L., 298, 323
Riehm, J. P., 324

Riley, C. T., 298
Riley, H. P., 391, 419
Riley, P. A., 201, 222
Rind, D., 18, 43
Ringo, J. M., 331, 339, 343, 364–66
Rising, J. D., 350, 364, 380, 419
Ritchie, M. G., 163, 164, 169, 172, 174, 277, 285, 293, 298, 383, 406, 412, 428, 431, 435
Robbins, P. S., 295, 298, 325, 364
Roberts, D. G., 43
Robertson, A., 150, 155
Robertson, H. M., 422, 438
Robertson, J. G. M., 335, 364
Robertson, T. S., 461
Robson, T., 257
Roelofs, W. L., 265, 269, 271, 273, 275, 276, 278, 280, 281, 283, 284, 287, 289, 290, 292, 293, 295–99, 304, 306, 311, 312, 314–17, 322, 324, 325, 337, 351, 352, 364, 388, 425, 433
Rogers, C. R., 225, 237
Rohlf, F. J., 407, 419, 424, 439
Rollenhagen, T., 359
Romer, A. S., 58, 60, 69
Ronsseray, S., 237
Roote, J., 173
Rosen, D. E., 407, 419
Rosenblum, S. L., 358
Ross, Q. E., 41
Ross, R., 90, 92, 101
Rossiter, A., 74, 89
Rothenberg, K., 248, 258
Rothschild, G. H. L., 341, 365
Roughgarden, J., 192, 195, 197, 199, 201, 202, 207, 216, 219, 222, 223
Rubin, G. M., 237
Russell, E. S., 240, 241, 258
Ryan, M. J., 283, 298, 329, 336, 337, 347, 365, 382, 419, 423, 424, 426, 429–31, 433, 435, 438, 439, 472, 477

Sabelis, M. W., 331, 354, 359
Sadawa, N., 436
Saedler, H., 233, 236
Sagan, D., 321, 325
Sage, R. D., 162, 174
Sakuma, M., 461
Salceda, V. M., 134
Sale, P. F., 200, 213, 223
Saltzman, B., 41
Sanborn, A. F., 408, 419

Sanderson, N., 169, 172, 174, 419
Sappington, T. W., 338, 365
Saura, A., 342, 343, 362
Saussure, F. de, 86, 88, 89
Saxena, R. C., 443, 460, 462
Scanlon, J. E., 90, 100
Schilcher, F. von, 365
Schildberger, K., 415
Schlyter, F., 298
Schneider, D., 419
Schneider, H., 416
Schneiderman, A. M., 319, 325
Schnitker, D., 43
Schoener, T. W., 223
Schremmer, F., 383, 419
Schueler, F. W., 380, 419
Schull, J., 8, 43
Schwartz, J., 438
Schweber, S. S., 223
Scott, G., 425, 436
Searcy, W. A., 335, 365, 429, 439
Searle, J., 169, 174
Sebeok, T. A., 263, 299
Seger, J., 198, 199, 221
Selander, R. K., 134
Sergeeva, M. V., 419
Shackleton, N. J., 42, 43
Shambaugh, G. F., 461
Shani, A., 292, 356
Shannon, C. E., 267, 299
Sharma, V. P., 101, 236, 237
Sharp, B. L., 94, 101
Sharp, J. L., 27, 363
Shaver, G. R., 205, 219
Shaw, D. D., 186, 189, 233, 235, 237
Shaw, K. C., 342, 363, 442, 462
Shaw, P. W., 295
Sheldon, P. R., 50, 52, 53, 56
Sheppard, P. M., 172, 360
Shewell, G. E., 419
Shorey, H. H., 325
Short, L. L., 404, 419
Sibatani, A., 71–74, 77–82, 84, 85, 89
Silverstein, R. M., 362
Simberloff, D., 213, 223
Simmons, J. A., 378, 419, 423, 439
Simmons, P., 409, 419
Simon, C. M., 52, 55, 358
Simon, H. A., 3, 6, 40, 43, 303, 325
Simpson, G. G., 57, 59, 60, 69, 186, 189, 417
Sjögren, M., 297

Slatkin, M., 219, 272, 299
Slayter, R. O., 219
Slessor, K. N., 299, 363
Slobodchikoff, C. N., 219, 222, 224
Slooten, E., 437
Smith, D. B., 149, 150, 152–54
Smith, K. G., 408, 420
Smith, M. F., 349, 364
Smith, R. H., 220
Smith, R. L., 295
Smith, S. L., 341, 363, 424, 430, 438
Smith, W. J., 263, 267, 278, 281, 299
Sneath, P. H. A., 406, 419
Snir, R., 294
Sokal, R. R., 407, 419, 424, 439
Solis, R., 438
Soper, R. S., 383, 419
Southwood, T. R. E., 223
Spangler, H. G., 299
Spencer, H. G., 67, 68, 103, 121, 142, 145, 146, 148, 150, 151, 155, 158, 159, 171, 174, 178, 185, 188, 189, 246, 258, 277, 299, 331, 332, 348, 349, 363, 365, 377, 382, 417, 419, 431, 433–34, 437
Spiess, E. B., 130, 137, 334, 365
Spiess, L. D., 334, 365
Spieth, H. T., 150, 156, 331, 365
Spooner, J. D., 383, 419
Sprey, T. E., 77, 88
Sreng, I., 283, 299, 325, 365
Sreng, L., 299
Stanley, S. M., 24, 43, 44, 56, 69
Starck, M., 41
Stark, B. P., 365
Starr, C., 104, 122
Stearns, S. C., 15, 28, 43, 203–5, 220
Stebbins, G. L., 69, 155, 476
Stenesh, J., 313, 325
Stephen, R. O., 408, 415
Stevens, P. M., 324
Stewart, K. W., 342, 365
Stewart, T., 362
Stockel, J., 292
Stoneking, M., 98
Stoner, G., 347, 357
Stratton, G. E., 174
Strickland, A. F., 295
Strobel, E., 234, 237
Strong, D. R., 201, 221, 223
Strubing, H., 442, 462
Stumm-Zollinger, E., 323
Subbarao, S. K., 101, 237

Subchev, M. A. I., 296, 314, 323, 325, 360
Suga, N., 407, 416, 423, 437
Sullivan, B. K., 341, 365, 423, 439
Sutcliffe, A. J., 18, 43
Sutherland, L., 435
Swellengrebel, N. H., 90, 91, 101
Sylvester-Bradley, P. C., 69
Symes, C. B., 92, 101
Szabo, T., 436
Szczytko, S. W., 365
Szöcs, G., 292, 296, 322, 323, 360

Taggart, R., 104, 122
Takabatake, I., 361, 436
Takamura, T., 364
Takanashi, E., 364
Takemon, Y., 76, 79, 89
Tamaki, Y., 265, 299
Tang, J. D., 266, 299
Tang, X.-H., 295, 298, 325, 365
Tanida, K., 80, 89
Taper, M. L., 375, 420
Tattersall, I., 59, 66, 69
Taylor, C. R., 6, 43
Taylor, D. H., 437
Taylor, O. R., 131, 137, 285, 296, 338, 365
Teal, P. E. A., 266, 299, 326
Telford, S. R., 363
Templeton, A. R., 3, 9, 13, 14, 21, 23, 24, 25, 28, 35, 36, 37, 38, 43, 62, 67, 128, 137, 142, 143, 144, 146, 147, 148, 149, 150, 151, 152, 153, 154, 156, 263, 299, 339, 343, 356, 373, 376, 377, 381, 382, 410, 413, 420, 468, 477
Teplow, D. B., 324
Teramoto, L. T., 131, 135, 335, 339, 358
Theobald, F. V., 90, 92, 100, 101
Thibaud, G. R., 189
Thoday, J. M., 339, 365
Thomson, D. R., 92, 93, 94, 100, 101, 276, 299, 489
Thornhill, R., 124, 137, 263, 267, 299, 376, 420
Throckmorton, L. H., 142, 156
Tiivel, T., 224
Tilley, S. G., 341, 365
Tilman, D., 197, 204, 223
Tinbergen, N., 307, 326
Todd, J. L., 281, 299
Tomlinson, I. P. M., 361
Tompkins, L., 343, 365
Toms, R., 433, 439

Toolson, E. C., 408, 415
Tòth, M., 292, 296, 322, 323, 360
Toure, Y. T., 95, 102
Tracey, M. L., 41, 292, 356
Tsacas, L., 132, 137
Tumlinson, J. H., 299, 326, 363
Turelli, M., 332, 357
Turner, A., 29, 57, 66, 69, 316
Tuttle, M. D., 329, 365
Tyrell, D., 419

Ueshima, N., 180, 190
Uetz, G. W., 170, 174
Umesao, T., 72, 89
Underwood, A. J., 192, 197, 201, 216, 217, 223
Usinger, R. L., 175, 179, 180, 181, 186, 190

Vaisnys, R., 44
Vakenti, J. M., 276, 299
Val, F. C., 132, 137, 358, 373, 416
Van der Pers, J. N. C., 278, 279, 280, 297, 298, 300, 324, 362
Van Eeden, G. J., 101
Varela, F. J., 254, 255, 258
Vargo, A., 462
Vasantha, K., 101, 237
Velasco, M. V., 462
Veltman, C. J., 361
Verrell, P. A., 303, 326, 341, 365, 366, 377, 420
Vessey, S. H., 252, 258
Vigneault, G., 170, 174
Villet, M., 377, 420, 423, 424, 425, 426, 428, 429, 430, 433, 439
Vincent, C., 299
Vincke, I. H., 92, 102
Virdee, S. R., 160, 167, 174
Voerman, S., 314, 322
Vogt, W. G., 391, 420
Volney, W. J. A., 341, 362
von Helversen, O., 424, 435
Vrba, E. S., 17, 18, 19, 24, 27, 28, 31, 32, 33, 34, 41, 42, 43, 44, 47, 49, 56, 59, 66, 68, 69, 88, 101, 137, 142, 174, 186, 189, 190, 222, 258, 263, 264, 272, 277, 286, 294, 296, 298, 300, 303, 325, 364, 418, 437, 438, 439, 462, 466, 473, 474, 475, 477

Waage, J. K., 346, 366, 375, 420
Waddington, C. H., 242, 259

Wagner, G., 41
Wagner, W. E., 336, 337, 365, 382, 419, 423, 439
Wake, D. B., 41, 391, 420
Waldrop, B., 289, 300
Waldrop, M. M., 76, 89
Walgebach, C. A., 326
Walker, L. R., 212, 213, 223
Walker, T. J., 342, 345, 346, 366, 375, 377, 420
Wallace, A. R., 174
Wallschlager, D., 424, 439
Walpole, D. E., 175, 177, 179, 180, 183, 185, 186, 190
Walter, G. H., 193, 194, 195, 196, 198, 200, 201, 204, 205, 206, 207, 209, 210, 211, 212, 213, 215, 217, 221, 223, 224
Wang, Y.-B., 416
Waring, G. L., 377, 391, 415
Waser, P., 425, 439
Wasserman, M., 345, 346, 366, 375, 381, 406, 413, 420
Watanabe, A., 409, 414
Watanabe, T. K., 339, 343, 345, 361
Watson, G. F., 344, 346, 348, 362
Watson, M., 90, 99, 102
Watt, A. S., 211, 224
Weary, D., 430, 435, 439
Weaver, W., 267, 299
Webb, S. F., 357
Webb, T., 18, 42
Weber, T., 258, 408, 415, 420
Webster, G., 241, 258, 259
Webster, R., 266, 293
Webster, T. P., 341, 345, 366
Wei, K.-J., 44
Weinberg, E. S., 47, 95, 236
Weir, J., 362
Weissenbacher, B. K., 173, 188, 363
Wells, K., 425, 429, 439
Wells, M. M., 342, 361, 366, 383, 420
Wenner, A. M., 259
Westberg Smith, J., 43
West-Eberhard, M. J., 125, 137, 330, 331, 335, 366, 376, 420, 429
Wever, E. G., 419, 439
Wheeler, Q. D., 26, 44, 292
Whelan, P., 100
White, C. S., 55, 158, 256, 259, 267, 295, 297, 323, 324, 328, 377, 382, 416, 420
White, M. J. D., 145, 156, 172, 173, 182,

White, M. J. D. (*continued*)
  188, 190, 263, 300, 362, 379, 406, 417, 420, 455, 462
White, P. R., 306, 322
Wholers, D., 436
Wiens, J. A., 201, 203, 207, 224
Wiese, L., 10, 44
Wiese, W., 10, 44
Wijk, E., 297
Wilczynski, W., 365, 423, 424, 430, 437, 438, 439, 472, 477
Wiley, E. O., 44, 59, 63, 69, 70, 186, 190
Wiley, R. H., 278, 281, 283, 300, 425, 426, 429, 438, 439
Wilke, C. M., 339, 365
Wilkinson, P., 237
Williams, E. F., 341, 345, 359
Williams, G. C., 8, 9, 11, 44, 61, 70, 157, 174, 178, 182, 183, 190, 200, 204, 209, 224, 242–46, 253, 256, 259, 264, 300, 327, 366
Williams, J. L. D., 415
Williams, K. S., 408, 420
Williamson, P. G., 286, 300
Wilson, A. C., 31, 42, 67, 98, 174
Wilson, D. S., 204, 224
Wilson, E. O., 263, 300, 375, 381, 412
Wilson, M. R., 415, 453–55, 460, 461, 463
Wisotzkey, R. G., 129, 135
Witzgall, P., 270, 300, 311, 326
Wohlers, D. W., 415
Wolda, H., 373, 420, 421
Wolf, W. A., 275, 299, 322

Wood, D., 339, 343, 364, 366
Wood, D. L., 308, 326
Woodger, J. H., 253, 254, 259
Woodruff, D. S., 67, 172, 362, 380, 391, 403, 417, 421
Wool, D., 338, 360
Woolfenden, G. W., 109, 110, 111, 112, 115, 121, 122
Worthly, L. H., 293
Wright, J. W., 99, 418
Wright, S., 144, 148, 151, 156
Wu, C.-I., 418

Xu, M.-L., 416

Yager, D. D., 307, 324
Yan, L., 416
Yanev, K. P., 420
Yoon, J. S., 154, 343, 358
Yoro, T., 83, 89
Young, A. M., 378, 382, 383, 390, 392, 405, 407–9, 413, 416, 419, 421, 423, 427, 428, 436, 439
Young, D., 426, 439
Yusof, O. M., 375, 409, 421, 447, 448, 463

Zachar, Z., 233, 234, 235
Zalucki, M. P., 211, 212, 222, 224
Zanger, C. D., 413
Zimmerman, B. L., 347, 366
Zimmerman, H., 43
Zouros, E., 170, 173, 174, 366, 457, 463

Subject Index

Acer campestre, 453
Acris, 329, 346, 348, 352, 430; *A. crepitans*, 329, 341, 346, 348, 365, 430; *A. gryllus*, 341, 346, 348, 417, 438
Adalia, 336, 361; *A. bipunctata*, 336, 338, 342, 361
adaptation: for acoustic communication, 439; as change, 207, 468, 470, 471, 476; the concept in ecology, 193–95, 199, 217; current utility, 243, 246; defined, 242; devices, 182, 202, 206; and diversity, 203; economic, 473; to environmental conditions, 157, 329, 333, 355; as evolution, 204; increased, 129; "isolating mechanisms" as, 245; local, 430; to a new habitat, 264; niche, 14; as an operational approach, 106; patterns of, 221; peaks, 150, 466, 476; physiological, 211; recognizing an, 243, 244, 246; sexual selection and, 136; of the SMRS, 433–34; of species, 200, 208–10, 214, 215
Adoxophyes orana, 268, 296
Aedes, 346, 362, 363; *albopictus*, 342, 346, 362, 363
Aesculus hippocastanum, 453
Agapornis, 250
Agrotis segetum, 268, 292, 296, 297, 313, 322, 324, 326, 341, 360, 362
Alebra, 453, 454, 461; *albostriella*, 453, 454; *coryli*, 453; *wahlbergi*, 453, 454
Allonemobius, 415; *fasciatus*, 346, 415; *socius*, 342, 346
allopatric populations, 20, 23, 28, 119, 120, 171, 366, 458, 473
allozymes, 162
Alnetoidea alneti, 454
Alnus glutinosa, 454
Amalda, 47, 48, 55
ambient environment, 123, 129, 134
amelioration, 159, 168–71
Amphisalta: cingulata, 392, 409; *zelandica*, 392
anagenesis, 154, 220, 477
Ancilla, 55
Anobium punctatum, 322

Anolis, 358; *brevirostris*, 341, 345, 366; *distichus*, 337, 341, 345, 358; *marmoratus*, 341
Anopheles, 55, 90–92, 95–102, 225, 226, 235–37; *abranchiae*, 97; *annularis*, 231; *annulipes*, 99; *arabiensis*, 101; *atroparvus*, 97; *confusus*, 227; *costalis*, 92, 99; *coustani*, 99; *culicifacies*, 101, 236, 237; *dirus*, 98; *farauti*, 100; *funestus*, 97, 99, 236; *gambiae*, 55, 92, 95, 97–102, 235, 237; *hughi*, 97; *letabensis*, 97; *maculatus*, 95, 101, 227, 236; *maculipennis*, 91; *marshallii*, 97, 100; *merus*, 93; *parensis*, 227; *stephensi*, 235, 236; *superpictus*, 235; *tangensis*, 100; *theileri*, 231; *vaneedeni*, 227; *ziemanni*, 95
antelope, 17, 33, 41
Anura, 334, 340, 345, 362, 364, 417, 418, 421, 436, 438, 439
Aphelocoma c. coerulescens, 110, 121
aphids, 10, 22, 79, 80
Aphis fabae, 10
Aphytis, 419
apomixis, 39
apomorphic characters, 29, 227, 231
Archips, 293, 322, 346, 358; *argyrospilus*, 290, 293, 322, 346, 358; *mortuanus*, 312; *rosanus*, 276; *semiferanus*, 278
arctiid moths, 296, 306, 324
Argyrotaenia, 292, 336; *velutinana*, 267, 292, 298, 311, 325, 336, 337
armyworms, 306
arrhenotoky, 9, 39
asexual clones, 27
asexual species, 14
Auchenorrhyncha, 416, 420, 441, 459, 462
autapomorphic characters, 31, 32, 34
automixis, 8
autopoiesis, 255
autosomal modifiers, 131
autosomes, 131, 170, 172, 174, 227

balanced polymorphism, 127, 128
barriers to gene flow, 159, 327, 355
Baryspira, 55

493

Barytettix, 346, 358
bedbugs, 179, 180, 184–90
behavioral antagonists, 282, 288, 289
behavioral evolution, 133; explanation for, 171
beliefs, 62, 205, 242, 246, 247, 291, 308, 391
Belocephalus sabalis, 345
Betula, 450, 460; *pendula*, 450; *pubescens*, 450
biogeography, 20, 42, 43, 137, 413
biological control, 46
biological function, 126, 238
biological species concept. *See* isolation concept
biosociology, 72
biotypes, 369, 443, 460
Blackburnium, 361
body size, 347, 407, 424, 428, 430, 431, 438
bottlenecks, 127–29, 134, 142, 149, 150
brown planthoppers, 442
Bufo, 345, 365, 417, 438, 439; *americanus*, 345, 417; *amphimixis*, 8, 39; *compactilis compactilis*, 345; *microscaphus californicus*, 345; *punctatus*, 345; *valliceps*, 439; *woodhousii*, 341, 345, 365; *woodhousii fowleri*, 417
bush crickets, 408, 415, 417

cabbage looper, 273, 281, 287, 296
Cacoecimorpha, 270; *pronubana*, 270, 300, 326
Caenorhabditis elegans, 75
Caledia, 189; *captiva*, 233
Calopteryx, 346, 366, 420; *maculata*, 346
canonical discriminant analysis, 447, 454
Castanea sativa, 454
causal processes, 470
charcter displacement, 194, 207, 220, 276, 277, 291, 334, 348–50, 353, 360, 366, 375–77, 380, 406, 414, 417, 420
character states, 26, 29, 33, 34, 391, 406
Chen hyperborea, 119
Chlamydomonas, 10, 12, 44; *moewusii*, 10
Choristoneura, 296, 362; *occidentalis*, 341; *retiniana*, 362; *rosaceana*, 276, 296, 299
Chorthippus, 163, 172, 174, 298, 335, 336, 354, 357, 358, 364; *biguttulus*, 334, 344, 364; *brunneus*, 335, 336, 338, 342, 357, 358, 364; *parallelus*, 159, 163, 172–74, 298, 342, 359; *p. erythropus*, 172, 173; *yersini*, 342

chromosome homologies, 97
chromosome inversions, 96, 236
Chrysopa downsei, 415
Chrysoperla, 361, 366, 420; *plorabunda*, 361; *cataphractica*, 418
cicadas. *See Kikihia*
Cicadetta, 418; *sinuatipennis*, 418, 438
Cimex, 175, 182, 187, 189; *hemipterus*, 175, 177, 179–82, 187, 189; *lectularius*, 175, 182, 187, 189
clades, 10, 11, 12, 17, 21, 29, 30, 33–37, 231, 232, 428
cladogenesis, 264
cladograms, 30, 33, 34
Clarkia, 185, 417; *biloba*, 185
climate, 41, 43, 48, 50, 213, 220
clines, 162–64, 167, 172, 407, 413
Cnemidophorus, 413, 418
Cnemidopyge, 53
coadaptation: and mutual tuning, 457, 458; in the SMRS, 10, 22, 124, 143, 146, 149, 245, 255, 286, 317, 329, 333, 422, 423, 425, 431, 432
coadapted gene pools, 143, 146
Cochliomyia hominovorax, 342, 360
Coelopa frigida, 335, 359
coevolution, 219
coexistence, 93, 180, 195, 199–202, 212, 218, 348, 351
Colcondamyia auditrix, 419
Colias, 131, 137, 365; *eurytheme*, 137, 296, 338, 365; *philodice*, 137
colonization, 142, 220, 326
common ancestry, 103, 104, 109, 114, 117, 226, 371
community: ecology, 191–93, 197, 199, 201, 202, 204, 205, 210, 216, 217, 222, 223, 224; regularity, 205; structure, 191–93, 197, 198, 200–207, 210, 213, 214, 218, 221, 222, 223
competition. *See* interspecific competition; intraspecific competition; male-male competition
competitive exclusion, 193, 195, 196, 199, 200, 204, 207, 217, 219, 221, 222
complex systems, 3, 4, 6, 8–10, 35, 37, 38, 40, 76, 128, 303
congeners, 77, 78
connectivity as an operational statement, 111–20, 121. *See also* operationalism
Conocephalus, 298
contingencies, 210, 246
convergence, 135, 289

Corylus avellana, 453, 454
Cretaceous, 18, 43
Crinia, 417; *glauerti*, 346; *insignifera*, 346
Ctenopseustis, 273, 285, 294, 296, 313, 323; *obliquana*, 273, 294–96, 313
Cuvier, Georges, 238, 240–42, 249, 253, 257
Cystosoma, 405, 407, 413, 416, 419, 421, 436, 439; *saundersii*, 405, 407, 413, 416, 419, 421, 436, 439

Dacus, 414; *tryoni*, 420
Dalbulus, 415, 442, 461
Damaliscus, 34; *niro*, 34
Damalops, 34
Darwinian fitness, 135
Darwinism, 56, 68, 69, 89, 103, 248, 249, 259, 293, 420. See also evolutionary theory
DDT, 92, 94, 101, 175, 179, 180, 186, 214
definitions, 3, 8, 80, 103, 106, 312, 417; operational definitions, 107, 117
demes, 42, 130, 148, 473
demography, 112
Dendroctonus brevicomis, 307
density-dependence, 432
Desmognathus, 365; *ochrophaeus*, 341, 365, 366
differential reproductive success, 130, 472, 476
Dikraneura, 459
discontinuities in nature, 48, 465, 466, 475; explanations for, 96
discriminant analysis, 48, 447, 451, 453, 454
discriminant function analysis, 42, 296
disequilibrium, 129, 152, 169, 219
dispersal, 109, 113, 159, 160, 248, 348, 350
disruptive selection, 365
DNA fingerprinting, 112, 249, 257
dominance, 131
Doppler shift, 425
Drosophila: *ananassae*, 360; *arizonae*, 348, 352; *athabasca*, 349; *equinoxialis*, 342, 356; *e. caribbinensis*, 356; *funebris*, 235; *gaucha*, 340; *grimshawi*, 363; *heteroneura*, 31, 132, 352, 356; *immigrans*, 360; *littoralis*, 343, 361; *melanogaster*, 28, 130, 132, 134, 136, 137, 173, 234–37, 307, 322, 325, 331, 335, 336, 340, 352, 354, 356, 359–66, 430, 436, 460; *mercatorum*, 142, 361, 430, 436; *miranda*, 342; *mojavensis*, 170, 173, 174, 336, 346, 348, 362, 366, 420; *nasuta*, 136; *obscura*, 359, 362; *pallidosa*, 360; *paulistorum*, 334, 340, 346, 352, 359, 362; *pavani*, 340; *persimilis*, 365, 460; *pseudoobscura*, 130, 134, 136, 142, 338, 345, 356, 362–64; *pugionata*, 132; *silvestris*, 31, 130–32, 135, 137, 153, 355, 352, 353, 358; *simulans*, 170, 172, 338, 360, 361, 364, 366; *virilis*, 131, 136
dynamic systems, 4, 125

Ecdyonurus, 73
ecological processes: adaptation, 473; complexity, 193; stasis, 475
ecophenotypes, 28; variation, 15
electrophoretic distance, 349
Eleuthrodactylus coqui, 341, 347, 430
emergence, 6, 8, 76, 78, 83, 89, 388, 437, 474; operaitonal concept of, 10
Empoasca, 462
endothermy, 408
Enallagma, 438
Ensantina eschscholtzi, 412, 420
environmental change, 22, 23, 205, 317, 330, 473
Epeorus, 73
Ephemera, 76, 89; *strigata*, 76
Ephestia cautella, 268, 292, 337, 356
Ephippiger, 342, 357, 359
Epiphyas postvittana, 323
epistasis, 135, 143, 146, 147, 160, 167
epistemology, 3, 30, 73
Erythroneura, 459
essence, 39, 74, 121, 381
ethology, 378
Euchorthippus, 344, 364
Euphydryas editha, 220
European corn borer, 275, 287, 292, 295, 296, 298, 317, 324, 325, 351, 365
Euscelis, 462
Eve-population, 26
evolutionary by-products, 200, 225, 264
evolutionary theory, 44, 55, 71, 73, 88, 89, 412, 466, 467, 471, 475, 477. See also Darwinism
extinction, 11, 20, 134, 151, 158–61, 171, 184, 185, 235, 465, 474
extinct species, 27, 28, 32, 54

F1 sterility, 168
fecundity, 123, 189, 460, 472
fertile hybrids, 132, 181, 275
fertility, 349, 376

fertilization systems: in *Amalda*, 47; in bedbugs, 185, 187; and biological groupings, 54; characters and, 64, 66; common to individuals, 20, 39, 124, 178; confusion with other signalling systems, 349; to diagnose species, 22; divergence, 24, 264; facilitating successful mating, 464; function of, 382; and gene recombination, 183, 186; interactions, 9, 10; internal dynamics, 37; in motile animals, 180; natural selection maintains, 23; origination of new, 27; and phenotypic similarity, 14; SMRS as a subset of, 40, 47, 78, 381; stabilizing selection on, 12, 28. *See also* postfertilization; prefertilization
Fidicina rana, 436
fine-tuning, 282, 290, 291
fireflies, 383, 417
foraging, 78, 335
fossil record, 18, 28–33, 45, 47, 53–56, 58–60, 63–66, 68, 69, 103, 473
fossils, 28, 34, 44, 45, 48, 53, 56, 59, 63, 65, 69, 477
founders, 84, 104, 121, 135, 141–56, 431, 437, 439, 457, 458
functional interactions, 210, 213, 215
functionalism, 242, 258
function and history, 468

Galago, 68, 437; *crassicaudatus*, 437
Gambusia affinis, 415
Gasterosteus aculeatus, 341, 347, 363
Gastrophryne olivacea, 341
genealogy, 6, 11, 23, 31, 33
gene exchange, 40, 181, 334, 340, 351
gene pools, 127, 128, 143, 146, 149, 150, 154, 170, 184, 206, 272, 371, 422, 434
generalized selection argument, 244–45
genetic assimilation, 242
genetic drift, 14, 21, 38, 129, 133, 155, 160, 161, 330, 354, 376, 431, 433, 467, 468, 470, 473
genetic transilience, 143, 149, 150, 153
genetic variability, 123, 125–26, 127–34, 143, 155, 184, 283, 284, 309, 328, 332, 362
Geocrinia laevis, 344, 417
geographic isolation, 466
geographic variation, 273–77, 329, 330, 339–41, 344, 345, 352, 357, 359, 361, 362, 430–32, 435, 446, 458, 472
geography, 419, 466

Geospiza, 419
Gerris odontogaster, 338
good genes, 355
Grapholitha, 354; *molesta*, 268, 293, 322, 341
Gryllus: bimaculatus, 294, 435; *campestris*, 438; *commodus*, 417; *firmus*, 160; *integer*, 345; *oceanicus*, 417; *pennsylvanicus*, 160; *rubens*, 342
guilds, 195, 197, 198, 199, 204
guppies, 335, 347, 352, 353, 357, 360, 361
gynogenesis, 40

habitat: choice, 74, 161, 254; difference, 23, 37; segregation, 77; specificity, 15, 17, 18, 21–23, 34, 36, 37, 426
habituation, 281, 306
hair pencils, 307
Hamilton-Zuk hypothesis, 355
Haplochromis, 31
haplotypes, 143
Hawaiian crickets, 364, 418
Hawaiian *Drosophila*, 135–37, 142, 150, 154, 358, 363, 416
Heliothis, 299, 307, 323, 325; *zea*, 307, 325
heritability, 17, 131, 145, 269, 270, 284, 294, 336, 356, 358
heterogeneity, 134
heterogonic lineages, 10, 22
heterozygote superiority, 130
hierarchies, 6, 11, 40, 75, 430, 464, 471
Hirundo rustica, 363
holism, 6, 82, 83, 85
holospecia, 75, 76
Homalopteon, 54
homologies, 11, 25, 97, 106, 227, 240, 241, 258
homologous DNA sequences, 234
hormones, 250
host preference polymorphisms, 450
hybrids: interspecific, 419, 447; laboratory, 373; as offspring, 96, 184, 347, 348, 377; putative, 367, 380, 392, 396, 405, 410; songs, 373, 389, 409; swarms, 233, 396, 406, 415; zones, 159–64, 166–69, 172–74, 298, 363, 391, 418
Hydropsyche, 80, 89
Hyla: arborea, 346; *chrysoscelis*, 346; *cinerea*, 335, 360; *ewingi*, 340, 344, 346, 348, 362, 417; *eximia*, 341; *meridionalis*, 346; *versicolor*, 337, 341, 346, 436
Hyperolius marmoratus, 436

imagination, 246–48, 256
immanence, 85
inbreeding, 142, 147–49, 284, 433
incidental effects, 243, 264, 376
inevitability, 191, 194, 200–203, 208, 209, 212, 215, 245, 349, 465
insect communication, 415
insecticide resistance, 93, 95, 99, 445
interbreeding, 20, 32, 33, 34, 39, 40, 103, 104, 106–8, 119–21, 177, 373, 375, 376, 378–80, 388, 414, 443, 450, 451, 465, 469
intergeneric mating, 378
intersexuality, 318, 323
interspecific competition, 192–96, 199–202, 205, 207, 212, 213, 219, 224, 276, 278, 279, 291
interspecific mating, 180, 189, 190, 282, 377, 409
intraspecific competition, 130, 203, 291, 331
intraspecific variation, 208, 274, 347
introgression, 159, 161, 171, 179, 184, 189, 371, 373, 380, 389, 391, 396, 403, 404, 406
invariability, 309
inversions, 95–98, 130, 225, 226, 227, 231–33, 235, 236
Ips pini, 338, 342, 362, 363
Iris, 419
irreversible lineage branching, 4, 25, 37
isolation concept, 13, 125, 272, 373, 375, 381; adaptive utilization of the environment, 183; anomalies, 441, 469; *Anopheles gambiae* species complex, 94, 95, 97; botanists long opposed to, 470; and cicadas (*Kikihia*), 367, 377–82; conceptual difficulties, 411; and gene flow, 404; and hybrids, 410, 445; and mating, 267; and punctuated equilibrium, 64; and recognition concept, 66–67, 193, 263, 264, 440–41, 475

Javasella, 442, 462

katydids, 383, 419
Kikihia (cicadas), 367–421; *cutora*, 367, 368, 370, 373, 375, 392, 396, 405; *c. cumberi*, 367, 368, 369, 370, 371, 374, 377, 384, 395–411; *c. cutora*, 367, 368, 371, 374, 397, 402; *laneorum*, 367, 369, 371, 374–76, 384, 392, 393, 396–411;

muta, 378, 409; *m. muta*, 392, 394; *rosea*, 378; *scutellaris*, 383–84, 388, 392; *subalpina*, 367, 368, 371–77, 392, 396–411
Kobus, 17

lacewings, 361, 366, 420
larch budmoth, 313, 323
Laspeyrisia, 351; *pomonella*, 351, 364
Laupala, 346, 349
leafhoppers, 415, 421, 430, 431, 440, 441, 446, 450, 453, 454, 457–61, 463
leafroller moths, 273, 292, 295, 311, 323
Leersia, 442–45, 460, 462; *hexandra*, 442, 460–62
Lembeja brunneosa, 418
life history traits, 197
lifestyle partitioning, 73, 77–80, 84, 89
lineage branching, 4, 23, 34, 36, 37
linguistic theory: Chomsky, 81; langue/parole, 79, 81–82; Saussure, 79, 80–81
Lipara lucens, 342, 361
Litoria: ewingi, 340, 344, 346, 348, 362, 417; *paraewingi*, 340, 346
living fossils, 56
Lymantria dispar, 307, 318, 323

macroevolution, 66, 67, 155, 219, 412
Macrosteles fascifrons, 462
Magicicada, 378, 379, 388, 413; *cassini*, 413; *septendecim*, 413, 420
malaria: transmission, 57, 90–102, 179, 189, 225, 235, 236, 258; vectors, 96, 97, 101, 236
male-male competition, 355, 356
Manduca, 294, 319, 324; *sexta*, 289, 294, 300, 319, 324
Maoricicada, 379, 413, 436
Mastomys, 42
mate choice, 124, 333, 340, 346, 361, 363, 429, 430, 436, 438, 439, 443, 451, 460; experiments, 186, 333, 443, 449, 451; female choice, 127, 294, 335, 347, 361. *See also* interspecific mating
Meimuna kuroiwae, 418
meiosis, 39, 41
Melampsalta cruentata, 413
Milankovitch cycle curve, 18
mimicry, 172
mitochondrial DNA, 17, 33, 41, 98, 418
monophyly, 4, 25–27, 31, 35–37

morabine grasshoppers, 173, 181, 182, 184, 188, 190
morphology: change, 31, 45, 65; evolution, 133, 155; stasis, 50
Muellerianella, 342, 346, 357, 435, 458, 459; *brevipennis,* 342, 346; *fairmairei,* 342, 346
Musca domestica, 142, 342, 356
mutations: in adaptation of populations, 245; of beneficial gene, 108; in biosynthetic pathway of pheromone, 287; emergent properties, 8; evolution of biodiversity, 76, 77; and heterozygosity, 145; major gene moves to fixation, 128; neutral, 21; polygenic, 133, 332
mutual specific discrimination, 77

narrative explanations, 247, 252, 256
negative heterosis, 157–59, 167, 171, 185
neobiology, 76
neontology, 33, 57, 58, 63, 66
Nephotettix, 446–50, 460, 461, 462; *apicalis,* 461; *impicticeps,* 461; *nigropictus,* 342, 446–50, 460, 462; *virescens,* 342, 446–50, 460, 462
niche, 14, 38, 40, 196, 199, 264, 279
Nilaparvata, 354, 442–46; *lugens,* 336, 338, 340, 342, 357, 358, 431, 435, 442, 443, 446, 460, 461, 462
noisy environments, 291
nonadaptive, 330
norms of reaction, 15, 28, 32
Notus, 459

Oecanthus, 342, 346
Ogygiocarella, 53
Okanagana rimosa, 419
Omocestus, 364
Oncopsis, 450–52, 460, 461; *flavicollis,* 450, 460
Oncotympana maculaticollis, 416
oneness, 79
ontology, 3, 36, 41, 464, 476
operationalism, 103, 105–6, 122. *See also* connectivity as an operational statement
operational thinking, 106
organisms as systems, 6
Oriental fruit moth, 268, 292, 293, 297, 307, 311, 322, 365
Orocharis, 345
Oryx gazella, 6, 8
Ostrinia nubilalis, 270, 275–77, 280, 283, 284–90, 292, 296, 298, 317, 325, 351
outcrossing, 26, 36

paleontology, 28, 41, 57, 67, 69, 70
Panaxia dominula, 335, 360
Papilio dardanus, 172
paracentric inversions, 96, 97, 226, 227
parapatry, 158, 159, 178, 185, 189, 235, 340
paraphyly, 26, 27, 31, 35
parasites, 90–92, 108, 215, 225, 226, 329, 354
parasitoids, 357
Parmularius angusticornis, 34
parthenogenesis, 8–11, 22, 40, 142, 418
Partula, 363; *mooreana,* 346; *suturalis,* 328, 346, 348, 361
Parus, 109; *atricapillus,* 119; *major,* 437
Passerina, 430
Pauropsalta, 376, 414
PBAN (pheromone-biosynthesis-activating neuropeptide), 306, 319
Pectinophora: gossypiella, 268, 294, 296, 336, 337, 341, 358–60; *scutigera,* 365
"penis at twelve" syndrome, 318
perturbations, 255, 318
phenotype, 11, 12, 15, 127, 264, 267, 269, 286, 291; variation, 15, 39, 131, 332, 336
phonotaxis, 294, 357, 360, 382, 407, 413, 435
Photinus, 383, 417
Phragmatobia fuliginosa, 296, 324
Phthorimaea operculella, 269, 298
phyletic change, 45, 54
phyletic gradualism, 42, 55, 62, 65, 68, 323
pink bollworm moth, 268, 283, 294, 296, 358–60
Pinna bicolor, 221
Planotortrix, 273, 274, 294, 295, 312, 313, 323; *excessana,* 273, 295, 313, 323
plate tectonics, 18
pleiotropy, 31, 130, 136, 376, 433, 457, 458
pleistocene, 18, 32, 34, 44, 47, 59, 68, 110
plesiomorphic characters, 29, 34
Plethodon vehiculum, 341, 364
Plusia chalcites, 294
Podisma pedestris, 168, 173
Poecilia reticulata, 335–37, 341, 347, 361
Poeciliidae, 382, 435
Polia pisi, 316
Polistes, 315, 325
polyandry, 109, 257
polygenes, 128, 131, 352
polygenic inheritance, 131, 409, 444
polygenic traits, 136, 362, 416, 437

polyploidy, 24
polytene chromosomes, 96, 97, 226, 227, 235, 236
population bottlenecks, 128, 129, 142, 149
population genetics, 41, 95, 123, 143, 144, 146, 149, 154, 181, 184, 185, 225, 296
positive assortative mating, 95, 96, 98, 150, 178, 180, 181, 187, 440, 443, 445
postfertilization, 12–14, 21, 24, 25, 36, 37, 39. *See also* fertilization systems
Praomys: coucha, 42; *natalensis,* 42
predation, 111, 193, 199, 200, 212, 222, 347, 349, 353, 357, 360, 365, 414
predators, 145, 215, 329, 335, 354, 383, 445
prefertilization, 13, 23, 25, 36, 37. *See also* fertilization systems
primates, 31, 69
probabilism, 213, 217, 223
process thinking, 253
proteins, 21, 318
protoidentity, 74, 75, 77, 78, 84
Prunella modularis, 249, 257
Pseudacris: feriarum, 345; *nigrita,* 345, 360; *triseriata,* 345, 362
Pseudaletia unipuncta, 306
Pseudophryne, 418, 421
Pseudoplusia includens, 282, 297, 316, 324
Pteronarcella badia, 342, 365
Pterophylla, 356; *camellifolia,* 342, 363
Ptilonorhynchus violaceus, 357
punctuated equilibrium, 45, 46, 55, 61–66, 264, 286
purpose in evolution, 104, 157, 243–45, 264, 464
Pygosteus pungitius, 10
pyrethrum, 94, 101
Pyrrharctia isabella, 296, 324

quantitative characters, 128, 129, 145, 332, 335, 336
Quercus robur, 453

Rana: calmitans, 438; *catesbiana,* 333; *pipiens,* 437
Ranidella riparia, 329, 363
recognition concept, 22–32, 141, 158, 177, 178, 191, 303, 317, 321, 381, 382; and bedbugs, 179–81, 187; and the biological species concept (*see* recognition concept: and the isolation concept); and cicadas (*Kikihia*), 388–90; cross-attraction, 315; development of, 95–98, 244, 286; and ecology, 194, 205–11, 213, 214, 218; fossil record, 47, 58, 64–66; and function, 244; heuristic value, 382, 410, 411; and isolating mechanisms, 314; and the isolation concept, 66–67, 193, 263, 264, 440, 441, 475; and mating, 162, 309; and negative heterosis, 157; operational definition, 184; at organismal level, 11; pheromone function, 266, 271, 321; postfertilization divergence, 37; predictions, 310; as a relational concept, 27, 178, 198, 199, 207, 381, 382; and secondary sexual characters, 340; species definition, 57, 382, 466; and stasis, 64; and syngamy, 185; system among organisms, 35; taxic approach, 45, 46; taxonomic species concept, 46, 47, 53, 54; transformational approach, 45; typological species concept, 13, 39; wider application, 151
recombination: and flush-crash theory, 143; and genetic associations, 167; during period of relaxed selection, 150; rates, 8; and sexual selection, 128, 134; from the species' gene pool, 206; types, 149
Redunca, 17
refugia, 159
reinforcement: aids, 169; and allopatric speciation, 375, 441; alternative outcomes, 167; characters affecting, 158; in cicadas (*Kikihia*), 410; and divergence, 347, 378; Dobzhansky's model, 466, 467, 475; evidence against, 277, 377; examples, 291, 406, 410; extinction as an alternative, 159; frequency, 157; and habitat change, 24; in hybrid zones, 164, 166, 351; and hybridization, 161; and isolation, 36, 37, 158, 170, 171, 381; rejecting, 246; in sympatry, 178, 348
repeatability, 269, 270
replicators, 474
reproductive isolation: in bedbugs, 180; and boundaries of isolation species, 28; as a consequence, 13, 467, 473; and cross-attraction, 315; in ecology, 193; ensuring, 316; ethological isolation, 96, 98, 155, 171, 275, 335, 350, 351, 359, 363; function, 13, 37, 382; intrinsic, 125; and niche specialization, 196; and postfertilization divergence, 37; postmating, 244; premating, 41, 163, 170–72, 334, 351, 357, 412; and the recognition concept,

reproductive isolation (*continued*) 157–58; selection for, 362, 475; and species definition, 373, 379
reproductive processes: adaptations, 473, 475; character displacement, 277, 291, 334, 348–50, 353, 375, 377, 406, 417; cohesion, 469–71; communities, 465–67, 469, 471, 473–75; discontinuities, 465, 475; entities, 470; modes of, 8, 9; networks, 108; reproductive success, 124, 130, 249, 253, 272, 283, 349, 471–73, 476
resource partitioning, 195, 196, 200, 207, 212, 219, 223
Ribautodelphax, 354, 359, 458, 459
rice plant, 442, 461

Scaphiosus bombifrons, 345
Sceloporus undulatus, 345
Schizocosa, 174
secondary contact, 4, 37, 158, 159, 167, 171, 277, 282, 290, 291, 375, 381, 467
secondary sexual characters, 127, 128, 131, 133, 330, 335, 340
selection: pressures, 149, 196, 204, 290, 327, 328, 335, 354, 355, 376; runaway, 127, 130, 134, 330, 354, 359, 363; viability, 155, 330, 332, 353
sex-linked genes, 131, 325, 365
sexual coadaptation, 124
sexual differentiation, 316–19, 321
sexual dimorphism, 428
sexual environment, 123, 134
sexuality, 8, 10, 41, 316, 321, 322
sexual selection: and the biological species concept, 267; directional, 331, 472; and divergence, 272, 376, 458; and female choice, 126–27, 163–64, 335, 347; founder-flush models, 141–43, 151; function of, 126; and genetic variability, 126, 127–34; intersexual selection, 328, 330, 437; and mating systems, 291, 309; meaning, 471–74; modes of, 355; polygenic traits, 128; stabilizing, 334–35, 472, 476; variation in SMRS signals, 430; and viability selection, 353, 354
sexual systems, 10, 11, 13, 35, 124, 321, 322
shifting balance theory, 144
sister-taxa, 32
Sitophilus zeamais, 326
Smith, Adam, 248

sociobiology, 77, 248, 472
Sophophora, 236
Sorex araneus, 169, 174
speciation: allopatric model, 45, 68, 159, 195, 432; causes, 23, 37, 157; and ethological barriers, 171; flush-crash, 144, 149–53; founder effect, 141, 439; founder events, 135, 141, 143–53; founder-flush, 141–43, 149–53, 155; founder populations, 143, 148, 149, 457, 458; gradual divergence, 376; hybrid inviability, 158; modes of, 155; patterns, 52–54; peripatric, 144, 146, 153; by polyploidy, 24; reinforcing selection, 13, 105, 161, 296 (*see also* reinforcement); rules of, 135, 154; sympatric, 158, 317, 450, 452
species: abundance, 197; as adaptive units, 202; ancestral, 25; complexes, 91, 95, 273, 278, 295, 364, 457; cryptic (*see* species: sibling or cryptic); daughter, 27, 35; diagnosis, 20; divergence, 23; extinction, 474; field for gene recombination, 177, 181, 183, 184, 186, 188, 380, 412, 440; genetical, 95, 179, 180, 182, 186, 194, 210, 380, 381; identification, 25, 89, 94, 98, 309, 371; metaspecies, 25, 33; morphospecies, 55, 59, 68, 442; nature of, 61, 63, 90, 173, 256, 418, 420, 464, 468, 475, 476; packing, 202; preferred habitat, 264, 271, 329, 332, 333; ring, 420; selection, 471, 473–75, 477; as self-defining systems, 55, 297, 416; semispecies, 279, 334, 340, 359, 362, 391; sibling or cryptic, 32, 42, 96, 99, 100, 101, 158, 173, 174, 222, 236, 237, 273, 275–78, 293–95, 314, 322, 323, 331, 358, 366, 371, 375–77, 414, 452–54, 460, 468, 473; status, 9, 11, 14, 15, 21, 23, 26, 31, 32, 108, 124, 126, 179, 180, 277, 278, 373, 380, 389, 390, 392, 396, 410, 411
species concepts, 15, 32, 37, 41, 45, 53, 90, 97, 103, 107, 121, 177, 187, 413, 465, 476; allospecies, 120; biological species (*see* isolation concept); cohesion concept, 3, 13, 14, 21, 25–28, 31, 35–38, 201, 212, 315, 468–71, 474; isolation concept (*see entry* isolation concept); operational problems, 27–28; phylogenetic species concept, 11, 21, 23–27, 29, 30, 32–34, 38, 39, 42, 470; polytypic species, 467;

recognition concept (see entry recognition concept)
Specific-Mate Recognition Systems (SMRSs): acoustic signals, 298, 333, 340, 346, 409, 425–27, 429, 431, 434, 441, 442, 444, 448, 451, 453, 455, 457–61; antagonistic interactions, 282, 283, 289, 314, 316, 321; auditory information, 383; auditory physiology, 426; auditory signals, 31, 32, 307, 310, 361, 378, 379, 383, 405, 406, 410, 415, 421, 426–28, 435–39, 461; calling signals, 442, 445, 446, 462; calling song, 294, 363, 407, 421, 435, 439, 454; changes in, 277, 279, 329, 423, 430, 432, 433, 468; chemical signals, 264–66, 278, 284, 297, 306, 315, 316, 321, 322; chemical "noise," 286; cross-attraction, 276, 279, 282, 314–16, 321, 414; genetic coupling, 415, 428, 435, 436, 450, 461; geographic variation in signals, 432, 458; male response thresholds, 270, 280; mating displays, 377; modalities, 53, 263, 265, 305–7, 310; modification of, 467; multicomponent pheromone blends, 265, 267, 271, 279, 280, 282, 285, 290, 291; olfactory signals, 10, 31, 124, 284, 285, 289, 292, 294, 296, 298, 306, 307, 324; pheromone blends, 268, 274, 275, 278, 280, 281, 282–84, 287, 290–91, 299, 306, 316, 325, 329, 336, 337, 341, 351, 364; pheromone plumes, 289; pheromone response, 294, 295, 297, 324, 358; pulse repetition frequency, 329, 361, 427, 442; receptor cells, 282, 314, 316, 383; recognition mechanisms, 37; sex pheromones, 267, 275, 278, 283–85, 292–300, 304–7, 308, 310–13, 318, 319, 322, 323–25, 356, 364, 414; signal receivers, 267, 285, 286, 305, 321, 352, 353, 355, 383, 388, 406, 410, 422, 425–27, 429–32, 434, 438, 450, 457, 477; signal-to-noise ratio, 429; song structure, 376, 379, 392; stability, 125, 331, 405; stimulus-response elements, 10; structural coupling, 255; tuning of, 282, 290, 291, 327, 376, 434, 457; variation in, 149, 267, 309, 423, 430–32, 434, 458
speciose groups, 31
Spizella pusilla, 438
Spodoptera littoralis, 342, 357
stasis, 50, 53, 62, 64–66, 194, 273, 286, 300, 332, 475
stenotopic, 24
sterility, 68, 93, 95–97, 160, 167, 168, 170, 172–74, 177, 181–86, 189, 222, 418
stickleback fish, 10, 23, 255, 307, 363
structuralism, 72, 73, 77–79, 80–82, 84–89, 221, 256, 259
sumiwake, 77
symplesiomorphous, 470, 475
Synanthedon pictipes, 341
synapomorphy, 10
syngameon, 14, 40
syngamy, 8, 39, 41, 151, 185, 264, 271, 286, 333, 382, 434
system habitat specificity, 22
syutaisei, 74

Teleogryllus, 131, 134, 361, 436, 459; *commodus*, 344, 361; *gryllus*, 459; *oeceanicus*, 344, 361, 346
teleology, 212, 245
tension zones, 159, 161, 168, 169, 171, 173, 350
teretiusculus Shales, 52
terminal taxa, 25, 26
Thomomys, 349, 350, 364; *bottae*, 349, 350
Tibicen: chiricahua, 415; *duryi*, 415; *linei*, 388
Tilapia, 42
Tilia, 453; *cordata*, 455
tortricid, 278, 290, 296, 312, 313, 322
transilience, 143, 146, 149, 150, 153
transmitter, 457
transposable genetic elements, 129, 135, 136, 233, 234, 235, 236
tree of life, 13, 23, 24, 29
triassic, 18
Tribolium castaneum, 338, 360
Trichoplusia ni, 88, 273, 293, 296, 297, 299, 316, 324
trophic levels, 205, 211, 216
Trypanosoma, 234
turnip moth, 268, 273, 296, 297, 313, 324, 360
twoness, 79
Typhlocyba, 459
Typhlocybidae, 459, 462
typical habitat, 160, 388

Uca, 341, 359; *pugnax*, 346; *rapax*, 346
ultimate cause, 253

unique habitat specificity, 18, 21, 36, 37
Uperoleia rugosa, 364

vacant niche, 196
Vanessa urticae, 317
vector-borne diseases, 95
vicariance, 19, 20, 24, 43

wallflower effect, 354

X chromosomes, 227
Xiphophorus nigrensis, 336, 337, 435

Yponomeuta, 294, 296, 323; *padellus,* 294, 323; *rorellus,* 294, 323

Zahavi Handicap principle, 355
Zea mays, 233
Zeiraphera, 351; *diniana,* 313, 322, 351, 356
zones of overlap, 340, 361, 375, 412, 415, 416
Zygothrica, 132, 135